American Statistical Association

1999
Proceedings

of the

Section on Physical and Engineering Sciences

Papers presented at the

1999 Spring Research Conference on Statistics in Industry and Technology,
Minneapolis, Minnesota, June 2–4, 1999

Annual Meeting of the American Statistical Association,
Baltimore, Maryland, August 8–12, 1999,

1999 Fall Technical Conference,
Houston, Texas, October 14 & 15, 1999

all under the sponsorship of
the Section on Physical and Engineering Sciences

American Statistical Association ■ 1429 Duke Street, Alexandria, VA 22314

The papers and discussions in this Proceedings volume are reproduced exactly as received from the authors. None of the papers has been submitted to a refereeing process. However, the authors have been encouraged to have their papers reviewed by a colleague prior to final preparation. These presentations are presumed to be essentially as given at the Joint Statistical Meetings in Baltimore or other conferences where indicated. The 1999 Proceedings volumes are not copyrighted by the Association; hence, permission for reproduction must be obtained from the author, who holds the rights under the copyright law.

Authors in these Proceedings are encouraged to submit their papers to any journal of their choice. The ASA Board of Directors has ruled that publication in the Proceedings does not preclude publication elsewhere.

American Statistical Association
1429 Duke Street
Alexandria, Virginia 22314

PRINTED IN THE U.S.A.

ISBN 1-883276-96-9

TABLE OF CONTENTS

Invited Papers by Topic

Contributed Papers by Topic

4. **Applications of Statistics to Vehicle Emission Studies**
    Organizer: *Tim Coburn*, Abilene Christian University
    Chair: *Robert L. Mason*, Southwest Research Institute

5. **Recent Advances in Response Surface and Regression Methods**
    Chair: *Paul Stober,* SmithKline Beecham Pharmaceuticals

6. **Statistics Application in Army R & D**
    Organizer: *Barry Bodt*, US Army Research Laboratory

10. Statistical Inference and Modeling
Chair: *Nicole Lazar*, Carnegie Mellon University

11. Times Series and Related Topics
Chair: *Teri Crosby*, Oronite/Chevron Chemical Company

Contributed Papers-Poster Sessions

Papers Presented at the
1999 Spring Research Conference
On Statistics in Industry and Technology
Minneapolis, Minnesota, June 2 - 4, 1999

Papers Presented at the
1999 Fall Technical Conference
Houston, Texas, October 14 - 15, 1999

KIEFER ORDERING OF SECOND-DEGREE MIXTURE DESIGNS FOR FOUR INGREDIENTS

Norman R. Draper[1], U Wisconsin–Madison, Berthold Heiligers[2], U Magdeburg,
Friedrich Pukelsheim[1], U Augsburg
Berthold Heiligers, Faculty of Mathematics, University of Magdeburg,
D–39016 Magdeburg, Germany

Key Words: Complete class results, Exchangeable designs, Kiefer design ordering, Kronecker product, Majorization, Scheffé canonical polynomials, Weighted centroid designs.

1 Introduction

Many practical problems are associated with the investigation of mixture ingredients $t_1, t_2, \ldots, t_m$ of m factors, with $t_i \geq 0$ and being further restricted by $\sum_{i=1}^m t_i = 1$. The definitive text Cornell (1990) lists numerous examples and provides a thorough discussion of both theory and practice. Early seminal work was done by Scheffé (1958, 1963) in which he suggested (1958, page 347) and analyzed canonical model forms when the regression function for the expected response $y = y(t)$ is a polynomial of degree one, two, or three. We shall refer to these as the S-polynomials, or S-models; for example, the second-degree S-polynomial has the form

$$y(t) = \sum_{1 \leq i \leq m} \beta_i\, t_i + \sum_{1 \leq i < j \leq m} \beta_{ij}\, t_i t_j. \quad (1)$$

In this paper, we use the alternative representation of mixture models introduced in Draper and Pukelsheim (1998b, 1999). Our versions are based on the Kronecker algebra of vectors and matrices, and give rise to homogeneous model functions and moment matrices. We refer to the corresponding expressions as the K-models, or K-polynomials. We emphasize, however, that our results on the Kiefer ordering of experimental designs for second-degree mixture models are the same whether the S-model or the K-model is employed.

Our notation is the same as in the previous papers Draper and Pukelsheim (1998b, 1999). We

consider multifactor experiments, for m deterministic ingredients that are assumed to influence the response only through the percentages or proportions in which they are blended together. For $i = 1, \ldots, m$, let $t_i \in [0,1]$ be the proportion of ingredient i in the mixture. As usual, we assemble the individual components to form the column vector of experimental conditions, $t = (t_1, \ldots, t_m)'$. It ranges over the experimental domain $\mathcal{T}$, the standard probability simplex in the space $\mathbb{R}^m$. Let $\mathbf{1}_m = (1, \ldots, 1)' \in \mathbb{R}^m$ be the unity vector, whence $\mathbf{1}'_m t = t_1 + \cdots + t_m$ is the sum of the components of a vector t. Therefore, in our case, the experimental domain is $\mathcal{T} = \{t \in [0,1]^m : \mathbf{1}'_m t = 1\}$.

Under experimental conditions $t \in \mathcal{T}$, the experimental response Y_t is taken to be a scalar random variable. Replications under identical experimental conditions, or responses from distinct experimental conditions are assumed to be of equal (unknown) variance σ^2, and uncorrelated. When the regression function is a second-degree K-polynomial, the expected response takes the form

$$\mathrm{E}[Y_t] = y(t) = \sum_{i=1}^m \sum_{j=i}^m t_i t_j \theta_{ij} = (t \otimes t)'\theta. \quad (2)$$

An experimental design τ on the experimental domain $\mathcal{T}$ is a probability measure having a finite number of support points. If τ assigns weights $w_1, w_2, \ldots$ to its points of support in $\mathcal{T}$, then the experimenter is directed to draw proportions $w_1, w_2, \ldots$ of all observations under the respective experimental conditions. We associate with τ its (second-degree K-) moment matrix,

$$M(\tau) = \int_{\mathcal{T}} (t \otimes t)(t \otimes t)'\, \mathrm{d}\tau. \quad (3)$$

The Kronecker square $(t \otimes t)$ in (2) repeats the mixed products $t_i t_j = t_j t_i$, $i < j$, and thus overparameterizes the quadratic response function y, (while the corresponding Scheffé setup (1) is based on a minimal parameterization). As a consequence, (2) is a non-parsimonious representation of (1), as,

[1] Supported in part by the Alexander von Humboldt-Stiftung, Bonn, through a Max–Planck–Award for cooperative research.

[2] Support from the Deutsche Forschungsgemeinschaft and from the American Statistical Association for participating in the JSM 99 is gratefully acknowledged.

1

for $t \in \mathcal{T}$,

$$t'\beta = (t'\beta) \otimes (t'\mathbf{1}_m) = (t \otimes t)'(\beta \otimes \mathbf{1}_m)$$
$$= (t \otimes t)'\theta.$$

Nevertheless we propose to utilize the K-model (2) instead of (1), since the Kronecker algebra is powerful enough to outweigh the overparameterization, and, more importantly, in the K-model the moment matrices from (3) have all entries homogeneous of degree four. This homogeneity is a distinctive advantage over the S-model for which some of the entries of the moment matrix are homogeneous of degree two, others of degree three, and the rest of degree four. A closely related emphasis on proper standardization is put forward by Dette (1997) though the motivation, rescaling the experimental domain, is different. In our work, the experimental domain $\mathcal{T}$ is the simplex and stays fixed.

Given an arbitrary design τ, we obtain an exchangeable (permutation invariant) design $\overline{\tau}$ by averaging over the permutation group,

$$\overline{\tau} = \frac{1}{m!} \sum_{R \in \mathrm{Perm}(m)} \tau \circ R^{-1}.$$

If the original design τ itself is exchangeable then it is reproduced, $\overline{\tau} = \tau$. Otherwise the average $\overline{\tau}$ is an improvement over τ, in that it exhibits more symmetry and balancedness. In terms of matrix majorization (relative to the congruence action that is induced on the moment matrices M), the moment matrix of the averaged design $\overline{\tau}$ is majorized by the moment matrix of τ, $M(\overline{\tau}) \prec M(\tau)$. The terminology "is majorized by" is standard, even though for design purposes the emphasis is reversed: $M(\overline{\tau})$, being more balanced, is superior to $M(\tau)$. As a consequence, the design $\overline{\tau}$ yields better values than τ, under a large class of optimality criteria (Pukelsheim 1993, page 349). For a recent review of invariance and optimality of polynomial regression designs see Gaffke and Heiligers (1996).

Symmetry and balancedness have always been a prime attribute of good experimental designs, and comprise the first step of the Kiefer design ordering. The second step concerns the usual Loewner matrix ordering. Lemma 1 in Heiligers (1991) and Theorem 2 in Heiligers (1992) imply that any second-degree mixture design which is not a weighted centroid design can be improved upon, in the sense of the Kiefer ordering, by a weighted centroid design. Here we show *how* to derive the improving design from the properties of the starter design.

We shall restrict ourself to the K-model (2) with $m = 4$ ingredients, only. We show that the class

of weighted centroid designs is minimal complete, see Theorem 4. As a consequence, the search for optimal designs may be restricted to weighted centroid designs, for most criteria. For particular criteria, this was observed already by Kiefer (1959, 1975, 1978), and Galil and Kiefer (1977). Related results on Kiefer ordering completeness of rotatable designs on the ball are reviewed by Draper and Pukelsheim (1998a). The setting of Cheng (1995) is slightly different, in that his permutations act on the regression vector $x = t \otimes t$, rather than on the experimental conditions t.

While for models with two or three factors Kiefer comparability of exchangeable moment matrices is described by one parameter only, cf. Draper and Pukelsheim (1999), the corresponding result in the four factor model is more complicated as it involves two dependent parameters, see Lemma 1, below. This is also true for $m \geq 5$ factors—these models, however, have to cope with some ambiguity introducing additional complications. Setups with five or more ingredients are discussed in Draper, Heiligers and Pukelsheim (1998); that paper also provides a complementary, geometric view of the present Complete Class Theorem 4.

Our approach is an extension of Draper and Pukelsheim (1998a, 1999). In view of the initial symmetrization step it suffices to search for an improvement in the Loewner ordering sense, among exchangeable moment matrices only. We first aim at finding necessary and sufficient conditions for two exchangeable second-degree K-moment matrices to be comparable in the Loewner matrix ordering. The "Comparison Lemma" 1 provides conditions in terms of two moment inequalities, in the spirit of Theorem 2 of Heiligers (1992). Weighted centroid designs effectively remove one degree of freedom, see the "Characterization Lemma" 2. Given a first, poor design we then construct a second, better design, in the "Existence Lemma" 3. The complete class result is stated in Theorem 4.

2 Four Factors

The four-ingredient second-degree model features all possible moments of order four,

$$\mu_4 = \int_{\mathcal{T}} t_i^4 \, d\overline{\tau}, \qquad \mu_{31} = \int_{\mathcal{T}} t_i^3 t_j \, d\overline{\tau},$$

$$\mu_{22} = \int_{\mathcal{T}} t_i^2 t_j^2 \, d\overline{\tau}, \qquad \mu_{211} = \int_{\mathcal{T}} t_i^2 t_j t_k \, d\overline{\tau},$$

$$\mu_{1111} = \int_{\mathcal{T}} t_i t_j t_k t_\ell \, d\overline{\tau},$$

where the subscripts $i, j, k, \ell = 1, \ldots, 4$ are pairwise distinct and where $\overline{\tau}$ is some exchangeable design on the simplex $\mathcal{T}$. The associated K-moment matrix $M = M(\overline{\tau})$ is of the generic form

$$M = \mu_4 V_4 + \mu_{31} V_{31} + \mu_{22} V_{22} + \mu_{211} V_{211} + \mu_{1111} V_{1111}$$

with 0-1 matrices V_i of order 16×16, indicating the position of the moments μ_i in M. As usual, let e_i denote the i-th Euclidean unit vector in $\mathbb{R}^4$ with i-th component one and zeros elsewhere. For a concise representation of V_i we use the 16×1 Euclidean unit vectors $e_{ij} = e_i \otimes e_j$ having a single one as the i-th block's j-th element, for $i, j = 1, \ldots, 4$,

$$V_4 = \sum_i e_{ii} e_{ii}',$$

$$V_{31} = \sum_{i,j}{}' (e_{ii} e_{ij}' + e_{ij} e_{ii}' + e_{ii} e_{ji}' + e_{ji} e_{ii}'),$$

$$V_{22} = \sum_{i,j}{}' (e_{ii} e_{jj}' + e_{ij} e_{ij}' + e_{ij} e_{ji}'),$$

$$\begin{aligned} V_{211} = \sum_{i,j,k}{}' (&e_{ii} e_{jk}' + e_{jk} e_{ii}' + e_{ij} e_{ki}' + \\ &+ e_{ji} e_{ik}' + e_{ij} e_{ik}' + e_{ji} e_{ki}'), \\ &+ e_{ji} e_{ik}' + e_{ij} e_{ik}' + e_{ji} e_{ik}'), \end{aligned}$$

$$V_{1111} = \sum_{i,j,k,\ell}{}' e_{ij} e_{k\ell}'.$$

The sign $\sum'$ means that the summation is restricted to pairwise distinct subscripts. The rank of M is at most 10, implying M has at least six nullvectors.

The simplex restriction entails $\mathbf{1}_{16}' M \mathbf{1}_{16} = \int (\mathbf{1}_4' t)^4 \, d\overline{\tau} = 1$. That is, the elements of M sum to one, $4\mu_4 + 48\mu_{31} + 36\mu_{22} + 144\mu_{211} + 24\mu_{1111} = 1$. Moreover, the lower order moments are functions of the fourth order moments, i.e.,

$$\mu_3 = \mu_4 + 3\mu_{31},$$
$$\mu_{21} = \mu_{31} + \mu_{22} + 2\mu_{211},$$
$$\mu_{111} = 3\mu_{211} + \mu_{1111};$$

$$\begin{aligned} \mu_2 &= \mu_3 + 3\mu_{21} \\ &= \mu_4 + 6\mu_{31} + 2\mu_{22} + 6\mu_{211}, \end{aligned}$$

$$\begin{aligned} \mu_{11} &= 2\mu_{21} + 2\mu_{111} \\ &= 2\mu_{31} + 2\mu_{22} + 10\mu_{211} + 2\mu_{1111}. \end{aligned}$$

Now we consider two exchangeable designs η and $\overline{\tau}$ possessing identical moments of order three. Then the moments of order two are equal, too. For the fourth order moment differences we get, with $\gamma =$ $\mu_4(\eta) - \mu_4(\overline{\tau})$ and $\delta = \mu_{1111}(\eta) - \mu_{1111}(\overline{\tau})$,

$$\begin{aligned} \mu_4(\eta) - \mu_4(\overline{\tau}) &= \gamma, \\ \mu_{31}(\eta) - \mu_{31}(\overline{\tau}) &= -\tfrac{1}{3}\gamma, \\ \mu_{22}(\eta) - \mu_{22}(\overline{\tau}) &= \tfrac{1}{3}\gamma + \tfrac{2}{3}\delta, \qquad (4) \\ \mu_{211}(\eta) - \mu_{211}(\overline{\tau}) &= -\tfrac{1}{3}\delta, \\ \mu_{1111}(\eta) - \mu_{1111}(\overline{\tau}) &= \delta. \end{aligned}$$

Of course, there is an infinity of ways to parameterize the two degrees of freedom in (4). We find γ and δ a natural choice to work with. Using the indicator matrices V_i, the moment matrices of η and $\overline{\tau}$ differ by

$$\begin{aligned} \Delta &= M(\eta) - M(\overline{\tau}) \\ &= \gamma V_4 - \frac{\gamma}{3} V_{31} + \frac{\gamma + 2\delta}{3} V_{22} - \frac{\delta}{3} V_{211} + \delta V_{1111}. \end{aligned}$$

This decomposition has five terms although there are only two degrees of freedom, γ and δ.

There are, however, simpler representations for Δ involving just two matrices A and B, say. A convenient choice is $A = V_4 - \frac{1}{3} V_{31} + \frac{1}{3} V_{22}$, i.e.,

$$A = \frac{1}{3} \begin{pmatrix}
3 & -1 & -1 & -1 & -1 & 1 & 0 & 0 & -1 & 0 & 1 & 0 & -1 & 0 & 0 & 1 \\
-1 & 1 & 0 & 0 & 1 & -1 & 0 & 0 & 0 & 0 & 0 & 0 & 0 & 0 & 0 & 0 \\
-1 & 0 & 1 & 0 & 0 & 0 & 0 & 0 & 1 & 0 & -1 & 0 & 0 & 0 & 0 & 0 \\
-1 & 0 & 0 & 1 & 0 & 0 & 0 & 0 & 0 & 0 & 0 & 0 & 1 & 0 & 0 & -1 \\
-1 & 1 & 0 & 0 & 1 & -1 & 0 & 0 & 0 & 0 & 0 & 0 & 0 & 0 & 0 & 0 \\
1 & -1 & 0 & 0 & -1 & 3 & -1 & -1 & 0 & -1 & 1 & 0 & 0 & -1 & 0 & 1 \\
0 & 0 & 0 & 0 & 0 & -1 & 1 & 0 & 0 & 1 & -1 & 0 & 0 & 0 & 0 & 0 \\
0 & 0 & 0 & 0 & 0 & -1 & 0 & 1 & 0 & 0 & 0 & 0 & 0 & 1 & 0 & -1 \\
-1 & 0 & 1 & 0 & 0 & 0 & 0 & 0 & 1 & 0 & -1 & 0 & 0 & 0 & 0 & 0 \\
0 & 0 & 0 & 0 & 0 & -1 & 1 & 0 & 0 & 1 & -1 & 0 & 0 & 0 & 0 & 0 \\
1 & 0 & -1 & 0 & 0 & 1 & -1 & 0 & -1 & -1 & 3 & -1 & 0 & 0 & -1 & 1 \\
0 & 0 & 0 & 0 & 0 & 0 & 0 & 0 & 0 & 0 & -1 & 1 & 0 & 0 & 1 & -1 \\
-1 & 0 & 0 & 1 & 0 & 0 & 0 & 0 & 0 & 0 & 0 & 0 & 1 & 0 & 0 & -1 \\
0 & 0 & 0 & 0 & 0 & -1 & 0 & 1 & 0 & 0 & 0 & 0 & 0 & 1 & 0 & -1 \\
0 & 0 & 0 & 0 & 0 & 0 & 0 & 0 & 0 & 0 & -1 & 1 & 0 & 0 & 1 & -1 \\
1 & 0 & 0 & -1 & 0 & 1 & 0 & -1 & 0 & 0 & 1 & -1 & -1 & -1 & -1 & 3
\end{pmatrix},$$

and $B = \frac{2}{3} V_{22} - \frac{1}{3} V_{211} + V_{1111}$, i.e.,

$$B = \frac{1}{3} \begin{pmatrix}
0 & 0 & 0 & 0 & 0 & 2 & -1 & -1 & 0 & -1 & 2 & -1 & 0 & -1 & -1 & 2 \\
0 & 2 & -1 & -1 & 2 & 0 & -1 & -1 & -1 & -1 & -1 & 1 & -1 & -1 & 1 & -1 \\
0 & -1 & 2 & -1 & -1 & -1 & 1 & 2 & -1 & 0 & -1 & -1 & 1 & -1 & -1 \\
0 & -1 & -1 & 2 & -1 & -1 & 1 & -1 & -1 & 1 & -1 & -1 & \cdot 2 & -1 & -1 & 0 \\
0 & 2 & -1 & -1 & 2 & 0 & -1 & -1 & -1 & -1 & -1 & 1 & -1 & -1 & 1 & -1 \\
2 & 0 & -1 & -1 & 0 & 0 & 0 & 0 & -1 & 0 & 2 & -1 & -1 & 0 & -1 & 2 \\
-1 & -1 & 1 & 1 & -1 & 0 & 2 & -1 & -1 & 2 & 0 & -1 & 1 & -1 & -1 & -1 \\
-1 & 1 & 1 & -1 & -1 & 0 & -1 & 2 & 1 & -1 & -1 & -1 & -1 & 2 & -1 & 0 \\
0 & -1 & 2 & -1 & -1 & -1 & 1 & 2 & -1 & 0 & -1 & -1 & 1 & 1 & -1 & -1 \\
-1 & -1 & 1 & 1 & -1 & 0 & 2 & -1 & -1 & 2 & 0 & -1 & 1 & -1 & -1 & -1 \\
2 & -1 & 0 & -1 & -1 & 2 & 0 & -1 & 0 & 0 & 0 & 0 & -1 & 0 & 2 \\
-1 & 1 & -1 & -1 & 1 & -1 & -1 & -1 & -1 & 0 & 2 & -1 & -1 & 2 & 0 \\
0 & -1 & 2 & -1 & -1 & 1 & -1 & -1 & 1 & -1 & -1 & 2 & -1 & -1 & 0 \\
-1 & -1 & 1 & 1 & -1 & 0 & 2 & -1 & 1 & -1 & -1 & -1 & -1 & 2 & -1 & 0 \\
-1 & 1 & -1 & -1 & 1 & -1 & -1 & -1 & -1 & 0 & 2 & -1 & -1 & 2 & 0 \\
2 & -1 & -1 & 0 & -1 & 2 & -1 & 0 & -1 & -1 & 2 & 0 & 0 & 0 & 0 & 0
\end{pmatrix}.$$

The matrices A and B have the same rank six, both possessing three non-zero eigenvalues (with respective multiplicities one, two, and three):

$$\lambda_1[A] = \tfrac{8}{3}, \qquad \lambda_1[B] = \tfrac{8}{3},$$
$$\lambda_2[A] = \tfrac{2}{3}, \qquad \lambda_2[B] = \tfrac{14}{3},$$
$$\lambda_3[A] = \tfrac{4}{3}, \qquad \lambda_3[B] = -\tfrac{4}{3}.$$

Moreover, the eigenspaces $\mathcal{E}_i$ associated with the i-th eigenvalue of A and of B coincide; upon setting $u_{ijk\ell} = (e_i - e_j) \otimes (e_k - e_\ell)$ we have

$$\mathcal{E}_1 = \mathrm{span}\big\{\mathbf{1}_{16} - 4\textstyle\sum_{i=1}^{4} e_{ii}\big\},$$
$$\mathcal{E}_2 = \mathrm{span}\big\{(u_{1234} + u_{3412}),\, (u_{1324} + u_{2413})\big\},$$
$$\mathcal{E}_3 = \mathrm{span}\big\{(u_{1212} - u_{3434}),\, (u_{1313} - u_{2424}),$$
$$(u_{1414} - u_{2323})\big\}.$$

In summary, the representation for $\Delta = M(\eta) - M(\overline{\tau})$ takes the form

$$\Delta = \gamma A + \delta B, \tag{5}$$

and has eigenvalues

$$\begin{aligned}
\lambda_1[\Delta] &= \tfrac{8}{3}(\gamma + \delta), \\
\lambda_2[\Delta] &= \tfrac{2}{3}(\gamma + 7\delta), \\
\lambda_3[\Delta] &= \tfrac{4}{3}(\gamma - \delta), \\
\lambda_4[\Delta] &= 0.
\end{aligned} \tag{6}$$

The comparison of two exchangeable moment matrices in the Loewner ordering is now reduced to the comparison of individual moments. Let $\mu_{(3)} = (\mu_2, \mu_{11}, \mu_3, \mu_{21}, \mu_{111})'$ be the vector of moments up to and including order three.

LEMMA 1.   *Let η and $\overline{\tau}$ be two exchangeable designs on the simplex $\mathcal{T}$. Then we have, with $\gamma = \mu_4(\eta) - \mu_4(\overline{\tau})$ and $\delta = \mu_{1111}(\eta) - \mu_{1111}(\overline{\tau})$,*

$$M(\eta) \geq M(\overline{\tau})$$

if and only if

$$\mu_{(3)}(\eta) = \mu_{(3)}(\overline{\tau}), \quad \text{and} \quad -\tfrac{1}{7}\gamma \leq \delta \leq \gamma.$$

PROOF.   For the direct part, assume that $\Delta = M(\eta) - M(\overline{\tau})$ is nonnegative definite. Then $(\mathbf{1}_4 \otimes \mathbf{1}_4)'\Delta(\mathbf{1}_4 \otimes \mathbf{1}_4) = 0$ forces $\Delta(\mathbf{1}_4 \otimes \mathbf{1}_4) = 0$. This implies equality of second order moments. Now we get

$$(e_1 \otimes \mathbf{1}_4)'M(\eta)(e_1 \otimes \mathbf{1}_4) = \int_{\mathcal{T}} t_1^2 \, d\eta = \mu_2$$
$$= \int_{\mathcal{T}} t_1^2 \, d\overline{\tau} = (e_1 \otimes \mathbf{1}_4)'M(\overline{\tau})(e_1 \otimes \mathbf{1}_4).$$

This yields $\Delta(e_1 \otimes \mathbf{1}_4) = 0$, that is, $\int (t \otimes t)t_1 \, d\eta = \int (t \otimes t)t_1 \, d\overline{\tau}$. Hence the third order moments of η and $\overline{\tau}$ are equal as well, giving $\mu_{(3)}(\eta) = \mu_{(3)}(\overline{\tau})$. By (6), nonnegative definiteness of Δ entails nonnegativity of $\gamma + \delta$, $\gamma - \delta$ and $\gamma + 7\delta$, that is, $-\tfrac{1}{7}\gamma \leq \delta \leq \gamma$. For the converse part, equality of third order moments implies the representation (5). According to (6), the assumption on γ and δ immediately implies $\Delta \geq 0$.   □

By Lemma 1, comparability of exchangeable designs η and $\overline{\tau}$ ensures that the difference $\Delta = M(\eta) - M(\overline{\tau})$ lies in $\mathcal{D} = \big\{\gamma A + \delta B : \gamma \geq 0, -\tfrac{\gamma}{7} \leq \delta \leq \gamma\big\}$, which is a two-dimensional subcone in the set of nonnegative definite 16×16 matrices. Thus, for any fixed $\overline{\tau}$, the moment matrices of those exchangeable designs η which improve upon $\overline{\tau}$ w.r.t. the Kiefer ordering are obtained by intersecting the set $\mathcal{M}$ of all moment matrices with the shifted cone $M(\overline{\tau}) + \mathcal{D}$; Figure 1 illustrates this geometry.

Of course, most of the designs η associated with $\mathcal{M} \cap (M(\overline{\tau}) + \mathcal{D})$ can be further improved upon by a third exchangeable design ζ, which then is also a further improvement on the starter design $\overline{\tau}$. We are particularly interested here in describing those exchangeable designs η which improve upon $\overline{\tau}$ and which can not be further improved upon. These turn out to be among the weighted centroid designs, introduced next.

There are four *elementary centroid designs*: the vertex points design η_1, the edge midpoints design η_2, the face centroids design η_3, and the overall centroid design η_4,

$$\begin{aligned}
\eta_1(e_i) &= \tfrac{1}{4} && \text{for } 1 \leq i \leq 4, \\
\eta_2\big(\tfrac{1}{2}(e_i + e_j)\big) &= \tfrac{1}{6} && \text{for } 1 \leq i < j \leq 4, \\
\eta_3\big(\tfrac{1}{3}(e_i + e_j + e_k)\big) &= \tfrac{1}{4} && \text{for } 1 \leq i < j < k \leq 4, \\
\eta_4(\overline{e.}) &= 1;
\end{aligned}$$

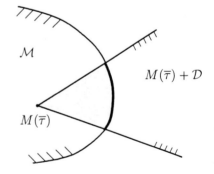

FIG. 1. Moment matrices of exchangeable designs which improve upon $\overline{\tau}$ w.r.t. the Kiefer ordering. The gray area describes the "improvement area" $M(\overline{\tau}) + \mathcal{D}$. The thick line in the east boundary of $M(\overline{\tau}) + \mathcal{D}$ is associated with the weighted centroid designs which improve upon $\overline{\tau}$.

Figure 2 illustrates these designs. The moments of order four of these designs are

$$\mu_4(\eta_1) = \tfrac{1}{4}, \qquad \mu_4(\eta_2) = \tfrac{1}{32},$$
$$\mu_4(\eta_3) = \tfrac{1}{108}, \qquad \mu_4(\eta_4) = \tfrac{1}{256},$$
$$\mu_{31}(\eta_1) = 0, \qquad \mu_{31}(\eta_2) = \tfrac{1}{96},$$
$$\mu_{31}(\eta_3) = \tfrac{1}{162}, \qquad \mu_{31}(\eta_4) = \tfrac{1}{256},$$
$$\mu_{211}(\eta_1) = 0, \qquad \mu_{211}(\eta_2) = 0,$$
$$\mu_{211}(\eta_3) = \tfrac{1}{324}, \qquad \mu_{211}(\eta_4) = \tfrac{1}{256},$$
$$\mu_{1111}(\eta_1) = 0, \qquad \mu_{1111}(\eta_2) = 0,$$
$$\mu_{1111}(\eta_3) = 0, \qquad \mu_{1111}(\eta_4) = \tfrac{1}{256};$$

we have $\mu_{31}(\eta_j) = \mu_{22}(\eta_j)$, for all $j = 1, 2, 3, 4$.

For weights $\alpha_1, \alpha_2, \alpha_3, \alpha_4 \geq 0$ summing to one, the design $\eta = \alpha_1\eta_1 + \alpha_2\eta_2 + \alpha_3\eta_3 + \alpha_4\eta_4$ is called a *weighted centroid design*.

LEMMA 2. *Let $\bar{\tau}$ be an exchangeable design on the simplex $\mathcal{T}$. Then we have*

$$\mu_{31}(\bar{\tau}) \geq \mu_{22}(\bar{\tau}),$$

with equality if and only if $\bar{\tau}$ is a weighted centroid design.

PROOF. The function

$$\psi(t_1, t_2, t_3, t_4) = \sum_{i<j} t_i t_j (t_i - t_j)^2$$

is nonnegative on the simplex $\mathcal{T}$, and integrates under $\bar{\tau}$ to $12(\mu_{31} - \mu_{22})$. This proves $\mu_{31} \geq \mu_{22}$. We have $\int \psi \, d\bar{\tau} = 0$ if and only if every support point $t = (t_1, t_2, t_3, t_4)'$ of $\bar{\tau}$ satisfies $t_i t_j (t_i - t_j)^2 = 0$. Because of exchangeability, $\bar{\tau}$ must then be a weighted centroid design. $\square$

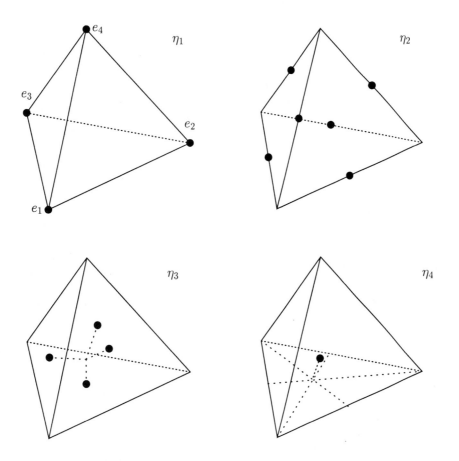

FIG. 2. *Elementary centroid designs $\eta_1, \eta_2, \eta_3, \eta_4$ in a three-dimensional space. The individial designs assign equal weights to the associated support points, represented by the dots. The weights total 1 in each case.*

Let $\eta = \alpha_1\eta_1 + \alpha_2\eta_2 + \alpha_3\eta_3 + \alpha_4\eta_4$ be a weighted centroid design. When we calculate the difference of line two and three in (4), the contribution of η vanishes due to $\mu_{31}(\eta) = \mu_{22}(\eta)$. Suppressing the dependence on $\overline{\tau}$ of the remaining moments, we get

$$\mu_{31} - \mu_{22} = \frac{2}{3}(\gamma + \delta). \qquad (7)$$

From this, we determine γ in terms of δ and the moments of $\overline{\tau}$, i.e., $\gamma = \frac{3}{2}(\mu_{31} - \mu_{22}) - \delta$. The restrictions $-\frac{1}{7}\gamma \le \delta \le \gamma$ provide initial bounds for δ,

$$-\frac{1}{4}(\mu_{31} - \mu_{22}) \le \delta \le \frac{3}{4}(\mu_{31} - \mu_{22}). \qquad (8)$$

In order to find a set of weights for $\eta = \alpha_1\eta_1 + \alpha_2\eta_2 + \alpha_3\eta_3 + \alpha_4\eta_4$ to improve upon $\overline{\tau}$, we refer to (4) and equate fourth order moments,

$$
\begin{aligned}
\mu_4 + \gamma &= \tfrac{1}{4}\alpha_1 + \tfrac{1}{32}\alpha_2 + \tfrac{1}{108}\alpha_3 + \tfrac{1}{256}\alpha_4, \\
\mu_{31} - \tfrac{1}{3}\gamma &= \tfrac{1}{96}\alpha_2 + \tfrac{1}{162}\alpha_3 + \tfrac{1}{256}\alpha_4, \\
\mu_{211} - \tfrac{1}{3}\delta &= \tfrac{1}{324}\alpha_3 + \tfrac{1}{256}\alpha_4, \\
\mu_{1111} + \delta &= \tfrac{1}{256}\alpha_4;
\end{aligned}
\qquad (9)
$$

The solutions are, inserting $\gamma = \frac{3}{2}(\mu_{31} - \mu_{22}) - \delta$,

$$
\left(\alpha_1, \alpha_2, \alpha_3, \alpha_4\right)' =
4\begin{pmatrix} 1 & -3 & 3 & -1 \\ 0 & 24 & -48 & 24 \\ 0 & 0 & 81 & -81 \\ 0 & 0 & 0 & 64 \end{pmatrix}
\begin{pmatrix} \mu_4 + \frac{3}{2}(\mu_{31}-\mu_{22}) - \delta \\ \frac{1}{2}(\mu_{31}+\mu_{22}) + \frac{1}{3}\delta \\ \mu_{211} - \frac{1}{3}\delta \\ \mu_{1111} + \delta \end{pmatrix}.
$$

In final terms we obtain

$$
\begin{aligned}
\alpha_1 &= 4\left(\mu_4 - 3\mu_{22} + 3\mu_{211} - \mu_{1111} - 4\delta\right), \\
\alpha_2 &= 48\left(\mu_{31} + \mu_{22} - 4\mu_{211} + 2\mu_{1111} + 4\delta\right), \\
\alpha_3 &= 108\left(3\mu_{211} - 3\mu_{1111} - 4\delta\right), \\
\alpha_4 &= 256\left(\mu_{1111} + \delta\right).
\end{aligned}
\qquad (10)
$$

In addition to the initial bounds (8), the requirements $\alpha_j \ge 0$ in (10) enforce further bounds on δ. Overall, we get the range $\delta_{\min}(\overline{\tau}) \le \delta_{\max}(\overline{\tau})$ where

$$
\delta_{\max}(\overline{\tau}) = \min\left\{ \frac{3}{4}(\mu_{31} - \mu_{22}), \; \frac{3}{4}(\mu_{211} - \mu_{1111}), \right.
$$
$$
\left. \frac{1}{4}(\mu_4 - 3\mu_{22} + 3\mu_{211} - \mu_{1111})\right\},
$$

and

$$
\delta_{\min}(\overline{\tau}) = -\min\left\{ \frac{1}{4}(\mu_{31} - \mu_{22}), \; \mu_{1111}, \right.
$$
$$
\left. \frac{1}{4}(\mu_{31} + \mu_{22} - 4\mu_{211} + 2\mu_{1111})\right\}.
$$

The following lemma shows that $\delta_{\min}(\overline{\tau}) \le 0 \le \delta_{\max}(\overline{\tau})$. In particular, $\delta = 0$ is always a feasible choice. The lemma says that, for every exchangeable design $\overline{\tau}$, there indeed exists a weighted centroid design $\eta(\delta)$ improving upon $\overline{\tau}$.

LEMMA 3. *Let $\overline{\tau}$ be an exchangeable design on the simplex $\mathcal{T}$, with fourth order moments $\mu_4, \mu_{31}, \mu_{22}, \mu_{211}, \mu_{1111}$. Then we have $\delta_{\min}(\overline{\tau}) \le 0 \le \delta_{\max}(\overline{\tau})$, and for every $\delta \in [\delta_{\min}(\overline{\tau}), \delta_{\max}(\overline{\tau})]$ the weighted centroid design $\eta(\delta) = \alpha_1\eta_1 + \alpha_2\eta_2 + \alpha_3\eta_3 + \alpha_4\eta_4$, with weights from (10), satisfies*

$$M(\eta(\delta)) \ge M(\overline{\tau}),$$

with equality if and only if $\delta = 0$ and $\overline{\tau} = \eta(0)$.

PROOF. The relation $\alpha_1 + \alpha_2 + \alpha_3 + \alpha_4 = 4\mu_4 + 48\mu_{31} + 36\mu_{22} + 144\mu_{211} + 24\mu_{1111} = 1$ is the simplex restriction formula. In order to show that the weights α_j are nonnegative, we start with the special case $\delta = 0$.

Clearly, we have $\alpha_4 = 256\mu_{1111} \ge 0$. We also have $\alpha_3 = 324(\mu_{211} - \mu_{1111}) \ge 0$, since the nonnegative function $162 t_1 t_2 (t_3 - t_4)^2$ integrates to α_3. For α_2, the inequality $\mu_{31} \ge \mu_{22}$ from Lemma 2 yields

$$
\begin{aligned}
\alpha_2 &\ge 96\left(\mu_{22} - 2\mu_{211} + \mu_{1111}\right) \\
&= 24\int (t_1 - t_2)^2 (t_3 - t_4)^2 \, d\overline{\tau} \ge 0.
\end{aligned}
$$

For α_1, we use the symmetric function

$$
\begin{aligned}
\psi(t_1, t_2, t_3, t_4) &= t_1^4 + t_2^4 + t_3^4 + t_4^4 \\
&- 2t_1^2 t_2^2 - 2t_1^2 t_3^2 - 2t_1^2 t_4^2 - 2t_2^2 t_3^2 - 2t_2^2 t_4^2 - 2t_3^2 t_4^2 \\
&+ t_1^2 t_2 t_3 + t_1^2 t_2 t_4 + t_1^2 t_3 t_4 + t_1 t_2^2 t_3 + t_1 t_2^2 t_4 \\
&+ t_2^2 t_3 t_4 + t_1 t_2 t_3^2 + t_1 t_3^2 t_4 + t_2 t_3^2 t_4 + t_1 t_2 t_4^2 \\
&+ t_1 t_3 t_4^2 + t_2 t_3 t_4^2 - 4 t_1 t_2 t_3 t_4.
\end{aligned}
$$

Because of homogeneity, ψ is nonnegative on the simplex $\mathcal{T}$ if and only if it is nonnegative on the quadrant $[0, \infty)^4$. In the interior $(0, \infty)^4$, the gradient vanishes only along the equiangular line, $t_1 = t_2 = t_3 = t_4$, where ψ attains the minimum value zero. By continuity, ψ stays nonnegative on all boundaries. This ensures $\alpha_1 = \int \psi \, d\overline{\tau} \ge 0$.

Hence, in the special case when $\delta = 0$, the weights α_j are nonnegative, and $\delta_{\min}(\overline{\tau}) \le 0 \le \delta_{\max}(\overline{\tau})$. Generally then, as long as δ stays in the range $[\delta_{\min}(\overline{\tau}), \delta_{\max}(\overline{\tau})]$, the weights α_j remain nonnegative. Therefore the weighted centroid design $\eta(\delta)$ is well-defined. It fulfills $\mu_{(3)}(\eta(\delta)) = \mu_{(3)}(\overline{\tau})$. We verify equation (7), whence the bounds on the range of δ secure $-\frac{1}{7}\gamma \le \delta \le \gamma$. Now Lemma 1 yields $M(\eta(\delta)) \ge M(\overline{\tau})$.

If equality holds then $\Delta = 0$ in (5). Hence $\gamma = \delta = 0$ and, from (7), $\mu_{31} = \mu_{22}$. By Lemma 2, $\overline{\tau}$ is a weighted centroid design. Denote the weights of $\overline{\tau}$

by $\beta_1, \beta_2, \beta_3, \beta_4$. That $\eta(0)$ and $\bar{\tau}$ are identical now follows from

$$
(\alpha_1, \alpha_2, \alpha_3, \alpha_4)'
$$
$$
= 4 \begin{pmatrix} 1 & -3 & 3 & -1 \\ 0 & 24 & -48 & 24 \\ 0 & 0 & 81 & -81 \\ 0 & 0 & 0 & 64 \end{pmatrix} \begin{pmatrix} \mu_4 + \frac{3}{2}(\mu_{31} - \mu_{22}) \\ \frac{1}{2}(\mu_{31} + \mu_{22}) \\ \mu_{211} \\ \mu_{1111} \end{pmatrix}
$$
$$
= (\beta_1, \beta_2, \beta_3, \beta_4)' \qquad \square
$$

Our examples share a common characteristic: Let $\epsilon(t_1, t_2, t_3, t_4)$ be the one-point design with support point $t = (t_1, t_2, t_3, t_4)' \in \mathcal{T}$. By averaging over all permutations we obtain the exchangeable design $\bar{\epsilon}(t_1, t_2, t_3, t_4)$, assigning equal weight to the distinct permutations of t. Because of exchangeability we may assume the components of t to be ordered, $t_1 \geq t_2 \geq t_3 \geq t_4$. In Examples 1 and 2, t depends on a real support parameter r.

EXAMPLE 1. Lemma 3 is illustrated using the one-parameter family of exchangeable designs $\tau_r = \bar{\epsilon}(\frac{1}{2} - r, \frac{1}{2} - r, r, r)$ for $r \in [0, \frac{1}{4}]$, assigning equal weight $1/6$ to each of the six permutations of $(\frac{1}{2} - r, \frac{1}{2} - r, r, r)'$. This family includes the edge midpoints design, $r = 0$, and the overall centroid design, $r = \frac{1}{4}$. The improving design $\eta(0)$ from Lemma 3 has weights

$$
\begin{aligned}
\alpha_1(r) &= \tfrac{1}{2} r(1 - 2r)(1 - 4r)^2, \\
\alpha_2(r) &= (1 - 6r + 12r^2)(1 - 4r)^2, \\
\alpha_3(r) &= 27 \, \alpha_1(r), \\
\alpha_4(r) &= 64 \, r^2(1 - 2r)^2.
\end{aligned}
$$

The bounds $\delta_{\min}$ and $\delta_{\max}$ for δ are conveniently expressed in terms of the preceding weights as

$$
\delta_{\max}(\tau_r) = \tfrac{1}{16}\alpha_1(r),
$$
$$
\delta_{\min}(\tau_r) = \begin{cases} -\tfrac{1}{48}\alpha_1(r) & \text{for } r \in [\tfrac{1}{4} - \tfrac{\sqrt{3}}{8}, \tfrac{1}{4}], \\ -\tfrac{1}{256}\alpha_4(r) & \text{for } r \in [0, \tfrac{1}{4} - \tfrac{\sqrt{3}}{8}]. \end{cases}
$$

The edge midpoints design has $\alpha_2(0) = 1$, and the overall centroid design has $\alpha_4(\frac{1}{4}) = 1$, as one would expect. With $\delta = 0$, the value of γ is $\frac{1}{8}\alpha_1(r)$. $\quad\square$

EXAMPLE 2. Our second example is provided by the designs $\tau_r = \bar{\epsilon}(1 - 3r, r, r, r)'$ for $r \in [0, \frac{1}{3}]$, assigning weight $1/4$ to the four permutations of $(1 - 3r, r, r, r)'$. This family includes the vertex points design, $r = 0$, the overall centroid design, $r = \frac{1}{4}$, and the face centroids design, $r = \frac{1}{3}$. The weights of the improving design $\eta(0)$ from Lemma 3 are

$$
\begin{aligned}
\alpha_1(r) &= (1 - r)(1 - 3r)(1 - 4r)^2, \\
\alpha_2(r) &= 12 \, r(1 - 3r)(1 - 4r)^2, \\
\alpha_3(r) &= 81 \, r^2(1 - 4r)^2, \\
\alpha_4(r) &= 256 \, r^3(1 - 3r).
\end{aligned}
$$

With $\delta = 0$, the value of γ is $\frac{1}{32}\alpha_2(r)$. The bounds $\delta_{\max}$ and $\delta_{\min}$ are

$$
\delta_{\max}(\tau_r) = \begin{cases} \tfrac{1}{16}\alpha_1(r) & \text{for } r \in [\tfrac{1}{4}, \tfrac{1}{3}], \\ \tfrac{1}{432}\alpha_3(r) & \text{for } r \in [0, \tfrac{1}{4}]; \end{cases}
$$
$$
\delta_{\min}(\tau_r) = \begin{cases} \tfrac{1}{192}\alpha_2(r) & \text{for } r \in [\tfrac{1}{8}, \tfrac{1}{3}], \\ \tfrac{1}{256}\alpha_4(r) & \text{for } r \in [0, \tfrac{1}{8}]. \end{cases}
$$

The graphs of $\delta_{\max}(\tau_r)$ and $\delta_{\min}(\tau_r)$ are shown in Figure 3. $\quad\square$

EXAMPLE 1: $\tau_r = \bar{\epsilon}(\frac{1}{2} - r, \frac{1}{2} - r, r, r)$

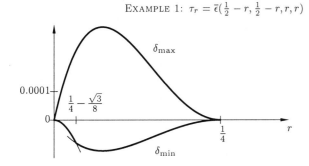

EXAMPLE 2: $\tau_r = \bar{\epsilon}(1 - 3r, r, r, r)$

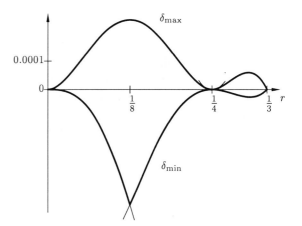

FIG. 3. Range $[\delta_{\min}(\tau_r), \delta_{\max}(\tau_r)]$, for various designs τ_r. The range contains the moment parameter δ. It thus reflects how many weighted centroid designs $\eta(\delta)$ improve upon the given exchangeable design τ_r.

In both examples, each of the two terms of which $\delta_{\min}$ is the minimum is needed, for some r. This is not so for $\delta_{\max}$. In Example 1, the three terms of which $\delta_{\max}$ is the minimum are identical, so any one of them suffices. In Example 2, the minimum in the definition of $\delta_{\max}$ is determined by the second and the third terms. In other examples, though, the first term may become relevant. For instance, this happens in the family of designs τ_r supported by the 12 permutations of $(0.4 - r, 0.3, 0.3, r)'$, for $r \in [0, 0.4]$. In general, each of the three terms of which $\delta_{\max}$ is defined to be the minimum has a role to play.

In conclusion, the complete class result for the Kiefer ordering in the case of four ingredients takes the following form. The class is formed by all convex combinations of the vertex points design η_1, the edge midpoints design η_2, the face centroids design η_3, and the overall centroid design η_4. The theorem refers to K-models and S-models alike.

THEOREM 4. *In the four-ingredient second-degree mixture model, the set of weighted centroid designs*

$$\mathcal{C} = \{\alpha_1 \eta_1 + \alpha_2 \eta_2 + \alpha_3 \eta_3 + \alpha_4 \eta_4 :$$
$$(\alpha_1, \alpha_2, \alpha_3, \alpha_4)' \in \mathcal{T}\}$$

is convex, and constitutes a minimal complete class of designs for the Kiefer ordering.

PROOF. Completeness follows as in Theorems 6.4 and 7.4 of Draper and Pukelsheim (1999). For minimal completeness, the last paragraph in the proof of Lemma 3 shows that any two weighted centroid designs η and τ satisfy $M(\eta) \geq M(\tau)$ only if $\eta = \tau$. $\square$

We append some remarks elucidating the various roles played by the simplex $\mathcal{T}$. In its primal meaning, as the experimental domain that underlies mixture models, $\mathcal{T}$ is exchangeable. In its dual meaning, as a set parameterizing the convex complete class $\mathcal{C}$ of Theorem 4, $\mathcal{T}$ is, of course, not exchangeable: α_1 belongs to the vertex design η_1, α_2 to the edge midpoints designs η_2 which is different, etc.

This helps in understanding the distinct geometric properties of the solutions of the system (10). As δ varies over its range $[\delta_{\min}(\bar{\tau}), \delta_{\max}(\bar{\tau})]$, let $\alpha(\delta) = (\alpha_1, \alpha_2, \alpha_3, \alpha_4)'$ denote the weight vector that uniquely solves (10). Evidently we have

$$\alpha(\delta) = \alpha(0) + \delta d,$$

where $d = 16(-1, 12, -27, 16)'$. This means that the set of all such weight vectors forms a segment on the line through $\alpha(0)$ in the direction given by d. The design $\bar{\tau}$ enters in that it determines the base point, $\alpha(0)$, and the length of the segment, $\delta_{\max}(\bar{\tau}) + \delta_{\min}(\bar{\tau})$. The direction d, however, always stays the same, because it has to be understood relative to the fixed orientation of the parameter space $\mathcal{T}$. In the geometry of these weight vectors, the range for δ ensures that $\alpha(0) + \delta d$ stays in the parameter space $\mathcal{T}$, besides securing (8).

The geometry translates into the moment space. For a given value of δ, let $\mu(\eta(\delta)) = (\mu_4(\eta(\delta)), \mu_{31}(\eta(\delta)), \mu_{211}(\eta(\delta)), \mu_{1111}(\eta(\delta)))'$ denote the fourth order moment vector of the weighted centroid design $\eta(\delta)$ of Lemma 3. From (9) we obtain the line segment

$$\mu(\eta(\delta)) = \mu(\eta(0)) + \delta e,$$

where again the direction $e = \frac{1}{3}(-3, 1, -1, 3)'$ does not depend on $\bar{\tau}$. The present line segment assembles those moment vectors $\mu(\eta(\delta))$ that improve upon $\bar{\tau}$, in the Loewner ordering sense of having $M(\eta(\delta)) \geq M(\bar{\tau})$. In the geometry of these moment vectors, the bounds on δ secure (4), and also imply that $\mu(\eta(0)) + \delta e$ lies in the moment polytope spanned by $\mu(\eta_j)$ for $j = 1, 2, 3, 4$.

The structure of the moment polytope, in the general case of five or more ingredients, is discussed in more detail in Draper, Heiligers and Pukelsheim (1998).

References

CHENG, C.-S. (1995). Complete class results for the moment matrices of designs over permutation-invariant sets. *Annals of Statistics* **23** 41–54.

CORNELL, J.A. (1990). *Experiments with Mixtures, 2nd Ed.* Wiley, New York.

DETTE, H. (1997). Designing experiments with respect to 'standardized' optimality criteria. *Journal of the Royal Statistical Society Series B* **59** 97–110.

DRAPER, N.R., B. HEILIGERS and F. PUKELSHEIM (1998). Kiefer ordering of simplex designs for second-degree mixture models with four or more ingredients. *Annals of Statistics* (submitted).

DRAPER, N.R. and F. PUKELSHEIM (1998a). Polynomial representations for response surface modelling. In: *New Developments and Applications in Experimental Design—Selected Proceedings of a 1997 Joint AMS-IMS-SIAM Summer Conference* (N. Flournoy, W.F. Rosenberger, W.K. Wong, eds.), *Institute of Mathematical Statistics Lecture Notes–Monograph Series* **34** 199–212.

DRAPER, N.R. and F. PUKELSHEIM (1998b). Mixture models based on homogeneous polynomials. *Journal of Statistical Planning and Inference* **71** 303–311.

DRAPER, N.R. and F. PUKELSHEIM (1999). Kiefer ordering of simplex designs for first- and second-degree mixture models. *Journal of Statistical Planning and Inference* **79** 325–348.

GAFFKE, N. and B. HEILIGERS (1996). Approximate designs for polynomial regression: Invariance, admissibility, and optimality. In: *Handbook of Statistics* **13** (S. Ghosh and C.R. Rao, eds.) 1149–1199. Elsevier, New York.

GALIL, Z. and J.C. KIEFER (1977). Comparison of simplex designs for quadratic mixture models. *Technometrics* **19** 445–453. Also in: Kiefer (1985) 417–425.

HEILIGERS, B. (1991). Admissibility of experimental designs in linear regression with constant term. *Journal of Statistical Planning and Inference* **28** 107–123.

HEILIGERS, B. (1992). Admissible experimental design in multiple polynomial regression. *Journal of Statistical Planning and Inference* **31** 219–233.

KIEFER, J.C. (1959). Optimum experimental designs. *Journal of the Royal Statistical Society Series B* **21** 272–304. Also in: Kiefer (1985) 54–101.

KIEFER, J.C. (1975). Optimal design: Variation in structure and performance under change of criterion. *Biometrika* **62** 277–288. Also in: Kiefer (1985) 367–378.

KIEFER, J.C. (1978). Asymptotic approach to families of design problems. *Communications in Statistics- Theory and Methods* **A7** 1347–1362. Also in: Kiefer (1985) 431–446.

KIEFER, J.C. (1985). *Collected Papers III—Design of Experiments.* (L.D. Brown, I. Olkin, J. Sacks, H.P. Wynn, eds.). Springer, New York.

PUKELSHEIM, F. (1993). *Optimal Design of Experiments.* Wiley, New York.

SCHEFFÉ, H. (1958). Experiments with mixtures. *Journal of the Royal Statistical Society Series B* **20** 344–360.

SCHEFFÉ, H. (1963). The simplex-centroid design for experiments with mixtures. *Journal of the Royal Statistical Society Series B* **25** 235–251.

Norman R. Draper
Department of Statistics
University of Wisconsin–Madison
Madison WI 53706-1685, USA
E-mail: `Draper@Stat.Wisc.Edu`

Berthold Heiligers
Fakultät für Mathematik
Otto-von-Guericke Universität
D-39016 Magdeburg, Germany
E-mail: `Heiligers@Mathematik.Uni-Magdeburg.De`

Friedrich Pukelsheim
Institut für Mathematik
Universität Augsburg
D-86135 Augsburg, Germany
E-mail: `Pukelsheim@Uni-Augsburg.De`

SIMULATION MODELS FOR SPOUTED BED ENGINEERING SYSTEMS

Carolyn B. Morgan and Morris H. Morgan, III , Hampton University
C. Morgan, Hampton University, Hampton, VA 23668

KEY WORDS: Modeling, prediction, simulation, residence time distribution

Abstract

The task of modeling the cycle time or residence time distribution for many engineering systems still remains a major challenge. The knowledge of the system cycle or residence time is critical for characterizing, controlling and predicting the behavior of spouted bed reactors. In previous investigations, it has been shown that the cycle time distribution (CTD) and its moments, when coupled with renewal theory concepts, are successful for analyzing such systems. Morgan and Morgan have demonstrated the utility of the Gamma Distribution for analyzing spouted bed coaters. In this paper simulation results are presented for a spouted bed electrolytic reactor. These results are used to develop both a predictive model for engineering scale up purposes and a mechanistic appreciation of the role of the overall system residence time distribution, reactor contact time and subsystem holding times on performance.

1. Background

Many manufacturing (aircraft, circuit board, machinery, metal finishing, etc.), dental, xray, and photography businesses generate metals which must be recovered from aqueous waste streams. Recent engineering research studies have demonstrated that Spouted Bed Electrolytic Reactors (SBER) are a potential cost effective "pollution prevention" technology for "point source" recovery and recycling of these metals. SBER systems employ no chemicals, generate no solid waste, and yield highly purified metal recycle. In these systems, conductive (metallized or metallic) particles serve as cathodes for electrolytic recovery. The challenge is to develop models to improve system performance and to reduce cost, equipment size and maintenance requirements. This paper describes the results of several SBER experimental runs performed to simulate metal depositions from silver (Ag), aluminum (Al), gold (Au), copper (Cu), and nickel (Ni) aqueous solutions. A predictive SBER model for identifying optimum operating conditions for each metal recovery application was developed.

2. Description of Spouted Bed Electrolytic Reactor

A schematic of the spouted bed electrolytic reactor apparatus and the "closed loop" flow system is given in Figure 1. The SBER consists of a draft tube area located in the center of the bed and the annular region surrounding the draft tube. In the SBER, the cathode is a spouted bed of conductive particles and the anode is located in the liquid above the particle bed. Liquid flow is initiated at the bottom of a conical vessel, either directly into the bed of particles or via a draft tube. In the electrolytic recovery process the metal is reduced from its ionic to its metallic state by the passage of current through the metal ion-containing solution.

Prior to entering the SBER, the aqueous liquids are pre-mixed in a holding tank. This fluid enters the reactive bed via the bottom inlet and circulates between the draft tube and the annulus. The distributor cone is used to direct all the particles disengaging from the fountain area to the periphery of the bed. A distributor cone was used in some runs to eliminate particle stagnation along the bed bottom.

As in Calo's study, three different SBER designs or configurations were used in the experimental program - SBER with draft tube and distributor cone, SBER with draft tube and no distributor cone, and SBER with no draft tube and no distributor cone. For the simulations, the additional reactor system design parameters studied were the bed diameter, particle diameter, current levels and flow rates. Simulation runs were conducted at the following levels for each of these variables.

PRIMARY VARIABLES	LEVELS	UNITS
solid particle diameters	3 and 6	mm
bed diameter	4, 7.5 and 12	inches
current levels	2.5, 5, 7.5 and 10	amperes
flow rates	6.3, 14, 21.2, 33.33	liters/min

Metal concentrations in solution were measured before and after each run by atomic absorption spectroscopy (AA), and the total amount of metal recovered was determined by the difference. Thus, the response variable and the other design variables were directly measurable and controllable.

3. Overview of The Modeling Strategy

To date no mechanistic or predictive models are available for determining the interrelationship between the response variable (concentration) and the system design factors, i.e. current, bed geometry, particle size, and internal system flow rates. In the present study, regression techniques were used to estimate the intrinsic deposition factor for this electrolytic precipitation process. This approach allows for design extrapolation to new conditions since it has a mechanistic undergirding. In fact, it is demonstrated that prediction of performance for untested metal solutions can be ascertained from our model. This model links the metal deposition to the current, amperage and critical subsystem times (SBER reactor and the holding tank) via Faraday's Law.

4. Engineering-Based Modeling Strategy

As part of the modeling strategy, efforts were focused on the development of a model which could be employed to describe the entire "closed loop" system. The starting point for model development is the following system conservation equation shown below

$$\tau_h \frac{dC_{out}(t)}{dt} + C_{out}(t) = C_{out}(t - \tau_r) - k\tau_r$$

where C_{out} – metal concentration time t

τ_r – reactor processing time

k – Faraday reaction constant

τ_h – holding time in the tank

I.

An approximate solution to this equation results when the reaction time is much less than the holding tank residence time. Under such conditions, the following linear model results for metal deposition in the SBER.

$$C(t) = C_0 - kt$$

$$where \quad k = \frac{k_d \times \tau_r}{metallization \ factor \times \tau_h}$$

$$k_d = a_0 \times amperes$$

II.

Thus, the metal deposition process as suggested by Faraday's Law analogy is a zero-order reaction process. This means that metal deposition can be directly related to the product of current and total processing time as depicted in Equation III.

$$C(t) = C_0 - \frac{a_0 \times amperes \times \tau_r \times time}{metallization \ factor \times \tau_h}$$

III.

Table 1.

Metallization Factor Estimation

Element	Charge Transfer	Molecular Weight	Metal. Factor x 10-6
Aluminum	3	27	111.1
Copper	2	63.5	31.5
Zinc	2	65	30.7
Nickel	2	58.7	34.1
Gold	3	197	15.2
Lead	2	207	9.66
Silver	1	107.9	9.27

The data from the simulation runs were analyzed using the SPLUS 2000 software. A model was developed to relate amperes to kd. Both linear and nonlinear models were investigated. The no intercept model was explored. Mechanistically, such a model should be appropriate for electrochemical deposition processes. Residual analyses and other graphical and non-graphical diagnostic evaluations were conducted to assess the appropriateness of the resulting model.

The resulting model is show and plotted in Figure 2. There are two salient features that arise from the modeling exercise. Equation III can be used to unravel the relationship between metal charge characteristics and deposition as highlighted in Figure 2. In addition, the effect of system amperage change on deposition is also revealed. When comparing Equation III with actual experimental data, it is also apparent that this linear behavior persists for long contact times and only at low concentrations is there any evidence of a backstripping reaction. Thus, it appears that Equation III can be used by designers to extrapolate to new untested conditions.

The estimated model was used to predict concentration as a function of time for the silver, copper, nickel and aluminum aqueous solutions assuming 10 amps and an initial concentration of 700 PPM. The modeling results are presented in Figure 3. These results are in close agreement with the actual results reported by Calo and others.

5. Concluding Remarks

The simulation results demonstrate that Faraday's Law is appropriate for describing depositions for SBER applications. The metallization factor can be used to identify a simple model for predicting the dependency of concentration on material charge characteristics. The mixer and reaction holding times are variables that appear explicitly in the model. For reactor times much less that the tank holding time, a simple linear relationship arises. Volumetric flow rates are imbedded in τ_r and τ_h. The series of figures provided highlight the effect of critical design variables on SBER performance.

References

1. Calo, J.M. "Point Source" Metals Recovery Via Spouted Bed Electrolytic Reactors (SBER), Annual Report submitted to Lucent Technologies Foundation, October 1977.
2. Eichenlaub, N., Freedman, G. and Noh, K., "The Silver Project," a student report to EN194/S16, Spring 1996, Division of Engineering, Brown University.
3. Hradil, G. and E. Hradil, "Electrolytic Recovery in Conjunction with Chemical Precipitation for Metal Recycling from Waste Streams," presented at AESS/EPA Conf. on Pollution Prevention and Control, Feb. 13-15, Orlando, FL.
4. Huber et. al., U.S. Patent #5164091 (1992).
5. Kammel, R. and H. W. Liber, U.S. Patent #4123340 (1978).
6. Kim, S. J. and H. Littman, *Can. J. Chem. Eng.* **65,** 723 (1987).
7. Morgan, III, M.H., H. Littman and B. Sastri, *Can. J. Chem. Eng.* **66**, 735 (1988).
8. Morgan, C.B., and Morgan III, M.H., "Statistical Mixing Parameters for a Down Flow Coal Liquefaction Reactor, " *Proceedings, Section on Physical and Engineering Sciences*, American Statistical Association, 1998.
9. Spearot, R.M., *Metal Products and Machinery: Proposed Effluent Guidelines -An Update*, Metal Finishing, Elsevier, p. 28, August 1994.

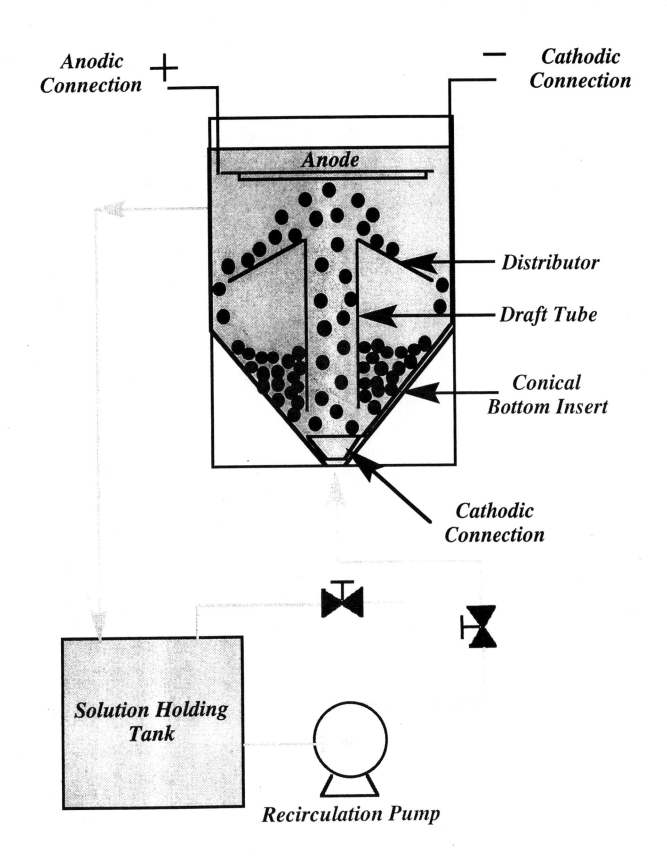

Figure 1. Schematic of spouted bed electrolytic reactor (SBER) apparatus and flow system.

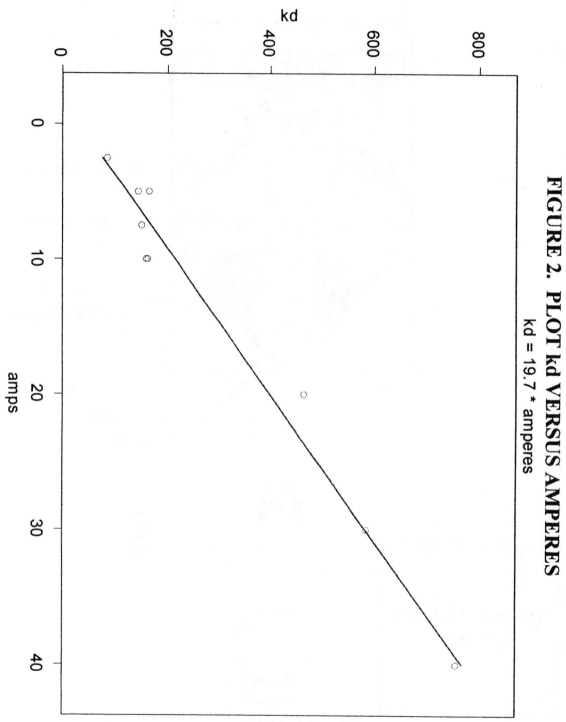

FIGURE 2. PLOT kd VERSUS AMPERES

kd = 19.7 * amperes

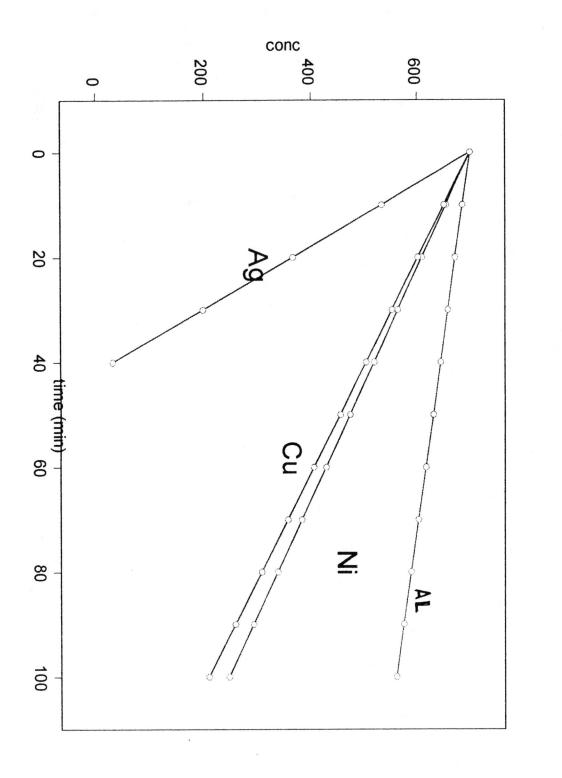

FIGURE 3. PLOT PREDICTED CONCENTRATION VS TIME

10 Amps, Initial Conc 700 ppm

COMPARATIVE STATISTICAL ANALYSIS OF ATMOSPHERIC OBSERVATIONS AND MODELING

Alexander Gluhovsky, Ernest Agee, Purdue University
Alexander Gluhovsky, Dept. of Statistics, Purdue University, West Lafayette, IN 47907

Key Words: Time Series, Data Analysis, Filtering
Random Fields

1. INTRODUCTION

Most of our understanding of atmospheric dynamics comes from observations and modeling. Both activities provide large amounts of data from which statistical characteristics of meteorological elements are computed. There are problems in a statistical treatment of these data, and the paper deals with the statistical reliability of atmospheric data analysis and modeling.

One problem is that real time series are not stationary, since meteorological fields may exhibit varied behavior over different regions. This is important, because in theory, i.e. in statistical fluid mechanics as formulated by Kolmogorov, a fluid dynamical variable is considered a random field in 4D space-time, and statistical characteristics are defined, accordingly, as ensemble averages.

In practice, estimates for these statistical characteristics are computed from data by space or time averaging. Such approximations of ensemble averages are valid for ergodic random fields. These are necessarily homogeneous, or stationary for observational data which are 1D (the behavior of a meteorological element is recorded along one spatial or temporal variable). Again, observational data are not stationary.

In modeling, a topic of intense development is *large eddy simulation* (LES), data are 3D, and horizontal fields can often be considered homogeneous.

To deal with nonstationarity, Gluhovsky and Agee (1994) suggested a procedure that permits selection of intervals within the record, where the time series can be considered stationary.

But even if data are stationary or homogeneous, there is *another problem* discussed in Section 2: record lengths from observations and domains in LES are typically inadequate for accurate estimation. To obtain an acceptable accuracy, the length of the record must be much larger than the integral scale of the process under study (that defines roughly the distance over which noticeable correlation is preserved). A technique based on band-pass filtering is suggested in Section 3 to deal with the typical situation when only one record of a fixed length is available of a process with the integral

scale too large to permit an accurate estimation. The technique makes possible rigorous comparisons between LES models and atmospheric measurements: one obtains quantities that can be reliably estimated and compared. It also provides a solid basis for explaining the well-known discrepancy between statistical characteristics of the vertical velocity in large eddy simulations of planetary boundary layers and field observations (Agee and Gluhovsky 1999). Section 4 describes the advantages of box area measurements.

2. RECORD LENGTH CONSIDERATIONS

Suppose that we have a record of length T of the stationary process W_t (t here does not necessarily denote time, it may be the position on a line) with mean $\mu = <W_t>$ and autocovariance sequence $B(\tau) = \left\langle (W_t - \mu)(W_{t+|\tau|} - \mu) \right\rangle$, and that the parameter of interest is μ, which is estimated with the sample mean

$$m_T = T^{-1} \sum_{t=1}^{T} W_t . \qquad (2)$$

If the covariance function satisfies certain regularity conditions, then m_T is $AN(\mu, \ \sigma^2/T)$, $\sigma^2 = \sum_{\tau=-\infty}^{\infty} B(\tau)$ as $T \to \infty$, (Brockwell and Davis 1991), so confidence intervals for μ can be constructed.

For σ^2 we used a consistent estimate (Politis and Romano 1993)

$$\hat{\sigma}_b^2 = \frac{b}{T-b+1} \sum_{i=1}^{T-b+1} \left(\frac{1}{b} \sum_{t=i}^{i+b-1} W_t - m_T \right)^2 ,$$

based on the values of the statistic computed over $T-b+1$ sub-samples $\{W_i, W_{i+1}, ..., W_{i+b-1}\}$. The size b of the sub-sample was chosen so as to maximize $\hat{\sigma}_b$, which gives a conservative estimate of σ resulting in a conservative estimate of the confidence interval length.

The variance of the estimator (2) is given by (Lumley and Panofsky 1964, Yaglom 1987)

$$\sigma_{m_T}^2 = \frac{\sigma^2}{T} = \frac{\sum_{\tau=-\infty}^{\infty} B(\tau)}{T} \approx 2B(0)\frac{I}{T} , \qquad (3)$$

where $I = B^{-1}(0)\left[\dfrac{1}{2}B(0) + \sum_{\tau=1}^{\infty} B(\tau)\right]$ is the integral scale of the process. Thus, the larger the integral scale of the process, the less accurate is our estimation.

This also means that if an acceptable level ε is picked for the relative error $\sigma_{m_{T_1}}/\mu$, then the length T of the averaging interval should be at least

$$T = 2\frac{B(0)}{\mu^2}\frac{I}{\varepsilon^2}.$$

Again, the larger the integral scale of the process, the longer the record should be to achieve the same accuracy.

Similarly, record lengths can be determined for higher moments and other characteristics. For example, by assuming W_t to be Gaussian (which is often not the case) and considering W_t^2 and W_t^4 instead of W_t, Lumley and Panofsky (1964) found for estimating, with relative error ε, the second and the fourth moments, respectively:

$$T = 4\frac{J}{\varepsilon^2} \quad \text{and} \quad T = \frac{64}{3}\frac{J}{\varepsilon^2}.$$

Lenschow et al. (1994) found that the record length for estimating with a relative error of 10% the fourth moment of a Gaussian characteristic in the atmosphere with integral scale $I = 0.1$ km is $T = 90$ km, and $T = 400$ km for a more realistic, non-Gaussian one.

We studied data from Project LESS (Lake-Effect Snow Studies). This field program was conducted in the winter of 1983-84 over Lake Michigan during the outbreak of a polar continental air mass over the Great Lakes region. Two research aircraft flew over Lake Michigan at five levels between west and east coasts, and produced five vertically stacked levels of boundary layer data. Figure 1 presents the records of the vertical velocity at four levels. We subjected the data to our test for stationarity and were able to compute (from stationary data) statistical characteristics of the vertical velocity and their confidence intervals (Gluhovsky and Agee 1994).

The results indicated that one can extract three regions over Lake Michigan where statistical properties of vertical velocity exhibit statistically significant differences. The existence of these regions reflects changing physical conditions in the convective boundary layer.

We failed to find intervals larger than 30 km where data can be considered stationary but we did find that to achieve acceptable accuracy much longer records are needed.

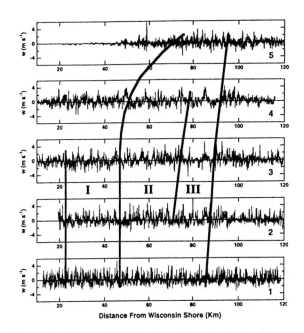

Figure 1. Plots of 20-Hz aircraft vertical velocity measurements from Project LESS

The third and the fourth column in Table 1 pres[ent] sufficient record lengths for the estimation of vertical velocity variance and the vertical velocity skewness for various segments of stationary data from Project LESS. These demonstrate that the situation is not completely safe even for the estimation of variance and it becomes hopeless for the estimation of skewness.

Table 1. Record lengths, $T^{(V)}$ and $T^{(S)}$, for estimation with 10% accuracy of the vertical velocity variance and skewness, respectively, and the side of the square, $\tilde{T}^{(S)}$ that provides the same accuracy for the estimation of skewness.

LEVEL	DATA SEGMENT	$T^{(V)}$	$T^{(S)}$	$\tilde{T}^{(S)}$
4	$50 - 75$	58	2995	40
	$75 - 92$	133	1120	35
3	$20 - 45$	20	7570	52
	$45 - 75$	27	1725	27
	$75 - 90$	64	10888	75
2	$20 - 45$	26	470	12
	$55 - 70$	23	659	16
	$70 - 85$	50	181	8
1	$20 - 45$	15	478	9
	$50 - 85$	37	170	5

3. COMPARISONS BETWEEN ATMOSPHERIC OBSERVATIONS AND LARGE-EDDY SIMULATIONS

What would be sensible in a typical situation with only one record of length T too short to estimate accurately statistical characteristics of a stationary process with certain integral scale? Note that it is usually large-scale (low-frequency) components that make the integral scale too large for estimation with acceptable accuracy at given record length. It may be practical to filter out larger-scale components of the process and study its smaller-scale part that can sometimes be accurately estimated. One immediate benefit of this approach is the possibility to make rigorous comparisons between LES models and atmospheric measurements.

To illustrate how this works, we have also modeled the above Project LESS situation with LES. Figure 2 presents profiles for the variance of the vertical velocity: profile I (dashed line) is computed from LES, profile II (solid line) - from the field data LESS. Then we removed larger-scale components. In fact we did more. Besides limited domains (5x5x2 km) that eliminate effects of larger wavelength structures, LES are also limited by a finite grid size. Thus, they are intrinsically band-pass filters, and to compare oranges with oranges we band-passed the field data.

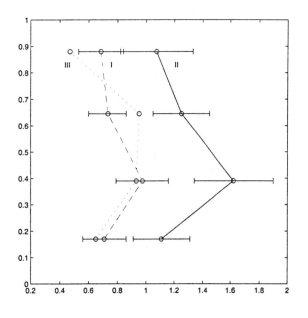

Figure 2. Vertical velocity variance.

New profile III (dotted line in Figure 2) from the filtered data agrees much better with that from LES (profile I) than profile II computed from the original data.

Nature does not have domains and grids. Because of inadequate domain size in current LES models, larger scales are underrepresented in the frequency spectrum in contrast to field observations, where larger scales contribute considerably to the formation of the vertical velocity variance profile. This probably explains, particularly for the two lowest levels, the discrepancy between statistical characteristics computed from atmospheric measurements and from modeling.

4. BOX AREA MEASUREMENTS

Wyngaard (1983) noted that averaging over an area is much more promising. He considered a characteristic with integral scale $I = 0.5$ km that required a record of length $T = 1000$ km to achieve 10% accuracy and suggested that averaging over 25 km square provided the same accuracy.

Indeed, if W is an *isotropic* random field ($B(\tau)$ depends only on the length of the vector τ) and its mean is estimated over a square of side $\tilde{T}$ with

$$m_{\tilde{T}} = \frac{1}{\tilde{T}^2} \sum_{t_1, t_2 = 1}^{\tilde{T}} W_{t_1, t_2},$$

then (Yaglom 1987)

$$\sigma_{m_{\tilde{T}}}^2 = \frac{\tilde{\sigma}^2}{\tilde{T}^2}, \quad \tilde{\sigma}^2 = \sum_{\tau_1 = -\infty}^{\infty} \sum_{\tau_2 = -\infty}^{\infty} B(\tau_1, \tau_2). \quad (4)$$

For an isotropic field, its statistical characteristics may also be estimated from measurements along any straight line. The variance of the estimate of the mean is then given by (3). By comparing (3) and (4), one obtains

$$\tilde{T} = \sqrt{\left(\sum_{\tau_1 = -\infty}^{\infty} \sum_{\tau_2 = -\infty}^{\infty} B(\tau_1, \tau_2) \middle/ \sum_{\tau = -\infty}^{\infty} B(\tau) \right)} \, T. \quad (5)$$

For fields with model covariance functions,

$$B_1(\tau) = e^{-\alpha \tau} \quad \text{and} \quad B_2(\tau) = e^{-\alpha \tau^2}, \quad (6)$$

Eq. (5) results in, respectively,

or
$$\tilde{T} = \sqrt{(\pi / \alpha) T} = \sqrt{\pi I T},$$
$$\tilde{T} = \sqrt{(\pi / \alpha)^{1/2} T} = \sqrt{2 I T}. \quad (7)$$

In the case considered by Wyngaard (1983), Eqs. (7) give, respectively, $\tilde{T} \approx 39.6$ km or $\tilde{T} \approx 31.6$ km.

The fact that averaging over a square with a side of 30 km may give reliable statistics indicates encouraging prospects for LES, and even for future atmospheric measurements using lidars.

5. SUMMARY AND CONCLUSIONS

This study has addressed critical statistical problems of atmospheric data analysis:

1) The testing technique for stationarity permits computing statistical characteristics with their confidence intervals thus providing estimate of their accuracy.

2) For *stationary* data, estimates can be made for record lengths required to achieve acceptable accuracy. These prove discouraging for now prevailing aircraft measurements.

3) Estimates of horizontal domains required to provide reliable inference show promise for large-eddy simulations and future box area atmospheric measurements (with lidars). It appears that horizontal domains of 30 km x 30 km should be adequate for most purposes.

6. ACKNOWLEDGMENTS

This work was supported by the Mesoscale Dynamics Meteorology Program, the National Science Foundation, under grants ATM-9813687 and ATM-9523572 awarded to Purdue University.

7. REFERENCES

Agee, E., and A. Gluhovsky, 1999: Further aspects of large eddy simulation model statistics and inconsistencies with field data. *J. Atmos. Sci.*, **56**, 2948 – 2950.

Gluhovsky, A., and E. Agee, 1994: A definitive approach to turbulence statistical studies in planetary boundary layers. *J. Atmos. Sci.*, **51**, 1682-1690.

Kendall, M, and A. Stuart, 1961: *The Advanced Theory of Statistics, Vol.2*. Hafner, New York.

Lenschow, D. H., J. Mann, and L. Kristensen, 1994: How long is long enough when measuring fluxes and other turbulence statistics? *J. Atmos. Oceanic Technol.*, **11**, 661-673.

Lumley, J. L., and H. A. Panofsky, 1964: *The Structure of Atmospheric Turbulence*. Interscience, 239 pp.

Politis, D. N., and J. P. Romano, 1993. On the sample variance of linear statistics derived from mixing sequences. *Stoch. Proc. Appl.*, **45,** 155-167.

Wyngaard, J. C., 1983: Lectures on the planetary boundary layer. *Mesoscale Meteorology –Theories, Observations and Models*. D. K. Lilly and T. Gal-Chen, Eds., D. Reidel Publishing Company, 603-650.

Yaglom, A., 1987: *Correlation Theory of Stationary and Related Random Functions*. Springer, 526 pp.

Bayesian Analysis of Optical Speckles in Light Diffusive Medium with An Embedded Object

Tak D. Cheung and Peter K. Wong

Department of Physics and Chemistry, Queensborough Community College, City University of New York, 222-05 56th Avenue, Bayside, NY 11364

Key Words: Bayesian decision procedure, object detection, inhomogeneous turbid medium, photon migration

ABSTRACT

Light diffusive medium was characterized using Bayesian analysis on the transmitted speckled laser profiles. Plastic tubes were inserted into a medium to imitate non-uniformity analogous to the presence of ducts in breast tissues. A calcium carbonate bead was used as the embedded object to mimic the presence of a small calcification in breast tissues. The lateral positions of the tubes were treated as known variables, while the depth position probability was estimated with additional data from sample rotation experiment. Point estimation of the median of the depth probability was used as the loss function in the Bayesian decision procedure. The presence of discontinuity in the loss function was interpreted as the presence of an embedded object.

1. INTRODUCTION

Propagation of electromagnetic waves in turbid media has been studied extensively in the last 20 years.[1] At the moment, many scientists are studying the use of laser imaging for early breast tumor detection. Earlier techniques include time-resolved detection of the ballistic and snake photons, using ultra-fast high power lasers in straight path geometry, and statistical modeling of late arrival photons. The most recent techniques include (a) the feasibility study of a 3-minute in-vivo time domain optical mammography on two patients who already have confirmed tumors[2], and (b) the application of the approximate extended Kalman filter, a recursive Bayesian minimum variance estimator, to frequency domain optical tomography in a simulated noise situation[3]. All these ultra-fast laser techniques aiming at replacing X-ray mammography have not, however, proven their definitive advantages over the latter, let alone being competitive with the emerging digital x-ray mammography.

In this paper, we address the issue of incorporating low power laser optical imaging into X-ray imaging for additional information via the Bayesian decision procedure. The feasibility of in-vivo optical imaging simultaneously with x-ray mammography depends on the statistical method used in modeling the limited optical data captured during x-ray mammography.

2. EXPERIMENT AND ANALYSIS

The turbid medium consists of an oil colloidal system containing hollow plastic tubes simulating breast tissue with ducts. Standard X-ray imaging was used to identify the lateral position of the plastic tubes. Transmitted laser profiles were used to measure the diffusive properties of the colloidal system[4]. The transmitted profile carries optical speckles, which are very common with coherent light sources. The phase information is retained due to the high degree of coherence in monochromatic laser light. The colloidal system has a mean free path of 0.5 mm under He-Ne irradiation and is 20 mean free path thick. Two uniform tubes of 2-mm diameter (4 mean free path length) were implanted with random direction (Figures 1 and 2). A calcium carbonate bead of size 0.5 mm was used as the embedded object and is located midway between the two tubes. A low power 3-mw He-Ne laser focused to 0.1 mm in diameter and a digital camera was used to measure the transmitted speckled profiles. Refractive index matching fluid was used to minimize the boundary effect.

Determination of the Mean Free Path Length

Although the mean free path (mfp) is controlled in the sample preparation, in-vivo application necessitates the characterization of the given diffusive medium from the speckled profiles. The variable formed by the ratio of the log of the transmitted profile peak to the log of the intensity of a nearby pixel follows a Cauchy distribution if the log of pixel data is transformed to the standard $N(0,1)$ distribution. We assume that σ represents material environment fluctuation and therefore is the same for nearby pixels in the standard transformation $(x-\mu)/\sigma$.

Let x_1 and x_2 denote data from pixels 1 and 2 and obey $N(0,1)$ distributions.

Let $r = x_1/x_2$, then $P(r) = 1/[(1+r*r)*p_i]$

A general Cauchy is $b/p_i*[b*b+(x-a)*(x-a)]$ with full width half maximum $= 2b$ and peak at $x=a$

Prob $(r_k|a,b) = b/pi*[(b*b+(x_k-a)*(x_k-a)]$

Prob $(r_k|a,b) = \Pi\ b/pi*[b*b+(x_k-a)*(x_k-a)]$ for k = 1 to 10 (assume there are 10 fluctuation data points)

Prob $(a,b|r_k,model) =$ Prob $(r_k|a,b)*$Prob $(a,b|model)$

$L = \log_e$Prob $(a,b|r_k,model)$

$L =$ Const $- \Sigma\log_e[b*b +(r_k-a)*(r_k-a)]$ for k= 1 to 10

We assume uniform pdf for Prob$(a,b|model)$

Starting with $\mu_1 = x_1$ average and $\mu_2 = x_2$ average,

find σ such that L is maximum for $b=1$ and $a =0$.

Then reiterate until L is maximum for μ_1, μ_2, and σ

The final values of μ_1 and μ_2 are the log intensity of pixels 1 and 2. The long-winged Cauchy distribution precludes the use of least square estimation.

After the speckled profile is smoothed out by the Bayesian Theorem procedure, the transmission profile can then be matched to those curves generated by photon migration simulation to determine the mean free path.

Photon migration statistics on uniform turbid medium without embedded plastic tubes was reported by us earlier[5]. It will be described here briefly for easy reference. Monte Carlo simulation based on the diffusion random walk statistics with the incorporation of Snell's law at the sample boundary yielded comparable good fit to literature data fitted with transport equation analytical solutions, provided that the sample has a thickness more than 10 mean free path length[6]. The flight of a photon is modeled as a random walk in 2 dimensions. Values of mean free path and penetration depth where the flux starts to randomize are inserted into the simulation. A weighting factor is included to allow for absorption.

Determination of Non-uniformity

The non-uniformity posed by the plastic tube can be written as the product of three probabilities: $P(x)P(y)P(z)$. $P(x)$ and $P(y)$ are probabilities (basically flat-top) extracted from straight path X-ray imaging. $P(z)$ contains the tube's depth information. Its probability is estimated from two different experiments, namely, taking data from both sides by reversing the geometry of laser/camera. This is also equivalent to 180° sample rotation. If the tube is in the middle, then the two transmitted profiles should be identical. Since the mfp was determined from the previous pixel ratio Bayesian procedure, Monte Carlo simulation can be used to decide how much of the medium must be added to the sample so that the profiles will be the same when measured from either side (A & B configurations). The amount that must be added is proportional to the tube depth (Figure 3).

Two models are used. Model 1 assumes a flat top distribution for Prob(tube depth position) for the possible values of position. Model 2 calculates a distribution profile with Prob(depth) inversely proportional to (profile area A–profile area B) (Figure 4).

Detection Modeling

The direct Monte Carlo simulation of non-uniform turbid medium with embedded objects is not possible due to the limit of the pseudo-random generator.[7] Other detection method has to be used. The detection scheme is quite straight-forward based on the concept of loss function.

Choose z' such that
Min Exp(Loss(z',z))=$\int$Loss$(z',z)P(z|data,model)dz$
Loss$(z',z) = (z'-z)^2$ quadratic loss (posterior point estimation of the mean)
  $= |z'-z|$ absolute loss (posterior point estimation of the median)
  $= 0$ (if $z' = z$), or $=1$ (if $z' \neq z$), zero-one loss (posterior point estimation of the mode)
Deviation on the otherwise smooth loss versus vertical distance curve indicates the presence of embedded objects (Figure 5). The loss function mode can be used as a risk value once the embedded object is detected.

3. RESULTS

The reduction of the speckled profile to a smoothed profile through Bayesian Theorem procedure gives a mean free path error of about 20% when the smooth profile is compared to the Monte Carlo simulation curves. The error in mfp determination is less than 10% if the smoothed profile is compared to the experimentally generated profiles smoothed by frequency averaging using a dye laser. This latter small error value is probably not useful in vivo because it takes too much time to collect the data.

It was found that the loss function $|z-z'|$ that corresponds to the point estimation of the median of P (tube depth|data, model) is most useful. Samples with tubes/bead near the interior give the best result, probably due to less boundary effect and that the light is truly randomized after the first mfp.

4. CONCLUSION

Other than the configuration with tubes near the laser flux randomization depth, the proposed algorithm based on Bayesian decision is effective in choosing the bead size. The effectiveness of reducing the speckled profiles to smooth profiles enables development of this procedure in vivo with x-ray mammography.

The information in optical measurement as modeled in this study would provide additional supplement to X-ray data, especially for early recognition of small calcifications, which tend to evade detection by the X-ray procedure, but contribute to optical scattering.

The power of the Bayesian decision procedure lies in its capacity to incorporate addition information without the normality assumption. Whether this algorithm tends towards false positive or false negative remains as an important clinical issue. In summary, this study shows that the diffusion information from various measurements in a non-uniform turbid medium could be modeled via the Bayesian decision procedure and used for detecting a hidden object.

5. ACKNOWLEDGMENT

The laser equipment used in this experiment was acquired through a VATEA grant. One of us (PKW) acknowledges the New York City Alliance for Minority Participation for a SEMRAP award. We appreciate the contribution to experimental results by Mr. Morgan Chen.

6. REFERENCES

1. P. de Vries, D. Coevorden, and A. Lagendijk, "Point Scatters for Classical Waves," *Review of Modern Physics*, **70**, pp.447-466, 1998

2. D. Grosenick, H. Wabnitz, H. Rinneberg, K. Moesta and P. Schlag, "Development of a Time Domain Optical Mammograph and First in-vivo Applications," *Applied Optics*, **38**, pp.2927-2943, 1999.

3. M. Eppstein, D. Dougherty, L. Troy, and E. Sevick-Muraca, "Biomedical Optical Tomography Using Dynamic Parameterization and Bayesian Conditioning on Photon Migration Measurements," *Applied Optics* , **38**, pp.2138-2150, 1999.

4. Tak D. Cheung, Peter K. Wong and Eva Chan, "Detection of objects embedded in a non-uniform turbid medium," *Proccedings of the Int. Soc. Optical Eng.(SPIE)*, **3459**, pp. 192-195, 1998.

5. E. Chan, T. Cheung, and P. Wong, "Statistical modeling of photon diffusion in turbid medium with applications to bioimaging", *Proceedings of the 156th American Statistical Association*, Section on Physical and Engineering Sciences (SPES), pp.181-187, Chicago, August, 1996, Ed. American Statistical Association, Alexandria, VA.

6. J. Li, A. Lisyansky, T. Cheung, D. Livdan, and A. Genack, "Transmission and surface profile intensity," *Europhysics Letters*, **22**, pp 675-679, 1993.

7. E. Chan, T. Cheung, and P. Wong, "Optical Speckles and Embedded Object Detection in Turbid Media," *Proceedings of the 157th American Statistical Association*, Section on Physical and Engineering Sciences (SPES), pp.31-36, Anaheim, CA, August, 1997, Ed. American Statistical Association, Alexandria, VA.

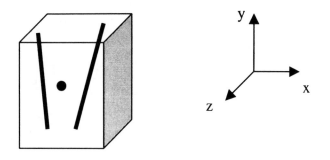

Figure 1 : Geometry of the sample.
The tubes are in random directions. The bead is midway between the tubes with different configurations given by different y values.

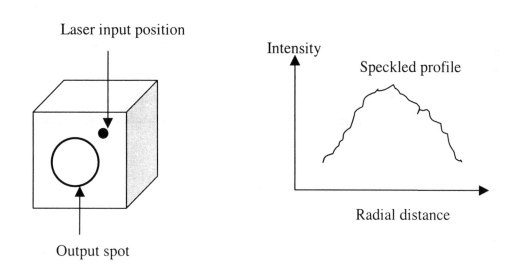

Figure 2: Detection Geometry.
The 1-dimensional cross-section of the output spot is used to generate the speckled transmitted profile. (Usually it would take 100 curves of slightly different wavelengths to average out the speckles to get a smooth profile.) In order to do in-vivo application, we only used one frequency. There are two ways to get profile curves. (a) We assume angular symmetry within a 2-mm spot. (b) We move the laser input point by a very small amount to get another speckled profile, assuming material environment fluctuation to be identical in small scale. In this paper, (a) is used. The intended resolution is 1 mm and 10 profiles are generated for each laser input position.

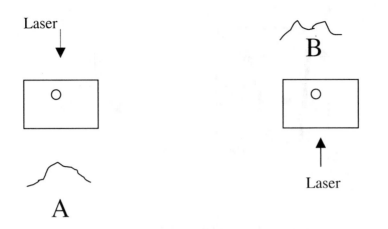

Figure 3 A: Sample rotation for configurations A & B.

 Add a new layer thickness t such that profiles A and B are identical

Let the tube depth be z , then $t + z = 20 - z$ (sample is 20 mfp thick)

Figure 3 B: Diagram illustration of adding a new layer geometry.

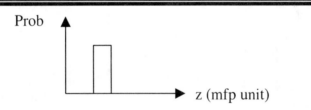

Figure 4 a: Typical flat-top distribution for Prob(tube depth z). Error in mfp determination may not jusitfy a profiled distribution. (3 to 4 mfp wide) Model 1

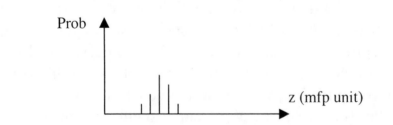

Figure 4 b: Typical profiled distribution for Prob(tube depth z) assuming that the error in mfp determination is due mostly to systemic error. (2 to 3 mfp half width) Model 2

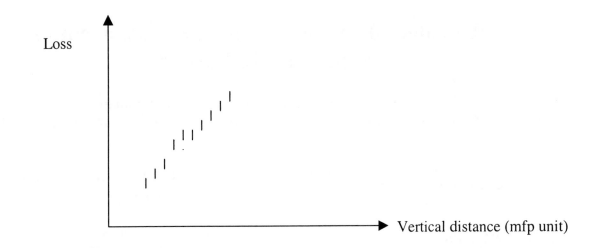

Figure 5 a: Typical loss function versus vertical distance using model 1. Discontinuity is less pronounced as compared to figure 5 b.

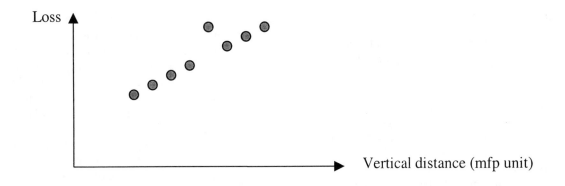

Figure 5 b: Typical loss function versus vertical distance using model 2. Although the discontinuity is more pronounced, the general use of model 2 profiled distribution is still under investigation.

Repeated Measures Models for Assessment of Metal Loss in Culverts

Peter Bajorski, Rochester Institute of Technology
Center for Quality and Applied Statistics, 98 Lomb Memorial Dr., Rochester, NY 14623

Key words: repeated measures, censored data, maximum likelihood, metal thickness, culverts, reliability.

1. Introduction

This paper is based on a project dealing with a problem of an optimal engineering design of metal culverts. A culvert is a pipe that is put under a roadway for draining purposes. The question to be answered in this project was: What is the minimum metal thickness required for pipes used for culverts, so that they can serve the required number of years?

The proper design for the required service life (usually 50 years) is very important because if a culvert fails and needs to be replaced, the road above it, which is usually planned for 50 years of service life, also needs to be redone, which is very costly.

It is known that metal pipes used as culverts lose their thickness due to corrosion and attrition caused by sand, stones, and water. Metal loss rate may vary a lot from culvert to culvert due to variability of environmental conditions. The purpose of this project is to build a stochastic model for metal loss as a function of time and then calculate the required thickness that would withstand 50 years before the thickness is reduced to zero. The proposed model will contain some parameters, which need to be estimated based on available data. An established statistical methodology for parameter estimation is the maximum likelihood method, which seeks parameters' values that maximize likelihood of the results that were obtained in the sample. This methodology was used in the project, and it led to complex optimization problems. Once the model was established, it was used for estimation of the metal thickness required, so that 90 percent of culverts would last at least 50 years.

Notice the indirect application of optimization in engineering design here, in contrast to the more traditional applications where some engineering parameters are directly optimized.

2. Stochastic Models for Metal Loss in Culverts

For the purpose of this study, some of the culverts investigated in 1977 study were re-inspected in 1996. The inspection team was able to obtain metal thickness data on 42 culverts according to a predefined sampling scheme (metal thickness was calculated as the average of eight measurements along the culvert). After preliminary analysis of the collected information, data obtained from five culverts were excluded from further analysis because of conflicting or incomplete data.

Investigating the same culvert at two points in time (1977 and 1996 studies) provides us with the advantage of observing how metal loss changes as a function of time. Figure 1 shows that the metal loss rate increases as a function of time. Each of the 37 lines in Figure 1 represents one culvert, and each line consists of two line segments. The first one represents the time period from the time when the culvert was build (age zero) to 1977 when the first study was performed. The second segment represents the period from 1977 to 1996, except for culverts that were 100% perforated before 1996. For these cases, the age of the culvert at the known time of 100% perforation is used in Figure 1. For 31 out of 37 culverts, the slope of the second segment is larger than the slope of the first one. This indicates that metal loss rate increases as culverts become older.

Inspection of Figure 1 suggests the following model for the metal loss M:

$$M = a \cdot X + b \cdot X^2$$

where X is the culvert age, and a, b are unknown constants which characterize metal loss rate of a specific culvert. The metal loss rate is not constant in this model but is equal to $(a + 2 b \cdot X)$. Since the constants a, b vary over the population of culverts, it is assumed that they are random variables with a joint normal distribution with parameters:

$$E(a) = m_1, \ E(b) = m_2, \ Var(a) = \sigma_1^2, \ Var(b) = \sigma_2^2,$$
$$Cov(a,b) = K.$$

The model can now be written in the form:

(1) $\qquad M = m_1 \cdot X + m_2 \cdot X^2 + \epsilon_1 \cdot X + \epsilon_2 \cdot X^2$

where $\epsilon = (\epsilon_1, \epsilon_2)$ is normal N(0, D), $D = \begin{bmatrix} \sigma_1^2 & K \\ K & \sigma_2^2 \end{bmatrix}$

Let us now introduce some notation for incorporation of the data available. Let

- M_{i1} - metal loss (in the i-th culvert) [in mils] at the time of the 1977 study, and
- M_{i2} - metal loss (in the i-th culvert) [in mils] at the time of the 1996 study.

Model (1) can now be written in the form:

(2) $\qquad M_{ij} = m_1 X_{ij} + m_2 X_{ij}^2 + \epsilon_{i1} X_{ij} + \epsilon_{i2} X_{ij}^2$,

$\qquad\qquad\qquad i = 1,...,n, \;\; n=37, \;\; j=1,2,$

where X_{ij} is the culvert age during the 1977 study (j=1) or 1997 study (j=2), m_1, m_2 are constants that characterize average increase in metal loss over the whole population of culverts, ϵ_{i1}, ϵ_{i2} are random variables that characterize the individual culvert (with the variability across the population due to variety in environmental conditions). Model (2) can be written in the following matrix form

(3) $\qquad\qquad y_i = X_i (m + \epsilon_i),$

where $y_i = (M_{i1}, M_{i2})^T$, $X_i = \begin{bmatrix} X_{i1} & X_{i1}^2 \\ X_{i2} & X_{i2}^2 \end{bmatrix}$, $m = (m_1, m_2)^T$,

$\epsilon_i = (\epsilon_{i1}, \epsilon_{i2})^T$, and
ϵ_i's are independent normal random vectors N(0, D), where D is a positive semidefinite 2×2 (symmetric) variance-covariance matrix of unknown parameters (defined before) to be estimated. Model (3) is well known in statistical literature (specifically in repeated measures theory).

The maximum likelihood estimation of the unknown five parameters can be performed by maximization of the log-likelihood function

(4) $\qquad \left(-\frac{1}{2}\right) \sum_{i=1}^{n} \left[\ln|S_i| + (y_i - X_i m)^T S_i^{-1} (y_i - X_i m) \right]$

where $S_i = X_i D X_i^T$, and $|S_i|$ is the determinant of S_i.

The model considered so far assumes that all observed metal thicknesses are positive. However, several thicknesses observed in 1996 are equal to zero, indicating that entire metal was lost, at least along the "line of worst metal loss", where metal thickness measurements were taken. For those culverts, the time when the thickness was reduced to zero is unknown, although we know that it occurred before 1996. Assuming that the perforation occurred in 1996, gives an estimation biased towards longer than actual service life. To overcome this problem, methods for censored data need to be used.

The assumption of censoring does not change the general structure of the model (the formula (3)). However, the distribution of the observed data changes (see [2]), and consequently the log-likelihood function (4) is now in the form:

$$\left(-\frac{1}{2}\right) \sum_{i=1}^{n} \left\{ \delta_i \left[\log|S_i| + (y_i - X_i a)^T S_i^{-1} (y_i - X_i a) \right] \right.$$

(5)

$$\left. + (1-\delta_i) \; (-2) \; logB \right\}$$

where

$$B = \left(1 - \Phi\left(\frac{M_{i2} - m_{cond}}{\sigma_{cond}} \right) \right) \cdot \phi\left(\frac{M_{i1} - m_1}{S[1,1]} \right) \cdot \frac{\sqrt{2\pi}}{S[1,1]}$$

$$m_{cond} = m_2 + \left[\frac{(M_{i1} - m_1) \; S_i[1,2]}{S_i[1,1]} \right]$$

$$\sigma_{cond} = S_i[2,2] - \frac{S_i[1,2]^2}{S_i[1,1]}$$

$S_i[k,l]$ is the (k, l) element of the matrix S_i, $\delta_i = 1$ for non-censored observations, and $\delta_i = 0$ for censored observations, and Φ, ϕ are the cumulative distribution function and the density function of the standard normal distribution, respectively. The derivation of this formula requires calculation of conditional distributions.

3. Optimization Methods Used

An iterative procedure for finding m and D which maximize (4) is given in [1]. This procedure converges very quickly in the case of our data set. Unfortunately, it cannot be used to maximize (5). It was, therefore, necessary to use general optimization techniques to find the maximum likelihood estimate for the censored observations.

Notice that both the censored-data problem to maximize (5) and the non-censored problem (4) are constrained non-linear problems in five dimensions $(m_1, m_2, \sigma_1^2, \sigma_2^2, K)$, with the following constraints:

$$\sigma_1^2 > 0 \, , \, \sigma_2^2 > 0 \, , \sigma_1^2 \cdot \sigma_2^2 - K^2 > 0 \, .$$

These constraints are necessary because of probabilistic interpretations. The two parameters σ_1^2, σ_2^2 are variances and have to be nonnegative (and can not be zeros to have non-degenerate distributions in (4) and (5)). The same applies to the last condition, which states that the determinant of the variance-covariance matrix D needs to be positive. Because of this probabilistic interpretation, we know that the optimum must be inside the feasibility region rather than on the boundary. That is, none of the constraints will become active at the optimal solution point (even approximately).

In order to translate this constrained problem to an unconstrained one, the infinite barrier method was used. Zero-order optimization methods were used because the objective function was highly non-linear with respect to some variables. The first method was the Powell's Method, and the second was the Evolutionary Optimization Method ("moving grid" method) with some improvements. Both methods were giving consistent results.

4. Numerical Results

The numerical results are given in Table 1 for both models: (4) without censoring and (5) with censoring.

Table 1.

Model	without censoring	with censoring
m_1	1.293	0.744
m_2	0.05019	0.08418
σ_1^2	2.24062	2.27057
σ_2^2	0.00317349	0.00599793
K	-0.0609418	-0.0846084

The goodness of fit of the model to the data was verified by calculation of residuals. It was checked that residuals are normal (as assumed in the model) and no patterns were identified which could suggest departure from the assumed model.

Using the stochastic model (3), one can describe the distribution of metal loss over the population of culverts at a specific age X. This distribution is normal with mean

$$\mu = m_1 \cdot X + m_2 \cdot X^2$$

and variance

$$S^2 = \sigma_1^2 \cdot X^2 + 2 K \cdot X^3 + \sigma_2^2 \cdot X^4 \, .$$

According to that distribution, a specified proportion p of culverts (e.g., p= 0.9 , i.e., 90%) at age X would have metal loss less than M_p and proportion (1-p) of culverts would have metal loss larger than M_p, where $M_p = \mu + z_p \cdot S$, and z_p is the p order quantile from the standard normal distribution. Selected values of M_p (in mils) , for culverts at the 50 years of age, are given in Table 2 (for both models).

Table 2.

p	M_p [in mils] for the model without censoring	M_p [in mils] for the model with censoring
0.1	61	58
0.25	122	148
0.5	190	248
0.75.	258	348
0.9	320	438

The metal loss M_p as a function of age (X) for the model (4) without censoring is plotted in Figure 2. The solid line represents p=0.5, the two dotted lines p=0.25 and p=0.75, and the dashed line p=0.1 and p=0.9. Figure 3 shows the metal loss M_p for the model (5) with censoring.

5. Summary and conclusions

The main purpose of this project was to identify the minimum metal thickness required for pipes used for culverts, so that they can serve the required number of years. The usually required service life is 50 years. Hence the required thickness for this time period is at least 320 mils according to the model without censoring, up to 438 mils if the model with censoring is assumed (see Table 2). As calculated in the previous section, 90% of such culverts should withstand that time without 100% perforation at the line of worst metal loss, which is judged to be a sufficient safety measure. The formulas shown in Section 4 enable calculation of the required thickness for service life periods other than 50 years, if needed.

References

[1] N. Laird, N. Lange, and D. Stram. "Maximum Likelihood Computations With Repeated Measures: Application of the EM Algorithm." Journal of the American Statistical Association. March 1987, Vol. 82, No. 397, pp. 97-105.

[2] J.F. Lawless, *Statistical Models and Methods for Lifetime Data*, John Wiley & Sons, 1982.

Figure 1. Metal loss of culverts as a function of age.

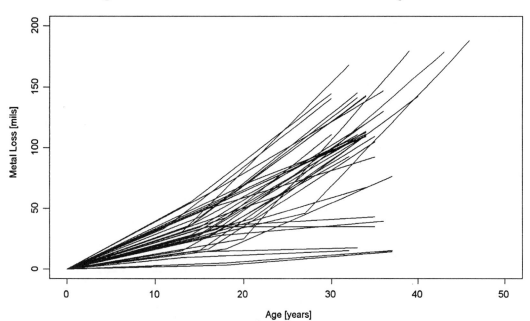

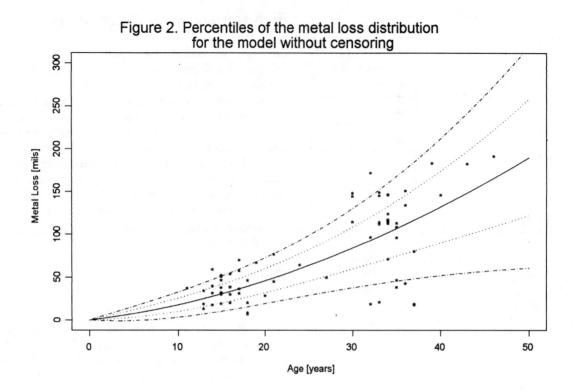

Figure 2. Percentiles of the metal loss distribution for the model without censoring

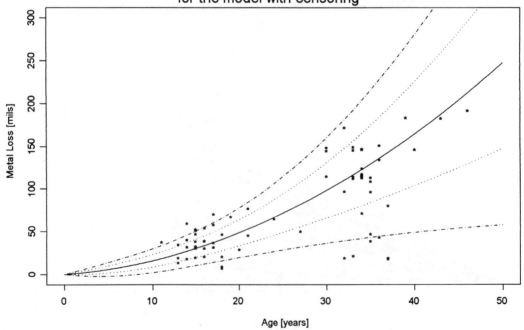

Figure 3. Percentiles of the metal loss distribution for the model with censoring

TWO-LEVEL FACTORIAL EXPERIMENT FOR CRACK LENGTH MEASUREMENTS OF 2024-T3 ALUMINUM PLATES USING TEMPERATURE SENSITIVE PAINT AND ELECTRODYNAMICS SHAKERS

David Banaszak, Gary A. Dale Air Force Research Laboratory
David Banaszak, AFRL/VASS-Bldg 24C, 2145 Fifth St. Ste 2, Wright Patterson AFB, OH45433

Key Words: Fatigue, Experimental Design, Crack Length, Measurement, Aluminum, Shakers, Temperature Sensitive Paint

Abstract

The Air Force Research Laboratory (AFRL) investigates instrumentation and techniques for measuring and analyzing dynamics data. To support the high cycle fatigue requirements of the sustainment focus area of AFRL; laboratory engineers conducted a research effort to assess the reduction in sonic fatigue crack growth rate in secondary aircraft structures with durability patches. To understand the crack growth process, AFRL is investigating crack growth rate measurement techniques and crack generation techniques for aluminum plates fatigued by out-of-plane excitation. Recently, AFRL demonstrated a visual crack measurement system (VCMS) using temperature sensitive paint. The system allows engineers to view and measure the crack length versus cycles of a plate being fatigued at resonance by an electrodynamics shaker. Over 24 factors impacting fatigue crack growth rates should be evaluated using a screening experiment. During 1999, AFRL engineers performed a 3 factor full two-level factorial experiment with 2024-T3 aluminum plates. The factors are plate size (180x85mm and 360x170mm), plate thickness (1 mm and 3 mm), and patching (patch/no patch). This paper will discuss the experiment and the latest crack length versus test cycles curves derived using the VCMS.

1. Background

The Air Force needs to increase lifetimes of an aging aircraft fleet. For sonic fatigue, there is a need to dynamically measure crack growth rates before and after application of a durability patch. The sustainment technical area needs a technique to make high cycle fatigue measurements on materials representative of aircraft skin structures from 1 to 3mm thick.

Unfortunately, there is no known standard procedure for dynamically measuring crack growth rates in secondary structures due to out-of-plane excitation. Banaszak, Dale, Jordan and Watkins (1999) demonstrated a new technique using a visual crack measurement system (VCMS).

Callinan (1997) states that typical acoustic induced cracks occur along a line of rivets or run parallel to the rivet line and may turn into the center of a panel. This effort demonstrates how to dynamically grow cracks and measure crack growth rates using electrodynamics shakers to provide out-of-plane excitation at resonance in different plates. Since most fatigue cracks on aging aircraft start at rivet heads; a bolt and two washers are used to simulate a rivet head in an aircraft panel.

There are numerous control factors affecting crack growth rates, so it is reasonable to use a design of experiment approach to evaluating crack growth rates. Designed experiments help to determine the optimum combination of control factors for desired responses. Yurkovich (1994) gives an example of an experimental design using a L32 Orthogonal Array for reducing the number of computer runs; for evaluating 7 variables, from 128 to 32 runs. For this project, we want to have at least a resolution V experiment. Box (1978) notes that, a design of resolution V does not confound main effects and two-factor interactions with each other, but does confound two-factor interactions with three-factor interactions, and so on.

AFRL desired a 4 factor 2-level factorial experimental design to validate the usefulness of the new crack measurement technique and determine the impact of the control factors on crack growth rate for 2024-T3 aluminum. Due to resource limitations, the final experiment was a 3 factor full factorial experiment with replications. For this experiment two plate sizes (180x85 and 360x180 mm) and two plate thickness (1mm and 3.175 mm); each with and without durability patches, are tested using a VCMS to measure crack length and determine dynamic crack growth curves. Durability patches are applied to cracks of 25mm length to determine if the patch arrests the crack. The crack growth curves will be similar to those in ASTM methods and static fatigue tests by Schubbe (1997).

2. Experimental Constraints

Obtaining stress levels in experimental items to induce cracks require use of a high force level shaker with armature weights larger than the test structure. According to Tustin (1974), "It is dangerous, moreover, to test loads that weigh more than half the weight of the armature; excessive cross-talk, even broken flexures, may result. In an email, Tustin noted that the fixture weight should be between 1 and 2 times specimen weight for frequencies less than 2000 Hz. A small shaker severely limits the experiment to small, lightweight test structures. Also, excessive side loads can damage the shaker. Thus a larger shaker is highly desirable for experiments with larger, heavier structures.

The mass of the 4 plates range from 43 to 533 grams. For a small modal shaker with an armature mass of less than 181 grams (.4 pounds) only small structures can be tested. The larger MB Dynamics Model C10E shaker has an armature weight of 8.2 kilograms (18 pounds). With a head plate attached the C10E shaker has an effective armature mass of 11.3 kilograms (25 pounds). For heavier plates, the larger shaker is required. Then Newton's second law (F=ma) implies a more constant force for a fixed acceleration input level.

The VCMS using temperature sensitive paint (TSP), developed by AFRL, views crack growth while shaking the plate with out-of-plane excitation. Innovative Scientific Solutions, Inc. (ISSI) developed the TSP used on the plates. As show in Figure 1, the VCMS includes a digital camera recording frames on a personal computer while the crack grows. Temperature sensitive paint is applied to the plate to enhance crack visibility and get a temperature profile of the plate while it is cracking. Liu (1997) discusses the use of temperature sensitive paint. As the plate starts to crack, stress concentrations at the crack tip are observed as temperature increases. The camera views the topside of the plate and patches are on the bottom side of the plate.

3. Preliminary Experiments with a Small Shaker

During June-October 1998, AFRL and ISSI conducted preliminary experiments with small lightweight pure Aluminum plates. These experiments demonstrated that cracks could be initiated and grown in the laboratory. These experiments aided in understanding the possible control factors in formulating a designed experiment for testing the effect of durability patches for several different control factors. The preliminary experiments demonstrated that cracks could be initiated from pre-drilled holes on small plates vibrated at the first bending mode. The first bending mode is illustrated in Figure 2. This phase was limited to small plates due to using a small shaker with a rated output of 111.2 Nts (25 pounds) without air-cooling.

Small, pure aluminum plates, approximately 85mm x 180 mm x 1.3 mm (7.09375" x 3.375" x .050"), were cracked with a manually swept sine excitation at the first resonant mode. A 6.35mm (1/4in) hole was drilled in the plate for attachment to the shaker. For some plates, washers were added above and below the plate to simulate rivet heads. AFRL generated cracks in nine plates. The quality factors (Q) were measured and cracks were generated. Test times to crack ranged from 6 to 95 minutes, depending on the shaker input level. Plates with washers had symmetrical cracks propagating from the washer's edge. Transducers were

not used to maintain plate balance. The mode was tuned by visually observing the beam tip and adjusting the frequency to get maximum displacement.

During October 1998, a prototype VCMS was used to successfully get crack length versus time, cycles versus time and crack length versus cycles. A view of the growing crack on the computer monitor is shown in Figure 3. Based on this prototype VCMS, the instrumentation system in Figure 1 was designed for future experiments.

During September of 1998, AFRL experimenters obtained a MB Dynamics model C10E 1200 pound shaker. This larger shaker allows using a more realistic set of control variables than using a small modal shaker. Control variables are based on the experimental design control factors detailed in the next section. The pilot experimental design is based on the checklist recommended by Dean and Voss (1999).

4. Experimental Design

Objective: The primary objective of the experiment is to measure number of cycles to generate given crack lengths for combinations of the sources of variation or control factors below. This will help to estimate fatigue life of materials used on aging aircraft. Many control factors and responses will be important. A large number of decisions need to be made at each stage of the planning process. The steps are not independent. At any stage, it may be necessary to go back and revise some of the decisions. That is, the experimental design process is an iterative process. The following experimental design results from consultation with numerous vibration and acoustic experts in AFRL, and experience with the preliminary experiments above.

Potential Control Factors: Where possible, factors with two levels are used to try to reduce the number of test combinations. The number of control factors and response measurements are narrowed down to reduce the number of required experimental units (plates). For these experiments, a potential list of control factors and their levels are listed in Table I. Other variation sources were considered nuisance variables.

Rationale for selecting the above control factors in Table I follow. The materials are the same used by Schubbe (1997) for tension/compression tests on primary structures. The small plate size was based on the preliminary experiments. The large plate size is double the small plate size and allows generation of cracks of lengths similar to Schubbe (1997), who used a specimen of 508 mm x 153 mm with thickness greater than 3.175 mm. Plate thickness represents typical aircraft secondary structure. Patch factors are being studied. The patch factors selected covered 25mm long cracks. The initial crack size is an attempt to make the crack grow in a desired direction. The preliminary experiments indicate that this can be set to 0. The hole and washer diameters were selected to be representative of a typical aircraft panel with skin, rivets and stiffeners. A plate with actual

Table I Potential Control Factors

Material: **Al 2024-T3 (Unclad)**, Al 7075-T6 (Unclad)
*****Plate Area: 180 x 85 mm (7.09 x 3.35 ")**, 360 x 170 mm (14.17 x 6.69")
*****Plate Thickness: 1.0 (.040")**, 3.175 mm (.125")
*****Patch Status: No Patch, Patch**
Patch Type: Boron/Epoxy, **Fiberglas (Durability)**
Patch Sides: **1 sided**, 2 sided
Patch Size: **Small (40x70mm)**, Large(80x80mm)
Patch Thickness: 4 plies, **8 plies**
Patch Damping: 0, **.05**
Patch Cure Temp: **120 F**, 180 F
Initial Crack Size(a0): **0 mm** , 25.4 mm(1")
Crack Size When Patched: a0, **a0+25.4mm(1")**
Fastener: **Bolt**, Rivet, Bond
Hole Diameter (Rivet): 3.175(.125"), **6.35 mm (.250")**
Washer Diameter (Rivet Head): **11.1 mm (7/16")**, 12.7 mm (1/2")
Shaker Input for Fixed Strain: 3gs for Small Plates, 6gs for Large plates
Excitation Location: **Center of Beam**, End of Beam
*****Excitation: Sine Track**, Narrow Band Random
Vibration Mode: **Bending Mode 1:1** or Bending Mode 1:2
Operation Temp: -55 C, **25 C (Room Temperature)**, 125 C
Configuration: **Bolt & Washer in Hole**, Rivet & Stiffener, Bonded Stiffener
Boundary Conditions: **Clamped**, Clamped-Clamped, Free-Free
Data Technique: **Empirical** , Analytical (FEA or AFGROW)
Excitation Type: In Plane, **Out-of-Plane**
Metal Treatment: None, **Alodine**
Stop Drill: Yes, **No**
Crack Type: Top, Middle, Bottom
Plate Side: Top, Bottom

Note: **Bold** = Fixed Values. * => 4 Factor Full Factorial Experiment.

rivets will be more realistic. The shaker acceleration input was selected based on the preliminary experiments and cracking practice plates to get approximately equal strain on both plate areas based on beam equations. Also, a reasonable short fatigue time is desired for these pilot experiments.   During the pilot experiment, the experimenters used sine excitation and tried to manually track the change in resonance as the crack grew. In the real world, panels are usually excited by random noise; which excites the structure as the crack propagates and the resonance changes. Narrow band random inputs may require a larger shaker than the C10E.   Thus sine excitation was used for this experiment. The excitation location is the center of the balanced beam to induce symmetric cracks on both sided of the attachments. The first bending mode grows cracks quickly.   Room temperature is used to cut down the number of variables. A convenient configuration and boundary condition was chosen from numerous possibilities. The configuration is a double cantilever plate with a center hole and mounted on a shaker as shown in Figure 2. This configuration

simulates multiple fatigue cracks growing at the edge of rivet heads (simulated by washers) and can be expanded to larger plates with multiple holes at predetermine spacing to more closely duplicate the conditions in Callinan (1997).

The experimental units will be plates of given material and dimensions indicated in Table I. Material and configuration may be considered as blocking factors and covariates. Experimental units are tested in as random order an order as possible. To test all combinations of the potential two level control factors requires 2^{28} = 268,435,456 experimental runs.

Realistic Control Factors: To reduce the number of experimental runs, some factor levels can be fixed at one level as show in Table I.   AFRL vibration and acoustics experts decided that for the first phase of this experiment, a maximum of 4 control factors should be used as indicated by an * in Table I.   A full factorial completely randomized design will require 2^4 or 16 possible test runs. Since random noise excitation requires a larger shaker, the factors were narrowed down to three: area, thickness and patch status. Thus a 3 factor full factorial experiment requires 2^3 or 8 runs.   Several replications of the control factors resulted in a total of 20 runs. A new control factor, metal treatment, was identified at the start of the experiment when the TSP did not adhere to aluminum

Desired   Response   Measurements: Potential   response   measurements   for analysis of variance (ANOVA) include the measurements in Table II.   For this paper a typical ANOVA for the first response factor bending mode frequency (F0) is included.

5. Instrumentation

As shown in Figure 1, the shaker input signal, from the signal generator, and sound level meter (SLM) output signal was recorded on tape.   Two-strain gages measured strain near the plate's edge.   One accelerometer (A1) measured the shaker input and one accelerometer (A2) measured the plate response.   The time code of the tape recorder and digital camera's personal computer was synchronized before each run.

While the plate coated with TSP is vibrating and illuminated with a blue light, a 1024 by 1024 pixel, 16bit digital camera acquires, averages and stores tif images on a PC hard drive.   The camera resolution is better than the resolution 512x512 used by Redner (1990). The growing crack is viewed on the PC monitor.   A scale over the monitor helps to determine the crack length as the experiment proceeds. After the run the tape recorder data and camera data are merged into a computer file using the time code from each system. With the data, the

Table II Desired Response Measurements

N0   -Number of Cycles to Crack Size = 0 mm
N25 -Number of Cycles to Crack Size = 25 mm
N50 -Number of Cycles to Crack Size = 50 mm
N75 -Number of Cycles to Crack Size = 75 mm
N100 -Number of Cycles to Crack Size = 100 mm
N125 -Number of Cycles to Crack Size = 125 mm
N150 -Number of Cycles to Crack Size = 150 mm
F0   -Resonant Frequency at Crack Size = 0 mm
F25 -Resonant Frequency at Crack Size = 25 mm
F50 -Resonant Frequency at Crack Size = 50 mm
Q0   - System Q at Crack Size = 0 mm
Q25 - System Q at Crack Size = 25 mm
Q50 - System Q at Crack Size = 50 mm
Resonant Frequency versus Time
Cycles versus Time
Crack Size versus Time
Crack Size versus Cycles
υε (Microstrain out) by highest stress intensity

experimenters can plot crack size versus time, cycles versus time and finally crack length versus cycles by using several computer programs.

Since no vibration controller is available for this experiment, the vibration system was an open loop system. It is recommended that a digital vibration control system with sine tracking and narrow band random

Table III Run Data with F0 Responses

ID-Type	Alodine	Area	Thickness	Patch	F0
10-S1	No	180x85	1 mm	D	112
1A-S1	Yes	180x85	1 mm	D	110
1B-S1	Yes	180x85	1 mm	No	110
1C-S1	Yes	180x85	1 mm	D	112
1D-S1	Yes	180x85	1 mm	No	112
20-S3	No	180x85	3 mm	D	333
2A-S3	Yes	180x85	3 mm	D	330
2B-S3	Yes	180x85	3 mm	D	334
2C-S3	Yes	180x85	3 mm	No	336
2D-S3	Yes	180x85	3 mm	No	336
30-L1	No	360x170	1 mm	D	28
3A-L1	Yes	360x170	1 mm	D	27
3B-L1	Yes	360x170	1 mm	No	27
3C-L1	Yes	360x170	1 mm	D	27
3D-L1	Yes	360x170	1 mm	No	27
40-L3	No	360x170	3 mm	D	82
4A-L3	Yes	360x170	3 mm	D	82
4B-L3	Yes	360x170	3 mm	D	82
4C-L3	Yes	360x170	3 mm	No	82
4D-L3	Yes	360x170	3 mm	No	83

capability be used on future experiments. The control system should be able to record the transducer signals.

6. Experiment Procedure

Twenty plates were cracked as shown in the run data in Table III. The input acceleration on the shaker was 3gs for large plates and 6 gs for small plates to get an estimated constant stress concentration at the washer's edge. Several practice plates were cracked to determine the shaker input levels for a reasonable test time.

The following process was followed for each run:
(1) Weigh and measure each plate.
(2) Alodine all plates.
(3) Apply TSP to all plates.
(4) Record low level random noise records for Q measurement using 3-dB power bandwidth method.
(5) Confirm bending mode using low level sine.
(6) Activate tape recorder and VCMS.
(7) Set shaker input acceleration level.
(8) Use VCMS to view crack.
(9) Terminate when desired crack length is reached.
(10) Apply patches to 12 plates pre-cracked to 25 mm.
(11) For patched plates repeat steps 1-10.
(12) Derive Crack Length versus Cycles from camera.
(13) Derive Cycles versus Time from recorder data.
(14) Merge data to get Crack Length versus Cycles.
(15) Enter response data in SAS® JMP® for ANOVAs.

Data were recorded on tape during the entire test duration. The digital camera saved up to 1000 images per run. A shift in frequency resonance was monitored to help determine crack initiation. A manual log was kept to indicate date, start time, end time, frequency of resonance, transducer outputs, temperature, and configuration data.

7. Analysis

The sinusoidal signal input into the shaker was played back from the tape recorder into a PC where a computer program generated frequency versus time and cumulative cycles versus time for input into an Excel spreadsheet. The camera images were stored as tif frames. An image-processing program was used to tabulate the pixel location of the crack tip for individual frames to help determine crack length versus time. Figure 4 shows typical images versus time as the crack grew with time. The cycles versus time and crack length versus time data were merged to generate a final crack length versus cycles as shown in Figure 5. This curve has a shape similar to crack growth curves using in-plane excitation. In addition ANOVA tables and interaction profile plots will be produced to determine which set of factors have the most significant effect on crack growth rate. A typical ANOVA table and interaction profile plot (for the area and thickness interaction) for the frequency response at zero crack size (F0) was generate by SAS® JMP®

software and is shown in Figure 6. As seen in the ANOVA table, the Area, Thickness and Area-Thickness interaction are the significant factors. The interaction plots shows that the small (185x85mm), thick (3mm) plate has the highest resonant frequency (F0). Also frequency increases with thickness faster for small plates than for large plates (360x170mm). Area-Patch and Thickness-Patch interactions are insignificant as expected since testing and analysis of patched plates is not complete.

8. Summary and Conclusions

The visual crack growth system using temperature shows promise in measuring crack length versus cycles for out-of-plane excitation. A smoother crack length versus cycles curve requires use of a more automated shaker control system. So far, preliminary analysis of variance was completed for the response variable F0 (frequency with a crack length of 0 millimeters). Measurements and analysis on patched plates are needed for completion of the experiment. This preliminary analysis confirms that area and thickness of the plates is a significant factor determining F0. Also, there is interaction between area and thickness for the resonant frequency F0. This is expected from conventional mechanical engineering. The patch is not yet a significant factor since patched plates still need to be tested to investigate all of the desired response variables. Of primary interest will be the effect of patching on number of cycles for given crack length. Future investigations should include use of random noise excitation and a screening experiment that takes all the potential control factors into consideration.

Acknowledgements

The authors thank Neal Watkins, Ph.D. and Jeffrey Jordan, Ph.D. for helping perform the preliminary and pilot experiments. Special thanks to Henry Baust for computer program support to acquire images, to Gary Clinehens for optical operational support and Mike Fabian for vibration control support during the pilot experiments. Thanks to Maj. John Lassiter, whose Air Force Reservists project initiated this experiment and to Howard Wolfe, Ph.D. for reviewing this paper.

References

Banaszak, David, Dale, Gary, Jordan, Jeffrey D., Watkins, A.Neal (1999), "An Optical Technique for Detecting Fatigue Cracks in Aerospace Structures", 18th International Congress on Instrumentation in Aerospace Simulation Facilities, Toulouse, France, June.

Box, George E.P. (1978), Statistics for Experimenters, Wiley.

Callinan, R.J., Galea, S.C. and Sanderson, S. (1997), "Finite Element Analysis of Bonded Repairs to Edge Cracks in Panel Subjected to Acoustic Excitation", Composite Structures Vol. 38, No. 1-4 pp.649-660.

Dean, Angela and Voss, Daniel (1999), Design and Analysis of Experiments, Springer-Verlag, NY, NY.

Liu, T., Campbell, B.T., Burns, S.P.and Sullivan, J.P. (1997), "Temperature and Pressure Sensitive Luminescent Paints in Aerodynamics", App. Mech. Rev. Vol. 50 No. 4, April.

Redner, A. S., Voloshin, A.S., and Nagar, A. (1990), "Fatigue Crack-Growth Measurement Using Digital Image Analysis Technique," Applications of Automation Technology to Fatigue and Fracture Testing, ASTM STP 1092, Philadelphia, PA, pp133-142.

Schubbe, Joel J. (1997), Thickness Effects on a Cracked Aluminum Plate with Composite Patch Repair., AFIT/DS/ENY/97-4.

Tustin, Wayne (1974), Environmental Vibration and Shock-Testing, Measurement, Analysis and Cables Course Notes, Revised.

Yourkovich, Rudy (1994), "The Use of Taguchi Techniques with the ASTROS Code for Optimum Wing Structural Design", presented at the 35th AIAA Structure, Structural Dynamic and Material Conference, Hilton Head, SC.

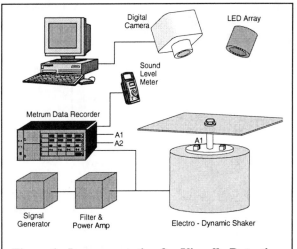

Figure 1. Instrumentation for Visually Detecting Cracks Induced by Out-of-Plane Excitation.

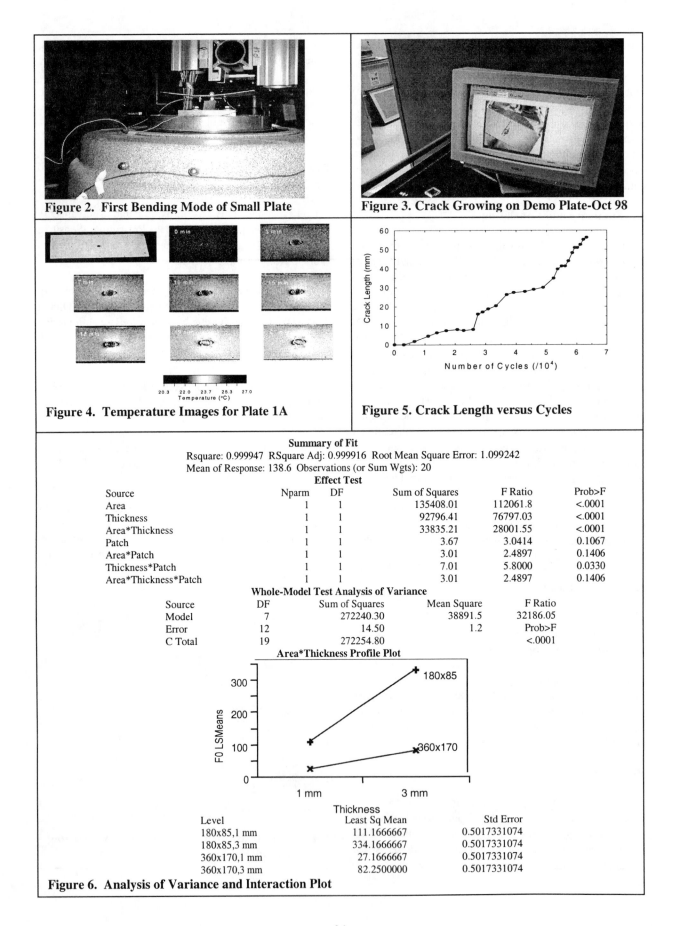

Figure 2. First Bending Mode of Small Plate

Figure 3. Crack Growing on Demo Plate-Oct 98

Figure 4. Temperature Images for Plate 1A

Figure 5. Crack Length versus Cycles

Summary of Fit
Rsquare: 0.999947 RSquare Adj: 0.999916 Root Mean Square Error: 1.099242
Mean of Response: 138.6 Observations (or Sum Wgts): 20

Effect Test

Source	Nparm	DF	Sum of Squares	F Ratio	Prob>F
Area	1	1	135408.01	112061.8	<.0001
Thickness	1	1	92796.41	76797.03	<.0001
Area*Thickness	1	1	33835.21	28001.55	<.0001
Patch	1	1	3.67	3.0414	0.1067
Area*Patch	1	1	3.01	2.4897	0.1406
Thickness*Patch	1	1	7.01	5.8000	0.0330
Area*Thickness*Patch	1	1	3.01	2.4897	0.1406

Whole-Model Test Analysis of Variance

Source	DF	Sum of Squares	Mean Square	F Ratio
Model	7	272240.30	38891.5	32186.05
Error	12	14.50	1.2	Prob>F
C Total	19	272254.80		<.0001

Area*Thickness Profile Plot

Level	Least Sq Mean	Std Error
180x85,1 mm	111.1666667	0.5017331074
180x85,3 mm	334.1666667	0.5017331074
360x170,1 mm	27.1666667	0.5017331074
360x170,3 mm	82.2500000	0.5017331074

Figure 6. Analysis of Variance and Interaction Plot

SPATIAL STRUCTURE OF SATELLITE OCEAN COLOR DATA

Montserrat Fuentes, North Carolina State University
Scott C. Doney, National Center for Atmospheric Research
David M. Glover, Scott McCue, Woods Hole Oceanographic Institution.
Montserrat Fuentes, Stat. Dept., Box 8203, Raleigh, NC 27595, fuentes@stat.ncsu.edu

Key Words: Anisotropy, Kriging, Non-stationarity, Ocean Color, Spatial Statistics, Variogram.

Abstract:

We determine the spatial scales of satellite chlorophyll images using the semivariogram from geostatistics. The local correlation patterns derived from the semivariograms are both anisotropic (varying with direction) and non-stationary (varying in space and time). The resulting zonal (East-West) and meridional (North-South) distributions of chlorophyll variance and ranges of spatial correlation are then compared to physical, mesoscale properties for the North Atlantic ocean such as the Rossby deformation radius and the eddy kinetic energy.

1. Introduction

As part of a longer term marine ecosystem modeling project, we examine the spatial structure of satellite derived ocean color data for the North Atlantic Ocean. Ocean color is considered a proxy for surface layer phytoplankton chlorophyll concentrations, and the large-scale ocean color field is governed by the seasonal distributions of light, nutrients and upper ocean mixing. On the so-called mesoscale (approx. 10–200 km), ocean color variability is modulated by biological sources and sinks (e.g. phytoplankton growth, zooplankton grazing), ocean flow, and mixing. The spatial correlation function of ocean color provides a useful measure for quantifying these biological-physical interactions and discriminating among theoretical models. It is also a necessary component of future work to objectively analyze the ocean color images using the best linear, unbiased estimates from kriging.

M. Fuentes acknowledges the support of the NCAR Geophysical Statistics Project, and S. Doney and D. Glover are supported in part by NASA SeaWiFS Grants W-19,223 and NAG-5-6456.

The SeaWiFS satellite ocean color data used in this study are described in Section 2. In Section 3, we discuss in detail example calculations of the semivariogram and model parameters for a single $5° \times 5°$ box in the middle of the North Atlantic. The North Atlantic ocean correlation patterns are presented and analyzed in Section 4, followed by conclusions and some final remarks in Section 5.

2. SeaWiFS Ocean Color Data

Phytoplankton are small, generally single-celled organisms that grow in the sea by converting sunlight into chemical energy by the process of photosynthesis. Chlorophyll, one of the main light harvesting pigments within phytoplankton, absorbs sunlight in the blue and red spectral ranges. The amount or concentration of chlorophyll in sea water is a primary factor in determining the depth penetration and spectral quality of light in the upper ocean. Ocean color instruments such as the Sea-viewing Wide Field of view Sensor (SeaWiFS; [McC98]) rely on this change in the spectral composition of radiance to quantify the chlorophyll concentration (mg m^{-3}). The SeaWiFS instrument measures eight spectral bands in the visible and near-infrared. Concentrations of chlorophyll are calculated from ratios of radiances of different bands [O'R98]. Only a small fraction of the light observed by the satellite is upwelled from below the ocean surface, the remainder being reflected sunlight from the surface, atmospheric gases or aerosols. In addition to the required atmospheric corrections, the SeaWiFS data also are processed to remove pixels with land or clouds before geophysical variables such as chlorophyll are computed.

The SeaWiFS instrument became operational in September of 1997. Its sun-synchronous, polar orbit (altitude 705 km, orbit period approximately 100 minutes, image swath width roughly 1500 km) provides near global coverage every two days with a nominal resolution of about 1 km when the sensor is directly overhead. The SeaWiFS data are reported at different processing levels: from raw spacecraft telemetry Level-0 to geophysical variables on standard grids Level-3 and at different spatial resolutions. The high resolution Local Area Coverage (LAC), with nominal resolution of 1km×1km, is broadcast continually by the spacecraft. The Global Area Coverage (GAC), with a nominal resolution of 4km×4km, is a continuous, global sub-sampled version of the LAC and is stored aboard the spacecraft for later transmission to ground stations. A daily, Level 3, resolution product is produced by averaging all of the GAC data from a single day after binning the data on to an equal area grid (81 km^2). This binned grid is difficult to work with because the number of longitudinal grid cells varies with latitude. Thus, the primary data used in this study are projections of the daily binned product onto a global, equal angle grid ($2\pi/2048$) with a nominal resolution of 9km×9km, referred to as standard mapped images. When higher resolution is required, we use data from the Bedford Institute of Oceanography, Nova Scotia station processed to Level-2 chlorophyll concentration still in the form of satellite images (i.e. with geometric distortions).

Geostatistical measures give a complete description of the spatial distribution of Gaussian, stationary processes. Unfortunately, neither of these two assumptions hold completely for ocean chlorophyll. Spatial trends, in particular, can produce spurious effects in the semivariogram [Cla79, Cre93]. Two steps are taken to normalize the chlorophyll data prior to analysis. First, the natural log transformation ln(Chl) is applied to the data following the arguments of Campbell [Cam95] and others that oceanic bio-optical variability is approximately lognormal. Second, the large-scale spatial patterns in chlorophyll were removed by subtracting from the daily chlorophyll a smoothed version of the monthly average, and we will call the resulting measurements *daily anomalies*. Specifically, monthly mean for each pixel were calculated on the log scale and then the resulting image was smoothed using a two-dimensional kernel with a 0.5 degree bandwidth. The resulting daily anomalies are merged into two day blocks, the time it takes the sensor to collect global coverage. Due to the presence of clouds and other ocean-atmosphere effects (e.g. aerosols, sun-glint), the number of samples that go into the composite for any particular pixel can vary widely, and thus the composite is a biased estimator of the mean field.

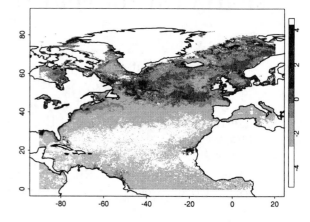

Ocean Color in June 1998

Figure 1: Natural log of the composite SeaWiFS chlorophyll concentration (mg m^{-3}) for June, 1998 over the North Atlantic ocean.

3. Ocean color semivariograms

3.1 Spatial analysis for the ocean chlorophyll

The *semivariogram* measures the local spatial variation of a random field by describing how sample data are related with distance and direction. The semivariogram function γ for a process Z is defined as:

$$\gamma(\mathbf{v}) = \frac{1}{2}\text{var}\{Z(\mathbf{x} + \mathbf{v}) - Z(\mathbf{x})\} \quad (1)$$

where $\mathbf{v}$ is the vector distance (considering direction) separating two points in space. The traditional semivariogram estimate $\hat{\gamma}$ suggested by Matheron [Mat71] is:

$$\hat{\gamma}(\mathbf{v}) = \frac{1}{2N(\mathbf{v})} \sum_{N(\mathbf{v})} (Z(\mathbf{x}_i) - Z(\mathbf{x}_j))^2 \quad (2)$$

where $N(\mathbf{v})$ are the number of data pairs $Z(\mathbf{x}_i)$ and $Z(\mathbf{x}_j)$ separated by $\mathbf{v}$. Most semivariograms are defined through several parameters; namely, the *nugget effect*, *sill*, and *range*.

- *nugget effect:* represents micro-scale variation and/or measurement error. It is estimated from the empirical semivariogram as the value of $\gamma(h)$ as $h \to 0$.

- *sill:* the value of $\gamma(h)$ for $h \to \infty$ representing the variance σ^2 of the random field.

- *range:* a scale parameter controlling the correlation (the precise definition depends on the form of γ).

Semivariograms using Matheron [Mat71] estimate were computed from the SeaWiFS Level-3 ocean color data for June, 1998 and are presented in Figure 2. The data for the semivariograms in Figure 2 are from a 5° × 5° box (470 km × 557 km) located at 30°W–35°W and 30°N–35°N in the open ocean North Atlantic in a region of relatively low chlorophyll concentrations (Figure 1). The size of this box was chosen after some experimentation to minimize the spatial heterogeneity of the ocean color data while retaining sufficient length-scales to resolve the full span of mesoscale (10km–200km) features. Separate semivariograms are estimated for the zonal (E-W) and meridional (N-S) directions as well as for the two diagonals (SW-NE and NW-SE), shown as different panels in Figure 2, to highlight any potential directional anisotropy. For a single month, the estimated semivariogram values versus distance (in km) are computed using the robust semivariogram form [Cre93] for 15 pairs of daily chlorophyll anomalies (we combine consecutive days to improve data coverage as explained in Section 2). The approximate bin size (nominally 9 km) differs with direction varying with the latitudinally dependent, Level-3 equal angle grid.

The boxplots in Figure 2 represent the resulting distribution for the 15 two-day pairs where the median value is shown as a horizontal line across the box and the mean as a circle. Although there is considerable scatter in the individual two-day pair semivariogram estimates at any particular distance, the mean and median values show general well defined curves with the semivariogram values increasing from about 0.15 near the origin to a uniform sill of about 0.35 beyond 100 km.

Quantitative estimates of the semivariogram parameters range, sill and nugget are derived by fitting an exponential model form [Cre93] to the median value for each boxplot (Table 1). A Gauss-Newton non-linear, least squares algorithm is used to find the set of model parameters that minimizes the sum of the squared residuals (SSR) between the response and the prediction. Approximate standard errors are estimated by linearizing the semivariogram model about the final convergence point and then applying traditional linear regression where the estimated standard deviations of the estimates are functions of the SSR.

The ocean color data for the 30°W–35°W and 30°N–35°N box (Figure 2) show a weak directional anisotropy (see Table 1), with a larger estimated range parameter in the meridional (N-S) direction (69 km) relative to either the zonal or diagonal directions. The gradient in the large-scale chlorophyll

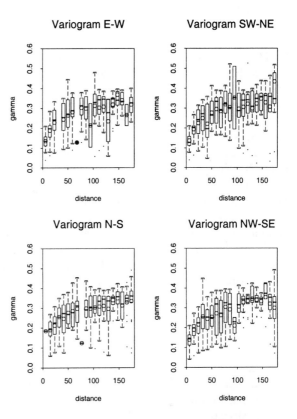

Figure 2: Semivariogram for the log chlorophyll anomalies for June 1998, in a region bounded by 30°W–35°W and 30°N–35°N in the middle of the North Atlantic Ocean. Each panel shows the estimated semivariogram in a different direction, and each boxplot represents the distribution of the semivariogram values for the anomalies at a fixed distance. The median value in each boxplot is shown as a line across the box, and the mean as a circle.

Table 1: Model fits of the semivariogram parameters for the ocean chlorophyll anomalies for a $5° \times 5°$ box in the North Atlantic using different directions, for the month of June, 1998. The table shows 95% confidence intervals for the nugget, sill and range.

Longitude	Latitude	Direction	nugget	partial sill	range	residual	n
30°W–35°W	30°N–35°N	N-S	0.15±0.04	0.21±0.02	69±32	0.014	16
30°W–35°W	30°N–35°N	NW-SE	0.13±0.06	0.23±0.04	63±48	0.028	16
30°W–35°W	30°N–35°N	E-W	0.09±0.12	0.22±0.10	32±26	0.034	15
30°W–35°W	30°N–35°N	SW-NE	0.11±0.06	0.24±0.06	47±28	0.028	16

field for this region is also oriented approximately N-S, and the larger range calculated in that direction may reflect a partial aliasing of the non-stationarity in the mean, despite the removal of the smoothed monthly field. The total sill parameter (partial sill plus nugget) is approximately the same in each direction, roughly .35, and the nugget effect is larger in the North-South direction, with a value of .15 versus around .11 in the other directions. The nugget, however, is not well defined from the 9km Level-3 data and is quite sensitive to the fit through one or two data points near the origin. The observed weak anisotropy is expected for an open ocean site away from strong boundary currents. There is more clear evidence of anisotropy in the coastal regions (not shown).

3.2 Effect of sampling resolution

In addition to position and direction, the semivariogram of a process $Z(\mathbf{x})$ can also depend on the spatial resolution of the sampling [Cla79]. Larger sample sizes tend to average out some of the smaller scale variability in the data, thus reducing both the total variance and sill as well as changing the shape of the semivariogram at short distances. The modification of the semivariogram with the resolution of the measurements is termed *change of support* or *regularization*.

Figure 3 shows a comparison of the N-S and E-W semivariograms computed from the Level 3 standard mapped image data ($\sim$9km$\times$9km) and the Bedford station Level 2 LAC data ($\sim$1km$\times$1km) for June for the box 30°W–35°W and 45°N–50°N. The total variance of the daily standard mapped image data is approximately 1/3 to 1/4 that of the original LAC data because of spatial averaging. The sub-sampling of the data from the fine resolution LAC grid to the medium resolution GAC grid ($\sim$4km$\times$4km) should not change the observed variance or semivariogram. Depending on the exact geometry of an individual satellite swath, the number of GAC points going into a single equal area, binned product grid cell can vary from approximately 4 to 9. Given typical ranges of 30–50 km, the expected reduction in variance (excluding the nugget) is about 30%. In comparison, note that the variance would be expected to drop by a factor of about 6 for spatially uncorrelated data.

Temporal averaging on the binned product grid can contribute an additional reduction of variance in those regions, mostly subpolar, where more than one pass of a region is available for a single day. Finally, the variance will be reduced by the conversion of the binned product to the equal angle, standard mapped image grid. The transformation involves a simple linear mapping between the two grids, but because of the distorted geometry of the binned product at high latitudes the remapping can lead to biases particularly in the zonal semivariogram estimates. The expected difference between the Level 2 and Level 3 standard mapped image data is difficult to quantify exactly because of the number of processing steps and the heterogeneity of the original sample coverage, and we are currently investigating how to calculate the semivariograms either from the original GAC data or from the daily binned product, which includes variance estimates. Other possible complications include overestimating the level 2 variance from inclusion of outliers because of subtle differences in the data processing algorithms for level 2 and level 3 data.

4. Spatial patterns for the North Atlantic ocean

The variance and scales of the ocean color data are determined by the underlying physical and biological dynamics that govern phytoplankton distributions. One of the goals of our research is to see whether we can relate the length scales of ocean chlorophyll, defined by the range paramter of the fitted exponential semivariogram models, to patterns of physical variables. The analysis presented here is preliminary but points towards a complicated system with a number

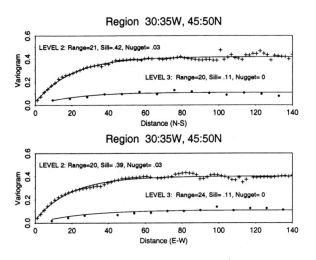

Figure 3: Comparison of the semivariograms for high resolution (Level 2) and low resolution (Level 3) satellite ocean chlorophyll for the region 30°W–35°W, 45°N–50°N for June, 1998 in the North–South and East–West directions. Model parameters are estimated using non-linear least squares.

of competing effects.

The variation of the ocean color semivariogram with latitude from 25°N to 50°N is presented in Figure 4 for a set of 5° × 5° boxes along 35°W–30°W. Following the analysis above, each plot shows an exponential model fit to the median N–S semivariogram values for the 15 pairs of anomalies in June, 1998. The range parameter decreases with latitude, going from about 69 km at 30–35°N to about 25 km at 50–55°N. The range is not well defined for the lowest latitude box 25–30°N, as shown by the observed nearly linear semivariogram and an extremely large standard error of 588 km on the model range fit. However, the standard error on the model range fit for the other boxes in Figure 4 is approximately 10 km. We observed a similar though slightly smaller reduction in the estimated range from the eastern to the western part of the North Atlantic Ocean. The model estimated range parameter decreased from approximately 50 km in the eastern subtropics to about 32 km in the west.

Significant work has been done on characterizing the mesoscale spatial variability of other physical variables. For example, [Smi99] using high resolution, eddy-resolving numerical simulations of the North Atlantic, found that the zonal average physical eddy autocorrelation lenght-scales decreased from approximately 250 km at the equator to about 40 km at 60°N. Similar autocorrelation length-scales have been calculated from satellite observations us-

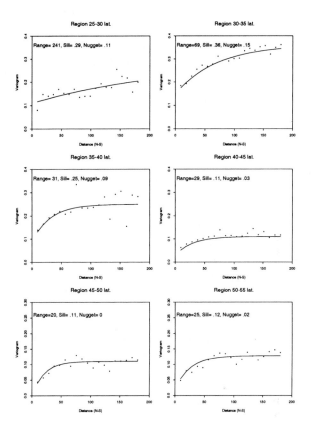

Figure 4: Semivariograms for the ocean chlorophyll anomalies in June 1998. Each panel shows the semivariogram values versus distance in the North-South direction for a 5° × 5° box along 35°W–30°W, with the latitude changing from panel to panel from 25°N to 55°N. Model parameters are estimated using non-linear least squares.

41

ing TOPEX/Poseidon altimeter sea surface height data poleward of 15°N and from Advance Very High Resolution Radiometer (AVHRR) sea surface temperature imagery for the eastern North Atlantic from 35°N to 60°N [Kra90]. Krauss et al. [Kra90] and Stammer [Sta97] further show that the physical eddy length scales outside of the tropics vary roughly linearly with the Rossby deformation radius, the horizontal scale where rotation becomes important relative to buoyancy and a key parameter governing turbulent ocean flow. These results are consistent with the hypothesis that the primary eddy generation mechanism in the ocean related to baroclinic instability of the density field.

5.   Conclusions and final remarks

The availability of routine, high spatial resolution satellite ocean color measurements provides a unique opportunity for characterizing the spatial and temporal scales of ocean biology. Complete knowledge of the time-space variability structure of ocean color provides a window into the mechanisms of marine biological–physical interaction and are a necessary component of future work to objectively analyze satellite chlorophyll images using spatial predictors, such as kriging. We present here a detailed geostatistical framework for estimating spatial length scales using the semivariogram and show preliminary calculations for a single month of the SeaWiFS data. The spatial decorrelation scales or ranges of autocorrelation are found to be both non-stationary and anisotropic. In addition to the expected poleward decrease in length-scale, as observed for physical variables, our analysis also suggested a significant east-west gradient that may be related to the elevated eddy kinetic energy in the Gulf Stream/North Atlantic Current system.

Further work is needed in a number of areas to clarify some of the questions raised here. Efforts are underway to expand our calculations to the full annual cycle and to the global domain to assess the robustness of the patterns we observe. We are also investigating the impact of the standard SeaWiFS data processing on variance and spatial autocorrelation scales. Finally, we have neglected in our treatment here any discussion of the temporal correlation scales. In addition to the obvious information gained on time-scape scales of biological variability, a number of researchers (e.g. *Chelton and Schlax* [Che91]) propose local, temporal (rather than spatial) objective analysis as a simpler, more straightforward method of interpolation of patchy satellite geophysical data such as ocean color.

References

[Cam95] Campbell, J.W. The lognormal distribution as a model for bio-optical variability in the sea. *Journal of Geophysical Research*, 100:13,237–13,254, 1995.

[Che91] Chelton, D. B. and M. G. Schlax. Estimation of time averages from irregularly spaced observations: with application to coastal zone color scanner estimates of chlorophyll concentration. *Journal of Geophysical Research*, 96:14,669–14,692, 1991.

[Cla79] Clark, I. *Practical Geostatistics*. Elsevier Applied Science Publishers, New York, USA, 1979.

[Cre93] Cressie, N. A. C. *Statistics for Spatial Data*. J. Wiley and Sons, New York, USA, 1993.

[Kra90] Krauss, W., R. Doscher, A. Lehmann, T. Viehoff. On eddy scales in the eastern and northern north atlantic ocean as a function of latitude. *Journal of Geophysical Research*, 95:18,049–18,056, 1990.

[Mat71] Matheron, G. The theory of regionalized variables and its applications. Fontainebleau, Les Cahiers du Centre de Morphologie Mathématique, 1971.

[McC98] McClain, C.R., M.L. Cleave, G.C. Feldman, W.W. Gregg, S.B. Hooker, and N. Kuring. Science quality seawifs data for global biosphere research. *Sea Technology*, 39:10–14, 1998.

[O'R98] O'Reilly, J.E., S. Maritorena, B.G. Mitchell, D.A. Siegel, K.L. Carder, S.A. Garver, M. Kahru, and C. McClain. Ocean color chlorophyll algorithms for seawifs. *Journal of Geophysical Research*, 103:24,937–24,953, 1998.

[Smi99] Smith, R. D., M. E. Maltrud, F. O. Bryan, and M. W. Hecht. Numerical simulation of the north atlantic at 1/10°. *Journal of Physical Oceanography*, submitted:–, 1999.

[Sta97] Stammer, D. Global characteristics of ocean variability estimated from regional topex/poseidon altimeter measurements. *Journal Physical Oceanography*, 27:1743–1769, 1997.

TABULATED TREATMENT PLANS, CONFOUNDED MODEL COEFFICIENTS, AND CROSSED-CLASSIFICATION BLOCK PARAMETERS FOR EXPANSIBLE TWO-LEVEL FRACTIONAL FACTORIAL EXPERIMENTS

Arthur G. Holms, Response Surfaces Consultant
9303 Newkirk Dr., Cleveland, OH 44130-4166

KEY WORDS; Experimental design; Response surfaces; Sequential experiments; Fractional factorials; Orthogonal blocks; Defining Contrasts.

Abstract

Plans include 4 through 7 independent variables on blocks of size 8. The first block has minimum aberration resolution 3 to estimate all first degree model coefficients. Adding equal sized blocks in stages permits fitting models with successively higher order terms. The stages terminate with at least resolution 5 to at least estimate all two-factor interaction coefficients. Tables for each stage show the Yates method order of observations. The tables also show the Yates method order of estimates of aliased sets of model coefficients, including those sets containing confounded crossed-classification block effect parameters.

1. INTRODUCTION

Holms (1998), henceforth abbreviated H98, cited many references to expansible experiments.

The information available at each stage of an expansion is measured by the resolution number, Box and Hunter (1961). As used here, a plan has resolution R if R is the number of letters in the shortest word in the group of defining contrasts. The condition that the coefficients of a first degree model be free of aliases with other first degree coefficients is that R = 3. The condition that the first degree coefficients also be free from aliases with two-factor interaction coefficients is that R = 4. The condition that two-factor interaction coefficients be free of aliases with first degree and with other two-factor interaction coefficients is that the experiment be a full replicate, or at least have R = 5.

For optimum seeking, Box and Wilson (1951) suggested using an R = 3 plan initially to estimate the coefficients of a first degree model in the method of steepest ascents. Such an experiment later might be expanded with equal sized blocks to an R = 5 plan as the factorial part of a central composite in their method of local exploration. In that case, the block is the smallest block that will insure that no first degree or two-factor interaction coefficients are confounded with block effects, Box and Hunter (1957).

Holms (1967), henceforth abbreviated H67, gave design tables for expansible experiments on 4 through 8 independent variables. The tables were constructed to be appropriate to the optimum seeking and response surface methods of Box and Wilson (1951), in that the plans are expansible from R = 3 to the full factorial or at least to R = 5. The tables listed the treatment assignments to the blocks of a sequence and listed the resulting observations from successive fractional stages in the proper order for Yates' method of coefficient estimation. The tables also listed the aliased sets of model coefficients in the order the sets are estimated by Yates' method. The tables of H67 identified which aliased sets of model coefficients are confounded with block effects but they gave no opportunity to benefit from a knowledge of a block effect structure.

To illustrate the parameterization of a crossed classification block effect structure, suppose treatments are performed in lab $i = 1$ and in lab $i = 2$. Suppose that in both labs, treatments are performed on material from vendor $j = 1$ and on nominally the same material from vendor $j = 2$. The difference between labs can be parameterized by μ_i and the difference between vendors by μ_j. An interaction μ_{ij} can be assumed to exist, because the block structure is a crossed structure. But suppose that the treatments in lab $i = 1$ are performed on materials from vendors $j = 1$ and $j = 2$, and that the treatments in lab $i = 2$ are performed on materials from two other vendors $k = 1$ and $k = 2$. Such a block structure is classified as vendors nested within labs and such a nested classification is outside of the present discussion.

Holms (1998) henceforth abbreviated H98 gave methods for the design of expansible experiments with multiple types of crossed classification block structures. The purpose now is to supplement the information in the tables of H67 with a display of what aliased sets of model coefficients are confounded with what types of block parameters using the methods enabled by H98. Such displays are useful when the experimenter has prior beliefs as to which block parameters are active. Model coefficients confounded with active block parameters would be excluded from the predictive model. But from the aliased sets of coefficients confounded with block parameters believed to be inactive, one coefficient from each such set could be included in the predictive model.

2. NOTATION AND BASIC CONCEPTS

The notation follows H98. Where p is the number of independent variables, any stage is a $1/2^q$ fraction of a full factorial done in m blocks each of size n_m to have resolution R. The number of factorial (hypercube) points is

$$n_{c} = 2^{p-q}_{m,R} = mn_m$$

The standard symbols for treatments, such as (1), a, b, ab, c, ac,..., also are used to represent the observations, y, resulting from such treatments. The standard symbols for contrasts, such as I, A, B, AB. C, AC,..., also are used as the subscripts of the coefficients of a model equation such as

$$E(Y) = \beta_I + \beta_A x_A + \beta_B x_B + \beta_{AB} x_A x_B + \beta_C x_C$$

$$+\beta_{AC} x_A x_C + \beta_{BC} x_B x_C + \beta_{ABC} x_A x_B x_C$$

The x_A, x_B, x_C, are the independent variables as transformed from natural units so that the treatment high levels are $+1$ and the low levels are -1. The contrasts then are the linear operators on the observations, which on dividing by n_c are the estimators of model coefficients.

The block structures will be indicated by $i, j,...$ where the integer $= 1$ if that type of block effect has not been introduced, and $= 2$ if that type of block effect has been introduced. The blocks of a particular plan will be named $i, j,....$ The group of defining contrasts associated with a particular stage of a sequence having plan $i, j,...$ will be named $C(i, j, ...)$.

3. RESPONSE SURFACE SEQUENCES

Sequences for response surface methods are here prescribed as follows.
1. The first block shall be a principal fraction (shall contain the treatment symbolized as (1), with all independent variables at their low levels). The first block shall be a minimum aberration plan (Fries and Hunter (1980)) within a sequence that meets all the other requirements here prescribed.
2. All blocks shall be equal in size with minimum $R = 3$ to insure orthogonality of block effects to first degree and two-factor interaction coefficient estimates, Box and Hunter (1957).
3. Any subsequent stage of a sequence shall be a doubling of the preceding stage that maintains orthogonality of coefficient estimates.

4. The completed sequence shall be the minimum design that is a full factorial or at least an $R = 5$ design.

Table 1. Table Numbers Listing Yates Order Properties of Stages of Sequences

Sequence number	p	Replicate	q	m	n_c	R	Table number
1	4	1/2	1	1	8	4	2
		2/2	0	2	16	full	3
2	5	1/4	2	1	8	3	4
		2/4	1	2	16	4	5
		4/4	0	4	32	full	6
3	6	1/8	3	1	8	3	8
		2/8	2	2	16	4	9
		4/8	1	4	32	4	10
		8/8	0	8	64	full	11
4	7	1/16	4	1	8	3	13
		2/16	3	2	16	3	14
		4/16	2	4	32	3	15
		8/16	1	8	64	7	16

Table 2. Yates Order Properties, Sequence no. 1, $p = 4$, 1/2 Rep. $I = ABCD$ $q = 1$, $m = 1$, $n_c = 8$, $R = 4$

i	y	Estimate	i	y	Estimate
1	(1)	β_I	1	cd	β_C
1	ad	β_A	1	ac	$\beta_{AC}+\beta_{BD}$
1	bd	β_B	1	bc	$\beta_{BC}+\beta_{AD}$
1	ab	$\beta_{AB}+\beta_{CD}$	1	abcd	β_D

Table 3. Yates Order Properties, Sequence no. 1, $p = 4$, 2/2 Rep. $q = 0$, $m = 2$, $n_c = 16$, full

i	y	Estimate	i	y	Estimate
1	(1)	β_I	2	d	β_D
2	a	β_A	1	ad	β_{AD}
2	b	β_B	1	bd	β_{BD}
1	ab	β_{AB}	2	abd	β_{ABD}
2	c	β_C	1	cd	β_{CD}
1	ac	β_{AC}	2	acd	β_{ACD}
1	bc	β_{BC}	2	bcd	β_{BCD}
2	abc	β_{ABC}	1	abcd	$\beta_{ABCD}+\mu_i$

Because the doublings to successive stages maintain the orthogonality of the coefficient estimators, the confounded block parameter estimators remain orthogonal to the unconfounded coefficient estimators over the doublings.

The defining contrasts that are deleted with the expansions that introduce the block parameters become estimators of the block parameters (H98). Furthermore all the blocks are at least $R = 3$ plans.

Thus the block parameters at any stage are confounded with coefficient estimates only for three-factor or higher degree coefficients, Box and Hunter (1957).

Blocks of size $n_m = 8$ with $R = 3$ can be achieved with $p < 8$. If $p = 8$, the minimum with $R = 3$ is $n_m = 16$. H67 shows reductions in n_c for $R = 5$ with p = 5 and 6 when n_m is increased from 8 to 16. Limited space precludes listing of $n_m = 16$ plans here.

Table 4. Yates Order Properties, Sequence no. 2, p = 5, 1/4 Rep. I = -BCD = ABCE = -ADE, q = 2, m = 1, n_c = 8, R = 3

i	y	Estimate	i	y	Estimate
1	(1)	β_I	1	cde	$\beta_C - \beta_{BD}$
1	ae	$\beta_A - \beta_{DE}$	1	acd	$\beta_{AC} + \beta_{BE}$
1	bde	$\beta_B - \beta_{CD}$	1	bc	$-\beta_D + \beta_{BC} + \beta_{AE}$
1	abd	$\beta_{AB} + \beta_{CE}$	1	abce	$\beta_E - \beta_{AD}$

Table 5. Yates Order Properties, Sequence no. 2, p = 5, 2/4 Rep. I = ABCE, q = 1, m = 2, n_c = 16, R = 4

i,j	y	Estimate	i,j	y	Estimate
1,1	(1)	β_I	2,1	d	β_D
1,1	ae	β_A	2,1	ade	β_{AD}
2,1	be	β_B	1,1	bde	β_{BD}
2,1	ab	$\beta_{AB} + \beta_{CE}$	1,1	abd	$\beta_{ABD} + \beta_{CDE}$
2,1	ce	β_C	1,1	cde	β_{CD}
2,1	ac	$\beta_{AC} + \beta_{BE}$	1,1	acd	$\beta_{ACD} + \beta_{BDE}$
1,1	bc	$\beta_{BC} + \beta_{AE}$	2,1	bcd	$\beta_{BCD} + \beta_{ADE} + \mu_i$
1,1	abce	β_E	2,1	abcde	β_{DE}

Table 6. Yates Order Properties, Sequence no. 2, p = 5, 4/4 Rep. q = 0, m = 4, n_c = 32, full

i,j	y	Estimate	i,j	y	Estimate
1,1	(1)	β_I	1,2	e	β_E
1,2	a	β_A	1,1	ae	β_{AE}
2,2	b	β_B	2,1	be	β_{BE}
2,1	ab	β_{AB}	2,2	abe	β_{ABE}
2,2	c	β_C	2,1	ce	β_{CE}
2,1	ac	β_{AC}	2,2	ace	β_{ACE}
1,1	bc	β_{BC}	1,2	bce	β_{BCE}
1,2	abc	β_{ABC}	1,1	abce	$\beta_{ABCE} + \mu_j$
2,1	d	β_D	2,2	de	β_{DE}
2,2	ad	β_{AD}	2,1	ade	$\beta_{ADE} + \mu_{ij}$
1,2	bd	β_{BD}	1,1	bde	β_{BDE}
1,1	abd	β_{ABD}	1,2	abde	β_{ABDE}
1,2	cd	β_{CD}	1,1	cde	β_{CDE}
1,1	acd	β_{ACD}	1,2	acde	β_{ACDE}
2,1	bcd	$\beta_{BCD} + \mu_i$	2,2	bcde	β_{BCDE}
2,2	abcd	β_{ABCD}	1,1	abcde	β_{ABCDE}

Table 7. Defining Contrasts, Sequence no. 3, p = 6

Block parameter	Defining contrasts		
	1/8 rep.	2/8 rep.	4/8 rep.
μ_0	I		
μ_i	-ABD		
μ_j	ACDE	ACDE	
μ_k	ABCF	ABCF	ABCF
μ_{ij}	-BCE		
μ_{ik}	-CDF		
μ_{jk}	BDEF	BDEF	
μ_{ijk}	-AEF		

Table 8. Yates Order Properties, Sequence no. 3, p = 6, 1/8 Rep. q = 3, m = 1, n_c = 8, R = 3

i	y	Estimate
1	(1)	β_I
1	adf	$\beta_A - \beta_{BD} - \beta_{EF}$
1	bdef	$\beta_B - \beta_{CE} - \beta_{AD}$
1	abe	$-\beta_D + \beta_{AB} + \beta_{CF}$
1	cef	$\beta_C - \beta_{BE} - \beta_{DF}$
1	acde	$\beta_{AC} + \beta_{BF} + \beta_{DE}$
1	bcd	$-\beta_E + \beta_{BC} + \beta_{AF}$
1	abcf	$\beta_F - \beta_{AE} - \beta_{CD}$

Table 9. Yates Order Properties, Sequence no. 3, p = 6, 2/8 Rep. q = 2, m = 2, n_c = 16, R = 4

i	y	Estimate
1	(1)	β_I
2	aef	β_A
2	bf	β_B
1	abe	$\beta_{AB} + \beta_{CF}$
1	cef	β_C
2	ac	$\beta_{AC} + \beta_{DE} + \beta_{BF}$
2	bce	$\beta_{BC} + \beta_{AF}$
1	abcf	β_F
2	de	β_D
1	adf	$\beta_{AD} + \beta_{CE}$
1	bdef	$\beta_{BD} + \beta_{EF}$
2	abd	$\beta_{ABD} + \beta_{BCE} + \beta_{CDF} + \beta_{AEF} + \mu_i$
2	cdf	$\beta_{AE} + \beta_{CD}$
1	acde	β_E
1	bcd	$\beta_{ABE} + \beta_{ADF} + \beta_{BCD} + \beta_{CEF}$
2	abcdef	$\beta_{BE} + \beta_{DF}$

4. TABULATED INFORMATION

The values of p and n_m uniquely identify the sequences. The values of q uniquely identify the stages of a given sequence. Table 1 lists the tables of Yates order properties for the stages of the sequences.

Table 10. Yates Order Properties, Sequence no. 3,
p = 6, 4/8 Rep. q = 1, m = 4, n_c = 32, R = 4

i,j	y	Estimate
1,1	(1)	β_I
2,2	af	β_A
2,1	bf	β_B
1,2	ab	$\beta_{AB} + \beta_{CF}$
1,2	cf	β_C
2,1	ac	$\beta_{AC} + \beta_{BF}$
2,2	bc	$\beta_{BC} + \beta_{AF}$
1,1	abcf	β_F
2,2	d	β_D
1,1	adf	β_{AD}
1,2	bdf	β_{BD}
2,1	abd	$\beta_{ABD} + \beta_{CDF} + \mu_i$
2,1	cdf	β_{CD}
1,2	acd	$\beta_{ACD} + \beta_{BDF}$
1,1	bcd	$\beta_{BCD} + \beta_{ADF}$
2,2	abcdf	β_{DF}
1,2	e	β_E
2,1	aef	β_{AE}
2,2	bef	β_{BE}
1,1	abe	$\beta_{ABE} + \beta_{CEF}$
1,1	cef	β_{CE}
2,2	ace	$\beta_{ACE} + \beta_{BEF}$
2,1	bce	$\beta_{BCE} + \beta_{AEF} + \mu_{ij}$
1,2	abcef	β_{EF}
2,1	de	β_{DE}
1,2	adef	β_{ADE}
1,1	bdef	β_{BDE}
2,2	abde	$\beta_{ABDE} + \beta_{CDEF}$
2,2	cdef	β_{CDE}
1,1	acde	$\beta_{ACDE} + \beta_{BDEF} + \mu_j$
1,2	bcde	$\beta_{BCDE} + \beta_{ADEF}$
2,1	abcdef	β_{DEF}

For the $n_c = 2^{p-q}$ observations, there are n_c estimates, each being an estimate of a set of 2^q of the full model aliased coefficients. The Yates order properties tables list the treatments in the order the associated observations must be listed to use the Yates method of coefficient estimation. The tables also list the estimated sets of aliased model coefficients in the order they are estimated by the Yates method. The methods for establishing the Yates orders for observations and estimates are given in H67. The methods follow the rules of Daniel (1956).

Except for I, negative signs are attached to all defining contrasts having an odd number of letters. This is consistent with making the first block a principal fraction, that is, having the first block contain the treatment with all independent variables at their low level.

Table 11. Yates Order Properties, Sequence no. 3,
p = 6, 8/8 Rep. q = 0, m = 8, n_c =64, full

i,j,k	y	Estimate	i,j,k	y	Estimate
1,1,1	(1)	β_I	1,1,2	f	β_F
2,2,2	a	β_A	2,2,1	af	β_{AF}
2,1,2	b	β_B	2,1,1	bf	β_{BF}
1,2,1	ab	β_{AB}	1,2,2	abf	β_{ABF}
1,2,2	c	β_C	1,2,1	cf	β_{CF}
2,1,1	ac	β_{AC}	2,1,2	acf	β_{ACF}
2,2,1	bc	β_{BC}	2,2,2	bcf	β_{BCF}
1,1,2	abc	β_{ABC}	1,1,1	abcf	$\beta_{ABCF} + \mu_k$
2,2,1	d	β_D	2,2,2	df	β_{DF}
1,1,2	ad	β_{AD}	1,1,1	adf	β_{ADF}
1,2,2	bd	β_{BD}	1,2,1	bdf	β_{BDF}
2,1,1	abd	$\beta_{ABD} + \mu_i$	2,1,2	abdf	β_{ABDF}
2,1,2	cd	β_{CD}	2,1,1	cdf	$\beta_{CDF} + \mu_{ik}$
1,2,1	acd	β_{ACD}	1,2,2	acdf	β_{ACDF}
1,1,1	bcd	β_{BCD}	1,1,2	bcdf	β_{BCDF}
2,2,2	abcd	β_{ABCD}	2,2,1	abcdf	β_{ABCDF}
1,2,1	e	β_E	1,2,2	ef	β_{EF}
2,1,2	ae	β_{AE}	2,1,1	aef	$\beta_{AEF} + \mu_{ijk}$
2.2.2	be	β_{BE}	2,2,1	bef	β_{BEF}
1,1,1	abe	β_{ABE}	1,1,2	abef	β_{ABEF}
1,1,2	ce	β_{CE}	1,1,1	cef	β_{CEF}
2,2,1	ace	β_{ACE}	2,2,2	acef	β_{ACEF}
2,1,1	bce	$\beta_{BCE} + \mu_{ij}$	2,1,2	bcef	β_{BCEF}
1,2,2	abce	β_{ABCE}	1,2,1	abcef	β_{ABCEF}
2,1,1	de	β_{DE}	2,1,2	def	β_{DEF}
1,2,2	ade	β_{ADE}	1,2,1	adef	β_{ADEF}
1,1,2	bde	β_{BDE}	1,1,1	bdef	$\beta_{BDEF} + \mu_{jk}$
2,2,1	abde	β_{ABDE}	2,2,2	abdef	β_{ABDEF}
2,2,2	cde	β_{CDE}	2,2,1	cdef	β_{CDEF}
1,1,1	acde	$\beta_{ACDE} + \mu_j$	1,1,2	acdef	β_{ACDEF}
1,2,1	bcde	β_{BCDE}	1,2,2	bcdef	β_{BCDEF}
2,1,2	abcde	β_{ABCDE}	2,1,1	abcdef	β_{ABCDEF}

One element of each of the aliased sets of model coefficients is a member of the Yates order sequence β_I, β_A, β_B, β_{AB}, β_C, β_{AC}, ... with 2^{p-q} coefficients in that sequence. The subscripts of the other coefficients of the aliased sets were identified by multiplying the subscripts of the Yates order coefficients by the defining contrasts to obtain the subscripts of the sets aliased with the Yates order coefficients (H98). The algebraic signs of these products were attached to the coefficients. A negative sign attached to a coefficient requires that the sign of a

Yates estimate be reversed if that coefficient is used in the predictive equation. Because the block parameters are not used in the predictive equation, algebraic signs for the block parameters are ignored.

Table 12. Defining Contrasts, Sequence no. 4, p = 7

Block parameter	1/16 rep.	2/16 rep.	4/16 rep.
μ_0	I	I	I
μ_i	-BCF		
μ_j	ABCG	ABCG	
μ_k	ACDF	ACDF	ACDF
μ_l	-ABCDEFG	-ABCDEFG	-ABCDEFG
μ_{ij}	-AFG		
μ_{ik}	-ABD		
μ_{il}	ADEG		
μ_{jk}	BDFG	BDFG	
μ_{jl}	-DEF	-DEF	
μ_{kl}	-BEG	-BEG	-BEG
μ_{ijk}	-CDG		
μ_{ijl}	BCDE		
μ_{ikl}	CEFG		
μ_{jkl}	-ACE	-ACE	
μ_{ijkl}	ABEF		

In the tables listing the aliased coefficient sets, if a first degree or a two-factor interaction coefficient is aliased with three-factor or higher degree coefficients, the aliased coefficients of degree higher than two are assumed negligible, and are not shown. In any set of aliased coefficients of minimum degree higher than two, merely the lowest degree coefficients of the aliased set are shown.

The display of the aliased coefficients allows the experimenter to choose, based on prior knowledge, one model coefficient from each aliased set that the experimenter believes the most likely to be important. That coefficient and associated term then is included in the predictive equation.

Any set of aliased coefficients confounded with a block parameter is shown with the block parameter. If the experimenter believes that a listed block parameter is important, then neither the parameter nor any of the confounded coefficients would be included in the predictive equation. If the experimenter believes that a listed block parameter is negligible, a confounded coefficient can be selected from the alised set for inclusion in the predictive equation.

Table 13. Yates Order Properties, Sequence no. 4, p = 7, 1/16 Rep. q = 4, m = 1, n_c = 8, R = 3

i	y	Estimate
1	(1)	β_I
1	adeg	$\beta_A - \beta_{BD} - \beta_{CE} - \beta_{FG}$
1	bdfg	$\beta_B - \beta_{AD} - \beta_{CF} - \beta_{EG}$
1	abef	$-\beta_D + \beta_{AB} + \beta_{CG} + \beta_{EF}$
1	cefg	$\beta_C - \beta_{AE} - \beta_{BF} - \beta_{DG}$
1	acdf	$-\beta_E + \beta_{AC} + \beta_{DF} + \beta_{BG}$
1	bcde	$-\beta_F + \beta_{BC} + \beta_{AG} + \beta_{DE}$
1	abcg	$\beta_G - \beta_{CD} - \beta_{BE} - \beta_{AF}$

Table 14. Yates Order Properties, Sequence no. 4, p = 7. 2/16 Rep. q = 3, m = 2, n_c = 16, R = 3

i	y	Estimate
1	(1)	β_I
2	aefg	$\beta_A - \beta_{CE}$
2	bg	$\beta_B - \beta_{EG}$
1	abef	$\beta_{AB} + \beta_{CG}$
1	cefg	$\beta_C - \beta_{AE}$
2	ac	$-\beta_E + \beta_{AC} + \beta_{DF} + \beta_{BG}$
2	bcef	$\beta_{BC} + \beta_{AG}$
1	abcg	$\beta_G - \beta_{BE}$
2	df	$\beta_D - \beta_{EF}$
1	adeg	$\beta_{AD} + \beta_{CF}$
1	bdfg	$\beta_{BD} + \beta_{FG}$
2	abde	$\beta_{ABD} + \beta_{CDG} + \beta_{BCF} + \beta_{AFG} + \mu_i$
2	cdeg	$\beta_{CD} + \beta_{AF}$
1	acdf	$\beta_F - \beta_{DE}$
1	bcde	$\beta_{ADG} + \beta_{ABF} + \beta_{CFG} + \beta_{BCD}$
2	abcdfg	$\beta_{DG} + \beta_{BF}$

If instead of using Yates' method, coefficients are estimated by matrix inversion, the model to be fitted can be formed by first selecting one coefficient from each of the tabulated 2^{p-q} sets of aliased coefficients. The model is then written with the associated terms.

REFERENCES

Box, G. E. P. and Hunter, J. S. (1957), "Multi-Factor Experimental Designs for Exploring Response Surfaces." *Annals of Mathematical Statistics*, 28, pp. 195–241.

Box, G. E. P. and Hunter, J. S. (1961), "The 2^{k-p} Fractional Factorial Designs," I. *Technometrics*, 3, pp. 311–351.

Box, G. E. P. and Wilson, K. B. (1951), "On the Experimental Attainment of Optimum Conditions." *Journal of the Royal Statistical Society*, Series B, 13, pp. 1–38, discussion pp. 38–45.

Table 15. *Yates Order Properties, Sequence no. 4,* $p = 7$, 4/16 *Rep.* $q = 2$, $m = 4$, $n_C = 32$, $R = 3$

i,j	y	Estimate
1,1	(1)	β_I
2,2	af	β_A
2,1	bg	$\beta_B - \beta_{EG}$
1,2	abfg	β_{AB}
1,2	cf	β_C
2,1	ac	$\beta_{AC} + \beta_{DF}$
2.2	bcfg	β_{BC}
1,1	abcg	$\beta_{ABC} + \beta_{BDF}$
2,1	df	β_D
1,2	ad	$\beta_{AD} + \beta_{CF}$
1,1	bdfg	β_{BD}
2,2	abdg	$\beta_{ABD} + \beta_{BCF} + \mu_i$
2,2	cd	$\beta_{CD} + \beta_{AF}$
1,1	acdf	β_F
1,2	bcdg	$\beta_{BCD} + \beta_{ABF}$
2,1	abcdfg	β_{BF}
1,2	eg	$\beta_E - \beta_{BG}$
2,1	aefg	β_{AE}
2,2	be	$-\beta_G + \beta_{BE}$
1,1	abef	$-\beta_{AG}$
1,1	cefg	β_{CE}
2,2	aceg	$\beta_{ACE} + \beta_{DEF} + \mu_j$
2,1	bcef	$-\beta_{CG}$
1,2	abce	$-\beta_{ACG} - \beta_{DFG}$
2,2	defg	β_{DE}
1,1	adeg	$\beta_{ADE} + \beta_{CEF}$
1,2	bdef	$-\beta_{DG}$
2,1	abde	$-\beta_{ADG} - \beta_{CFG}$
2,1	cdeg	$\beta_{CDE} + \beta_{AEF}$
1,2	acdefg	β_{EF}
1,1	bcde	$-\beta_{AFG} - \beta_{CDG} + \mu_{ij}$
2,2	abcdef	$-\beta_{FG}$

Table 16. *Yates Order Properties, Sequence no. 4,* $p = 7$, 8/16 *Rep.,* $I = -ABCDEFG$, $q = 1$, $m = 8$, $n_C = 64$, $R = 7$

i,j,k	y	Estimate	i,j,k	y	Estimate
1,1,1	(1)	β_I	2,2,2	fg	β_F
1,1,2	ag	β_A	2,2,1	af	β_{AF}
2,1,1	bg	β_B	1,2,2	bf	β_{BF}
2,1,2	ab	β_{AB}	1,2,1	abfg	β_{ABF}
2,1,2	cg	β_C	1,2,1	cf	β_{CF}
2,1,1	ac	β_{AC}	1,2,2	acfg	β_{ACF}
1,1,2	bc	β_{BC}	2,2,1	bcfg	$\beta_{BCF} + \mu_i$
1,1,1	abcg	β_{ABC}	2,2,2	abcf	$-\beta_{DEG}$
1,2,2	dg	β_D	2,1,1	df	β_{DF}
1,2,1	ad	β_{AD}	2,1,2	adfg	β_{ADF}
2,2,2	bd	β_{BD}	1,1,1	bdfg	β_{BDF}
2,2,1	abdg	$\beta_{ABD} + \mu_{ik}$	1,1,2	abdf	$-\beta_{CEG}$
2,2,1	cd	β_{CD}	1,1,2	cdfg	β_{CDF}
2,2,2	acdg	β_{ACD}	1,1,1	acdf	$-\beta_{BEG} + \mu_k$
1,2,1	bcdg	β_{BCD}	2,1,2	bcdf	$-\beta_{AEG}$
1,2,2	abcd	$-\beta_{EFG}$	2,1,1	abcdfg	$-\beta_{EG}$
1,2,1	eg	β_E	2,1,2	ef	β_{EF}
1,2,2	ae	β_{AE}	2,1,1	aefg	β_{AEF}
2,2,1	be	β_{BE}	1,1,2	befg	β_{BEF}
2,2,2	abeg	β_{ABE}	1,1,1	abef	$-\beta_{CDG} + \mu_{ijk}$
2,2,2	ce	β_{CE}	1,1,1	cefg	β_{CEF}
2,2,1	aceg	$\beta_{ACE} + \mu_{jk}$	1,1,2	acef	$-\beta_{BDG}$
1,2,2	bceg	β_{BCE}	2,1,1	bcef	$-\beta_{ADG}$
1,2,1	abce	$-\beta_{DFG}$	2,1,2	abcefg	$-\beta_{DG}$
1,1,2	de	β_{DE}	2,2,1	defg	$\beta_{DEF} + \mu_j$
1,1,1	adeg	β_{ADE}	2,2,2	adef	$-\beta_{BCG}$
2,1,2	bdeg	β_{BDE}	1,2,1	bdef	$-\beta_{ACG}$
2,1,1	abde	$-\beta_{CFG}$	1,2,2	abdefg	$-\beta_{CG}$
2,1,1	cdeg	β_{CDE}	1,2,2	cdef	$-\beta_{ABG}$
2,1,2	acde	$-\beta_{BFG}$	1,2,1	acdefg	$-\beta_{BG}$
1,1,1	bcde	$-\beta_{AFG} + \mu_{ij}$	2,2,2	bcdefg	$-\beta_{AG}$
1,1,2	abcdeg	$-\beta_{FG}$	2,2,1	abcdef	$-\beta_G$

Daniel, C. (1956), "Fractional Replication in Industrial Research" *Proceedings of the Third Berkeley Symposium on Mathematical Statistics and Probability*, Vol. V, University of California Press, 87–98.

Fries, A. and Hunter, W. G. (1980), "Minimum Aberration 2^{k-p} Designs," *Technometrics*, 22, 601-608.

Holms, A. G. (1967), "Designs of Experiments as Telescoping Sequences of Blocks for Optimum Seeking (as Intended for Alloy Development)". NASA TN D-4100.

Holms, A. G. (1998), "Design of Experiments as Expansible Sequences of Orthogonal Blocks With Crossed-Classification Block Effects. *Technometrics*, 40, 244-253.

EFFICIENCY JUSTIFICATION OF GENERALIZED MINIMUM ABERRATION

Ching-Shui Cheng, University of California-Berkeley
Lih-Yuan Deng, University of Memphis
Boxin Tang, University of Memphis
Lih-Yuan Deng, University of Memphis, Memphis, TN 38152 (dengl@msci.memphis.edu)

Key Words: Plackett-Burman design, confounding frequency vector, regular design, $\mathcal{E}(s^2)$, D_f criterion.

Abstract:

Deng and Tang (1999) first proposed the generalized minimum aberration criterion (GMA) for both regular and nonregular designs as a natural extension to the minimum aberration criterion (MA) for regular designs. In this paper, we establish a relationship between the GMA criterion and some design efficiency criteria. It is shown that the GMA criterion is supported by these design efficiency criteria. Several advantages of using the GMA criterion are also discussed.

1 Introduction

The most commonly used designs for factorial experiments are the two-level fractional factorial designs. *Regular designs*, often denoted by 2_R^{m-q}, are the most popular among two-level designs. Here m is the number of factors, $m - q$ is the number of independent factors, and R is the design resolution. A regular design is uniquely determined by its defining relation. Each term in the defining relation is called a *word*, and the number of factors in the word is called the *word-length*. The frequency distribution of the word-length in the defining relation is called the *word-length pattern* of the regular design. In fact R is the shortest word-length in the defining relation. We prefer a design with a higher design resolution. While the design resolution is an important criterion, it cannot differentiate designs with the same resolution. Fries and Hunter (1980) proposed a more refined *minimum aberration* (MA) criterion based on the word-length pattern to compare designs with the same resolution. More recent works on MA for regular 2_R^{m-q} designs can be found in Franklin (1984), Chen and Wu (1991), Chen (1992), Tang and Wu (1996), and H. Chen and Hedayat (1996).

A necessary condition for a regular design is that its run size must be a power of 2 (say, $4, 8, 16, 32, 64 \cdots$). This creates a major problem of using regular designs because the "gap" in run size increases exponentially. Nonregular designs can be taken from a Hadamard matrix of order n, $\mathbf{H}$, where all the entries in $\mathbf{H}$ are ± 1 and $\mathbf{H}'\mathbf{H} = n\mathbf{I}$. It is well-known that the order n has to be a multiple of 4, except when $n = 1, 2$. Hamada and Wu (1995) discussed several applications of nonregular designs. For more discussion on regular and nonregular designs, see Deng and Tang (1998). For nonregular designs, the MA criterion is not applicable and there is no systematic criterion until recently. Deng and Tang (1999) proposed the generalized resolution (GR) criterion and the generalized minimum aberration (GMA) criterion. When n is not a power of 2, no regular designs are available and GMA is a natural extension of the MA criterion. When n is a power of 2, there exist both regular and nonregular designs. In this case, GMA is the same as the MA criterion when applied to regular designs and the GMA criterion is applicable to both regular and nonregular designs. In Deng and Tang (1999), GR and hence GMA are justified from projection and from minimizing bias. In this paper, we build a connection between the GMA criterion and some of traditional design efficiency criteria. We give another justification of the GMA criterion by showing that it is supported by these model-dependent criteria.

In Section 2, we give a brief review of the GMA criterion as proposed by Deng and Tang (1999). Some notation is also introduced. In Section 3, we establish a relationship between the GMA criterion and some of the classical design efficiency criteria. Based on this relationship, we then show that the GMA criterion is supported by these design efficiency criteria. Several advantages of using the GMA criterion are also discussed.

2 Generalized minimum aberration

For a set of k vectors, $s = \{\mathbf{v}_1, \mathbf{v}_2, ..., \mathbf{v}_k\}$, define

$$j_k(s) = \sum_{i=1}^{n} v_{i1}v_{i2}...v_{ik}, \quad J_k(s) = |j_k(s)|, \quad (1)$$

where v_{ij} is the i-th component of $\mathbf{v}_j$. Throughout this paper, we use D to denote an $n \times m$ matrix chosen from a Hadamard matrix of order n. We also use s to denote a subset of columns taken from D. From the orthogonal property of Hadamard matrix, it is easy to see that $J_1(s) = 0$, $J_2(s) = 0$, for any s. If design D is regular, then $J_k(s)$ is either 0 or n. If D is nonregular, then the value of $J_k(s)$ can be between 0 and n. Deng and Tang (1998) prove some useful results concerning the possible values of $J_k(s)$.

2.1 Confounding frequency vectors

Let $\mathbf{A}_k(D)$ be the frequency distribution of $J_k(s)$ over all subsets s of k columns from design D. Deng and Tang (1999) proposed to use the *confounding frequency vector*

$$F_A(D) = [\mathbf{A}_1(D); \ldots; \mathbf{A}_m(D)] \quad (2)$$

to classify and rank various designs. Given two designs D_1 and D_2 with their confounding frequency vectors, $[\mathbf{A}_1(D_1); \ldots; \mathbf{A}_m(D_1)]$ and $[\mathbf{A}_1(D_2); \ldots; \mathbf{A}_m(D_2)]$, suppose r is the smallest value for which $\mathbf{A}_r(D_1) \neq \mathbf{A}_r(D_2)$. Obviously, we prefer a design with a smaller frequency count of the largest value of $J_r(s)$. If they are the same, continue to compare the frequency count of the next largest value of $J_r(s)$, and so on. We note that the $F_A(D)$ in (2) becomes the word-length pattern when D is a regular design.

Tang and Deng (1999) proposed another GMA criterion for ranking designs. Specifically, let

$B_k(D) = n^{-2} \sum_{|s|=k} j_k^2(s)$. Since design D is taken from a Hadamard design, we have $B_1(D) = B_2(D) = 0$. A new GMA criterion can be considered according to the confounding frequency vector

$$F_B(D) = [B_1(D); B_2(D); B_3(D); ...]. \quad (3)$$

According to our empirical study, GMA ranking based on $B_k(D)$ is quite consistent with the GMA ranking based on $\mathbf{A}_k(D)$. This observation can be easily explained by the fact that $B_k(D)$ is a function measuring the magnitude of $\mathbf{A}_k(D)$. As it will be shown later, there are some particular situations where both rankings are identical. While the GMA criterion based on $B_k(D)$ is simpler, the GMA criterion based on $\mathbf{A}_k(D)$ is "better" in terms of the classification ability. As shown in Deng and Tang (1998), only two or three leading terms of the $F_A(D)$ are needed to rank and/or classify various designs. A similar idea can be extended for the confounding frequency vector $F_B(D)$ as defined in (3).

2.2 Other notational preparation

Consider the following two models:

$$\begin{array}{lll} (I) & \mathbf{y} = \mathbf{X}_1\boldsymbol{\beta}_1 + \boldsymbol{\epsilon} \\ (II) & \mathbf{y} = \mathbf{X}_1\boldsymbol{\beta}_1 + \mathbf{X}_2\boldsymbol{\beta}_2 + \boldsymbol{\epsilon}, \end{array}$$

where $\mathbf{y}$ is a column vector of dimension n, $\mathbf{X}_1$ and $\mathbf{X}_2$ are two matrices of dimension $n \times m$ and $n \times f$, respectively, and $\boldsymbol{\epsilon}$ is an n-dimensional column error random vector with a zero mean and a variance-covariance matrix $\sigma^2 \mathbf{I}_n$. The matrix $\mathbf{X}_1$ is the $n \times m$ matrix selected from a Hadamard matrix of order n, and m is the number of factors studied. Model (I) is therefore the main effect model. The matrix $\mathbf{X}_2$ is the $n \times f$ matrix representing f interaction effects. It is common to assume that interactions of order 3 or higher can be ignored. Therefore, model (II) is the main effects plus f second-order interaction effects model.

In this paper, we refer to $D = \mathbf{X}_1$ as the *design* matrix and $\mathbf{X} = [\mathbf{X}_1, \mathbf{X}_2]$ as the *model* matrix. Since design matrix D is an orthogonal matrix, all the main effects are unconfounded among themselves. However, the main effects can be confounded with some of the two-factor interactions. The first objective in choosing design matrix D is that it can minimize these confounding effects. Secondly, if some of the two-factor interactions are present, we hope design matrix D will allow the identification and estimation of these effects. Finally, we hope such a design matrix will result in a "good" model matrix yielding efficient estimates.

For $\mathbf{X}_2$, it is convenient to denote its index of the entry by $(i, \mathbf{u})$, where $1 \leq i \leq n$ and $\mathbf{u}$ corresponds to the set of interaction columns in $\mathbf{X}_2$. For any two column index sets $\mathbf{u}$ and $\mathbf{v}$, we use $\mathbf{u} < \mathbf{v}$ to indicate that the column in $\mathbf{X}_2$ corresponding to the index $\mathbf{u}$ is placed before that corresponding to $\mathbf{v}$. In addition, we use $(\mathbf{u} \times \mathbf{v})$ to denote the set corresponding to a set containing either $\mathbf{u}$ or $\mathbf{v}$ but not both. That is, $(\mathbf{u} \times \mathbf{v}) = (\mathbf{u} - \mathbf{v}) \cup (\mathbf{v} - \mathbf{u})$. If $\mathbf{u}$ and $\mathbf{v}$ are two index sets corresponding to two-factor interactions, then there are only three possible values for the number of elements in $s = (\mathbf{u} \times \mathbf{v})$: $|s| = 4$, if $\mathbf{u}$ and $\mathbf{v}$ are disjoint, $|s| = 0$, if $\mathbf{u} = \mathbf{v}$, and $|s| = 2$ otherwise. The corresponding value for its j value is $j_4(s)$, n, and 0, respectively. Here we define $j_0(s) = n$, when s is an empty set. In this section, we use $|s|$ to denote the number of elements in a set s. In the next section, we also use $|\mathbf{Z}|$ to denote the determinant of a matrix $\mathbf{Z}$. Since its meaning should be clear from the context, we will use the same notation. Define

$$\mathbf{Z_{s_f}} = \frac{\mathbf{X}'\mathbf{X}}{n} = \begin{pmatrix} \mathbf{I}_m & \mathbf{Z}_{12} \\ \mathbf{Z}'_{12} & \mathbf{Z}_{22} \end{pmatrix}, \qquad (4)$$

where $\mathbf{I}_m$ is an identity matrix of order m, $\mathbf{Z}_{12} = \frac{1}{n}\mathbf{X}'_1\mathbf{X}_2$, $\mathbf{Z}_{22} = \frac{1}{n}\mathbf{X}'_2\mathbf{X}_2$. Note that $\mathbf{Z}_{12}$ is an $m \times f$ matrix and we denote its index of the entry by $(i, \mathbf{u})$, where $1 \leq i \leq m$ and $\mathbf{u} \in \mathbf{s}_f$. Here $\mathbf{s}_f$ is a set of two-factor interactions corresponding to the set of interaction columns in $\mathbf{X}_2$. In that case

$$\mathbf{Z}_{12}(i, \mathbf{u}) = \begin{cases} 0, & i \in \mathbf{u}, \\ z_3(s), & i \notin \mathbf{u}, s = \{i\} \cup \mathbf{u}, \end{cases} \qquad (5)$$

where $z_k(s) = j_k(s)/n$, and k is the number of elements in the set s. If the value k in $z_k(s)$ cannot be specified, we simply denote $z_k(s)$ in (5) as $z(s)$. The notation $z_k(s)$ and $z(s)$ are frequently used in this paper. For example, we can rewrite $B_k(D)$ as $B_k(D) = \sum_{|s|=k} z_k^2(s)$. In (4), $\mathbf{Z}_{22}$ is an $f \times f$ matrix and we denote its index of the entry in $\mathbf{Z}_{22}$ by $(\mathbf{u}, \mathbf{v})$,

where $\mathbf{u}, \mathbf{v} \in \mathbf{s}_f$. Obviously,

$$\mathbf{Z}_{22}(\mathbf{u}, \mathbf{v}) = \begin{cases} 1, & |\mathbf{u} \cap \mathbf{v}| = 2, \\ 0, & |\mathbf{u} \cap \mathbf{v}| = 1, \\ z_4(s), & |\mathbf{u} \cap \mathbf{v}| = 0, s = \mathbf{u} \cup \mathbf{v}. \end{cases} \qquad (6)$$

3 Design efficiency and GMA

3.1 D-efficiency and D_f criterion

From (4), the D-efficiency, given by the determinant of $\mathbf{Z_{s_f}}$, is

$$|\mathbf{Z_{s_f}}| = |\mathbf{D_{s_f}}|, \quad \text{where} \quad \mathbf{D_{s_f}} = \mathbf{Z}_{22} - \mathbf{Z}'_{12}\mathbf{Z}_{12}.$$

Here $\mathbf{D_{s_f}}$ is an $f \times f$ matrix and its entry indexed by $(\mathbf{u}, \mathbf{v})$ is

$$\mathbf{D_{s_f}}(\mathbf{u}, \mathbf{v}) = z(\mathbf{u} \times \mathbf{v}) - \sum_{i=1}^{m} z(\{i\} \times \mathbf{u})z(\{i\} \times \mathbf{v}).$$

When $f = 1$, we have

$$|\mathbf{Z_{s_1}}| = |\mathbf{Z}_{\{\mathbf{u}\}}| = 1 - \sum_{i=1}^{m} z^2(\{i\} \times \mathbf{u}). \qquad (7)$$

For a general value of f, $|\mathbf{Z_{s_f}}|$ has a complicated relationship involving a non-linear function of $z_3(s)$ and $z_4(s)$.

Clearly, the value of $|\mathbf{Z_{s_f}}|$ depends on the choice of two-factor interactions in $\mathbf{s}_f$. Since we do not know in advance which of the f two-factor interactions are present, it is natural to consider the average over the finite population of all two-factor interactions. As it can be expected, the calculation of the expected value needs some results in the theory of finite population sampling. We first review some of the terminology here. The sampling unit is indexed by $\mathbf{u}$ (or $\mathbf{v}$), the sample of size f is denoted by $\mathbf{s}_f$, and the population $\mathbf{P}$ consists of $F = \binom{m}{2}$ two-factor interactions. The total number of possible samples is $\binom{F}{f} = \binom{\binom{m}{2}}{f}$ which is also the number of possible models with f two-factor interactions. For an introduction to the area of finite population sampling, see Cochran (1977).

Since $|\mathbf{Z_{s_f}}|$ depends on the random sample $\mathbf{s}_f$, we define the D_f criterion as

$$D_f = E_{\mathbf{s}_f}(|\mathbf{Z_{s_f}}|).$$

For simplicity, we may drop the subscript s_f from $E_{\mathbf{s}_f}(\cdot)$ and denote $E(\cdot)$ as the expectation taken over the random sample, if its meaning is clear from the context. For $f = 1$, using (7), we have

$$D_1 = E[|\mathbf{Z}_{\{\mathbf{u}\}}|] = 1 - \sum_{i=1}^{m} \frac{1}{F} \sum_{\mathbf{u} \in \mathbf{P}} z^2(\{i\} \times \mathbf{u}). \quad (8)$$

From (5), we know that $z^2(\{i\} \times \mathbf{u})$ is either $z_3^2(s)$ or 0, depending on whether the value $|s| = |(\{i\} \times \mathbf{u})|$ is 3 or not. Summing over $1 \leq i \leq m$, $\mathbf{u} \in \mathbf{P}$ and counting the number of duplications, we have

$$\sum_{i=1}^{m} \sum_{\mathbf{u} \in \mathbf{P}} z^2(\{i\} \times \mathbf{u}) = 3 \sum_{|s|=3} z_3^2(s) = 3B_3(D). \quad (9)$$

Therefore, we have

$$D_1 = 1 - \frac{3B_3(D)}{F}.$$

For $f \geq 2$, no simple explicit formula for D_f is available.

3.2 $\mathcal{E}(s^2)$ and S_f^2 criteria

The $\mathcal{E}(s^2)$ criterion, proposed by Booth and Cox (1962), is the average of the square of the off-diagonal entries in the matrix $\mathbf{Z}_{\mathbf{s}_f}$. If $\mathcal{E}(s^2) = 0$, then $\mathbf{Z}_{\mathbf{s}_f}$ is an identity matrix achiving the maximum value of $|\mathbf{Z}_{\mathbf{s}_f}|$. If $\mathcal{E}(s^2)$ is "small," then the off-diagonal values of $\mathbf{Z}_{\mathbf{s}_f}$ are expected to be small. In that case, $|\mathbf{Z}_{\mathbf{s}_f}|$ tends to be larger. A more precise argument on the relationship between $\mathcal{E}(s^2)$ and $|\mathbf{Z}_{\mathbf{s}_f}|$ is given next.

Note that each of the diagonal elements of $\mathbf{Z}_{\mathbf{s}_f}$ is 1 and $tr(\mathbf{Z}_{\mathbf{s}_f}) = \sum_{i=1}^{m+f} \lambda_i = m + f$, where $tr(A)$ is the trace of the matrix A and $\lambda_1, \lambda_2, \cdots, \lambda_{m+f}$ are the eigenvalues of $\mathbf{Z}_{\mathbf{s}_f}$. Furthermore, we have

$$\mathcal{E}(s^2) = \frac{\sum_{i=1}^{m+f} (\lambda_i - \bar{\lambda})^2}{2\binom{m+f}{2}}, \quad (10)$$

where $\bar{\lambda} = \frac{\sum_{i=1}^{m+f} \lambda_i}{(m+f)} = 1$. Equation (10) follows from

$$tr(\mathbf{Z}_{\mathbf{s}_f}^2) = \sum_{i=1}^{m+f} \lambda_i^2$$

and

$$tr(\mathbf{Z}_{\mathbf{s}_f}^2) = 2\binom{m+f}{2} \mathcal{E}(s^2) + (m+f).$$

From (10), we can see that the problem of minimizing $\mathcal{E}(s^2)$ is equivalent to the problem of minimizing the variation of λ_i. It is well-known that $|\mathbf{Z}_{\mathbf{s}_f}| = \prod_{i=1}^{m+f} \lambda_i$ and the best solution of maximizing $\prod_{i=1}^{m+f} \lambda_i$ subject to the condition $\sum_{i=1}^{m+f} \lambda_i = m+f$ is that all λ_i are the same. Intuitively, the λ_i's with a small variance (hence, a small $\mathcal{E}(s^2)$) tends to yield a large value of $|\mathbf{Z}_{\mathbf{s}_f}| = \prod_{i=1}^{m+f} \lambda_i$. Therefore, a small value of $\mathcal{E}(s^2)$ is a good and simple indication of a large value of $|\mathbf{Z}_{\mathbf{s}_f}|$.

Because of the strong connection between $\mathcal{E}(s^2)$ and $|\mathbf{Z}_{\mathbf{s}_f}|$, $\mathcal{E}(s^2)$ can be used as a surrogate of D-efficiency $|\mathbf{Z}_{\mathbf{s}_f}|$. The major advantages of $\mathcal{E}(s^2)$ over $|\mathbf{Z}_{\mathbf{s}_f}|$ are as follows.

1. When $f > n' - m - 1$, where n' is the number of distinct runs, the matrix $\mathbf{Z}_{\mathbf{s}_f}$ becomes singular and hence its determinant is zero while $\mathcal{E}(s^2)$ can still be useful in comparing various designs. In fact, this is one of the major motivations behind the Booth and Cox's $\mathcal{E}(s^2)$ criterion for supersaturated designs, where design matrix D is always a singular matrix. For some recent developments in supersaturated designs, see Lin (1993a, 1993b), Wu (1993), Nguyen (1996), Cheng (1997), Tang and Wu (1997), and Deng, Lin and Wang (1994, 1999).

2. It is easier to compute $\mathcal{E}(s^2)$ than $|\mathbf{Z}_{\mathbf{s}_f}|$. Furthermore, $\mathcal{E}(s^2)$ is a linear function of $z_3^2(s)$ and $z_4^2(s)$ while $|\mathbf{Z}_{\mathbf{s}_f}|$ is a complicated function of $z_3(s)$ and $z_4(s)$ when $f \geq 2$. Hence, it should be easier to study the expected value of $\mathcal{E}(s^2)$ than that of $|\mathbf{Z}_{\mathbf{s}_f}|$.

Like the D_f criterion, we define the S_f^2 criterion as the expected value of $\mathcal{E}(s^2)$ over $\mathbf{s}_f$

$$S_f^2 = E[\mathcal{E}(s^2)].$$

As will be shown next in Proposition 1, S_f^2 can be written as a linear combination of $B_3(D)$ and $B_4(D)$. This result describes the relationship between S_f^2 and $B_3(D), B_4(D)$ and it gives another justification for the GMA criterion.

Proposition 1. Let D be a set of m column vectors selected from a Hadamard design of size n, and let f be any positive integer $\leq F = \binom{m}{2}$. Then we have

$$S_f^2 = \alpha B_3(D) + \beta B_4(D),$$

where

$$\alpha = \frac{3f}{\binom{m+f}{2}F}, \quad \beta = \frac{3f(f-1)}{\binom{m+f}{2}F(F-1)}, \quad \frac{\beta}{\alpha} = \frac{f-1}{F-1}.$$

Note that S_f^2 depends only on m, f, and $B_3(D), B_4(D)$, and not on n or Hadamard designs. Proposition 1 shows that $B_3(D), B_4(D)$ determines the behavior of S_f^2. Equation (10) also shows that the S_f^2 criterion is closely connected to the D_f criterion. Therefore, we should expect a major influence of $B_3(D), B_4(D)$ on the D_f criterion.

In the next subsection, we will show how Proposition 1 can be used to justify the claim that the GMA criterion should be consistent with the D_f criterion.

3.3 Relationship among D_f, S_f^2 and GMA

From Proposition 1, we show that S_f^2 is a linear combination of $B_3(D)$ and $B_4(D)$. From the GMA ranking procedure, we can view the GMA criterion as a weighted sum of $B_k(D)$ for $k = 3, 4, ..$

$$G = \alpha_3 B_3(D) + \alpha_4 B_4(D) + \alpha_5 B_5(D) + \cdots \quad (11)$$

The relative weights associated with $B_3(D)$ over $B_4(D)$ (α_3/α_4) and $B_4(D)$ over $B_5(D)$ (α_4/α_5) are large enough so that the leading term in equation (11) will dominate the criterion value G. As we can see from Proposition 1, the relative weight of $B_3(D)$ over $B_4(D)$ on S_f^2 is $(F-1)/(f-1)$, a function of f. Therefore, the degree of consistency among these criteria depends on f, the number of two-factor interactions entertained.

1. When $f = 1$, both D_f and S_f^2 depend only on $B_3(D)$ because

$$D_1 = 1 - \frac{3B_3(D)}{\binom{m}{2}}, \quad S_1^2 = \frac{3B_3(D)}{\binom{m}{2}\binom{m+1}{2}}. \quad (12)$$

Hence, the rankings based on D_1 and S_1^2 are exactly the same as the GMA ranking based on $B_3(D)$.

2. For a general f, the relative weight between $B_4(D)$ and $B_3(D)$ on the S_f^2 criterion is $(f-1)/(F-1) \leq 1$. Therefore, when $f = F$, the weight of $B_3(D)$ and $B_4(D)$ on S_f^2 are exactly the same while for small f, the effect of $B_3(D)$ on S_f^2 is more dominant than that of $B_4(D)$. From Proposition 1 and equation (11), the ranking based on the GMA criterion should be consistent with the ranking of S_f^2 (and D_f) especially when f is relatively small. According to (10), the $S_f^2 = E[\mathcal{E}(s^2)]$ criterion and the $D_f = E[|\mathbf{Z_{s_f}}|]$ criterion should be fairly consistent, and the GMA criterion should be consistent with the classical D_f criterion. This conclusion is supported by our extensive empirical study.

The D_f is an efficiency criterion but difficult to compute. As argued in subsection 3.2, the S_f^2 can be regarded as a cheap surrogate of D_f. Both D_f and S_f^2 are model dependent in that they rely on f, the number of two-factor interactions. In contrast, the GMA criterion is simple and free of model assumption. Furthermore, we provide some theoretical justification for the GMA rankings in terms of the design efficiency S_f^2 and D_f criteria. In addition, extensive empirical study, not reportrd here, on some designs provides additional support for the GMA criterion.

4 Concluding Remarks

We note that the GMA criterion for nonregular designs is a natural extension of the MA criterion for regular 2_R^{m-q} designs. GMA is useful for Hadamard matrices of size n which may not be of a power of 2. Other criteria such as the D_f and S_f^2 criteria are studied in this paper. We derive a simple formula for S_f^2 in terms of $B_3(D)$ and $B_4(D)$. Based on the formula, we argue that GMA should be consistent with the S_f^2 criterion. We also provide an extensive empirical evaluation showing that these criteria support GMA ranking. Due to space limitation, it will not be reported here.

5   References

Booth, K. H. V. and Cox, D. R. (1962). Some Systematic Supersaturated Designs *Technometrics* **4** 489-495.

Chen, H. and Hedayat, A.S. (1996). 2^{n-m} fractional factorial designs with (weak) minimum aberration. *Annals of Statistics* **24** 2536-2548.

Chen, J. (1992). Some results on 2^{n-k} fractional factorial designs and search for minimum aberration designs. *Annals of Statistics* **20** 2124-2141.

Chen, J. and Wu, C.F.J. (1991). Some results on s^{n-k} fractional factorial designs with minimum aberration or optimal moments. *Annals of Statistics* **19** 1028-1041.

Cheng, C.S. (1995). Some projection properties of orthogonal arrays. *Annals of Statistics* **23** 1223-1233.

Cheng, C. S. (1997). $E(s^2)$-Optimal Supersaturated Designs *Statistica Sinica*, **7** 929-939.

Cochran, W. G. (1977). Sampling Techniques. Third Edition. Wiley, New York.

Deng, L.Y., Lin, D.K.J. and Wang, J.N. (1994). Supersaturated designs using Hadamard matrices, IBM Research Report, #RC19470, IBM Watson Research Center.

Deng, L.Y., Lin, D.K.J. and Wang, J.N. (1999). On resolution rank criterion for supersaturated designs, *Statistica Sinica*, **9, No. 2** 605-610.

Deng, L. Y. and Tang, B. (1999). Generalized resolution and minimum aberration criteria for Plackett-Burman and other nonregular designs, *Statistica Sinica*, to appear.

Deng, L. Y. and Tang, B. (1998). Design selection and classification for Hadamard matrices using generalized minimum aberration criterion, submitted for publication.

Franklin, M.F. (1984). Constructing tables of minimum aberration p^{n-m} designs. *Technometrics* **26** 225-232.

Fries, A. and Hunter, W.G. (1980). Minimum aberration 2^{k-p} designs. *Technometrics* **22** 601-608.

Hamada, M. and Wu, C.F.J. (1995). Analysis of designed experiments with complex aliasing. *Journal of Quality Technology* **24** 130-137.

Hall, M. J. (1965). Hadamard matrix of order 20. Jet Propulsion Laboratory, Technical Report 1, 32-76.

Lin, D.K.J. (1993a). A new class of supersaturated designs. *Technometrics* **35** 28-31.

Lin, D.K.J. (1993b). Another look at first-order saturated designs: the p-efficient designs. *Technometrics* **35** 284-292

Lin, D.K.J. (1995). Generating systematic supersaturated designs. *Technometrics* **37** 213-225.

Lin, D.K.J. and Draper, N.R. (1992). Projection properties of Plackett and Burman designs. *Technometrics* **34** 423-428.

Nguyen, N. K. (1996). An algorithmic approach to constructiing supersaturated designs. *Technometrics* **38** 69-73.

Plackett, R.L. and Burman, J.P. (1946). The design of optimum multi-factorial experiments. *Biometrika* **33** 305-325.

Tang, B. (1994). Unbiased estimation for finite population parameters, *Scandivinian Journal of Statistics*, 91-95.

Tang, B. and Deng, L. Y. (1999). Minimum G_2 aberration criterion for non-regular fractional facotorial designs, *Annals of Statistics*, to appear.

Tang, B. and Wu, C.F.J. (1996). Characterization of minimum aberration 2^{n-k} designs in terms of their complementary designs. *Annals of Statistics* **24** 2549-2559.

Tang, B. and Wu, C. F. J. (1997). A method for constructing supersaturated designs and its $E(s^2)$ optimality. *The Canadian Journal of Statistics* **25, No. 2** 191-201.

Wang, J.C. and Wu, C.F.J. (1995). A hidden projection property of Plackett-Burman and related designs. *Statistica Sinica* **5** 235-250.

Wu, C. F. J. (1993). Construction of supersaturated designs through partially aliased interactions. *Biometrika* **80** 661-669.

STATISTICAL ISSUES IN EVALUATION OF FUEL EFFECTS ON EMISSIONS FROM LIGHT-DUTY VEHICLES IN LABORATORY CHASSIS DYNAMOMETER EXPERIMENTS

Jim Rutherford, Chevron
100 Chevron Way, Richmond, CA 94802

Key Words: Lognormal, Transformation, Replication, Repeat, Emissions

Abstract

Through review of literature and examples from recent laboratory studies, the author explores the following topics: (1) how to best apply and compensate for the apparent lognormality of emissions data, and (2) whether to include and how to make appropriate use of the information from repeat emissions tests.

It is common practice when data are distributed lognormally for analyses to be carried out in the log (transformed) space; and, if estimates are desired in terms of original units, results (for example, measures of central tendency and error bands) are subsequently "back-transformed." In the Auto/Oil studies, arithmetic means in original units were combined with errors estimated in log space to come up with a compromise approach. Statistical authors have considered approximations to estimators having optimal theoretical properties. Other scientists have argued that any estimators other than arithmetic means are inappropriate for environmental data.

Also, a consistent and effective approach is needed for the treatment of measurements from repeat emissions tests. The problem goes to the very heart of the concept of replication, and the frequent misunderstanding about the difference between replicate and repeat measurements. In the present setting, the answer has much to do with the economics of vehicle emissions testing, since if duplicate measurements were eliminated, considerable cost savings would accrue

Lognormal Distribution Discussion

"We can distinguish broadly two types of dependent variable, extensive and non-extensive. The former have a relevant property of physical additivity, the latter not." (Box and Cox (1964) p.213) "Hence, transformations can be applied freely to non-extensive variables. ... In a narrow technological sense, therefore, we are interested in the population mean of y, not some function of y. Hence we either analyze linearly the untransformed data or, if we do apply a transformation in order to make a more efficient and valid analysis, we convert the conclusions back to the original scale. Even in circumstances where, for immediate application, the original scale y is required, it may be better to think in terms of transformed values in which, say, interactions have been removed."(Ibid. p.214)

"If a transformation of the data is required, one can work completely with the transformed data for testing the equality of factor level means. On the other hand, it is usually desirable when making estimates to change the confidence intervals based on the transformed variable back to intervals in the original variable since it is ordinarily easier then to understand the significance of the results." (Neter and Wasserman (1974) p.508)

"Often there is nothing in particular to recommend the original metric in which the measurements happen to be taken. ... the data may be able to indicate that the model assumptions are more nearly met in a particular metric. The *analysis* should be conducted in that metric. After analysis the results may be transformed back and reported in whatever scale is most easily understood."(Box, Hunter, and Hunter (1978) p.240)

In my first analysis of motor vehicle emissions data over 20 years ago (Rutherford (1977)), I convinced many others and myself that the data were lognormally distributed. Based on education founded on several of the above quoted sources, over the intervening years I thought I understood how to handle the data correctly. I analyzed logarithms of the results. After drawing conclusions from standard normal theory analyses, I often transformed means, least squares means, intervals, and other predictors back to the original grams per mile scale by exponentiation.

I thought that arithmetic means should be better if we were trying to estimate total population quantities using a legitimate probability sample, with adequate sampling fraction, from the population. When investigating factors affecting emissions, such as phenomena affecting catalysts or fuel factors, we often are dealing with multiplicative effects. After reading Atchison and Brown (1959) and considering the bias of geometric means, I made a rule for myself. If I wanted to look at population totals in lognormal situations, I would consider arithmetic means. If I were looking for effects on emissions from an average vehicle, I would analyze in log space. If estimators in the original scale were desired, I would transform back by simple exponentiation.

As I became involved with the Auto/Oil Air Quality Improvement Research Program (AQIRP), I discovered that I was committed to a compromise version of my rule: "The mean values in these plots, for any complete matrix, are simply the arithmetic means of the fuel-car averages across all repeats. ... For those responses analyzed in logarithms, the expression RMSE*½*q/√N$_C$ is exponentiated and then multiplied by and divided into the mean to give the upper and lower limits for each fuel mean." (Painter and Rutherford (1992))

For vehicle emissions, Dr. Gunst (1998) has suggested an analysis that uses the log transformation while attempting to correct for bias when transforming back to the original scale. He refers to the Finney (1941) approximations to the minimum variance unbiased estimators for original scale mean and variance as described by Johnson, Kotz, and Balakrishnan (1994). Finney claimed these estimators should be adequate for the mean with n>50 and for the variance with n>100. The more modern and thorough presentation of the bias discussion in Crow and Shimizu (1988) reveals that a great deal of theoretical effort has gone toward improvement of this type of estimator for the lognormal situation.

In a discussion of a closely related application, Parkhurst (1998) made an argument for using arithmetic means for lognormal data. His paper includes an interesting simulation that provides evidence of superiority of simple arithmetic means for lognormal distributions when bias is the primary criterion for evaluation.

While sensitive to the label of naïveté

> ("*The Naïve Transformation.* The use of $\zeta_\alpha = \exp(\mu_\alpha)$ as a confidence limit for EX may be, or may once have been, common, although this writer has found an unambiguous reference to it in the methodological literature only as an undesirable alternative to a more sophisticated method (Patterson, 1966)." (Land (1972))),

I still believe my approach to investigating effects is appropriate. Even in the case where we are looking for best estimates of total emissions, since we are trying to assess the impact on the atmosphere, inaccurate sample frame, miniscule sample fraction, poor representation of real world conditions by dynamometer procedures, correction factors, and other sources of error probably have greater impact than any bias in the estimator due to transformation. Therefore, I am more concerned about using an appropriate error estimate to judge

significance of effects than I am about theoretical bias induced by a naïve transformation back to the original scale to estimate absolute means.

"If decisions can reasonably be made in terms of the new metric (transformed data) then the problem discussed here is circumvented." (Patterson (1966))

Lognormal Distribution Example

For illustration of various approaches, we will look at a subset of data from a segment of the AQIRP (Burns, et. al. 1995). This segment consisted of emissions tests of eight 1989 model year passenger cars and two 1989 Class 2 light-duty trucks. The vehicles were each tested using two fuels, the AQIRP industry average (Fuel A), blended to represent 1988 national average composition and one reformulated gasoline (C2) blended to meet California Phase 2 1996 regulatory requirements.

Four approaches were applied to this simplified data set. (1) Emissions in grams per mile were analyzed with no transformation. (2) Error estimates from the analysis of log transformed data were used with arithmetic means of the grams per mile data as in AQIRP. (3) After analysis of log transformed data, means and intervals were naïvely transformed via exponentiation back to the grams per mile scale. (4) After analysis of log transformed data, Finney type estimators were used to derive means and standard deviations in the grams per mile scale. The appendix explains these analyses more fully.

Comparisons of the results of these analyses of HC and NO$_x$ are shown in Figures 1 and 2. In this example we found that the pairs of analyses using arithmetic means gave very similar results and the Naïve and Finney estimates were very close to each other. There was a substantial difference between the two pairs of analyses. In fact, for NO$_x$, the latter analyses indicate significant fuel effects while the former do not. In this example, I believe that the merit of the lognormal transformation for determining significance of effects is revealed. The benefit of minimum variance unbiased estimators does not seem justified relative to the more cumbersome calculations and difficulty of explaining such estimators to other than mathematical statisticians.

I suspect that the differences within the two pairs of similar estimators are so small due to the small standard deviations relative to the sizes mentioned in the literature. On the other hand, the differences between the arithmetic means and those based on the geometric means are symptomatic of the diversity of the vehicles for which the fleet means are being estimated. In the historical perspective, the differences in grams per mile estimated by the different techniques are quite small. As cars become cleaner, it may matter less what form of estimators we use. However, the differences in estimates become important when multiplied by the large number of vehicles and miles driven in the real world fleet.

Repeats and Replicates Discussion

A very common approach to emissions testing involves what we will call "repeat" tests. A vehicle is tested repeatedly under one set of experimental conditions before changing to the next set of experimental conditions. For example, the vehicle might be tested repeatedly with one of the treatment fuels before switching to another fuel. In an earlier paper (Rutherford and Crosby (1993)), we discussed the efficiency of repeat tests relative to replicates. In addition to inefficiency, we observe that data from experiments with repeats are occasionally subject to inappropriate analyses. Degrees of freedom for estimating experimental error are not provided by additional tests unless the additional tests at least have the potential to derive from randomization. In other words, repeat tests can be averaged before any further analyses. Repeat tests do not represent experimental error against which effects should be judged. This seems intuitively obvious. Yet, analyses appear in the literature that are carried out in violation of this premise. A most egregious example from emissions testing concerns the ability of a vehicle to recover from fuels that have detrimental effects. If we select one vehicle, test it repeatedly with low sulfur fuel, then test it repeatedly with high sulfur fuel, then test it repeatedly with low sulfur fuel, we have only once tested a vehicle's response and ability to recover from high sulfur fuel. This one vehicle provides one degree of freedom for testing the effect under investigation.

We might also argue that the randomization justification for statistics from designed experiments as espoused by R. A. Fisher (see for example Box (1980) or Yates (1964)) is violated by considering repeats to be replicates. However, it has come to our attention that not everyone agrees with the merits of randomization (see for example Harville (1975)). However, we agree with Finney (1964) that Fisher led us to the point where "Today in truth the obligation lies upon the experimental scientist to show reason why randomization is unnecessary in any particular inquiry, and not upon the statistician to advance special reasons for its use" (p.328)

Repeats and Replicates Example

Most of the AQIRP experiments used repeat testing and the first step in analysis (after checking for "outliers") was averaging the repeats. The example discussed for Issue 1 happens also to be a good one for this issue. Each of the vehicles was tested with each of the fuels in two separate blocks of emissions tests. In the first block, each vehicle was tested with each fuel with two consecutive repeat tests with a possible third test if the first two were more divergent than a predetermined limit. In the second block the two fuels were again tested with repeat tests. For each vehicle, the sequence including both blocks was either A-C2-C2-A or C2-A-A-C2. We averaged the repeat tests within each vehicle, fuel, and block before the analyses described under Issue 1. We consider the two blocks of tests to provide replication.

Using the data from this example we can compare the appropriate experimental error with a pooled repeat error estimate to see how much the repeat error underestimates the appropriate experimental error. We compare repeat error with experimental error for this example using analysis of logarithms of emissions in Figure 3. We find that for HC and NO_x, repeat error underestimates experimental error by about 30%. For CO, underestimation is about 15%.

Conclusions

I believe I have raised more questions than I have answered in this paper. I have stated my approach and reasoning on two issues which have implications for analysis and design of emissions experiments.

For most vehicle emissions experiments, I would tend toward lognormal analyses with naïve back transformations in the absence of compelling reasons to do otherwise for a specific experiment. We have made great strides in convincing chemists and engineers to consider the implications of lognormality. I don't think we should push beyond what is beneficial and risk what we have gained.

Only in very special circumstances with very limited scope of inference could valid analyses use repeats as though they were replicates.

As a consulting statistician, I am always looking to improve my knowledge, skills, and judgement. If you can help me do that or have other comments, please contact me at the above address or via

Phone: (510) 242-3410 or
e-mail: jaru@chevron.com

Acknowledgements

I have had the pleasure of learning about emissions from more people over the years than I could hope to acknowledge in this short paper. I'll just say thank you.

I thank Chevron Products Company for their support of this work.

I acknowledge the assistance of Eric Burk in finding many, if not all, of my calculation errors.

References

Aitchison, J. and Brown, J.A.C. (1959). *The Lognormal Distribution*, Cambridge University Press.

Box, G. E. P., and Cox, D. R. (1964). An analysis of transformations, *Journal of the Royal Statistical Society,* Series B, **26**, 211-243.

Box, G. E. P., Hunter, G. H., and Hunter, J. S. (1978). *Statistics for Experimenters*, Wiley.

Box, Joan Fisher (1980). R. A. Fisher and the design of experiments, 1922-1926, *American Statistician* **34**, 1-7.

Burns, V. R., Rapp, L. A., Koehl, W. J., Benson, J. D., Hochhauser, A. M., Knepper, J. C., Leppard, W. R., Painter, L. J., Rippon, B. H., Reuter, R. M., and Rutherford, J. A. (1995). Gasoline reformulation and vehicle technology effects on emissions - The Auto/Oil Air Quality Improvement Research Program, *SAE Paper* **952509**.

Crow, E.L. and Shimizu, K. eds (1988). *Lognormal Distributions*, Marcel Dekker.

Finney, D. J. (1941). On the distribution of a variate whose logarithm is normally distributed, *Journal of the Royal Statistical Society,* Series B, **7**, 155-161.

Finney, D. J. (1965). Sir Ronald Fisher's contributions to biometric statistics, *Biometrics, 322-329.*

Gunst, R. F. (1998). Logarithmic power, *Quality Progress, October*, 101-107.

Harville, D.A. (1975). Experimental randomization: who needs it?, *The American Statistician* **29**, 27-31.

Johnson, N.L. , Kotz, S., and Balakrishnan, N. (1994). *Continuous Univariate Distributions Volume 1, Second Edition*, Wiley.

Land, C. E. (1972). An evaluation of approximate confidence interval estimation methods for lognormal means, *Technometriccs* **14**, 145- 158.

Neter, J. and Wasserman, W. (1974). *Applied Linear Statistical Models*, Richard D. Irwin, Inc.

Painter, L. J. and Rutherford, J. A. (1992). Statistical design and analysis methods for the Auto/Oil Air Quality Research Program, *SAE Paper **920319***.

Parkhurst, D. F. (1998). Arithmetic versus geometric means for environmental concentration data, *Environmental Science & Technology / News, February 1*, 92A-98A.

Patterson, R. L. (1966). Difficulties involved in the estimation of a population mean using transformed sample data, *Technometrics, **8***, 535-537.

Rutherford, J. A. (1977). Automobile exhaust emission surveillance - analysis of the FY1975 program, EPA-460/3-77-022.

Rutherford, J.A. and Crosby, T. (1993). The statistical value of duplicates in vehicle emissions testing", *JSM*, San Francisco.

Yates, F. (1964). Sir Ronald Fisher and the design of experiments, *Biometrics*, 307-321.

Appendix: Example 1 Analyses

Data for Example 1 consist of hydrocarbon (HC) and nitrous oxides (NO_x) grams per mile for the ten vehicles with the two fuels. Each vehicle was tested in two blocks with the two fuels, i.e., within a larger set of fuels each of the two fuels was put in each vehicle and two or more emissions tests were run before switching to another fuel. Similar testing with each vehicle and fuel was also repeated within another larger set of fuels. Data were either log transformed or left in grams per mile. Averages were computed for each vehicle/fuel combination within each block of fuels. There were 40 vehicle/fuel/block means going into the analyses. Calculations for predicted means and intervals for comparing these means between the two fuels for significance at $\alpha=0.05$ for the four methods are shown below.

(1) Analysis without transformation: comparison interval

$$= x \pm RMSE \times t_{.975,df} / 2\sqrt{10},$$

where x = simple arithmetic average and RMSE = root mean square error from least squares fit of non-transformed means.

(2) AQIRP compromise: comparison interval $= x$ multiplied or divided by $\exp(RMSE_l \times t_{.975, df} / 2\sqrt{10})$,

where $RMSE_l$ = root mean square error from least squares fit of means of log-transformed data.

(3) Naïve back transformation: comparison interval $= \exp(x_l \pm RMSE_l \times t_{.975,df}/2\sqrt{10})$, where x_l = mean of means of log-transformed data.

(4) Finney back transformation: comparison interval $= \exp(x_l) g(MSE_l/2) \pm \{\exp(2x_l) [g(2MSE_l)-g((df-1)MSE_l/df)]\}^{\frac{1}{2}} t_{.975,df} /2\sqrt{10}$,

where MSE_l = mean square error from least squares fit of means of log-transformed data, and $g(t) = \exp(t)\{1 - t(t+1)/10 + t^2(3t^2+22t+21)/600\}$.

Figure 1: HC Means and Comparison Intervals

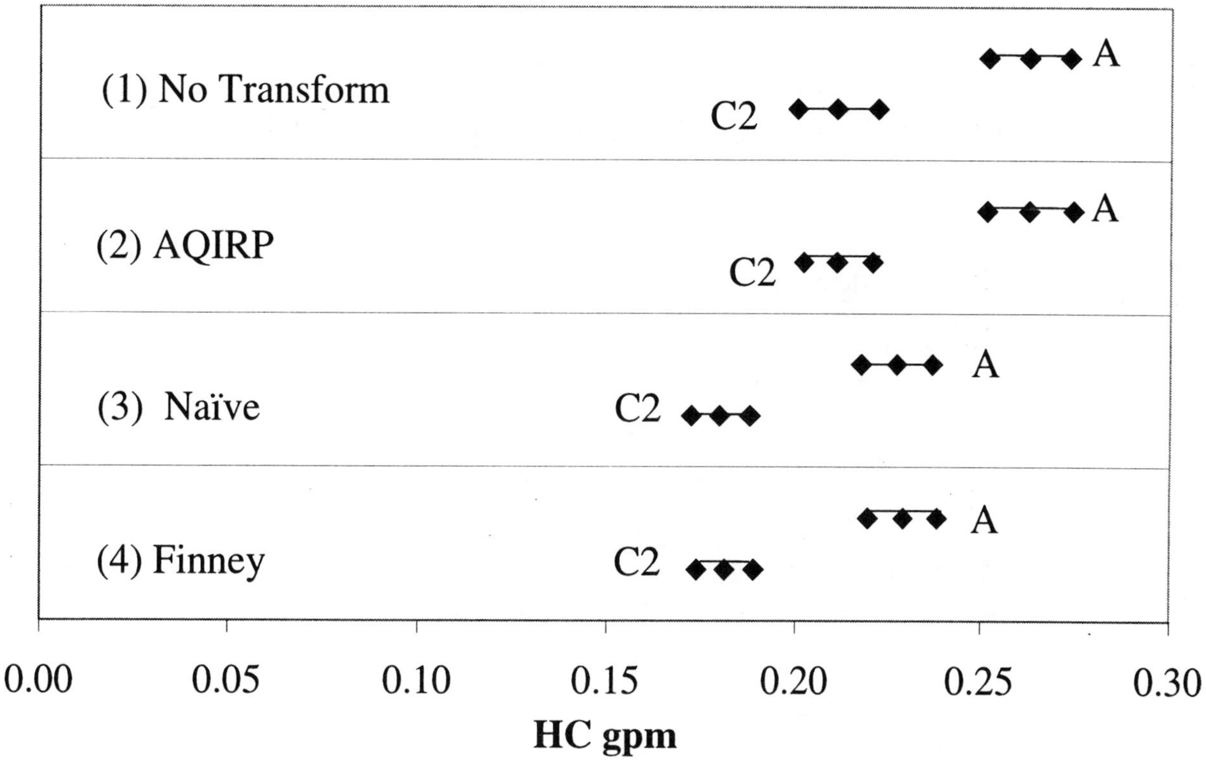

HC gpm

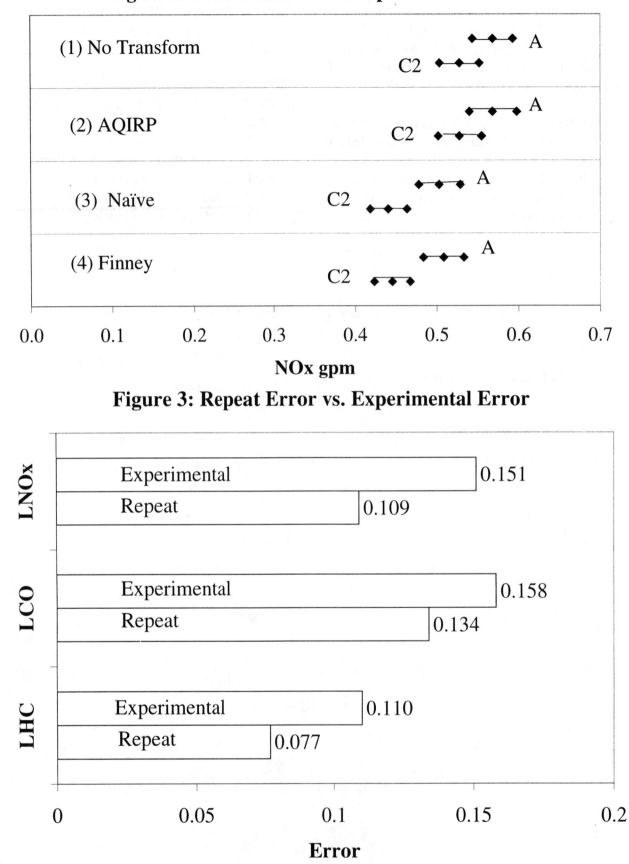

Figure 2: NOx Means and Comparison Intervals

(1) No Transform

A

C2

(2) AQIRP

A

C2

(3) Naïve

A

C2

(4) Finney

A

C2

NOx gpm

Figure 3: Repeat Error vs. Experimental Error

LNOx

Experimental 0.151

Repeat 0.109

LCO

Experimental 0.158

Repeat 0.134

LHC

Experimental 0.110

Repeat 0.077

Error

Confidence Regions for Constrained Optima in Response Surface Experiments with Noise Factors

Andrew M. Kuhn

Becton Dickinson Microbiology Systems

Mail Code 912

Sparks, MD 21152

max_kuhn@ms.bd.com

Key Words: Lagrangian multipliers, dual response model, robust parameter design.

1. Introduction

The importance of robust parameter design developed by Genichi Taguchi has proven to be useful in building quality into products or processes at the design phase. Quality is improved by minimizing the effect of variables which are uncontrollable outside experimental settings, called noise factors. Examples of such factors are environmental or operating conditions. In these experiments, variables which are controllable at all times ("control factors") may be used to make a process insensitive or robust to the effect of the noise factors.

Statistical criticisms have prompted alternatives to the analysis of Taguchi's signal–to–noise ratios. Many of these alternatives have adapted response surface methodology to simultaneously model the process mean and variance. A panel discussion in Nair(1992) provides an introduction to these methods and an excellent summary of both schools of thought. One advantage response surface alternatives have relative to the analysis of signal–to–noise ratios is that they inherit flexible and proven methods for modeling, interpreting and optimizing experimental data. Lucas (1994) illustrates these differences with an experimental dataset in which signal–to–noise ratios are analyzed and compared to direct modeling of the response using regression analysis. In the former case, control factor operating conditions can be found which are associated with large values of a specific type of signal–to–noise ratio chosen prior to data collection. In the latter case, the relationship between the process mean and variance can be investigated using contour plots, the quality of fit can be assessed using diagnostic plots and improved operating conditions can be found using existing methods for function optimization.

If a regression approach is taken, the framework of response surface methodology provides a variety of procedures to determine optimal operating conditions. For example, the model's stationary point can be estimated and characterized. In some cases, this location will fall outside of the experimental region and ridge analysis or numerical optimization routines can be used to optimize the response subject to constraints which force the optimum to be within the design region. While constrained optimization provides useful knowledge as to what operating conditions may be appropriate, it does not account for the variablity associated with these conditions. In many cases, a confidence region for the optimal factor settings illustrates that an area exists where the optimal conditions may lie and can aid in assessing the variablity surrounding the optimum.

2. Response Surface Models Using Noise Factors

Let the vector containing the c control factors in coded units be denoted as $\mathbf{x} = (x_1, x_2, \cdots, x_c)$ and the vector containing the n noise factors in coded units be $\mathbf{z} = (z_1, z_2, \cdots, z_n)$. It is assumed that the mean and variance of the noise factors are known and that the z_j can be centered so that the coded factors have mean $\mathbf{0}$ and covariance matrix $\mathbf{V}$. In this manuscript, it is assumed that the noise factors have zero covariance so that $\mathbf{V} = \text{diag}(\sigma_{z_1}^2, \cdots, \sigma_{z_n}^2)$. This assumption can be relaxed so that the methods given here can be extended to situations where $\text{Cov}[z_k, z_{k'}] \neq 0$.

If the experimental data are observed conditionally on the levels of the noise factors, a model can be developed that is a function of the control factors, noise factors, and control by noise interactions. This model can be any form supported by the experimental design, although for illustrative purposes, we assume that no quadratic noise factors or noise factor interactions are used in the model. The methods developed here assume the conditional model

$$y|\mathbf{z} = \beta_0 + \mathbf{x}'\boldsymbol{\beta} + \mathbf{x}'\mathbf{B}\mathbf{x} + \mathbf{z}'\boldsymbol{\gamma} + \mathbf{x}'\boldsymbol{\Delta}\mathbf{z} + \epsilon|\mathbf{z} \quad (1)$$

is used for the data, where

$$\boldsymbol{\beta}_{c\times 1} = \begin{bmatrix} \beta_1 \\ \vdots \\ \beta_c \end{bmatrix}, \quad \mathbf{B}_{c\times c} = \begin{bmatrix} \beta_{11} & \frac{1}{2}\beta_{12} & \cdots & \frac{1}{2}\beta_{1c} \\ \frac{1}{2}\beta_{12} & \beta_{22} & \cdots & \frac{1}{2}\beta_{2c} \\ & & \ddots & \vdots \\ sym & & & \beta_{cc} \end{bmatrix},$$

and

$$\boldsymbol{\gamma}_{n\times 1} = \begin{bmatrix} \gamma_1 \\ \vdots \\ \gamma_n \end{bmatrix}, \quad \boldsymbol{\Delta}_{c\times n} = \begin{bmatrix} \delta_{11} & \delta_{12} & \cdots & \delta_{1n} \\ \delta_{21} & \delta_{22} & \cdots & \delta_{2n} \\ \vdots & \vdots & \ddots & \vdots \\ \delta_{c1} & \delta_{c2} & \cdots & \delta_{cn} \end{bmatrix}.$$

The parameters β_j and β_{jj} are linear and quadratic terms associated with the jth control factor, respectively, β_{jk} is the interaction between the jth and k^{th} control factors, γ_l is the linear effect of the lth noise factor, and δ_{jl} is the interaction between the jth control factor and the lth noise factor. Note that $\boldsymbol{\Delta}$ may not be square. The model residuals, $\epsilon|\mathbf{z}$, are assumed to be independent and identically distributed normal random variables with mean 0 and constant variance σ^2.

Using identities for obtaining unconditional means,

$$\mathrm{E}\left[y\right] = \mathrm{E}_z[y|\mathbf{z}] = \beta_0 + \mathbf{x}'\boldsymbol{\beta} + \mathbf{x}'\mathbf{B}\mathbf{x}, \qquad (2)$$

where $\mathrm{E}_z[\cdot]$ denotes expectation with respect to the noise factor distribution. This model is free of the noise factors and represents the process mean of the response over the noise factor population. Given the assumption on the errors, $\mathrm{Var}\left[y|\mathbf{z}\right] = \sigma^2$, the process variance can be found to be

$$\begin{aligned} \mathrm{Var}\left[y\right] &= \mathrm{E}_z\left[\mathrm{Var}\left[y|\mathbf{z}\right]\right] + \mathrm{Var}_z\left[\mathrm{E}\left[y|\mathbf{z}\right]\right] \\ &= \mathrm{E}_z\left[\sigma^2\right] + \mathrm{Var}_z\left[\mathbf{z}'\boldsymbol{\gamma} + \mathbf{x}'\boldsymbol{\Delta}\mathbf{z}\right] \\ &= \sigma^2 + \boldsymbol{\alpha}(\mathbf{x})'\mathbf{V}\boldsymbol{\alpha}(\mathbf{x}). \end{aligned} \qquad (3)$$

where $\boldsymbol{\alpha}(\mathbf{x}) = \boldsymbol{\gamma} + \boldsymbol{\Delta}'\mathbf{x}$. Like the process mean, this model is not a function of the noise factors. The parameters contained in $\boldsymbol{\gamma}$ and $\boldsymbol{\Delta}$ determine the effect of the noise factors above and beyond the residual variance. Since $\boldsymbol{\Delta}$ is directly linked to the control factors in equation (3), the excess variance caused by the noise factors may be regulated by proper choice of the elements of $\mathbf{x}$. However, this can only occur when the control by noise interactions are nonzero.

3. Process Characteristics Without Constraints

Often, the calculation of the stationary point of a response surface model is considered a "black–box"

for determining optimal conditions. The researcher may desire to know if deviation from these settings can still result in acceptable performance, especially if the resulting operating conditions are impractical or expensive to initiate. Hence a gauge of the variability of the optimal operating conditions may be vital. In addition, the variability of the response at the stationary point can also help researchers understand how well operating conditions will perform when instituted.

The precision of the estimated parameters play a large role in the size of confidence regions for the optimal operating conditions. In fact, the variation associated with the parameter estimates is transfered directly to the variance of the stationary point. The use of confidence regions in these situations helps assess the precision of the estimated optimum relative to the error variance. If the precision of the model estimates is poor relative to the error variance, the area encompassing the stationary point will be large and indicate that the point estimate of the optimum may not be reliable.

For the process mean (2), the stationary point $\mathbf{x}_0$ is found by the solution to

$$\frac{d\mathrm{E}[y]}{d\mathbf{x}} = \boldsymbol{\beta} + 2\mathbf{B}\mathbf{x} = \mathbf{0} \qquad (4)$$

so that $\mathbf{x}_0 = -\frac{1}{2}\mathbf{B}^{-1}\boldsymbol{\beta}$. Following Box and Hunter (1954), a confidence region for the stationary point is obtained by constructing an estimator of equation (4), $\mathbf{d}(\mathbf{x}) = \widehat{\boldsymbol{\beta}} + 2\widehat{\mathbf{B}}\mathbf{x}$. Under normality assumptions on the residuals, the vector $\mathbf{d}(\mathbf{x})$ is normally distributed with mean $\mathbf{0}$ and variance $\mathrm{Var}\left[\mathbf{d}(\mathbf{x})\right]$. The elements of $\mathbf{d}(\mathbf{x})$ are a linear combination of model parameters so that $\mathrm{Var}\left[\mathbf{d}(\mathbf{x})\right]$ is a function of the elements of the covariance matrix of the model parameters, $\sigma^2(\mathbf{X}'\mathbf{X})^{-1}$, where $\mathbf{X}$ is the design matrix for the conditional model (1). A $100(1 - \alpha)\%$ confidence region for the stationary point of the mean model is given by the set of design locations satisfying

$$\frac{1}{c\,\widehat{\sigma}^2}\mathbf{d}(\mathbf{x})'\left\{\mathrm{Var}\left[\mathbf{d}(\mathbf{x})\right]\right\}^{-1}\mathbf{d}(\mathbf{x}) \leq F_{\alpha,c,N-p},$$

where N is the number of observations, p is the total number of parameters in the conditional model, σ^2 is estimated by the mean squared error and $F_{\alpha,c,N-p}$ is the upper α quantile of the F–distribution with c and $N - p$ degrees of freedom.

The quadratic form in the variance model (3) is the excess variance attributable to the noise factors. By minimizing this quadratic form, the effect of the noise factors in the process can be minimized. In fact, when σ^2 is constant with respect to the control factors, a region of minimum variance may be determined in a closed form solution. The quadratic form $\alpha(\mathbf{x})'\mathbf{V}\alpha(\mathbf{x})$ is minimized at a location $\mathbf{x}_0$ such that $\alpha(\mathbf{x}_0) = \mathbf{0}$, although the location of minimum variance may be a point, a line (or hyperplane) or may not exist at all, depending on the number of control and noise factors in the model.

The variance model (3) is a quadratic form which can be written in a notation similar to the mean model (2), i. e.

$$\sigma^2 + \alpha(\mathbf{x})'\mathbf{V}\alpha(\mathbf{x}) = \theta_0 + \mathbf{x}'\theta + \mathbf{x}'\Theta\mathbf{x} \qquad (5)$$

where $\theta_0 = \sigma^2 + \gamma'\mathbf{V}\gamma$, $\theta = 2\Delta\mathbf{V}\gamma$, and $\Theta = \Delta\mathbf{V}\Delta'$. Note that although Δ is not symmetric, Θ is symmetric by virtue of the symmetry of $\mathbf{V}$. When $\mathbf{V}$ is diagonal,

$$\theta_0 = \sigma^2 + \sum_{k=1}^{n} \sigma_{kk}^2 \gamma_k^2,$$

$$\theta_j = 2\sum_{k=1}^{n} \delta_{jk}\sigma_{kk}^2\gamma_k, \qquad j = 1\ldots c$$

and

$$\Theta_{jj'} = \sum_{k=1}^{n} \delta_{j'k}\delta_{jk}\sigma_{kk}^2 \qquad j,j' = 1\ldots c, \quad j < j'.$$

In this form, the variance model is not a linear function of the parameters given in the response model (1). For example, θ_0 is a function of γ_k^2. The location or region of minimum variance using equation (5) can be found to be the solution to the set of equations given by $\theta + 2\Theta\mathbf{x} = \mathbf{0}$ and it is not difficult to show that these equations are equivalent to $\alpha(\mathbf{x}) = \mathbf{0}$.

Myers and Montgomery (1995) and Myers, Kim, and Griffiths (1997) provide methods of obtaining confidence regions for the location of minimum variance. The procedure follows the development for the stationary point for the mean model. In this case, let $\mathbf{d}(\mathbf{x}) = \widehat{\gamma} + \widehat{\Delta}'\mathbf{x}$ be an estimator associated with the stationary point from the variance model constructed by inserting maximum likelihood estimates for the appropriate parameters. Again, this estimate is normally distributed with mean $\mathbf{0}$ and a covariance matrix which can be derived from $\sigma^2(\mathbf{X}'\mathbf{X})^{-1}$. A $100(1-\alpha)\%$ confidence region is given by the set of points satisfying

$$\frac{1}{c\,\widehat{\sigma}^2}\mathbf{d}(\mathbf{x})'\left\{\text{Var}\left[\mathbf{d}(\mathbf{x})\right]\right\}^{-1}\mathbf{d}(\mathbf{x}) \leq F_{\alpha,c,N-p}.$$

Note that neither the stationary point for the variance model or it's corresponding confidence region are functions of $\mathbf{V}$.

4. Confidence Regions for Process Characteristics Under Constraints

Similar to classical response surface methods for the mean, situations may arise when the stationary point of the process variance falls outside the experimental region. In this event, ridge analysis of the variance model may be helpful. Myers and Montgomery (1995) provide an example where the point of minimum variance within a circle of radius R from the design center is calculated using Lagrangian multipliers. The variation of the location of minimum variance under this constraint would provide a measure of the region of optimal conditions.

Taguchi classified optimization of the process mean into three categories: maximization, minimization or bringing the response to some target level t_0. For maximization or minimization, Vining and Myers (1990), denoted as VM, suggested that mean model should be optimized subject to the variance being fixed at some specified value v_0. Their method utilized the Lagrangian multiplier approach of Myers and Carter (1973) while Del Castillo and Montgomery (1996), denoted DCM, applied a gradient based routine that incorporates nonlinear constraints and Copeland and Nelson (1996) suggested the use of a simplex search procedure. In the last case, the authors modified this method so that variance is to be no greater than v_0. For "target is best" cases, VM propose that the variance model might be minimized while the mean model is constrained to be a value specified by the researcher, t_0. Here, Copeland and Nelson (1996), denoted CN, indicated that minimization of the process variance such that the squared deviation from target be no larger than some specified value, denoted T. These approaches are summarized in Table 1. When applied to a common dataset, these methods lead to similar results relative to the variation in the data. The Lagrangian multiplier approach of Myers and Carter is appealing since methods for constructing confidence regions associated with this approach have been studied.

Suppose that primary and secondary equations can be chosen to reflect the optimization goals. The primary function, $f_p(\mathbf{x})$, in maximization/minimization problems described in Table 1 is the mean model is given by equation (2) while the variance model (3) can the primary function for target is best cases and ridge analysis of the process variance. The secondary function, $f_s(\mathbf{x})$, is associ-

Table 1: Summary of Dual Response Model Optimization Procedures

Author	Case	Primary Function	Constraint
VM/DCM	Min	Min M	$V = v_0$
CN	Min	Min M	$V < v_0$
VM/DCM	Max	Max M	$V = v_0$
CN	Max	Max M	$V < v_0$
VM/DCM	Target	Min V	$M = t_0$
CN	Target	Min V	$(M - t_0)^2 < T$

M denotes $\mathrm{E}[y]$

V denotes $\mathrm{Var}[y]$

Table 2: Examples of Constrained Optimization Problems in Robust Design Experiments

Primary Function	Secondary Function Constraint	$\partial L / \partial \mathbf{x}$
$\mathrm{E}[y]$	$\mathrm{Var}[y] = v_0$	$(\mathbf{B} - \lambda\mathbf{\Theta})\mathbf{x} + (\boldsymbol{\beta} - \lambda\boldsymbol{\theta})/2$
$\mathrm{Var}[y]$	$\mathrm{E}[y] = t_0$	$(\mathbf{\Theta} - \lambda\mathbf{B})\mathbf{x} + (\boldsymbol{\theta} - \lambda\boldsymbol{\beta})/2$
$\mathrm{Var}[y]$	$\mathbf{x}'\mathbf{x} = R^2$	$(\mathbf{\Theta} - \lambda\mathbf{I}_c)\mathbf{x} + \boldsymbol{\theta}/2$

ated with the appropriate constraints such as requiring that the optimal location be within the experimental region or the variance be fixed at a value. Once these functions are defined, the constrained optimum can be found using Lagrangian multipliers by defining a function

$$L = f_p(\mathbf{x}) - \lambda\big(f_s(\mathbf{x}) - k\big)$$

where λ is the Lagrangian multiplier and k is a constant used in the constraint. To find the constrained optimum, L is differentiated with respect to the elements of $\mathbf{x}$ and set equal to $\mathbf{0}$. For a fixed value of λ, this equation can be solved for $\mathbf{x}$ and the primary and secondary functions are evaluated at $\mathbf{x}$. For example, when an optima is sought within a radius R of the design center, trial values of λ can be used to find one corresponding to R.

Table 2 provides summarizes the derivative of L for the approach of Vining and Myers (1990) and for ridge analysis of the variance. The are two differences between these equations their analogs used in unconstrained optimization. First, in each case $\partial L / \partial \mathbf{x}$ is a function of the noise factor variances $\mathbf{V}$. Secondly, these equations are not linear in the model parameters.

For two quadratic models, Stablein, Carter and Wampler (1983) constructed confidence regions for constrained optima by letting $\mathbf{d}_\lambda(\mathbf{x})$ be an estimator of $\partial L / \partial \mathbf{x}$ for a fixed value of λ constructed by replacing the appropriate model parameters by their maximum likelihood estimates. However, when the unconditional variance model is involved, the variance of $\mathbf{d}_\lambda(\mathbf{x})$ must be approximated since it's elements are not linear combinations of parameters. The use of statistical differentials (Johnson and Kotz, 1982), also referred to as the delta–method, can be used to compute such an approximation. If $\mathbf{d}_\lambda(\mathbf{x})$ is a system of d equations, let $\mathbf{L}$ be a $d \times p$ matrix of partial derivatives of $\mathbf{d}_\lambda(\mathbf{x})$ with respect to the p parameters so that

$$\mathrm{Var}\big[\mathbf{d}_\lambda(\mathbf{x})\big] \approx \sigma^2 \mathbf{L}(\mathbf{X}'\mathbf{X})^{-1}\mathbf{L}'.$$

For a specified value of λ, a $100(1 - \alpha)\%$ confidence region for the constrained optimum is given by the set of design locations satisfying

$$\frac{1}{D}\mathbf{d}_\lambda(\mathbf{x})'\Big[\mathbf{L}(\mathbf{X}'\mathbf{X})^{-1}\mathbf{L}'\Big]^{-1}\mathbf{d}_\lambda(\mathbf{x}) \leq F_{\alpha, d, N-p}. \quad (6)$$

5. Example: TV Signal Data

Myers and Montgomery (1995) describe an experiment to improve the reception and decoding of signals in televisions. The response is associated with the quality of the resulting image in decibels. The number of bits in an image (z_1) and the applied voltage (z_2) are difficult to control and are treated as noise factors. The two control factors, the number of tabs in a filter (x_1) and the sampling frequency (x_2), are used to increase the response while controlling the effect of the noise factors by reducing their effect on process variance.

The data can be analyzed using summary measures, such as the "larger–the–better" signal–to–noise ratio and the sample mean, which are computed for each of the nine control factor combinations. In this case, the noise factors are treated as replications used to compute the summary statistics. The results for this type of analysis are given in Figure 1. A "pick the winner" strategy where the control factor combination with the largest signal–to–noise ration and/or sample mean is often used to select new operating conditions, after confirmation experiments. Using this procedure, the low number of filter tabs (5) and the high sampling frequency (13.5 MHz) would be chosen. However, two of several difficiencies in this procedure are that control factor interactions are ignored and little is learned about the underlying process.

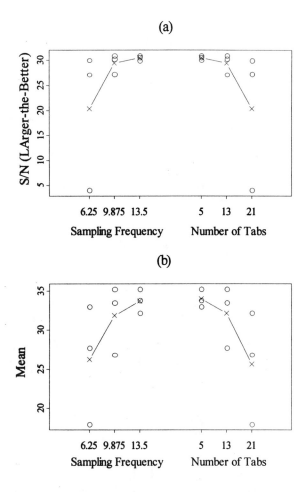

(a)

(b)

Figure 1: Summary Measure Analysis of TV Signal Data

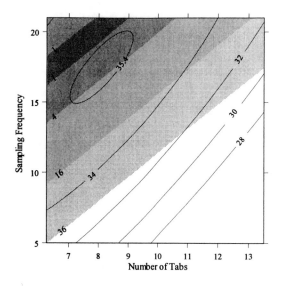

Figure 2: Contours of mean and variance in the example

A response surface approach may alleviate these problems. Following Myers and Montgomery, the variance of the two noise factors are assumed to be 1. A full second order model in the control factors, linear noise factor terms and all interactions between the control and noise factors were estimated in the conditional model. There is an indication that the control factor interaction is statistically significant in this experiment, a result that could not be seen in the summary measure approach.

From the parameter estimates, the process variance model contains elements $\widehat{\theta}_0 = 26.26$,

$$\widehat{\boldsymbol{\theta}} = \begin{bmatrix} 30.48 \\ -39.02 \end{bmatrix} \quad \text{and} \quad \widehat{\boldsymbol{\Theta}} = \begin{bmatrix} 9.13 & -11.60 \\ -11.60 & 14.98 \end{bmatrix}.$$

Figure 2 provides contour plots of the resulting mean and variance models where the solid lines correspond to the mean and shaded contours are associated with the process variance. Again, low numbers of filter tabs and high sampling frequencies are associated with improved picture quality and reduced variance.

The stationary point of the mean model occurs at 17.5 filter tabs and a 8.1 MHz sampling frequency and is associated with $\widehat{E[y]} = 35.470$ and $\widehat{\text{Var}[y]} = 2.69$. The location of minimum variance is given by a line. Myers and Montgomery (1995) provide a confidence region for this line.

To maximize the mean image quality subject to the variance being equal to a certain value, constrained optimization was performed. For values of λ ranging from 0 (unconstrained optimization) to 2, design locations were found by solving

$$\mathbf{d}_\lambda(\mathbf{x}) = \left(\widehat{\mathbf{B}} - \lambda\widehat{\boldsymbol{\Theta}}\right)\mathbf{x} + \frac{1}{2}\left(\widehat{\boldsymbol{\beta}} - \lambda\widehat{\boldsymbol{\theta}}\right) = \mathbf{0}$$

for $\mathbf{x}$. For values of λ near two, the optimal operating conditions begin to move along the line of minimum variance. A value of $\lambda = 0.7$ is associated with optimum process mean (35.3 decibels) while the variance is constrained to be 0.81, which corresponds to operating where sampling frequency is 8.2 MHz and the number of filter tabs is 20.12.

Using equation (6) with $\lambda = 0.7$, a 95% confidence region for the constrained optimum is given in Figure 3. This region is similar in shape to the constrained confidence region for the variance in Myers and Montgomery (1995, Figure 10.22). Their analysis and Figure 3 suggest that there is some flexibility in the choice of operating conditions, although neither contains the "pick the winner" conditions derived from Figure 1.

6. Discussion

The attractiveness of the response surface ap-

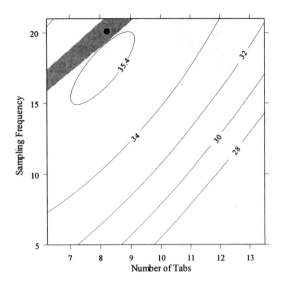

Figure 3: Contours of mean response, the constrained optima and the 95% confidence region (shaded) for the constrained optima when $\lambda = 0.7$

proach for the analysis of experiments containing noise factors is that a variety of well developed tools, such as confidence statements, can be applied to learn about the underlying process. Method such as sequential experimentation, lack of fit testing and canonical analysis are other examples of tools that may shed light on the process in addition to arriving at optimal operating conditions.

Myers and Montgomery note that the variance model in equation (3) is not an unbiased estimate of the process variance. The unbiased estimate is a function of the model mean squared error and can be used with the methods described here. However, the use of this estimate increases the complexity of the calculations since the forms of $\partial L/\partial \mathbf{x}$ in Table 2 are no longer appropriate.

When a Lagrangian multiplier approach is taken to optimize a function under constraints, the value of λ is a random quantity and has an associated variance. Peterson (1999) develops methods for constructing confidence regions that does not use Lagrangian multipliers. For situations where the constraints do not contain parameter estimates, this method can be used to obtain confidence regions for the process mean or variance under constraints involving the design region, such as in the third entry in Table 2, where the secondary function is given by $\mathbf{x}'\mathbf{x} = R^2$

7. References

Box, G. E. P. and Hunter, J. S. (1954), A confidence region for the Solution of a Set of Simultaneous Equations with an Application to Experimental Design. *Biometrika* **41**, 109–199.

Copeland, K. A. and Nelson, P. R. (1996), Dual Response Optimization via Direct Function Minimization. *Journal of Quality Technology* **28**, 331–336.

Del Castillo, E. and Montgomery, D. C. (1993), A Nonlinear Programming Solution to the Dual Response Problem. *Journal of Quality Technology* **25**, 199–204.

Johnson, S. K. and Kotz, N. L. (1982), Statistical Differentials *in Encyclopedia of Statistics* New York: Wiley.

Lucas, J. M (1994), How to Achieve a Robust Process Using Response Surface Methodology. *Journal of Quality Technology* **26**, 248–260.

Myers, R. H. and Carter, W. H. (1973), Response Surface Techniques for Dual Response Systems. *Technometrics* **15**, 301–317.

Myers, R. H., Kim, Y., and Griffiths, K. L. (1997), Response Surface Methods and the Use of Noise Factors. *Journal of Quality Technology* **29**, 429–44.

Myers, R. H., Khuri, A. I., and Vining, G. G. (1992),Response Surface Alternatives to the Taguchi Robust Parameter Design Approach. *The American Statistician* **46**, 131–139.

Myers, R. H. and Montgomery, C. M. (1995), *Response Surfaces Methodology: Process and Product Optimization Using Designed Experiments.* New York: Wiley.

Nair, V. N. et al (1992), Taguchi's Robust Parameter Design: A Panel Discussion. *Technometrics* **34**, 127–161.

Peterson, J. J. (1999), A General Approach to Inference for Optimal Conditions Subject to Constraints. Unpublished Manuscript.

Stablein, D. M., Carter, W. H. and Wampler, G. L. (1983), Confidence Regions for Constrained Optima in Response Surface Experiments. *Biometrics* **39**, 759–763.

Vining, G. G. and Myers, R. H. (1990), Combining Taguchi and Response Surface Philosophies: A Dual Response Approach. *Journal of Quality Technology.* **22**, 15–22.

A GENERAL APPROACH TO INFERENCE FOR OPTIMAL CONDITIONS SUBJECT TO CONSTRAINTS

John J. Peterson, SmithKline Beecham Pharmaceuticals, R&D, 709 Swedeland Road, King of Prussia, PA 19406-0939

Key Words: Confidence regions, Mixture Experiments, Response Surface Methodology, Stationary Points, Therapeutic Synergy

1. INTRODUCTION

In response surface experiments the investigator is often interested in making inferences about a minimizing (or maximizing) factor-level configuration within a specified, bounded region. In this paper I will discuss the minimizing point, the inference for the maximizing point being analogous.

An exact confidence region for the (unconstrained) stationary point for a response surface was given by Box and Hunter (1954) (referred to hereafter as BH). However, unconstrained estimation of a stationary point is often not well aligned with the practical needs to make inferences for a minimizing point within the region of the experimentation. In some cases the stationary point corresponds to a saddle point rather than a minimum in the experimental region. Even if stationary and minimizing points agree, it is sometimes the case that this point is outside of the region of experimentation.

Stablein, Carter, and Wampler (1983) (hereafter referred to as SCW) made an important first step in addressing the uncertainty of a optimal point within an experimenter's region, which by practicality must always be constrained. They modified the BH (unconstrained) stationary-point approach to address the stationary point of a Lagrange multiplier problem. For the quadratic response surface model subject to a quadratic constraint function, the work of Myers and Carter (1973) also allows one to identify the constrained minimizing point among multiple stationary points in the Lagrange multiplier system. However, the Lagrange multiplier approach has some technical difficulties which impose substantial restrictions upon both modeling of the response surface and the constrained inference region.

Carter et al (1982) introduced a key biomedical application of confidence regions by showing how they can be used to statistically assess the notion of "therapeutic synergism" (Venditti, et al. 1956, Mantel, 1974). In dose combination studies two treatments are said to be therapeutically synergistic if there exists a dose combination of both treatments that is superior to each of the best individual treatment doses. This notion can be extended to three or more treatments. This idea, of course, can be applied to any system where one is interested in assessing the "response enhancement" capability of two or more factors.

Carter et al (1982) proposed the idea of using a confidence region for the optimal dose combination as a way to test for therapeutic synergism. If the confidence region for the optimal dose combination excludes all zero dose treatment combinations, then there is statistically significant evidence that all of the treatments are therapeutically synergistic. This idea can be generally applied to situations where the unconstrained optimum is outside of the experimental region, provided a general approach to constructing confidence regions for constrained optima is available.

In the next section previous methodology on confidence regions associated with response surface optima will be reviewed. A general approach to inference for optimal factor conditions will be introduced in section 3. A discussion of the coverage probability for the confidence region proposed in section 3 is given in section 4. An example of this confidence region will be given in section 5. A summary discussion is given in section 6.

2. REVIEW.

For ease of introduction, it will be assumed in this section that the response surface model can be adequately expressed in the standard quadratic polynomial form,

$$Y = \beta_0 + \beta'x + x'Bx + e ,\qquad (2.1)$$

where Y is the response variable and e has a normal distribution with mean 0 and variance σ^2. Here, β_0 is the intercept term, x is a $k \times 1$ vector of factor levels, β is a $k \times 1$ vector of regression coefficients, and B is a $k \times k$ symmetric matrix of regression coefficients with i^{th} diagonal element equal to β_{ii} and the $(i, j)^{th}$ off-diagonal element equal to $\frac{1}{2}\beta_{ij}$. If x_0 is a stationary point of the response surface in (2.1), then H_0: $\beta + 2Bx_0 = 0$ is true. BH show that a $100(1-\alpha)\%$ confidence region for x_0 is the set of all x such that

$$\hat{\delta}_x{}' \hat{V}_x^{-1} \hat{\delta}_x \le k F(1-\alpha; k, \nu),$$

where $\hat{\delta}_x = \hat{\beta} + 2\hat{B}x$, $\hat{\beta}$ and $\hat{B}$ are least squares estimates β and B respectively, and $F(1-\alpha, k, \nu)$ is the upper $100(1-\alpha)^{th}$ percentile of the F-distribution with k and $v = (n-p)$ df. Here, $\hat{V}_x$ is the estimate of V_x, the variance of $\hat{\delta}_x$. Of course this approach can be applied to the stationary points of more general response surface functions but the related optimality criteria may be much more difficult to determine.

As stated above, SCW incorporated constraints into the BH approach by introducing Lagrange multipliers. Unfortunately, the Lagrange multiplier approach has several technical difficulties associated with it. As SCW point out, the Lagrange multiplier approach in general only addresses a constrained stationary point, not necessarily the minimum point. The Lagrange multiplier approach also allows only for equality constraints. For a fixed experimental region, the Lagrange multiplier is a function of the estimated model parameters. Usually, this function is not of closed analytic form, requiring as a practical matter that the estimate of the Lagrange multiplier value be treated as a constant. However, due to the actual sampling variability of the Lagrange multiplier, the quadratic form that defines the confidence region may not have an F or *chi-square* distribution in either small or large samples respectively (Peterson, 1999).

In the next section, an approach is introduced which avoids Lagrange multipliers and allows for more general modeling of both the response surface and the experimental region. This modeling flexibility can in turn help make for better statistical inferences.

3. A GENERAL APPROACH.

Throughout the rest of this paper I consider the use of a (parametrically) linear model for the response surface. I replace the last two terms of the deterministic part of the model in (2.1) with the more general form, $z(x)'\theta$, where $z(x)$ is a $p\times1$ vector-valued function of x, a $k\times1$ vector of independent variables, and θ is a $p\times1$ vector of regression coefficients. Let x_0 and $\eta(\theta)$ be defined as

$$z(x_0)'\theta = \min_{x \in R} z(x)'\theta = \eta(\theta)$$

and consider

$$H_0: \eta(\theta) - z(x)'\theta = 0$$

for some $x \in R$. If H_0 is true, then x is an optimal point for the response surface. If a test for H_0 can be constructed, then a $100(1-\alpha)\%$ confidence region for x_0 can be created from the set of all x-values such that H_0 is not rejected at level α. A test for H_0 can created if I can find a confidence interval for $\left(\eta(\theta) - z(x)\right)'\theta$.

If the confidence interval for $\left(\eta(\theta) - z(x)\right)'\theta$ does not contain 0, then we can reject H_0; otherwise we cannot reject H_0.

In order to find a confidence interval for $\left(\eta(\theta) - z(x)\right)'\theta$, consider the interval

$$[\min_{\theta \in C} \left(\eta(\theta) - z(x)'\theta\right), \max_{\theta \in C} \left(\eta(\theta) - z(x)'\theta\right)] \quad (3.1)$$

where C is the usual quadratic-form confidence region for θ,

$$C = \left\{\theta: \left(\hat{\theta} - \theta\right)'\hat{V}^{-1}\left(\hat{\theta} - \theta\right) \leq c_\alpha^2\right\}.$$

Here, $\hat{\theta}$ is an estimate of θ and $\hat{V}$ is an estimate of V, the variance covariance matrix of $\hat{\theta}$. The critical value, c_α^2, is the $100(1-a)\%$ upper percentile of an F-distribution distribution. The min-max confidence interval in (3.1) can be shown to be equivalent to an exact likelihood profile confidence interval (Clarke, 1987, Peterson, 1999). One can obtain a conservative confidence interval by choosing $c_\alpha^2 = pF(1-\alpha; p, v)$ (Rao, 1973, chap 7). But in this case I would be acting very conservatively. The choice of $c_\alpha^2 = F(1-\alpha; 1, v)$ is recommended by Clarke (1987) for estimating a confidence interval for a function of regression model parameters using the approach in (3.1). However, some discussion in section 4 will indicate that a critical value of $c_\alpha^2 = kF(1-\alpha, k, v)$ (where k is the dimension of R) is more appropriate. This critical value will be one I use in this paper.

Since $\left(\eta(\theta) - z(x)'\theta\right) \leq 0$ for all $x \in R$, it follows that we reject H_0 whenever the upper limit of the interval in (3.1) is less than zero. Hence we need only compute the upper limit in (3.1) to be able to test H_0.

For the confidence interval in (3.1) note that

$$\max_{\theta \in C}\left(\eta(\theta) - z(x)'\theta\right)$$

equals

$$\max_{\theta \in C}\left[\min_{w \in R}\left(z(w) - z(x)\right)'\theta\right] \quad (3.2)$$

Unfortunately, the 'min' term in (3.2) does not in general have a closed functional form. However,

$$\max_{\theta \in C}\left[\min_{w \in R}\left(z(w) - z(x)\right)'\theta\right]$$

is less than or equal to

$$\min_{w \in R}\left[\max_{\theta \in C}\left(z(w) - z(x)\right)'\theta\right]. \qquad (3.3)$$

The inequality above follows directly from a fundamental minimax result . See for example, Zangwill (1969, pp 45-46).

In this paper we consider (3.3) as an upper confidence bound to test H_0. This confidence bound is much more attractive from a computational standpoint. Since $\left(z(w) - z(x)\right)'\theta$ is linear in θ it follows that

$$\max_{\theta \in C}\left(z(w) - z(x)\right)'\theta = b_x(w), \text{ where}$$

$$b_x(w) = \left(z(w) - z(x)\right)'\hat{\theta} \; + $$

$$c_\alpha\left[\left(z(w) - z(x)\right)'\hat{V}\left(z(w) - z(x)\right)\right]^{1/2} \qquad (3.4)$$

Hence (3.3) equals the upper confidence bound,

$$\min_{w \in R} b_x(w). \qquad (3.5)$$

For testing H_0, note further that for a given x, we need only find one w for which $b_x(w) < 0$ in order to be able to reject H_0. A confidence region for x_0 can be obtained by inverting the hypothesis test for H_0 for each x in R. We can express this confidence region, C_{x_0}, as the set of all x-values for which the expression in (3.5) equals zero.

The confidence bound in (3.5) is intuitively appealing in that it is derived from a confidence band for a response surface for $\left(z(w) - z(x)\right)'\theta$. This form can also be modified to provide some insight from a hypothesis testing perspective. Note that if the bound in (3.5) is less than zero, then for some $w \in R$ we have

$$\frac{\left(z(x) - z(w)\right)'\hat{\theta}}{\left[z(x) - z(w)'\hat{V}\left(z(x) - z(w)\right)\right]^{1/2}} > c_\alpha \qquad (3.6)$$

So (3.6) implies that if we can find a point in R which corresponds to a point on the response surface that is statistically significantly less than the point which corresponds to x, then we reject x, i.e. reject H_0. Therefore C_{x_0} is simply the set of all x-points in R for which we can find no other w-points in R where $z(w)'\theta$ statistically significantly less than $z(x)'\theta$.

It may be possible that in some situations the inequality associated with (3.3) is strict. However, there are three points worth mentioning about this. The first point is given by the theorem below. (For a proof of this and other theorems in this section see Peterson, 1999.)

Theorem 1: If x_0 is unique for each $\theta \in C$, then the inequality in (3.3) becomes an equality.

While it can be difficult to determine analytically if x_0 is unique for all $\theta \in C$, one can reasonably assess this by looking at the contour plots of $z(x)'\hat{\theta}$ and the associated upper and lower confidence band surfaces with $c_\alpha^2 = kF(1 - \alpha; k, v)$. If all three contour plots show the same fundamental shape indicating a unique minimum then we can be reasonably sure that x_0 is unique for all $\theta \in C$. This is because the lower (upper) confidence bound is given by

$$\min_{\substack{(\max) \\ \theta \in C}} z(x)'\theta.$$

Such contour plots also given some insight into the uncertainty associated with the general shape of the response surface.

The second point deals with the quadratic model in (2.1) and is addressed by theorem 2.

Theorem 2. If (i) B is positive definite (p.d.) for all θ-values in C and (ii) R is a convex set, then the inequality in (3.3) becomes an equality.

It is possible to check the condition (i) for theorem 2 by doing a straightforward computation which is related to a lower confidence bound for the minimum eigenvalue of B, $\lambda_{\min}(B)$. (See Peterson, 1993 for details.) In addition, this condition for should be checked to help assess the nature of the stationary point for a quadratic model. If this condition holds then we know the response surface is convex in a statistically significant sense. The third point is that the familiar "confidence band" expression in (3.4) should not produce overly wide limits with $c_\alpha^2 = kF(1 - \alpha; k, v)$ as this critical value will usually be much less than the corresponding Scheffe' critical value with k replaced by p.

In order to test each point x in R to see whether or not it should be in the confidence region, one must search over R to see if there exits a w-point in R satisfying $b_x(w) < 0$. Without further investigation, this can be a computationally intensive procedure and prone to round-off errors if w is close to x. However, since $b_x(x) = 0$, it makes sense to begin a search of $b_x(w)$ for w in a close neighborhood of x. Some results below provide conditions that can be used to substantially reduce the number of computations and improve accuracy in small neighborhood of x.

The local behavior of $b_x(w)$ in a neighborhood of x cannot be assessed by computing the gradient vector of $b_x(w)$ evaluated at x since $b_x(w)$ is not differentiable at x. However, directional derivatives of $b_x(w)$ can be easily computed if $z(x)$ is differentiable with respect to x. The directional derivatives can be used to determine if there exists a w-point (local to x and inside of R) where $b_x(w)$ becomes negative. The directional derivative of $b_x(w)$ associated with the direction vector, d, is defined as

$$b_x'(w;d) = \lim_{h \to 0^+} \frac{\left(b_x(w+hd) - b_x(w)\right)}{h}.$$

If $b_x'(x;d) < 0$ then there exists a sufficiently small, positive h such that $b_x(x+hd) < 0$. If $z(x)$ is differentiable, then it is straightforward to show, taking limits, that

$$b_x'(x;d) = d'D(x)\hat{\theta} + c_\alpha \left[d'D(x)\hat{V}D(x)'d\right]^{1/2},$$

where $D(x)$ is the $k \times p$ matrix of derivatives of $z(x)$ with respect to x. Computing $b_x'(x;d)$ over the set

$$\mathcal{B} = \left\{d : d'd = 1, x + hd \in R \text{ for small } h > 0\right\}$$

provides a local check of $b_x(w) < 0$. If x is in the interior of R, then the number of computations to obtain a local check of $b_x(w) < 0$ can be reduced using the results of the following theorem.

Theorem 3. If x is in the interior of R, then $b_x'(x;d) < 0$ for some d if and only if

$$\hat{\theta}'D(x)'\left[D(x)\hat{V}D(x)'\right]^{-1}D(x)\hat{\theta} > c_\alpha^2. \qquad (3.7)$$

Note that (3.7) is the condition for rejecting a stationary x-point with respect to $z(x)'\theta$ for the BH confidence region. This makes sense in that rejecting a point in the interior of R as stationary is equivalent to rejecting that point as optimal.

The local check in (3.7) becomes a global check under the conditions of the theorem below.

Theorem 4. Suppose $z(x)'\theta = \beta_0 + \beta'x + x'Bx$. If x is in the interior of R and B is p.d. for all $\theta \in C$, then (3.7) holds if and only if $b_x(w) < 0$ for some w in R.

Using the above results I recommend the following three-step procedure for computing a confidence region for x_0.

Step 1. For x-values in the interior of R, check to see if x can be rejected by a local assessment using the BH criterion in (3.7).

Step 2. For x-values on the boundary of R, check to see if x can be rejected by a local assessment by searching for the first d-value on $\mathcal{B}$ such that $b_x'(x;d) < 0$.

Step 3. For any x-value not rejected by Step 1 or 2, search for the first w-value in R such that $b_x(w) < 0$.

Interestingly, the (linear approximation) delta method applied to the confidence region problem of this paper does not work as the linear term in the Taylor series approximation of $\eta(\theta) - z(x)'\theta$ collapses to zero under H_0. This happens because the gradient vector of $\eta(\theta)$ equals $z(x_0(\theta))$ where $x_0(\theta)$ is the optimal value of x for a given θ (Peterson, 1989).

4. COVERAGE PROBABILITY.

To assess the relationship of the coverage probability to the critical value, c_α^2, it is helpful to consider further a connection with the likelihood ratio test (LRT) statistic. Let $g(\theta)$ denote a function of θ. For the linear model defined in section 3, a likelihood-ratio based confidence interval for $g(\theta)$ is defined to be

$$\left\{g_0: \left[\max_{g(\theta)=g_0} L(\gamma) \Big/ \max L(\gamma)\right] \geq \tau_\alpha^2\right\}, \qquad (4.1)$$

where $\gamma = (\beta_0, \theta')'$ (Clarke, 1987) and L is the appropriate likelihood function. The set in (5.1) can be shown to equal $\left[\min_{\theta \in C} g(\theta), \max_{\theta \in C} g(\theta)\right]$ where $\tau_\alpha^2 = \left(v^{-1}c_\alpha^2 + 1\right)^{-n/2}$. See Peterson (1999) for a derivation. It follows then that for some x if $g(\theta) = \eta(\theta) - z(x)'\theta$, the LRT for $H_0: g(\theta) = 0$ rejects when $\max_{\theta \in C} g(\theta) < 0$.

In many cases where x is in the interior of R,

$$\left[\max_{\Theta_0} L(\gamma) \Big/ \max L(\gamma)\right] \approx \left[\max_{\Theta_0^*} L(\gamma) \Big/ \max L(\gamma)\right], \qquad (4.2)$$

where $\Theta_0 = \left\{\theta: \eta(\theta) - z(x)'\theta = 0\right\}$, and $\Theta_0^* = \left\{\theta: D(x)\theta = 0\right\}$. However, for x on the boundary of R, the '$\approx$' sign in (4.2) is better replaced by '$<$' as Θ_0 is a larger set, often of higher dimension, than Θ_0^*. To see this more easily, consider the simplified quadratic model, $\beta_0 + x'\beta + \frac{1}{2}x'x$, where $k = 2$. Let $R = \left\{x: -1 \leq x_i \leq 1, i = 1,2\right\}$. If x_0 is in the interior of R, then $\Theta_0^* = \Theta_0 = \left\{\beta: \beta = -x_0\right\}$, so the LRT has $c_\alpha^2 = 2F(1-\alpha, 2, v)$. If x_0 is such that $-1 \leq x_{10} \leq 1$, $x_{20} = 1$, then $\Theta_0^* \subset \Theta_0 = \left\{\beta: \beta_1 = x_{10}, \beta_2 \leq -1\right\}$, so the above c_α^2

will be conservative. Similarly, if $x_0 = (1,1)$, then $\Theta_0^* \subset \Theta_0 = \{\beta: \beta_1 \le -1, \beta_2 \le -1\}$, so c_α^2 will be even more conservative. But interestingly, if the data strongly support $x_0 = (1,1)$, then the confidence region will be small (i.e. the point $(1,1)$) even if $c_\alpha^2 = 2F(1-\alpha, 2, v)$ is "conservative".

More generally, if x_0 is in the interior of R and the data clearly support $z(x)'\theta$ convex over R, then the LRT statistics corresponding to Θ_0 and Θ_0^* in (4.2) should be very similar. If the data support x_0 on the boundary of R, then $c_\alpha^2 = kF(1-\alpha, k, v)$ will be conservative. The worst case scenario for less than nominal coverage might be as follows. Suppose x_0 is in the interior of R and the estimated response surface is convex but not statistically significantly so. In this case the distribution of the LRT statistic corresponding to Θ_0^* in (4.2) may be more skewed to the right than for the LRT statistic corresponding to Θ_0, as Θ_0 is more restrictive. This would result in somewhat less than nominal coverage probability. However, this situation is the best place for a worst case scenario as one perhaps should not compute a confidence region for an optimal configuration if the general shape of the response surface is in question.

5. AN EXAMPLE.

In this section, we apply the methodology in section 3 to an example from the literature. In this example, a constrained confidence region is computed for a convex response surface whose minimizing stationary point is far outside of the experimental region. The first example is developed from data from a preclinical cancer chemotherapy study illustrated in SCW. Here, it is sought to estimate the optimal dose combination of two treatments, 5-Fluorouracil (5-FU) and Teniposide (VM26) for treatment of leukemia in the B62DF1 mouse animal-model.

Sixteen different treatment combinations were used. The study end-point was survival time in days. The data was analyzed using the proportional hazards regression model. The standard quadratic model was fit to the data. Using the method in Peterson (1993), a 95% confidence interval for $\lambda_{\min}(B)$ is [1.09, 0.026]. Hence the response surface is statistically significantly convex. However, B is not p.d. everywhere on C (for $\alpha \le 0.1$) with asymptotic critical value, $c_\alpha^2 = \chi^2(1-\alpha; k), k = 2$. Therefore all three steps of the above computational procedure were used. For

step 2, the local searches were done on a circle of radius 0.05 centered at each x-value.

The unconstrained minimum (stationary point) of the fitted quadratic response surface yields a 5FU value of 428.9 mg/kg and a VM26 value of 36.1 mg/kg. As SCW point out, this shows that the unconstrained optimal combination estimate is far from the experimental region. As such, it would be unwise to try to make statistical inferences about optima so far from the experimental region. To estimate constrained optima it is natural to define the constraint region, R, to be the set of points within the convex hull of the design points (illustrated by the symbol '∎') shown in Figure 1. Using this constraint region, and the above quadratic model, the resulting confidence region is represented below as the gray area in Figure 1. The confidence region indicates that 5FU and VM26 are therapeutically synergistic with respect to the dose-region defined within the convex hull of the experimental treatment combinations.

6. DISCUSSION.

As can be seen from the example in the previous section, the approach in section 3 for estimating and testing optimal factor combinations is attractive in that general experimental regions and a wide variety of (parametrically) linear models can be employed. This allows for a much improved application of Carter et al's (1982) important idea of using confidence regions to test for therapeutic synergism. The generality in section 3 is also useful for constructing confidence regions for optima in mixture experiments. Here, the factors are proportions that are constrained to sum to one, with additional constraints often required. This produces a naturally constrained situation where optimal factor combinations can sometimes fall on the boundary of the experimental region. Furthermore, models for mixture experiments sometimes utilize exotic functions of the factor variables as covariates, instead of a standard quadratic model (Cornell, 1981, chap 6).

In addition, this approach can easily be generalized to other families of response surface models. Some examples of useful (parametrically) nonlinear response-surface models can be found in Nelder (1966), Mead and Pike (1975), Box and Draper (1987, chap 12), and Khuri and Cornell (1987, chap. 8). Since the confidence region inference is based upon confidence bounds for the mean response function, the parameter-effects nonlinearity associated with such inferences will be zero (Ratkowsky, 1983, chap 9). Applications also extend readily to generalized linear models such as Poisson or binary regression models.

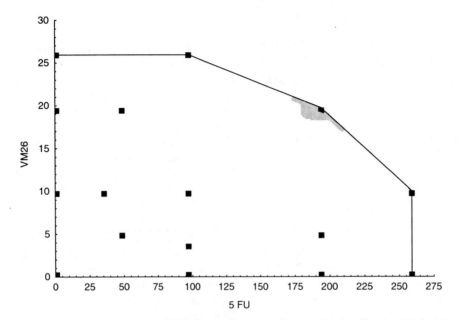

Figure 1. The Grey Area Represents a 95% Constrained Confidence Region for the Minimizing Combination of Chemotherapy Dose Levels of 5 FU and VM 26 within the experimental region. The experimental region is the convex hull of the points ('■') in the graph.

REFERENCES

Box, G. E. P. and Draper, N. R. (1987) *Empirical Model Building and Response Surfaces*, New York: John Wiley.

Box, G. E. P. and Hunter, J. S. (1954) "A Confidence Region for the Solution of a Set of Simultaneous Equations with an Application to Experimental Design", *Biometrika*, 41, 109-199

Carter, W.H., Wampler, G.L., Stablein, D. M. and Campbell, E.D. (1982) "Drug Activity and Therapeutic Synergy in Cancer Treatment" *Cancer Research* **42**, 2963-2971.

Clarke, G. P. Y. (1987) "Approximate Confidence Limits for a Parameter Function in Nonlinear Regression", *Journal of the American Statistical Association*, **82**, 221-230.

Khuri A. I. and Cornell, J. A. (1987) *Response Surfaces: Designs and Analyses*, New York: Marcel Dekker.

Mantel, N. (1974) "Therapeutic Synergism", *Cancer Chemotherapy Reports Part II* **4**, 147-149.

Mead R. and Pike D. J.(1975), "A Review of Response Surface Methodology from a Biometric Viewpoint", *Biometrics*, **22**, 803-851.

Myers R. H. and Carter, W. H., (1973) "Response Surface Techniques for Dual Response Systems", *Technometrics*, **15**, 301-317

Nelder, J. A. (1966) "Inverse Polynomials, a Useful Group of Multifactor Response Functions", *Biometrics*, **22**, 128-141.

Peterson, J. J. (1989) "First and Second Order Derivatives Having Applications to Estimation of Response Surface Optima", *Statistics and Probability Letters*, 8, 29-34.

Peterson, J. J. (1993) "A General Approach to Ridge Analysis with Confidence Intervals", *Technometrics*, **35**, 204-214.

Peterson, J .J. (1999), "A General Approach to Inference for Optimal Conditions Subject to Constraints", *Biostatistics and Data Sciences Technical Report*, January 26, 1999, King of Prussia, PA.

Rao, C. R. (1973) *Linear Statistical Inference and Its Applications*, New York, John Wiley.

Ratkowsky, D. A. (1983), *Nonlinear Regression Modeling: A Unified Approach*, New York: Marcel Dekker.

Stablein, D. L., Carter, W. H., and Wampler, G. L., (1983) "Confidence Regions for Constrained Optima in Response Surface Experiments", *Biometrics*, **39**, 759-763.

Venditti, J.M., Humphreys, S.R., Mantel, N. and Goldin, A. (1956) "Combined Treatment of Advanced Leukemia (L1210) in Mice with Amethopterin and 6-mercaptopurine", *Journal of the National Cancer Institute*, **17**, 631-638.

Zangwill, W. I. (1969) *Nonlinear Programming: A Unified Approach*, Englewood Cliffs, NJ, Prentice-Hall.

USE OF RESPONSE SURFACE METHODS IN CHEMICAL PROCESS OPTIMIZATION

Kenneth C. Syracuse, Douglas P. Eberhard, Anna M. Messinger
Kenneth C. Syracuse, Wilson Greatbatch Ltd., 10000 Wehrle Drive, Clarence, NY 14031

Abstract:
Implantable medical devices have increased both the quality of life and its longevity for a great many individuals. Pacemakers, defibrillators, neurostimulators, and drug delivery systems are included in this list. At the heart of these systems is the power system, the implanted battery or cell. Lithium/silver vanadium oxide (Li/SVO) is the battery chemistry of choice for high energy, implantable, pulsed current applications. Cells using this chemistry are successfully employed in the automatic implantable cardioverter defibrillator (AICD). Market forces, surgeon preference and patient morphology are combining to force these units to become smaller. Improved electronics have enabled these devices to reduce in size. As a result, the batteries are becoming smaller with as much current carrying capability as before.

A key cell component in achieving the smaller size and higher power in implantable power source technology is the cathode. Present Li/SVO cell construction utilizes pressed powder technology to form the cathode onto the current collectors (plates). Cathode plates (shielded by separator material) are interwoven with corresponding anodes to form the "cell stack" which is inserted into a case. The cell is filled with liquid electrolyte and hermetically sealed. While providing cells which meet present device requirements, alternatives are desirable for further product generations. A cathode sheeting process has been developed which reduces some of the challenges associated with dealing with particulates, and allows for operations under ambient conditions. The cathode sheeting technology[1] enables the design and manufacture of smaller cells with the same (or better) current carrying capabilities. Designed experiments and response surface models were used to identify and optimize the factors involved in the mechanical and processing aspects of the sheeting process.

Introduction:
Implantable medical devices have increased both the quality of life and its longevity for a great many individuals. Pacemakers, defibrillators, neurostimulators, and drug delivery systems are included in this list. While introduction of these devices increases the quality of life, pressure soon builds to make the devices more physiologically comfortable by making them smaller, thinner, and

lighter. At the heart of these systems driving the electronics is the power system -- the implanted battery or cell. Lithium/silver vanadium oxide (Li/SVO) is the battery chemistry of choice for high energy, implantable pulsed current applications such as the automatic implantable cardioverter defibrillator. As the battery accounts for a significant portion of the volume of an implantable defibrillator, size reduction of the battery is critical to the size reduction of the device.

Electrochemical cells or batteries have three basic components -- anode, cathode, and electrolyte. The chemistry of the battery determines its voltage. The surface area of the anode and cathode, to a large extent, determine the current that the battery can provide. Maintaining surface area while making the battery smaller implies that the cathode plate assembly must be made thinner. Alternative processing for future product generations was desired.

The cathode sheeting process.
Overview
The SVO sheet cathode process incorporates both chemical and mechanical processes as shown in Figure 1. The base material for SVO sheet cathode is silver vanadium oxide (SVO) produced from the reaction of silver nitrate and vanadium pentoxide. The next step involves mixing of SVO, conductive additives, binder, and solvent to produce the SVO sheet cathode mixture. The mixture is turned into pellets and processed into sheet cathode by passing them through three sets of rollers on the *primary* rolling mill. Finally, a *secondary* rolling mill, consisting of two sets of rollers, is used to attain model-specific product.

Rolling Mill Equipment
Primary and secondary rolling mills are used to transform pellets into sheet cathode and obtain the desired weight and thickness (see Figure 2, upper). The sheet on the reel is unrolled fed into the secondary rolling mill equipment (see Figure 2, lower). The coupons are "re-rolled" by passing them through two sets of rollers.

Each roll mill is configured with independent variable speed motors and gap adjustments. Therefore, as pellets are processed into sheet, the thickness of the sheet cathode is gradually. Proper gap settings are critical to obtain the required *basis* weights (defined in grams per square inch) and thickness. Finally, the sheet is collected, and forwarded to the punching operation.

---

[1]   Patent Numbers 5,435,874 and 5,571,640.

Identification of Key Variables

Sheet cathode designated for a specific cell model varies in surface area, and weight. Some cells incorporate a plate design consisting of multiple cathode "blanks" while others incorporate a wound element design consisting of a long rectangular sheet cathode. Both rolling mills are used when sheet is produced with the settings on the first mill fixed. Therefore, the remainder of the discussion pertains only to the secondary rolling mill operation.

The goal of the analysis to determine settings to maximize the cathode basis weight and minimize sheet thickness. A series of Phase 1 type studies identified three variables of primary interest; the relationship between gap settings, solvent content, and the direction of the sheet relative to the secondary rollers.

During the initial phases, a series of experiments was conducted to determine the limits for which mechanically sound material could be manufactured. For the development of gap parameters, settings on the secondary rolling mills were varied among five levels in combination. The solvent range was tested at 4 levels. Coupons were rotated $0°$, $90°$, $180°$, and $270°$ relative to the initial roll direction entering the secondary rolling operation to enhance mechanical integrity. All experiments involved processing the sheet cathode first through the fixed primary rolling mill.

Response Surface Modeling of Key Variables

Once the key variables had been identified and constraints determined, a central composite design experiment was conducted. In this instance rotatability was sacrificed for uniformity. The coupons were rotated prior to submission to the secondary mill.

Figure 3 shows the results of fitting a quadratic response surface[2] to the experiment involving the factors gap setting and solvent content for a given rotation. The model solution was determined to be outside the tested parameters. This is not surprising since the goal is to maximize one parameter (basis weight) while minimizing the thickness. Figure 4 shows a contour profile of the response with lower and upper limits. The plots enable selection of combinations of parameters for which the output will meet requirements. Results for the other degrees of rotation were similar as the primary effect of repositioning the coupon was determined to be mechanical.

While it is necessary to chemically duplicate the present cathode material, the final phase of these studies is the resultant output from actual cell discharge. Accordingly, a variety of cell types (models and

configurations) were constructed and placed on short and long term test.

Summary

A method has been developed which enables the reduction of thickness for cathode material for implantable grade power sources. While the chemical process to ensure the integrity of the material during the construction phase is important, the final test is performance. The sheet cathode possesses the necessary amount of cathode material to ensure adequate capacity and the mechanical strength to retain integrity during the manufacturing process. Short term testing has shown the sheet product to be a viable alternative. Long term tests are continuing.

The designed experiments and response surface modeling results enable the production operation to quickly determine new process settings as battery design requirements change.

Acknowledgement
The authors wish to thank Dr. Esther Sans Takeuchi for her careful review of this material.

---

[2]   Statistical analyses were performed using the JMP® software.

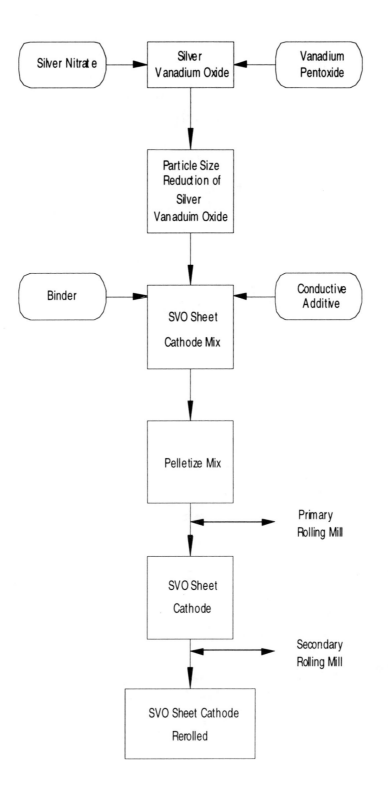

Figure 1 - SVO Sheet Cathode Process Flow Diagram

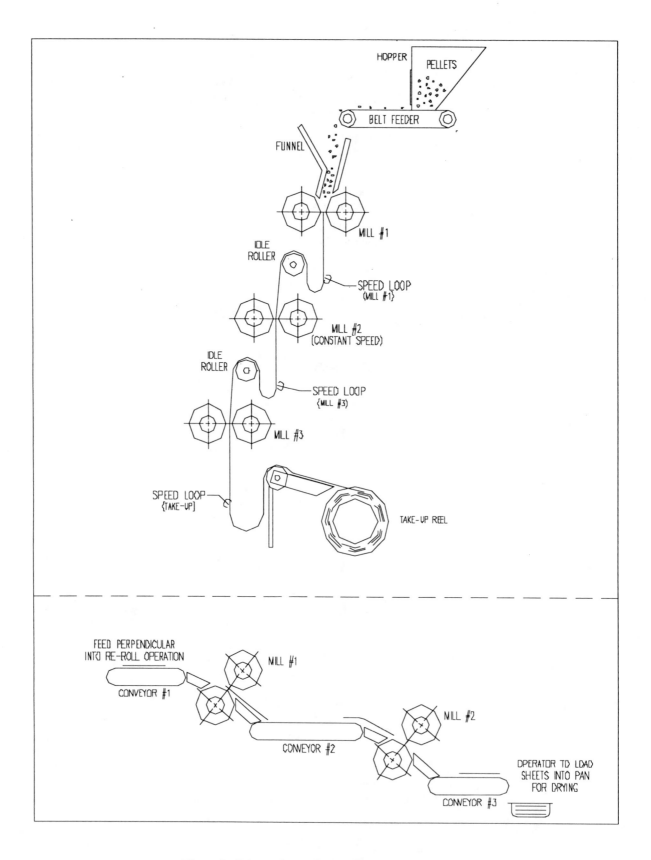

Figure 2 - Primary/Secondary rolling mill process

Response: Weight
Summary of Fit

RSquare	0.901345
RSquare Adj	0.900295

Response Surface

Coef	Solvent	Gap	Weight
Solvent	38.109264	0.3117531	-24.35182
Gap	?	-0.000953	-0.030324

Solution

Variable	Critical Value
Solvent	0.2304011
Gap	21.78322

Solution is a	SaddlePoint
Critical values outside data range	
Predicted Value at Solution	0.7128723

Contour Plots

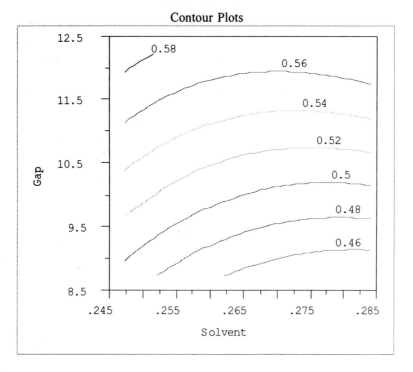

Figure 3 - Results from fitting a quadratic response surface

Response	Contour	Current Y	Lo Limit	Hi Limit
Weight	0.524	0.5241709	0.5	0.55

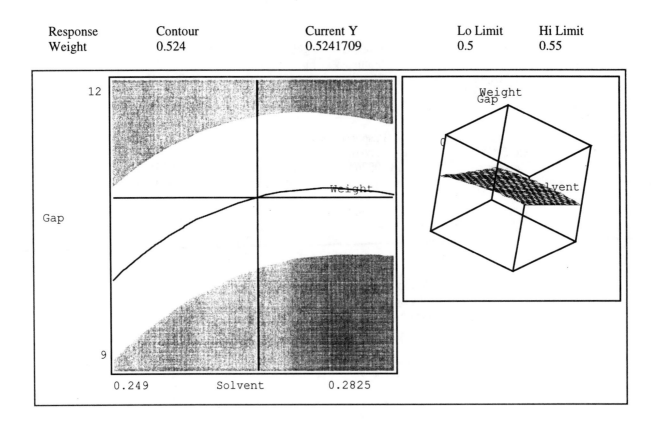

Figure 4 - Contour plot with output limits

Selection errors of stepwise regression in the analysis of supersaturated design

Shu YAMADA (shu@ms.kagu.sut.ac.jp)
Department of Management Science
Science University of Tokyo
Kagurazaka 1-3, Shinjuku, Tokyo 162-8601, Japan

Key Words: Active factors, Alias, Simulation study, Type I and II errors

Abstract:

Several papers consider constructing methods of supersaturated designs to assure low level of non-orthogonality between columns. As regarding data analysis for supersaturated design, stepwise selection method is usually applied to detect important factors under an assumption of the effect sparsity. This paper evaluates the Type II error in selection of stepwise regression via simulation study.

1 Introduction

Supersaturated design is a kind of fractional factorial design, in which the number of the experimental runs is greater than the number of the columns being assigned the factor effects. When experimentation is expensive and the number of factors being assigned in experiment is large, supersaturated design would be helpful, for example, screening problem to improve quality characteristic in industrial areas. The origin of supersaturated design was developed by Satterthwait (1959) as a random balance design. Booth and Cox (1962) formulated supersaturated design in a systematic manner and constructed some designs by a computer search.

After three decades, Lin (1993) has shown a simple and effective constructing method of supersaturated design. Supersaturated design became one of the hottest areas in the field of experimental design after the appearance of Lin (1993). Several papers consider constructing methods of supersaturated designs to assure low level of non-orthogonality between columns, such as Wu (1993), Deng, Lin and Wang (1994), Iida (1994), Lin (1995a), Li and Wu (1997), Nguyen (1996), Tang and Wu (1997), Cheng (1997) and Yamada and Lin (1997).

The author thanks Jun Harashima, graduate school student of Tokyo Metropolitan Institute of Technology, for his help of computer simulation.

In the above studies, they focus on construction of two-level supersaturated design, which consists of two-level columns. Recently another class of supersaturated design had been proposed by Yamada and Lin (1999), which is three-level supersaturated design consisting of three-level columns. A constructing method is shown by Yamada, Ikebe, Hashiguchi and Niki (1999). Extension to mixed-level supersaturated design had been done by Yamada and Lin (1998).

While there are several kinds of constructing method of supersaturated design, a few papers consider analysis methods for data collected by supersaturated design. Lin (1993) show an example of data analysis, where the data are generated by the half fraction of the real data listed in Williams (1968). In the example, stepwise regression and other methods are applied to the analysis. Lin (1995a) gave other examples of data analysis of supersaturated design. The examples suggest that stepwise regression may work well for the data analysis of supersaturated design.

However, several aspects needed to be examined still have remained due to many difficulties of the data analysis of supersaturated design, while the conclusions from the examples introduced in the above are meaningful. The discussion between Wang (1995) and Lin (1995b) can be regarded as an evidence of the difficulties. One of the aspects, we need to obtain, is error of selection in stepwise regression in the data analysis of supersaturated design. Westfall, Young and Lin (1998) consider the Type I selection error, which is selection of non-active factors as active factors. Furthermore, they derived a selection method to control the probability of the Type I error. The research would be helpful in terms of experimental cost reduction for additional experiments. Because, the selected factors would be used in additional experiments.

When we consider an improvement for quality characteristic, such as reduction of variation or adjusting average, we need to find active factors. From the viewpoint, the Type II error should be considered. In this paper, we focus on the Type II error in selection of stepwise regression. First, we evaluate the probability to detect active factors via simulation study by the reason of importance of the Type II error in selection. Furthermore, some guidelines for data analysis are discussed based on the results of the simulation study.

2 Selection errors of stepwise regression

Suppose that P factors, $x_1, \ldots, x_p, \ldots, x_P$, are assigned to a supersaturated design $\boldsymbol{X}$ with n runs, where each vector in the design consisting of equal numbers of -1s and 1s. Supersaturated design implies that the number of factors, P, is greater than the number of experimental runs, n. Let $\boldsymbol{Y}$ be an $n \times 1$ vector consisting of observations of a response variable y. We suppose that the true response function can be described by a linear regression model such that

$$\boldsymbol{Y} = \boldsymbol{1}\beta_0 + \boldsymbol{X}\boldsymbol{\beta} + \boldsymbol{\varepsilon}, \qquad \boldsymbol{\varepsilon} \sim N(\boldsymbol{0}, \sigma^2 \boldsymbol{I}_n), \quad (1)$$

where $\boldsymbol{1}$ and $\boldsymbol{\varepsilon}$ are $n \times 1$ vectors consisting of 1s and independent error terms, respectively and $\boldsymbol{I}_n$ is an $n \times n$ identity matrix. A factor x_p is called "active" when $\beta_p \neq 0$ in Equation (1), since the factor affect to the response y.

Consider data analysis of $\boldsymbol{Y}$ and $\boldsymbol{X}$ based on Equation (1). Due to the singularity of the matrix $\boldsymbol{X}^t \boldsymbol{X}$, usual least squares can not be applied to the data analysis of supersaturated design. Stepwise selection techniques are widely applied for data analysis of supersaturated design, such as forward selection of variables based on the F-statistic for the test of hypothesis $H_0 : \beta_p = 0$. For example, Lin (1993) applied the forward selection in regression based on the F-statistic for the data analysis of supersaturated design.

We also apply the forward selection in regression based on the F-statistic for the data analysis. The factor x_{s_1} is estimated as an active factor at the first stage, where

$$s_1 = \arg\max \left\{ F_p^1 \mid p \in \mathcal{P} \right\} \quad (2)$$
$$F_p^1 = RSS(p)/MSE(p), \quad (3)$$

and $\mathcal{P}$ is the set of suffices of all factors such that $\mathcal{P} = \{1, 2, \ldots, p, \ldots, P\}$. At the second stage, x_{s_2} is estimated as an active factor, where

$$s_2 = \arg\max \left\{ F_p^2 \mid p \in \mathcal{P} \setminus \{s_1\} \right\} \quad (4)$$
$$F_p^2 = RSS(p \mid s_1)/MSE(p, s_1). \quad (5)$$

By the similar manner, x_{s_k} is estimated as an active factor, where

$$s_k = \arg\max \left\{ F_p^k \mid p \in \mathcal{P} \setminus \{s_1, \ldots, s_{k-1}\} \right\} \quad (6)$$

and

$$F_p^k = \frac{RSS(p \mid s_1, \ldots, s_{k-1})}{MSE(p, s_1, \ldots, s_{k-1})}, \quad (7)$$

at the k-th stage of selection. According to the procedure, k factors are selected up to k-th stage and the selection procedure can be continued up to $(n-1)$-th stage. Let $\mathcal{Q}$ be the sets of suffices on the active factors such that $\mathcal{Q} = \{q_1, q_2, \ldots, q_Q\}$, where $|\beta_{q_1}| \geq |\beta_{q_2}| \geq \ldots \geq |\beta_{q_Q}|$. Furthermore, let $\mathcal{S}^k$ be the selected factors after the selection at k-th stage such that $\mathcal{S}^k = \{s_1, s_2, \ldots, s_k\}$. The most preferable selection is that all active factors are selected and there is no excessive selection, *i.e.* $\mathcal{S}^k = \mathcal{Q}$. The Type I error in selection implies that one or more non-active factors are selected, *i.e.* one or more suffices are in the set $\mathcal{S}^k$ and are not in the set $\mathcal{Q}$. On the other hand, the Type II error in selection implies that one or more suffices are not in the set $\mathcal{S}^k$ and are in the set $\mathcal{Q}$. In other words, the factors whose suffices are in the set $\{\mathcal{S}^k \cap \mathcal{Q}\}$ are detected active factors. Thus, we focus on the number of selected active factors, $| \mathcal{S}^k \cap \mathcal{Q} |$, in order to evaluate the Type II error of selection.

Since the set $\mathcal{S}^k$ is determined by random variable such as F_p^k, we will derive the probability to detect at least q $(1 \geq q \geq Q)$ active factors at k-th stage such that

$$P\left(| \mathcal{S}^k \cap \mathcal{Q} | \geq q \right). \quad (8)$$

For example, Equation (8) for $q = 1$ implies the probabilities that at least one and all active factors are detected at k-th stage. In general, the probability described by Equation (8) is utilized for the evaluation of the Type II error in the analysis of supersaturated design.

Furthermore, we also evaluate the probabilities to detect largest effect, second-largest effect, ..., Q-th largest effect such that

$$P\left(q_1 \in \mathcal{S}^k \right), P\left(q_2 \in \mathcal{S}^k \right), \ldots, P\left(q_Q \in \mathcal{S}^k \right).$$

3 Evaluations of errors in selection

3.1 Outline

For the evaluation of the probability to detect active factors in terms of the Type II error, we need to determine conditions of (1), ..., (6) as follows.

(1) Number of stages

As regarding the stopping rule of selection, we, first,

Table 1: The design matrices for the evaluation of the probability to detect active factors.

n	Num. of Columns	Design Matrix
8	35	Cheng (1997)[1]
12	66	Wu (1993)
16	71	Yamada and Lin (1997)
24	133	Yamada and Lin (1997)

select $n - 1$ factors out of P factors sequentially by the stepwise regression method. The behavior of the probability to detect active factors is examined with respect to the entrance stages. By an examination of the behavior, we, second, explore a stage number, in which almost active factors are detected efficiently. Note that the exploration procedure is essentially similar to the forward selection by F values.

(2) Number of runs and design n, X

Table 1 shows the design utilized in this paper. Cheng (1997) and Yamada and Lin (1997) had obtained essentially equivalent supersaturated design with $n = 8$ runs although their minimization criteria for construction are different. As regarding $n = 12$, we apply the design Wu (1993), since the number of columns in Wu (1993) is greater than the number in Lin (1993). The design for $n = 16$ and 24 are selected by the similar reasons of the numbers of columns.

(3) Number of assigned factors P

The number of assigned factors, P, are determined by

$$P = t(n - 1), \qquad (9)$$

where $t = 1, 2, 3$. Obviously, $t = 1$ implies saturated design and $t = 2, 3$ implies supersaturated designs. Comparisons between the results of $t = 1$ and $t = 2$ may derive the effects of supersaturated design.

(4) Number of active factors, Q

The number of active factors, Q, is selected from the set $\{1, \ldots, 7\}$. The condition $Q = 1$ or 2 or 3 is supposed the case of "effect sparsity", although $Q = 7$ is far from effect sparsity.

(5) Levels of effects, β
(6) Types of distribution of effects

We use two types of distribution on the effects of factors as follows: (a) constant type and (b) step type. In constant type, all factor effect are in the same level, i.e. $\beta_p = \beta_{\text{const}}$ for $p \in \mathcal{Q}$. On the

other hand, (b) step type implies that the levels of the Q active factors are distributed such as large, medium and small. For example, the levels of effects on three active factors are distributed 3, 2 and 1. In general, the largest level of factor effect is $\beta_{\max} = \max\{\beta_p \mid p \in \mathcal{Q}\}$. The second largest, k-th largest levels and Q-th largest levels are $\frac{Q-1}{Q}\beta_{\max}$, $\frac{Q-k+1}{Q}\beta_{\max}$ and $\frac{1}{Q}\beta_{\max}$, respectively.

The assignments of active factors are randomly determined. For example, 3 columns being assigned Q active factors are randomly determined from P columns.

After determination of conditions for all (1), ..., (6), we will derive probability to detect active factors, $P\left(\mid \mathcal{S}^k \cap \mathcal{Q} \mid \geq q\right)$ and so on. Specifically, after putting the design matrix X and n, the number of assigned factors P, the number of active factors Q and factor levels β, the model defined by Equation (1) is determined, such that $Y = \beta_0 + X\beta + \varepsilon$, $\varepsilon \sim N(0, \sigma^2 I_n)$. In the following, we put $\sigma^2 = 1$ without loss of generality for evaluation.

3.2 Some results

The probabilities to detect active factors are calculated by a Monte Calro simulation with 10,000 reputations. Figures 1 summarizes the probability of to detect active factors. Specifically, Figure 1 (a) shows the behavior of probability that at least one, at least two and three (all) factors are selected in sequential stages for $n = 8, 12, 16$ and 24 designs, i.e. $P\left(\mid \mathcal{S}^k \cap \mathcal{Q} \mid \geq q\right)$ $(1 \leq q \leq 3, 1 \leq k \leq n-1)$ for designs with $n = 8, 12, 16$ and 24 runs. On the calculation, other conditions are followings: the number of assigned factors, $P = t(n-1)$, $t = 3$, the number of active factors, $Q = 3$ and the levels of factors, $\beta_{\text{const}} = 3$. For example, the line "$(n, q) = (8, 2)$" indicates the behavior of probability to detect at least two out of three active factors in the sequential stages. At the second stage of selection, the probabilities of detect at least two out of three active factors are around 15% and 50% in $n = 8$ and $n = 12$ designs, respectively.

(1) Number of stages

We observe from Figure 1 (a) that the probability to detect active factors doesn't increase so much after fifth stage comparing to before fifth stage in most cases, although there are a few exceptions. In other words, there is no significant increasing of the probability after fifth stage. This result suggests that it may be sufficient to consider from first to fifth stages to detect active factors in many cases.

(2) Number of runs and design n, X

At the fifth stage in the design with $n = 24$ runs, the probabilities to detect at least one, at least two and all(three) active factors are around 0.96, 0.78 and 0.76, respectively, in Figure 1 (a). We may be able to expect that $n = 24$ supersaturated design would work well to detect all active factors under an assumption of threes active factors. The case of $n = 16$ is different from the case of $n = 24$. At the fifth stage, in the designs $n = 16$, the probabilities to detect at least one, at least two and all(three) active factors are around 0.94, 0.51 and 0.50, respectively, The levels of the probabilities in $n = 12$ are around same with $n = 16$. This result suggests that we may not able to detect all active factors in $n = 16$ or 12 design, although, detection of one active factor might be achieved. The case of $n = 8$ are more lower with comparing to $n = 24, 16, 12$. We may be able to detect only one out of three active factors, but may not be able to detect two or three factors.

From the viewpoint of practice, $n = 24$ design is recommended under the situation that the number of active factors is supposed to be around three and detection of all factors are requested. In such a situation, $n = 8$ design would not be recommended since the simulation results implies the poor performance of detection of all factors. On the other hand, when detection of one out of some active factors are requested and the cost for experiment is expensive, $n = 8$ supersaturated design may be acceptable since it may detect at least one factor out of some active factors.

(3) Number of assigned factors P

Figure 1 (b) displays the probability to detect at least one, at least two and three (all) active factors for the case of $P = t(n - 1)$, $t = 1, \ldots, 3$, where the design is $n = 12$, the number of active factors $Q = 3$ and the level of effect of active factors is $\beta_{\text{const}} = 3$. Although all active factors are detected perfectly (100%) in $t = 1$, the probability to detect active factors are less than 100% in $t = 2, 3$. The differences of the probabilities can be regarded as a demerit brought by supersaturated design compared to non-saturated design, while supersaturated design has a merit in terms of the number of factors, $i.e.$ more factors can be examined in an experiment comparing to non-saturated design. Thus, Figure 1 (b) implies a typical feature of supersaturated design concisely.

Let compare the cases of $t = 2$ and $t = 3$ for considerations of the effects of the number of assigned factors. Three (all) active factors are detected around 80% at the fifth stage in $t = 2$, where totally $P = t(n - 1) = 22$ factors are assigned to the design. The results suggest that we may be able to expect detection of all active factors in $t = 2$. On the other hand, when we assign $P = t(n - 1) = 33$ factors, the probability to detect all factors at the fifth stage is 52%. It may not be acceptable from practical viewpoint. By comparing the results of $t = 2$ and $t = 3$, we should not assign so many factors, such as $t > 3$, if detection of all active factors is requested. We may be able to conclude that $n = 12$ design would be appropriate to detect one or two out of two active factors for $P = 3(n - 1) = 33$, while $P = 2(n - 1) = 22$, $n = 12$ design would detect three out of three active factors.

(4) Number of active factors, Q

Figure 1 (c) shows the probability to detect active factors in the cases of $Q = 1, 2, 3$ and 5, where the design is $n = 12$, the number of active factors $P = t(n - 1)$, $t = 3$ and the level of effect of active factors is $\beta_{\text{const}} = 3$. In this simulation study, we found that the number of active factors significantly affects the probability to detect active factors. For example, all active factors are detected by 86% for $Q = 2$ at the fifth stage. On the other hand, all active factors are detected at the fifth stage by 52% and 2% for $Q = 3$ and $Q = 5$, respectively. Furthermore, in the case of $Q = 5$, the probability to detect active factors increases gradually, not rapidly after at the fifth stage. It means that we may not be able to detect all active factors when there are many active factors, such as $Q \geq 5$.

The reason of the above results would be explained by confounding relation of columns in supersaturated designs. In general, columns in supersaturated design are confounded each other. This property implies the effect of an active factor appears both the column of the active factor and the columns confounding the active factor column. When there are many active factors, total amount of the effects by confounding can't be negligible in non-active factor columns. As a result, such a non-active factor column is sometimes detected as an active factor column.

(5) Levels of effects β of factors

Figure 1 (d) displays the probability to detect active factors under $\beta_{\text{const}} = 1, 2, 3$ and 5, where $n = 12$, $P = t(n - 1)$, $t = 3$ and $Q = 3$. The results of $\beta_{\text{const}} = 2, 3$ and 5 are closed each other although the result of $\beta_{\text{const}} = 1$ is far from them. The results suggest that the levels of active factors affect the selection slightly when the levels are higher than a certain level such as $\beta_{\text{const}} \geq 2$.

(6) Types of distribution of effects

Due to the limiation of space of this paper, we briefly explained the results in the case of the step type of distributions of factor effects. We observed that the probability in the step type of effect levels is generally higher than the constant type. From the practical viewpoint, this results implies a positive aspect of supersaturated design, since the step type is supposed to be closed actual cases rather than the constant type. In other words, Figure 1 shows a pessimistic case to detect active factors. The other trends of the probability in the step type are similar to the results of the constant type.

4 Discussions

The results in this study is summarized as follows: In many cases, the probability to detect active factors are saturated around the fifth or sixth stage in the stepwise regression. In terms of the probability, the number of active factors is the most essential in the analysis of supersaturated design. The result suggests that supersaturated design would not work well if the number of active factors is large, such as five or more. In such a case, we may not be able to detect all active factors. On the other hand, effect sparsity holds, such as two active factors in $n = 12$ experiments, supersaturated design works well. Through this simulation study, we recommend to use supersaturated design to detect active factors when the number of active factors is supposed one, two or at most three.

References

[1] Booth, K. H. V. and Cox, D. R. (1962). Some systematic supersaturated designs. *Technometrics* **4** 489-495.

[2] Cheng, C. S. (1997). $E\left(s^2\right)$- optimal supersaturated designs. *Statistica Sinica* **7** 929-939.

[3] Deng, L. Y., Lin, D. K. J. and Wang, J. N. (1994). Supersaturated design using Hadamard matrix. *IBM Research Report* RC19470 IBM Watson Research Center.

[4] Iida, T. (1994). A construction method of two-level supersaturated design derived from $L12$. *Japanese Journal of Applied Statistics* **23** 147-153 (in Japanese).

[5] Li, W. W. and Wu, C. F. J. (1997). Columnwise-pairwise algorithms with applications to the construction of supersaturated designs. *Technometrics* **39** 171-179.

[6] Lin, D. K. J. (1993). A new class of supersaturated designs. *Technometrics* **35** 28-31.

[7] Lin, D. K. J. (1995a). Generating systematic supersaturated designs. *Technometrics* **37** 213-225.

[8] Lin, D. K. J. (1995b). Response to Wang (1995). *Technometrics* **37** 359.

[9] Nguyen, N. K. (1996). An algorithm approach to constructing supersaturated designs. *Technometrics* **38** 69-73.

[10] Satterthwaite, F. E. (1959). Random balance experimentation (with discussion). *Technometrics* **1** 111-137.

[11] Tang, B. and Wu, C. F. J. (1997). A method for constructing supersaturated designs and its $E\left(s^2\right)$ optimality. *Canadian Journal of Statistics* **25** 191-201.

[12] Wang, P. C. (1995). Comments to Lin (1993). *Technometrics* **37** 358.

[13] Williams, K. R. (1968). Designed experiments. *Rubber age* **August** 65-71.

[14] Westfall, P. H., Young, S. S. and Lin, D. K. L. (1998). Forward selection error control in the analysis of supersaturated design. *Statistica Sinica* **8** 101-117.

[15] Wu, C. F. J. (1993). Construction of supersaturated designs through partially aliased interactions. *Biometrika* **80** 661-669.

[16] Yamada, S., Ikebe, Y., Hashiguchi, H. and Niki, N., (1999). Construction of three-Level supersaturated design. *Journal of Statistical Planning and Inference* (accepted).

[17] Yamada, S. and Lin, D. K. J. (1997). Supersaturated designs including an orthogonal base. *Canadian Journal of Statistics* **25** 203-213.

[18] Yamada, S. and Lin, D. K. J. (1999). Three-level supersaturated design, *Statistics and Probability Letters* (accepted).

[19] Yamada, S. and Lin, D. K. J. (1998). Construction of mixed-level supersaturated designs. *Proceedings of Joint Statistical Meetings 1998* Section on Physical and Engineering Sciences 67-71.

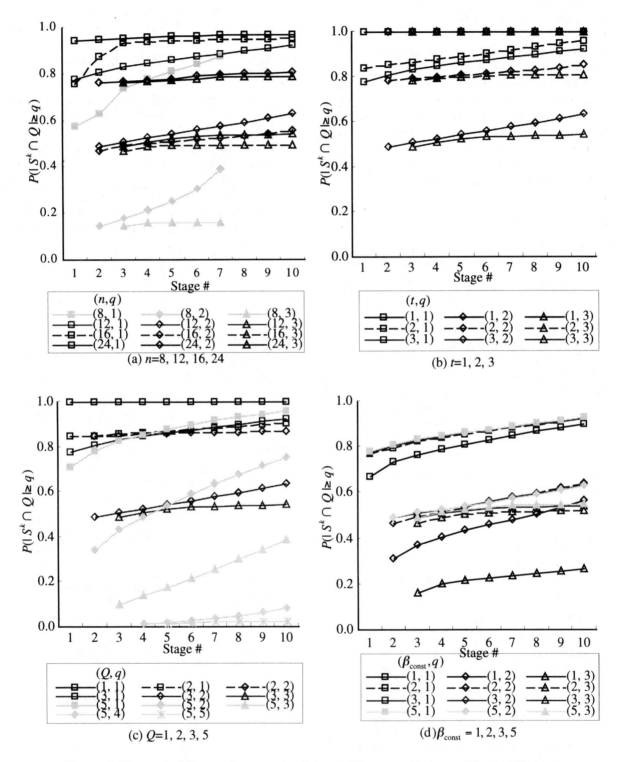

Figure 1: The probability to detect active factors: The constant type of factor effects.

CALIBRATION: A NONLINEAR APPROACH

Edna Schechtman, Ben Gurion University of the Negev,
Cliff Spiegelman, Texas A&M University
Contact-Author: Edna Schechtman, Department of Industrial
Engineering and Management, Ben Gurion University of the Negev,
Beer Sheva, Israel ednas@bgumail.bgu.ac.il

Key Words: Estimation, Confidence intervals, Least Squares, Inverse regression, radiocarbon.

1 Introduction

The paper focuses on two related problems. The first is estimating the x-intercept of a straight line, and the second is a calibration problem. In fact, the only difference is that rather than estimating x at a new observed value of $Y^* = 0$, we estimate x for a value $EY = 0$.

Calibration curves typically relate a standard unit x to an instrumental measurement Y by using data x_i and Y_i collected from a calibration experiment to estimate the calibration curve. After the calibration curve is estimated, new measurements Y_j^* are taken and the calibration curve is then used with the Y_j^* to estimate the associated standard values. More specifically, the notation and relationships are given by: $Y_i = \beta_0 + \beta_1 x_i + \sigma\epsilon_i$, for $i = 1, ..., n$, with the unknown parameters β_0 and β_1 and with the standard deviation σ, also unknown(See Scheffe (1973) and Osborne (1991)). The random variables ϵ are taken as unobservable noise and are independent and identically distributed. Initially we assume that they are normal random vectors. Typically, all of the unknown parameters are estimated only from the calibration experiment data and do not use the new, after calibration measurements. After the calibration experiment is completed, a new measurement $Y_i^* = \beta_0 + \beta_1 x^* + \sigma\epsilon^*$ is taken. Thus x^* can be estimated by point and interval estimates.

There are two major approaches to the calibration problem: the classical approach and the inverse regression approach. The inverse regression approach was first introduced by Krutchkoff (1967) and is based on the regression of X on Y, rather than Y on X. Krutchkoff's (1967) paper caused controversy, which will not be covered here. The interested reader will find an excellent review in Osborne (1991). The standard or classical formula was

We wish to thank ANRC and the NSF statistics and chemistry divisions for partial support.

derived by Eisenhart (1939), who obtained his point estimate by inverting the estimated regression line of Y on X, given by $\hat{Y} = b_0 + b_1 X$, where b_0 and b_1 are the least squares estimators of the intercept and the slope, respectively. Eisenhart also produced a confidence interval for x^*. While his formula is useful in many instances there are other cases where either it is not applicable or it gives too wide an interval.(This will happen when the slope of the regression line of Y on X is not significantly different from zero, at level α). We note that the classical approach does not use the point estimate as the center of the interval, but rather obtains the center of the interval from geometric arguments. Shukla (1972) obtained asymptotic expressions for the bias and mean square error (MSE) of the classical estimator, conditional on the event $|b_1| > 0$. Shukla and Datta (1985) studied the bias and MSE under the so-called truncation procedure that $H_0 : \beta_1 = 0$ is rejected for some α, but no confidence intervals are provided. Naszodi (1978) derived a new estimator which is approximately unbiased, is more efficient than the classical estimator and is consistent. His expression for the approximate bias is the same as in Shukla (1972).

In an interesting paper, Dahiya and McKeon (1991) present a method for getting confidence intervals for Naszodi's (1978) estimate. They give an explicit way to estimate the variance of the point estimate and construct a $(1 - \alpha)100\%$ confidence interval, based on the normal percentiles. Their intervals are shorter than all other methods that we have tried, but the coverage rate is much below the declared $(1 - \alpha)100\%$. By making a simple adjustment to the Dahiya and McKeon (1991) interval, namely replacing their recommended normal-percentile by a t- percentile we show, via simulation, that their intervals work well, have reasonable coverage rate, and are the narrowest of all the methods that we have tried.

In this paper, a different approach is suggested, with emphasis on confidence intervals of the form MLE $\pm t*$standard error. It is based on reparametrizing the linear model so that the unknown value of the standard measure becomes a parameter in a nonlinear regression model. Then, standard packages for nonlinear regression can easily provide a point estimator, as well as its standard error, which can then be modified and used as the basis for the confidence interval. We present a comparison of our nonlinear least squares approach to the standard frequentist intervals. Readers interested in Bayesian approaches should see Hunter and Lamboy (1981) and the extensive list of references contained therein. Our main emphasis here is on confidence intervals centered on the MLE, but for completeness, our simulation experiments will include the modified Dahiya and McKeon (1991) method. The structure of the paper is as follows: In section 2, the two approaches (classical and nonlinear) are introduced. Section 3 is devoted to an example while in section 4 we give some simulation results. Section 5 concludes the paper.

2 The two approaches: classical and nonlinear

We write the straight line model as follows:

$$Y_i = \beta_0 + \beta_1 x_i + \epsilon_i.$$

The x_i's are chosen predictors, measured without error, and the Y_i's are the responses. The errors in the Y_i's, denoted by ϵ_i's, are mutually independent having the same distribution as ϵ, which is assumed to be the normal distribution with mean zero and a constant variance σ^2.

The problems of interest in this paper are to estimate, by point and interval estimates:

1. The x-intercept, namely: $x^* = -\beta_0/\beta_1$, and
2. x for a single future value of Y, denoted by Y^*.

Since the solutions to the two problems are very similar, we shall develop the estimates for the intercept problem, and comment on the changes needed for the calibration problem, when appropriate.

The common procedure for the problem of estimating the intercept of a straight line, based on the least squares method, is (see Graybill (1976)):

1. Obtain the MLE of x^*, which is $-b_0/b_1$, where b_0, b_1 are the LS estimators of β_0, β_1, respectively.

2. Test $H_0 : \beta_1 = 0$ vs. $H_1 : \beta_1 \neq 0$ (size α test).

3. If H_0 is not rejected, no confidence interval exists (using this method).

4. If H_0 is rejected, then a $(1-\alpha)100\%$ confidence interval for x^* exists. The interval is centered at:

$$\frac{\bar{X} - b_1 \bar{Y}}{d}$$

and its length is given by:

$$LI_{LS} = (2t\hat{\sigma}/d)\sqrt{d/n + \bar{Y}^2/S_{xx}}$$

where $d = b_1^2 - t^2\hat{\sigma}^2/S_{xx}$, $\hat{\sigma} = $ the maximum likelihood (ML) estimator of the standard error, $S_{xx} = \sum(x_i - \bar{x})^2$, $t = $ the upper $(1 - \alpha/2)$-th percentile of the t-distribution with $(n - 2)$ degrees of freedom.

It has been noted by many (see Graybill(1976)) that these intervals are not $(1 - \alpha)100\%$ confidence intervals. The actual confidence levels are lower than $(1 - \alpha)100\%$ since the intervals do not always exist. A necessary and sufficient condition for their existence is $d > 0$, which is equivalent to rejecting the null hypothesis $H_0 : \beta_1 = 0$, at level α.

The corresponding notation and interval for the calibration problem is: Let Y^* be a new value of Y, and we wish to estimate x^*, where x^* is the value of x such that

$$E(Y^*) = \beta_0 + \beta_1 x^*.$$

The classical point estimator of x^* is given by

$$\hat{x}^* = (Y^* - b_0)/b_1.$$

A classical approximately $(1 - \alpha)100\%$ confidence interval for x^* is centered at

$$\bar{x} + b_1(Y^* - \bar{Y})/d,$$

and its length is given by

$$L_{LS} = (2t\hat{\sigma}/d)\sqrt{d(1 + 1/n) + (Y^* - \bar{Y})^2/S_{xx}}$$

We present an alternative approach to the calibration problem that can be employed, using nonlinear regression methods for a reparametrized model.

Schechtman, Spiegelman and Moran (1998) suggest to reparametrize the model and rewrite it as:

$$Y = \beta_1(X - \beta_2) + \epsilon,$$

where $\beta_2 = -\beta_0/\beta_1$ is the parameter of interest in the original problem. The model, as formulated now, is a nonlinear regression model, and nonlinear methods and computer algorithms can be used in order to estimate β_2, both by a point estimate and by an asymptotic $(1 - \alpha)100\%$ confidence interval. The interval obtained by the nonlinear approach is centered at $-b_0/b_1$ and its length is given by

$$LI_{NL} = 2t\hat{\sigma}\sqrt{b_1^2 S_{xx} + n\bar{Y}^2}/(b_1^2\sqrt{nS_{xx}}).$$

It is shown in Schechtman, Spiegelman and Moran (1998) that the interval for the intercept problem, obtained by the nonlinear approach is equivalent to applying the standard delta method to the ratio of MLE's and proved that the resulting confidence interval is shorter than the classical interval, for the same level of confidence. In what follows, we will show that the solution to the calibration problem can be written in a similar way. Let

$$Y^* = \beta_0 + \beta_1 x^* + \epsilon^*$$

where ϵ^* has the same distribution as the ϵ's and is independent of them, and define

$$W_i = Y_i - Y^* = \beta_1(x_i - x^*) + (\epsilon_i - \epsilon^*) \qquad (1).$$

Model (1) describes a nonlinear regression model, with parameters β_1 and x^*, and where the responses $W_1, W_2, ..., W_n$ are not independent. Therefore, modifications need to be made before one can use the standard nonlinear methods and computer algorithms to estimate and obtain intervals for x^*. Let

$$W_i = \beta_1(x_i - x^*) + \delta_i$$

be the nonlinear regression model, where $\delta_1, \delta_2, ..., \delta_n$ are normally distributed random variables with mean 0, variance $2\sigma^2$ and $cov(\delta_i, \delta_j) = \sigma^2$. The point estimator for x^*, which is the center of the interval for the nonlinear approach, is equivalent to the classical estimator, namely:

$$\hat{x}^* = (Y^* - b_0)/b_1.$$

The approximate variance(of the limiting distribution) of $\hat{x}^*$ is given by

$$\hat{\sigma}^2 \left(\frac{1}{b_1^2} + \frac{(b_1^2 S_{xx} + n(Y^* - \bar{Y})^2)}{b_1^4 n S_{xx}} \right)$$

and thus, the length of the interval, which is affected by the dependence of the error terms, is given by

$$L_{NL} = 2t\hat{\sigma} \sqrt{\frac{1}{b_1^2} + \frac{b_1^2 S_{xx} + n(Y^* - \bar{Y})^2}{b_1^4 n S_{xx}}}.$$

Note that these intervals always exist, whereas the classical approach does not always provide an interval. A closer look at the variance of $\hat{x}^*$ shows that the dependence among the error terms adds an extra term to the variance of the estimator of x^*, relative to the variance in the independent errors case for the intercept problem, which is derived in Schechtman, Spiegelman and Moran (1998). What it means, in practice, is that any standard statistical software for nonlinear regression can be applied with the dependent variable being $Y - Y^*$, and once the variance is obtained, the extra term, $\hat{\sigma}^2/b_1^2$, should be added. After adding this term, an interval can be constructed.

The advantage of using the nonlinear approach is that it provides shorter intervals, for the same confidence level, and the point estimate requires no explicit inversion. It can be shown that, ignoring the common factor $2t\hat{\sigma}$ and looking at the difference between the squares of the remaining components of the lengths, the difference between the length using nonlinear approach and classical approach can be expressed as:

$$1/b_1^2 - 1/d + [(b_1^2 S_{xx} + n(Y^* - \bar{Y})^2)/b_1^4 n S_{xx}$$

$$-1/d^2(d/n + (Y^* - \bar{Y})^2/S_{xx})].$$

The term in the squared brackets is negative as long as $d > 0$, as shown in Schechtman, Spiegelman and Moran (1998), and

$$1/b_1^2 - 1/d = (d - b_1^2)/db_1^2 = -t^2\hat{\sigma}^2/S_{xx} d b_1^2 < 0.$$

This completes the proof that as long as $d > 0$, that is: as long as the classical method has a solution, the length of the interval obtained by the nonlinear method will be shorter. It can be shown that the difference in length is a monotone increasing function of $\hat{\sigma}$. That means that whenever the data is more "noisy", the advantage of using the nonlinear approach is more noticeable.

3 An example

Carbon dating - a calibration example. The data is the carbon dating example, taken from Kromer et al. (1986). Carbon dating is an important calibration method used in archaeology, and is a widely accepted method for dating ancient artifacts. Carbon dating is based upon the simple idea that an uncommon isotope of carbon, carbon 14, is absorbed by all life forms. When the life form dies it no longer breathes in carbon 14 and the carbon 14 contained in its system decays, with a half life of 5730 years. Assuming that the amount of carbon 14 in the environment was approximately constant during the last 10,000 years, we can date an object by measuring the percentage carbon 14 that the study object has. Usually the radio carbon age based upon the 5730 years half life needs to be corrected by calibration because the amount of carbon 14 in the environment was not and is not exactly constant. The causes of this variation are beyond the scope of this paper. Interested readers may wish to consult Currie (1982). Below are two columns of data, from Kromer et al (1986). The first is the radio-carbon age of an artifact , Y, and the second, x, is an age determined by more accurate methods such as counting tree rings. It would be typical to produce a calibration curve for this type of data. The calibration curve would then be used to date an unknown artifact.

The first 21 observations were used (approximately 200 years), and are given in the following table:

Y	x
8199	7207
8271	7194
8212	7178
8211	7173
8198	7166
8141	7133
8166	7129
8249	7107
8263	7098
8161	7088
8163	7087
8158	7085
8152	7077
8157	7074
8081	7072
8000	7069
8150	7064
8166	7062
8083	7060
8019	7058
7913	7035

The above data was used in order to obtain a calibration curve and use it in order to get 95% confidence intervals by the classical method, the nonlinear approach, and Dahiya and McKeon's method (DM), at two different locations on the calibration curve: near the middle, and at the end. For this example, b_1=1.15, $\sqrt{MSE}$=68, and the value of the t-statistic is 3.82, with a p-value of .0012. The procedure was as follows: a data point was chosen and deleted from the data set, and then the three approaches were applied to the remaining 20 data values in order to obtain the center and a 95% confidence interval for the removed x value. The results are summarized below:

a) The point removed is (7194, 8271) (near the upper end of the x-variable). Center: 7299 (classical method), 7218 (nonlinear approach),7208 (DM), length: 435.35 (classical method), 310.40 (nonlinear approach), 276.2 (DM).

b) The point removed is (7107, 8249) (near the average of x). Center: 7234 (classical method), 7198 (nonlinear approach),7192 (DM), length: 322.17 (classical method), 265.66 (nonlinear approach),

246.55 (DM).

For both locations, the classical approach gave the longest intervals, and DM gave the shortest intervals. The nonlinear approach was slightly worse (longer) than DM. The intervals obtained by DM and by the nonlinear approach were centered closer to the true values of x than the classical approach.

An electrophysiological experiment, an intercept example, can be found in Schechtman, Spiegelman and Moran (1998).

4 Simulation results

The data generated in the simulation study, for both problems, follow the model: $Y = 2 + x + \sigma\epsilon$. One thousand samples of size 10 were generated for each case. (Similar trends, not reported here, were found for sample sizes 5 to 30). The x's were chosen to have a Uniform distribution on [0,1] and Uniform distribution on [0,10] for the calibration problem, and Uniform on [0,1] and U-shaped on [0,1] for the intercept problem. The performance of the different methods for the calibration problem was evaluated at two new values of x∗: in the middle of the x-range and towards the edge. The performance of the two methods for the intercept problem was evaluated at true intercept of .5, .25 and 0.

The mean zero error distributions used for the calibration problem were: normal, t-distribution with 4 degrees of freedom, and a centered lognormal distribution. For the intercept, only the normal distribution was used. Various values for the error variance were used and are listed in the selected tables of results. Tables 1, 2 and 3 show the results for the calibration problem, while table 4 is for the intercept problem. The intercept of the straight line, 2, is immaterial, since all of the estimators that we use are location invariant. It is known that an important parameter in calibration experiments is β_1/σ. We chose our values of σ so that the ratio β_1/σ covers a critical range of parameter values that are typically encountered in practice.

The performance of the three methods was compared by looking at the following statistics:

1. The center of the interval - mean(s.e.)

2. The number of samples for which the classical method has a solution.

3. The percent coverage.

4. The length of the interval - mean(s.e.)

For all the simulation results, and for both problems, when the ratio β_1/σ is large, there is theoretically little reason to choose the nonlinear or the modified DM methods over the classical one, except that the classical method fails to produce an interval in a small percentage of cases. For these cases, both the nonlinear approach, which is centered at the MLE, and the modified DM, which is centered at an unbiased estimator, give intervals, but sometimes those intervals are too long to be of practical use. (The MSE for the MLE is theoretically infinity, but the results are modal for most simulations that would be done with the parameter values that we chose). This is not true, however, when the error is at least moderately big, as in the carbon dating example. Then, the center of the interval obtained by the classical approach has a relatively large standard error. Clearly, Eisenhart (1939) did not intend for the center of the interval to be used as the point estimate, but in our experience, many applied people report intervals as the center ± a multiplier of the standard error, and use the center of the interval as their point estimate.

Table 1: $\epsilon \sim N(0, \sigma^2)$ Design: $x \sim U(0, 1)$
m = no. of samples for which LS has a solution.
true parameter $x^* = 0.1$

Confidence Interval Center

σ	LS	NL	DM
.25	-.763(9.50)	.0615(.336)	.097(.292)
.10	.069(.125)	.093(.116)	.097(.114)
.05	.092(.058)	.098(.057)	.099(.057)

Percent Coverage

σ	m	LS(of m)	NL(of 1000)	DM(of 1000)
.25	936	94.1	96.8	96.4
.10	1000	94.3	94.9	94.8
.05	1000	94.3	94.4	94.4

Length of Confidence Interval

σ	LS	NL	DM
.25	2.98(19)	1.44(.81)	1.26(.46)
.10	.528(.16)	.509(.144)	.502(.138)
.05	.253(.067)	.251(.065)	.250(.065)

Table 2: $\epsilon \sim$ log Normal, centered,
Design: $x \sim U(0, 1)$
m = no. of samples for which LS has a solution.
true parameter $x^* = 0.1$

Confidence Interval Center

σ	LS	NL	DM
.25	-.135(3.302)	-.39(13.89)	.1167(.277)
.10	.065(.327)	.095(.122)	.100(.114)
.05	.0926(.069)	.0988(.058)	.0999(.057)

Percent Coverage

σ	m	LS(of m)	NL(of 1000)	DM(of 1000)
.25	943	92.0	92.4	92.0
.10	993	92.4	92.2	92.1
.05	1000	92.5	92.4	92.4

Length of Confidence Interval

σ	LS	NL	DM
.25	1.74(6.48)	1491(30000)	.939(.545)
.10	.4425(.725)	.422(.421)	.4025(.302)
.05	.2076(.18)	.2029(.156)	.201(.150)

Table 3: $\epsilon \sim t_{(4)}$ Design: $x \sim U(0, 1)$
m = no. of samples for which LS has a solution.
true parameter $x^* = 0.1$

Confidence Interval Center

σ	LS	NL	DM
.25	-.286(2.14)	.081(.348)	.118(.266)
.10	.0755(.133)	.101(.111)	.1047(.109)
.05	.096(.057)	.101(.055)	.1024(.055)

Percent Coverage

σ	m	LS(of m)	NL(of 1000)	DM(of 1000)
.25	933	93.6	94.4	94.1
.10	999	93.9	93.8	93.7
.05	1000	94	93.8	93.9

Length of Confidence Interval

σ	LS	NL	DM
.25	1.94(4.33)	838(26457)	1.15(.482)
.10	.4996(.266)	.4788(.227)	.4695(.197)
.05	.239(.126)	.2357(.103)	.2346(.098)

Table 4: $\epsilon \sim N(0, \sigma^2)$ Design: $x \sim U(0, 1)$
m = no. of samples for which LS has a solution.
true intercept =0.25

Confidence Interval Center

σ	LS	NL
.707	-1.58(16.97)	.14(2.6)
.50	-.58(12.57)	.22(.038)
.05	.25(.001)	.25(.001)

Percent Coverage

σ	m	LS (out of m)	NL (out of 1000)
.707	448	94.4	95.3
.50	940	95.0	95.0
.05	1000	95.1	95.2

Length of Confidence Interval

σ	LS	NL
.707	4.47(34.0)	52.9(774.6)
.50	2.0(25.16)	.55(.59)
.05	.004(.0012)	.004(.0012)

When comparing the performance of the three approaches for the different error distributions for the calibration problem, we can see that the nominal level α, .05, is well met when the error distribution is normal by both the nonlinear procedure and DM. The classical procedure has slightly worse coverage rate. When the error distribution has a heavier tail both the coverage and the confidence interval width become smaller. Regardless of the error distribution the shortest interval are obtained by the DM procedure and the nonlinear confidence intervals are slightly longer. Thus the nonlinear intervals give greater coverage and shorter length than the classical intervals and the more complicated modified DM estimates have the best behavior in our simulation experiment.

We further checked, via simulations not shown here, the case of extrapolation in the intercept problem, as in the problem that initiated the study. The results are similar to those reported above: the classical and nonlinear approaches give similar results for small error variance, and the nonlinear is better both in terms of center and length of the interval for moderate error variances.

5   conclusions

We have presented an alternative, easy to compute, method for generating single use calibration confidence intervals and for estimating the x-intercept. The new method is based on reparametrizing the model and rewriting it as a nonlinear regression model. The new intervals are appealing from at least three different points of view. Scientists and engineers possessing a nonlinear curve fitting routine will find that the new intervals are easier to compute as the usual inversion is done painlessly. Secondly, our calibration and intercept confidence intervals are always shorter than those obtained by the standard intervals while still proving realistic coverage. When the measurement errors are moderately big, the difference in confidence interval width is noticeable. The nonlinear approach can easily be applied (some minor modifications are needed for the calibration problem), using any statistical package. And finally, the nonlinear intervals are centered at the MLE, which is the natural point estimator.

The method suggested by Dahiya and McKeon (1991), which is included in our simulation study (after some minor modification), performs well and gives the shortest intervals in the calibration problem, with realistic coverage rates. Therefore, we believe that scientists, who wish to use confidence intervals centered on estimates other than the MLE, would do well to consider Dahiya and McKeon's modified intervals. We recommend these intervals despite the fact that they are more complicated to obtain than the simple method that we present for MLE centered intervals.

6   References

Currie, Lloyd A. (1982). Nuclear and Chemical Dating Techniques: Interpreting the Environmental Record; American Chemical Society, no. 176; 516 pages.

Dahiya, R. C., and Mckeon, J. J. (1991) "Modified Classical and Inverse Regression Estimators in Calibration", Sankhya B., 53, 48-55.

Eisenhart, C. (1939)."The Interpretation of Certain Regression Methods and their Use in Biological and Industrial Research", Annals of Mathematical Statistics, 10, 162-180.

Graybill, F. A. (1976). *Theory and application of the linear model*, Wadsworth & Brooks/Cole, Pacific Grove CA.

Hunter, W. G. and Lamboy, W. F. (1981) "A Bayesian Analysis of the Linear Calibration Problem" (With Discussion), Technometrics, 23, 323-350.

Kromer, B., Rhein, M., Bruns, M., Schoch-Fischer, H., Munnich, K. O., Stuiver, M., and Becker, B. (1986) "Radiocarbon Calibration Data for the 6th to the 8th Millennia BC", Radiocarbon, 28, 954-960.

Krutchkoff, R.G. (1967) "Classical and Inverse Regression Methods of Calibration", Technometrics, 9, 425-439.

Naszodi, L. J.,(1978) "Elimination of the Bias in the Course of Calibration", Technometrics, 20, 201-205.

Osborne, C. (1991) "Statistical Calibration: a Review", International Statistical Review, 59, 309-336.

Schechtman, E., Spiegelman, C, and Moran, N. (1998) "Interval Estimate for the x-intercept of a Straight Line: A Nonlinear Approach", Communications in Statistics, Simulations and Computations , 27(4), 1171-1180.

Scheffe, H. (1973). "A statistical Theory of Calibration", Annals of Statistics, 1, 1-37.

Shukla, G. K. (1972) "On the Problem of Calibration", Technometrics, 14, 547-553.

Shukla, G.K. and Datta, P. (1985)"Comparison of the Inverse Estimator With the Classical Estimator Subject to a Preliminary Test in Linear Calibration", Journal of Statistical Planning and Inference, 12, 93-102.

SMOOTHING FOR SMALL SAMPLES WITH MODEL MISSPECIFICATION:
NONPARAMETRIC AND SEMIPARAMETRIC CONCERNS

James E. Mays, Virginia Commonwealth Univ., and Jeffrey B. Birch, VA Polytech. Inst. & State Univ.
James E. Mays: 1001 West Main Street, P.O. Box 842014, Richmond, VA 23284-2014

Key Words: Bandwidth, Misspecification, Model-robust.

1 INTRODUCTION

The basic regression problem consists of estimating the underlying function g for data of the form

$$y_i = g(X_i) + \varepsilon_i , \; i = 1, \ldots, n , \qquad (1.1)$$

where X is the (controlled) regressor variable, and the ε_i are independent, identically distributed random errors with mean zero and variance σ^2. Other regressors may be present in (1.1), but the single regressor case (with possible polynomial terms in the model) is focused on here. Emphasis is also placed on the small sample scenario. Parametric regression, such as ordinary least squares (OLS), assumes a known form for g, and then uses the data to estimate the parameters of this specified model. If this form for g is unknown or misspecified, then nonparametric regression should be used. These nonparametric *smoothing* procedures use only the data and a local weighting scheme to fit the data and estimate g. Data points close to the location being fit are given greatest weight, and the rate at which these weights decrease away from this location is determined by the *bandwidth* or *smoothing parameter*. Smaller bandwidths give less weight to points further away, essentially using fewer data points and resulting in a less smoothed fit. Thus, the choice of bandwidth is crucial in obtaining a "proper" estimate of g. The situation addressed in this article is when parametric and nonparametric procedures are combined to fit the data, specifically when there is some knowledge about the parametric form of g, but this form is not adequate throughout the entire range of the data. Such *semiparametric* procedures require the selection of a bandwidth for their nonparametric portions, and both familiar and new concerns regarding this selection are the focus of this work.

For any nonparametric fit, the chosen bandwidth b should strike the proper balance between the variance of fit, which is high when b is too small, and the bias (or squared bias) of fit, which is high when b is too large. These points are illustrated in Figure 1, which contains semiparametric fits to the tensile strength data of Montgomery and Peck (1992). Figure 1a uses a bandwidth that is too small, resulting in a fit that would change significantly for a different realization of the data (i.e., is too variable). Figure 1c uses a bandwidth that is too large, and oversmooths the data. Figure 1b uses a proper bandwidth (via PRESS**, to be

introduced in section 3.2) to give the desired fit. Based on these concerns, the minimization of a mean squared error criterion (or other global error criterion) is a logical starting point for bandwidth selection. Issues related to such an approach and numerous resulting techniques are discussed in Section 3, after an overview of the fitting techniques in Section 2. Section 4 introduces an optimality criterion for evaluation of techniques. These evaluations are carried out via specific examples in Section 5 and a small simulation study with concluding remarks in Section 6.

2 OVERVIEW OF TECHNIQUES
2.1 NONPARAMETRIC REGRESSION

A thorough discussion of kernel regression and other nonparametric procedures is given in Härdle (1990). Using the weighting scheme of Nadaraya and Watson, a kernel estimator at location X_i may be defined as

$$\hat{y} = \sum_{j=1}^{n} \left[K\left(\frac{X_i - X_j}{b} \right) \Big/ \sum_{j=1}^{n} K\left(\frac{X_i - X_j}{b} \right) \right] y_j = \sum_{j=1}^{n} h_{ij}^{(\text{ker})} y_j$$

[see Wand and Jones (1995)], where the kernel function $K(\text{u})$ is a decreasing function of $|\text{u}|$, and the weights $h_{ij}^{(\text{ker})}$ are the elements of the kernel "hat" matrix $\mathbf{H}^{(\text{ker})}$, allowing the vector of fits to be expressed as $\hat{\mathbf{y}}_{\text{ker}} = \mathbf{H}^{(\text{ker})} \mathbf{y}$. Other weighting schemes can be found in Priestley and Chao (1972), Gasser and Müller (1979), and Chu and Marron (1991). The choice of K is not critical in obtaining a proper kernel fit, and the simplified Normal (Gaussian) function $K(\text{u}) = \exp(-\text{u}^2)$ is used here. *Local polynomial regression (LPR)*, introduced by Cleveland (1979), addresses some of the inherent problems of kernel regression (including boundary bias), and is used as the nonparametric fitting procedure [see Fan and Gijbels (1996)]. Hastie and Loader (1993) provide more details on LPR while discussing this as a boundary bias correction, Mays and Birch (1996) present the form of the local polynomial hat matrix $\mathbf{H}^{(\text{LPR})}$, and Stone (1980, 1982) shows optimal convergence rates for LPR in a certain minimax sense. Local linear regression (*LLR*) is appropriate for this work, with asymptotic optimality properties and nice small sample properties given by Fan (1992).

2.2 SEMIPARAMETRIC (MODEL-ROBUST) REGRESSION

The two model-robust procedures which combine parametric (OLS) and nonparametric (LLR) fits are

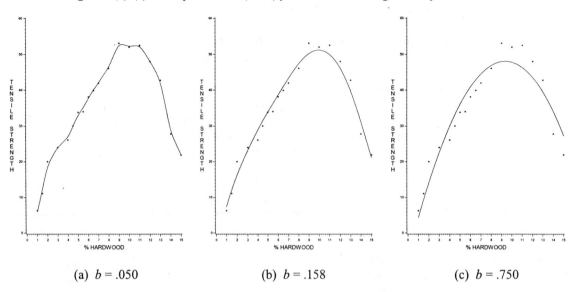

Figure 1 (a)-(c). Semiparametric (MRR) fits to Tensile Strength Data for various bandwidths.

(a) $b = .050$ (b) $b = .158$ (c) $b = .750$

briefly discussed here. Further details of the development and the establishment of the potential benefits of these techniques are given in Mays and Birch (1996). The first semiparametric method is *Partial Linear Regression* (*PLR*), an extension of the semiparametric method of Speckman (1988). Here the underlying function $g(X)$ is written as the sum of a parametric term and a nonparametric term in the partial linear model

$$y_i = \mathbf{x}_i'\boldsymbol{\beta} + f(X_i) + \varepsilon_i \quad (1 \leq i \leq n),$$

where $\mathbf{x}_i' = (X_i, X_i^2, X_i^3, \ldots)$ (no intercept term), $\boldsymbol{\beta}$ contains unknown parameters, and $f\colon \mathfrak{R} \to \mathfrak{R}$ is an unknown (smooth) regression function. The simultaneous estimates of $\boldsymbol{\beta}$ and $\mathbf{f} = (f(X_1), \ldots, f(X_n))'$ are

$$\hat{\boldsymbol{\beta}}_{PLR} = (\widetilde{\mathbf{X}}'\widetilde{\mathbf{X}})^{-1}\widetilde{\mathbf{X}}'\widetilde{\mathbf{y}},$$

$$\hat{\mathbf{f}}_{PLR} = \mathbf{H}_P^{(LPR)}(\mathbf{y} - \mathbf{X}_P\hat{\boldsymbol{\beta}}_{PLR}),$$

where $\mathbf{X}_P$ is the usual $\mathbf{X}$ matrix without a column of ones, $\widetilde{\mathbf{X}} = (\mathbf{I} - \mathbf{H}_P^{(ker)})\mathbf{X}_P$ and $\widetilde{\mathbf{y}} = (\mathbf{I} - \mathbf{H}_P^{(ker)})\mathbf{y}$ are the *partial* residuals for $\mathbf{X}$ and $\mathbf{y}$, respectively, after adjustment for the nonparametric component of X (via hat matrix $\mathbf{H}_P^{(ker)}$ from kernel smoothing on X), $(\mathbf{y} - \mathbf{X}_P\hat{\boldsymbol{\beta}}_{PLR})$ are the residuals from the parametric fit $\mathbf{X}_P\hat{\boldsymbol{\beta}}_{PLR}$, and $\mathbf{H}_P^{(LPR)}$ is the hat matrix from a local polynomial fit to these residuals (using weights from $\mathbf{H}_P^{(ker)}$). The vector of fits is then

$$\hat{\mathbf{y}}_{PLR} = \mathbf{X}_P\hat{\boldsymbol{\beta}}_{PLR} + \hat{\mathbf{f}}_{PLR} = \mathbf{H}^{(PLR)}\mathbf{y}, \quad \text{where}$$

$$\mathbf{H}^{(PLR)} = \mathbf{H}_P^{(LPR)} + (\mathbf{I} - \mathbf{H}_P^{(LPR)})\,\mathbf{X}_P(\widetilde{\mathbf{X}}'\widetilde{\mathbf{X}})^{-1}\widetilde{\mathbf{X}}'(\mathbf{I} - \mathbf{H}_P^{(ker)})$$

is the PLR hat matrix. Here bandwidth selection is important in smoothing the residuals, which contain the structure not captured by the parametric portion of the fit, including the lack of the intercept term.

The second semiparametric method is *Model Robust Regression* (*MRR*), developed in Mays and Birch (1996) as their *MRR2* method. The steps for this simple yet flexible procedure are as follows: (1) obtain a parametric fit to the raw data, say $\hat{\mathbf{y}}_{ols}$, (2) obtain the residuals $\mathbf{r}$ from this fit, and use a nonparametric procedure to smooth these residuals (say $\hat{\mathbf{r}} = \mathbf{H}_M^{(LLR)}\mathbf{r}$, where $\mathbf{H}_M^{(LLR)}$ is the local linear hat matrix for fitting the residuals), and (3) add back a portion of the residual fit to the original parametric fit, namely $\hat{\mathbf{y}}_{MRR} = \hat{\mathbf{y}}_{ols} + \lambda\hat{\mathbf{r}}$, where $\lambda \in [0,1]$. The mixing parameter λ increases from 0 to 1 as model misspecification increases, meaning that more of the nonparametric residual fit is needed to compensate for an insufficient parametric fit. This λ is chosen in a similar fashion as the bandwidth b for $\mathbf{H}_M^{(LLR)}$, and for MRR both λ and b contribute to smoothing considerations for obtaining a proper fit.

3. DATA-DRIVEN BANDWIDTHS
3.1 MOTIVATION AND CONCERNS

The following concerns for bandwidth choice are presented in the context of kernel regression, but extend to most other nonparametric methods. The most fundamental approach to bandwidth selection when estimating a function g is to minimize a global error criterion such as the *average squared error* (*ASE*)

$$d_A(b) = n^{-1}\sum_{i=1}^{n}\left[\hat{g}_b(x_i) - g(x_i)\right]^2 w(x_i)$$

or equivalent measure [Härdle (1990)].

Estimating ASE apart from a constant suggests minimizing the naïve estimate of prediction error

$$p(b) = n^{-1}\text{SSE} = n^{-1}\sum_{i=1}^{n}\left[y_i - \hat{g}_b(x_i)\right]^2 w(x_i),$$

92

which is biased towards choosing b too small (overfitting). This bias occurs because $p(b)$ relies too much on the individual data points, using them for both fitting and evaluating the fit. The "leave one out" criterion of cross-validation

$$\text{CV}(b) = n^{-1} \sum_{i=1}^{n} \left(y_i - \hat{y}_{i,-i} \right)^2 w(x_i) \qquad (3.1)$$

introduced by Clark (1975) addresses this problem. Here $\hat{y}_{i,-i}$ is the fit at x_i without using the point (x_i, y_i). Independent of n and w, CV(b) is the PRESS statistic

$$\text{PRESS} = \sum_{i=1}^{n} (y_i - \hat{y}_{i,-i})^2$$

[Allen (1974)]. This PRESS criterion has been observed in a wide variety of applications to overfit the data with a bandwidth too small.

A second approach to resolving the bias problem in $p(b)$ is to multiply $p(b)$ by a correction function $\Xi(n^{-1}b^{-1})$, which "penalizes" for small bandwidths. Rice (1984) and Härdle *et al* (1988) studied five prevalent forms of $\Xi(n^{-1}b^{-1})$: (*i*) Generalized Cross-validation $\Xi_{GCV}(n^{-1}b^{-1}) = [1 - n^{-1}b^{-1}K(0)]^{-2}$, (*ii*) Akaike's Information Criterion Ξ_{AIC}, (*iii*) Finite Prediction Error Ξ_{FPE}, (*iv*) Shibata's Ξ_S, and (*v*) Rice's own bandwidth selector $\Xi_T(n^{-1}b^{-1}) = [1 - 2n^{-1}b^{-1}K(0)]^{-1}$. These authors show that these penalized selectors behave differently for finite data simulations. The GCV selector appeared to perform best overall, but Rice argues that when considering squared error criteria for bandwidth selection, there is much less chance of encountering oversmoothing problems.

Alternative bandwidth selection methods include "plug-in" procedures based on the asymptotic expansion of ASE, and other asymptotic results or expressions. Arguments against the use of plug-in methods in favor of CV(b) in (3.1) or penalized functions are given in Härdle and Marron (1985) and Härdle *et al* (1988). Such asymptotic results are not deemed appropriate for the finite small sample cases in this work.

More computationally intensive procedures exist for choosing b, including bootstrapping, variable or local bandwidth selectors, and a more effective asymptotic plug-in method by Ruppert *et al* (1995). These more complex potential improvements are left for future study. Two new criteria based on concerns encountered while studying the new semiparametric approaches are now discussed.

3.2 PENALIZED PRESS CRITERIA

Einsporn and Birch (1993) developed a model-robust regression procedure called "Hatlink", which combines parametric and nonparametric fits, both to the data, in the proper proportions. In an attempt to improve upon the shortcomings of the bandwidth selectors described above, they also introduced a penalized PRESS criterion (called *PRESS**), given by

$$\text{PRESS*} = \frac{\text{PRESS}}{n - \text{tr}(\mathbf{H}^{(\text{ker})})} \, .$$

Here the denominator penalizes for small bandwidths. The idea behind PRESS* is to maintain the versatile performance of cross-validation, while also introducing extra protection against the common problem of overfitting. However, it is shown in Sections 5 and 6 that PRESS*, while correctly providing protection against small bandwidths when needed, often selects bandwidths that are much too large, especially for the model-robust procedures. Most alarming is that for the residual fits of PLR and MRR, PRESS* often chooses $b \geq 1$ (for the X's scaled between 0 and 1). For these cases, b is chosen as 1 to obtain convergence of the computer algorithm, which results in essentially fitting the mean. To resolve this problem, it is suggested that the user plot PRESS* versus bandwidth and then choose b at the point of the *first* local minimum or at the point where the PRESS* curve first starts leveling off (i.e., where the downward slope becomes significantly less). This idea follows closely the method used in ridge regression to choose the "shrinkage parameter" k [Myers (1990)]. This graphical approach involves some judgment from the user, but for most examples studied it has been clear how to choose b. Handling this technique for simulations is discussed in Section 6.

The shortcomings of PRESS* arise from its penalty for small bandwidths (i.e., for variance), but lack of protection against large bandwidths (bias). This has been found to be a common concern with the nonparametric residual fits of PLR and MRR. The proposed solution then is to add to PRESS* a penalty for large bandwidths that is comparable to the existing penalty $[n - \text{tr}(\mathbf{H})]$ for small bandwidths. Noting in PRESS* that $[n - \text{tr}(\mathbf{H}^{(\text{ker})})] \to 0$ as $b \to 0$, and $[n - \text{tr}(\mathbf{H}^{(\text{ker})})] \to n-1$ as $b \to 1$, it is desired to have the new penalty term approach 0 as $b \to 1$ and approach $n-1$ as $b \to 0$. This new penalty term is comprised of sums of squares error (SSE) terms, since it is known that SSE increases as b increases, and SSE is maximized when $b=1$ (when "fitting the mean" in kernel regression or fitting a simple linear regression line in LLR). Also, SSE $\to 0$ as $b \to 0$. Letting $\text{SSE}_{\text{mean}} = \text{SSE}$ with $b=1$, and $\text{SSE}_b = \text{SSE}$ for any candidate b, the expression $(\text{SSE}_{\text{mean}} - \text{SSE}_b)/(\text{SSE}_{\text{mean}})$ is between 0 and 1. This expression approaches 0 for $b \to 1$ and approaches 1 for $b \to 0$. Multiplying by $(n-1)$ then gives a penalty term that approaches 0 for $b \to 1$ and approaches $n-1$ for $b \to 0$, the penalty structure desired. In general, for selecting a parameter θ for any procedure with hat matrix $\mathbf{H}$, and defining SSE_{max} to be the maximum sum of squares error across all θ values, PRESS** may be defined as

$$PRESS^{**} = \frac{PRESS}{n - tr(\mathbf{H}) + (n-1)\frac{SSE_{max} - SSE_\theta}{SSE_{max}}} . \quad (3.2)$$

For kernel regression, $\mathbf{H} = \mathbf{H}^{(ker)}$, $SSE_{max} = SSE_{mean}$, and $\theta = b$. Note that $[n - tr(\mathbf{H})] \to n-2$ as $b \to 1$ for LLR, and thus one may adjust the $(n-1)$ multiplier in (3.2) accordingly. Results in this work are based on the general form of PRESS** in (3.2). As seen in Sections 5 and 6, PRESS** alleviates the concern over the occurrence of $b=1$ and usually improves the fits of PLR and MRR by choosing a slightly smaller bandwidth than PRESS* (or a slightly larger λ when bandwidths are chosen the same). One concern with PRESS** is that on rare occasions it is minimized by an inappropriately small b (e.g., .015, .03, ...). In these cases, the PRESS** curve starts out small over a narrow range of b, but then abruptly changes into an appropriate diagnostic curve. The initial small values of PRESS** are due to the instability of PRESS for small b's. The denominator in PRESS* eliminates this problem by approaching zero for small b, but the second penalty term in PRESS** prevents its denominator from going to zero in this case. Proper choice of starting values in the search routine for b (see Section 6), or plotting PRESS** versus b can prevent this choice of small b.

4. OPTIMALITY CRITERION

Mean squared error $[MSE = (bias)^2 + variance]$ of prediction is used to evaluate the bandwidth selectors in the next two sections. Mays and Birch (1996) give derivations and theoretical properties of the formulas for bias($\hat{\mathbf{y}}$) and var($\hat{\mathbf{y}}$) for each of the four fitting techniques considered here for the theoretical underlying model $\mathbf{y} = \mathbf{g}(X) + \boldsymbol{\varepsilon}$. Here any function $\mathbf{g}(X)$ may be expressed as $\mathbf{X}\boldsymbol{\beta} + \mathbf{f}(X)$, where $\mathbf{X}\boldsymbol{\beta}$ is some specified linear parametric form, and $\mathbf{f} = \mathbf{g} - \mathbf{X}\boldsymbol{\beta}$ is the remaining portion of $\mathbf{g}(X)$ (where $\mathbf{X}\boldsymbol{\beta}$ could be $\mathbf{X}_p\boldsymbol{\beta}_p$ for PLR). The bias and variance of an individual prediction $\hat{y}_o$ at any location $\mathbf{x}_o'$ within the range of the data can similarly be derived by considering the underlying model as $y_o = \mathbf{x}_o'\boldsymbol{\beta} + f(X_o) + \varepsilon$. These formulas are given in Mays and Birch (1996), along with the establishment of their accuracy via simulation studies.

These MSE formulas may be used to compute the *average MSE* (*AVEMSE*) over the specific data points, or an approximate *integrated MSE* (*INTMSE*) across the entire range of the data (using $\approx$ 500-1000 locations). AVEMSE is minimized to separately (not jointly) find the "optimal" bandwidth or mixing parameter, denoted by b_o or λ_o respectively, while INTMSE is used to evaluate the overall fits of the procedures. In deriving the MSE formulas, the bandwidth and mixing parameter for each procedure are taken to be *fixed* at these optimal b_o and λ_o, following the approach taken by Speckman (1988). AVEMSE, using only data points, is minimized instead of INTMSE because this more closely parallels the information available for data-driven selection methods, as mentioned in Härdle *et al* (1988).

5. EXAMPLES

The current work is motivated by such real data examples as the tensile data in Figure 1. This data has a strong underlying quadratic structure, but contains an interesting "peak" that deviates from this structure. The model-robust procedures PLR and MRR have been developed for situations where there is some knowledge about the parametric form of the model, but slight to moderate deviations from this model exist in the data. Mays (1995) studied the performance of the various bandwidth selectors for several generated (*small-sample*) data sets that exhibit deviations from an underlying polynomial structure. By comparing optimal fits from using using b_o and λ_o as described above, it was found that PLR and MRR showed consistently significant *potential* improvements over the individual procedures of OLS and LLR. By comparing MSE values of the data-driven fits to those of the optimal fits, Mays showed that PRESS** was successful in maintaining the potential benefits of PLR and MRR when using a data-driven selector. It was also shown that PRESS* and other criteria (in particular the GCV criterion found to work well for strictly nonparametric cases) were very inadequate selectors of the bandwidth.

One of the data sets studied came from the model

$$y = 2(X - 5.5)^2 + 5X + \gamma\left[10\sin\left(\frac{\pi(X-1)}{2.25}\right)\right] + \varepsilon \quad (5.1)$$

at ten evenly spaced X-values from 1 to 10, where $\gamma = .35$ and $\varepsilon \sim N(0,16)$. The sine function term adds a deviation from a quadratic model, mimicking the structure of the tensile data of Figure 1 (in this case giving a "dip" in the data). Other data sets included removing the error term from (5.1), taking replicates at each X location, using a cubic model to fit data from a sine curve with and without an error term, and other variations on the underlying quadratic structure with added sine deviation. These were all single data set examples (all small-sample), and now simulation results are presented for data from model (5.1).

6. SIMULATION RESULTS

The Monte Carlo simulation examples are based on model (5.1), where γ is the *misspecification parameter*, which is varied as 0, 0.25, 0.5, 0.75, and 1.0 to control the amount of deviation (misspecification) from a quadratic model. OLS should perform best for $\gamma=0$, and

LLR should perform well for larger γ. Of interest is how the model-robust methods perform in comparison across this range of misspecification. In addition to varying γ, the sample size n is also varied as $n=6$, 10, and 19. For all examples, X-values are taken at evenly spaced locations from 1 to 10. Through use of 500 simulated data sets for each combination of γ and n, Mays and Birch (1996) establish the accuracy of the MSE formulas used in Section 5 (by comparison to simulated MSE's), and they use INTMSE values to illustrate the significant potential benefits of using PLR or MRR. Here, interest is in how well the data-driven selectors PRESS* and PRESS** maintain these benefits, and if their performances are consistent with the observations of the previous section.

The b and λ values for each set of 500 runs are averaged to give the (mean) bandwidth and (mean) mixing parameter associated with each fitting technique, and these values are compared to b_o and λ_o for each example. To measure how close these data-driven fits are to the optimal fits, while also illustrating the benefits of the overall optimal fits for PLR and MRR versus OLS and LLR, the INTMSE values for the different fits are compared. Also reported is the number of times $b \geq 1$ occurred for each set of simulations. The plotting approach of visually selecting b from the PRESS* or PRESS** curve to eliminate these cases is not practical for the simulation runs. Instead, a three-point search routine is used to control for the cases of $b \geq 1$ for PRESS* or PRESS** and for b too small for PRESS**.

The results for PRESS*, given in Mays (1995), show that there is a significant problem with $b \geq 1$ for MRR and PLR, causing a loss in the benefits of these methods, as seen in significantly increased INTMSE values. However, for cases of $b_o \geq 1$ (for $\gamma=0$ and sometimes 0.25), PRESS* is appropriate in choosing $b \geq 1$. So, PRESS* seems to work fine for no (or very little) misspecification, but for the more important and practical cases of small to moderate misspecification, PRESS* consistently chooses bandwidths too large (also affecting the choice of λ).

Table 1 contains the results for PRESS** for $\gamma=0$, .5, and 1. The key result from this table is that for MRR and PLR, the INTMSE values are consistently "close" to optimal across all cases, maintaining the potential benefits of the model-robust methods. This is due primarily to the nearly eliminated problem of choosing $b \geq 1$. The cases of $b_o \geq 1$ for MRR or PLR do appear to be a problem at first glance. PRESS** selects b much smaller than one (due to the penalty for large b) and also selects a large λ value (a small λ would solve the bandwidth problem). Nonetheless, the INTMSE values are not greatly different from optimal due to the underlying parametric fits that are very adequate for these cases of no or minimal misspecification. The nonparametric fits are to residuals that contain very little structure, and litle is lost from varying b. MRR handles this a little better than PLR, as seen in the $\gamma=0$ cases. As n gets larger for these cases (still only 19, though), PRESS** selects b large enough for it to essentially behave as $b \geq 1$. PRESS** is very effective for MRR and PLR for the more practical cases of small to moderate misspecification.

Three additional observations can be made from Table 1. First, although MRR performs best in an overall sense, PLR is slightly better for the cases of $\gamma=1$ and the combinations of $n=19$ and $\gamma=.5$ and above (large misspecification). Second, for $n=19$ and $\gamma=.5$ and above, LLR performs best, as expected. However, there is no significant loss in performance when using MRR or PLR in these cases. Again, it is the entire range of misspecification that is of interest here, not just large misspecification, and it is in this respect that MRR and PLR prove beneficial. Lastly, the bandwidths chosen for LLR alone tend to remain too large for the smaller sample sizes, and the alternatives to PRESS** discussed in subsection 3.1 may be useful here.

In summary, the effective performance across all cases of MRR, with b chosen by PRESS**, has led to MRR being the recommended fitting technique whenever misspecification may be a concern [Mays and Birch (1996)]. This would provide adequate fits at either extreme of misspecification and would perform better than either OLS or LLR for any misspecifications in between. It would also eliminate the danger of incorrectly choosing between a parametric and nonparametric technique to use in a given fitting problem. These conclusions are based on b, λ, and INTMSE values directly from computer simulations, without user intervention (via the graphical aid) for the cases of $b \geq 1$ for MRR and PLR. If all inappropriate selections of $b \geq 1$ were ignored, PRESS* would still yield bandwidths that are too large, especially for larger misspecifications, but PRESS** would yield bandwidths extremely close to the optimal values for MRR and PLR for most cases. This lends even more evidence to the usefulness of MRR or PLR, with PRESS** (accompanied by the graphical aid) as the data-driven bandwidth selector.

REFERENCES

Allen, D. (1974) The Relationship Between Variable Selection and Data Augmentation and a Method for Prediction. *Technometrics*, **16**, 125-127.

Clark, R. (1975) A Calibration Curve for Radio Carbon Dates. *Antiquity*, **49**, 251-266.

Chu, C. and Marron, J. (1991) Choosing a Kernel Regression Estimator (with discussion). *Statistical Science*, **6**, 404-436.

Cleveland, W. (1979) Robust Locally Weighted Regression and Smoothing Scatter Plots. *Journal of the American Statistical Association*, **74**, 829-836.

Einsporn, R. and Birch, J. (1993) Model Robust Regression: Using Nonparametric Regression to Improve Parametric Regression Analyses. *Technical Report Number 93-5*, Dept. of Statistics, Virginia Polytechnic Institute and State University.

Fan, J. (1992) Design-adaptive Nonparametric Regression. *Journal of the American Statistical Association*, **87**, 998-1004.

Fan, J. and Gijbels, I. (1996) *Local Polynomial Modelling and Its Applications*. London: Chapman and Hall.

Gasser, T. and Müller, H. (1979) Kernel estimation of Regression Functions. In *Smoothing Techniques for Curve Estimation* (eds. Gasser and Rosenblatt). Heidelberg: Springer-Verlag.

Härdle, W. (1990) *Applied Nonparametric Regression*. New York: Cambridge University Press.

Härdle, W., Hall, P., and Marron, J. (1988) How Far are Automatically Chosen Regression Smoothing Parameters from Their Optimum? (with discussion). *Journal of the American Statistical Association*, **83**, 86-99.

Härdle, W. and Marron, J. (1985) Optimal Bandwidth Selection in Nonparametric Regression Function Estimation. *Annals of Statistics*, **13**, 1465-1481.

Hastie, T. and Loader, C. (1993) Local Regression: Automatic Kernel Carpentry (with discussion). *Statistical Science*, **8**, 120-143.

Mays, J. (1995) Model Robust Regression: Combining Parametric, Nonparametric, and Semiparametric Methods. *Ph.D. Dissertation*, Virginia Polytechnic Institute and State University

Mays, J. & Birch, J. (1996) Model Robust Regression: Combining Parametric, Nonparametric, and Semiparametric Methods. *Technical Report Number 96-7*, Dept. of Statistics, Virginia Polytechnic Institute and State University.

Montgomery, D. and Peck, E. (1992) *Introduction to Linear Regression Analysis*, 2nd edn. New York: Wiley.

Myers, R. (1990) *Classical and Modern Regression with Applications*, 2nd edn. Boston, MA: PWS-KENT.

Priestley, M. and Chao, M. (1972) Nonparametric Function Fitting. *J. Royal Stat. Soc. B*, **34**, 384-392.

Rice, J. (1984) Bandwidth Choice for Nonparametric Regression. *Annals of Statistics*, **12**, 1215-1230.

Ruppert, D., Sheather, S. J., and Wand, M. P. (1995) An Effective Bandwidth Selector for Local Linear Regression. *Journal of the American Statistical Association*, **90**, 432, 1257-1270.

Speckman, P. (1988) Kernel Smoothing in Partial Linear Models. *J. Royal Stat. Soc. B*, **50**, 413-436.

Stone, C. (1980) Optimal Rates of Convergence for Nonparametric Estimators. *Annals of Statistics*, **8**, 1348-1360.

Stone, C. (1982) Optimal Global Rates of Convergence for Nonparametric Regression. *Annals of Statistics*, **10**, 1040-1053.

Wand, M.P. and Jones, M.C. (1995) *Kernel Smoothing*. London: Chapman and Hall.

Table 1. PRESS** diagnostics for 500 Monte Carlo simulations. Optimal h_o, λ_o, and $INTMSE_o$ are in bold. The column "#b≥1" gives the number of times the bandwidth was chosen to be 1.

		OLS	LLR			MRR				PLR		
n	γ	*INTMSE*	b	#b≥1	*INTMSE*	b	λ	#b≥1	*INTMSE*	b	#b≥1	*INTMSE*
	0	6.384	.191	*0*	16.448	.145	.863	*0*	8.597	.118	*0*	10.453
		6.384	**.146**		**11.789**	**1**	**.016**		**6.384**	**1**		**6.391**
6	.5	18.028	.229	*0*	28.350	.132	.747	*0*	13.439	.126	*0*	13.669
		18.028	**.126**		**15.220**	**.140**	**.754**		**13.596**	**.140**		**13.659**
	1	52.959	.309	*0*	80.155	.125	.661	*0*	29.787	.123	*0*	23.969
		52.959	**.108**		**24.630**	**.110**	**.939**		**23.915**	**.110**		**23.408**
	0	4.105	.156	*0*	7.826	.138	.755	*14*	5.197	.123	*5*	6.592
		4.105	**.130**		**7.689**	**1**	**.016**		**4.105**	**1**		**4.110**
10	.5	14.956	.149	*0*	10.048	.133	.844	*14*	9.321	.139	*15*	9.275
		14.956	**.105**		**9.456**	**.118**	**.884**		**8.884**	**.118**		**8.867**
	1	47.509	.136	*0*	16.024	.109	.899	*5*	15.469	.116	*5*	14.854
		47.509	**.082**		**11.883**	**.083**	**1**		**11.675**	**.083**		**11.722**
	0	2.314	.128	*0*	5.079	.486	.670	*205*	2.322	.376	*110*	2.756
		2.314	**.113**		**4.622**	**1**	**.031**		**2.314**	**1**		**2.316**
19	.5	12.951	.108	*0*	5.691	.144	.908	*23*	6.724	.123	*5*	5.856
		12.951	**.089**		**5.695**	**.099**	**.996**		**5.348**	**.099**		**5.381**
	1	44.861	.082	*0*	7.089	.078	.976	*0*	7.441	.079	*0*	7.278
		44.861	**.068**		**7.089**	**.069**	**1**		**6.979**	**.070**		**7.010**

A STATISTICAL APPROACH TO CORPUS GENERATION

Ann E.M. Brodeen, Frederick S. Brundick, U.S. Army Research Laboratory
Malcolm S. Taylor, University of Maryland
Contact-Ann Brodeen, ARL, AMSRL-IS-CI, APG, MD 21005 (annb@arl.mil)

Key Words: Bootstrap, Time Series, Linguistics

1. Introduction

The impetus for this work lies in military application. The Forward Area Language Converter (FAL-Con) is a portable, field-operated, translation system designed to assist in intelligence collection. It enables an operator with no foreign language training to convert a foreign language document into an approximate English translation for an assessment of military relevance. The principal components of FALCon are an optical scanner, an optical character recognition (OCR) module, and a machine translation (MT) module. In order to assign a performance measure to the FALCon system, measures of effectiveness of the components must be developed and then aggregated into an overall measure. The focus of this paper is limited to evaluation of the OCR module.

A current procedure for determining a quantitative measure of the efficacy of an OCR product is as follows: A selection of carefully prepared source-language documents, called groundtruth, is stored in the computer; hardcopy of the same document set is then scanned into bitmap images; the OCR software partitions a gross bitmap image into homogeneous zones that are processed according to content. For zones that are identified as text, specialized scoring software then compares the OCR output against the corresponding groundtruth to produce accuracy statistics, usually including percentage agreement for both words and characters, and a confusion matrix.

A central database of groundtruth documents, accepted as a baseline, would enable the evaluation of OCR products to proceed from a common benchmark. Unfortunately, such a database does not exist, making the comparison of OCR software more difficult and any conclusions drawn more tentative. Fundamental questions regarding sample size requirements, and suitable document composition for such a database, remain to be addressed.

Collection of a corpus that is sufficient for evaluation of an OCR product is likely to remain, even in the best of circumstances, a burdensome task. Access to a sufficient number of source-language documents, representative of the document classes of interest may not be feasible and, even if obtained, the expensive and time-consuming process of preparing groundtruth remains. To address this problem, we are proposing a statistical approach to corpus generation based on a small set of source-language documents. Coincident with the statistical inquiry, substantial work involving language transliteration must be accomplished.

2. Time Series Model

Consider the passage of Serbian text shown in Figure 1. Every character—letters, punctuation marks, even inter-word spaces—is represented numerically in the computer. The set of character and numeric equivalents (the mapping) is called a codeset. For a specific language, the codeset representation may not be unique. Russian, for example, has 4 commonly used 8-bit encodings and some Asian languages even more [3]. A numeric representation of the Serbian text in Figure 1 for a particular codeset assignment is shown in Figure 2.

In Figure 2, the first 80 letters (emboldened) of the Serbian text are displayed. The vertical dashed lines mark the location of inter-word spaces, which have been removed, along with most punctuation, to facilitate our methodology. The x-axis indexes the order of occurrence of the characters in the text and the corresponding codeset values are plotted along the y-axis. If we allow a situation in which the characters are processed sequentially, then we can assign to each character an associated time epoch, and Figure 2 can be considered as a time series representation of the first 80 letters. The scale of measurement for the y-axis is nominal; an alternative codeset, if appropriate, would lead to a different graphic representation, but with no attendant loss (gain) of information.

In attempting to generate a corpus, we would like a core of authentic documents to serve as a basis from which to generate additional pseudo-documents. An analogous situation, arising in the

"Нецес остати пусто Невесиње равно,но цес бити оно сто си вазда било:расадник Српства и колевка лава!" "Србија мора постати велико радилисте и родилисте!" Ово су само два изватка из скорасњих говоранција Вука Драсковица. Анахроницна, срцепарајуца реторика, примерена политицарима осамнаестог века, представља данас најбољи пример лази-говора или језика-маске. А онај ко на себе стави маску лази-говора, пре или касније, изабраце и лаз као основни политицки принцип. Нико није изрекао толико лази, подвала и лазних доказа о Косову као г. Драсковиц и његова телевизија. Раније смо ту анахроницну реторску маску примали као некакав његов особењацки избор, као сто примамо нецији цудацки стил у одевању. Требало је за његову реторику реци оно сто је одувек и била - да је обицан киц.

Figure 1: Serbian Text

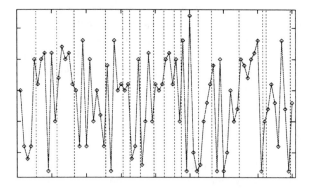

Figure 2: Time Series Representation

analysis of time series data collected as part of a clinical study, has been described and addressed using the bootstrap [1, 2].

3. Bootstrap Application

In this section we present an abridged description of the bootstrap procedure, modified for application to the textual model. Notice the time series has an inherent structure: the time series represents a block of text—it is not a random sequence. Moreover, the words themselves are subject to lexical constraints and, hence, the patterns that they assume in the codeset representation have meaning. These word patterns are, however, interrupted with great frequency; the inter-word spaces play the role of interventions in time series modeling. As a consequence, the time series has local structure contributed by the word patterns but little in the way of global structure due to the high frequency of interventions. A fundamental requirement for a bootstrap procedure applied to these data is that the fidelity of the overall structure be maintained.

Denoting the time series as a sequence of ordered pairs $(x_1, y_1), (x_2, y_2), \ldots, (x_n, y_n)$, we be-

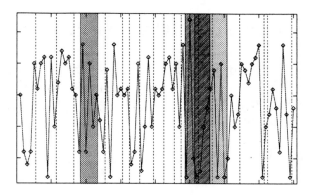

Figure 3: Intermediate Results

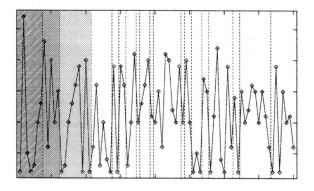

Figure 4: Bootstrapped Time Series

gin the bootstrap procedure by choosing a random location within the time series, say (x_r, y_r). Starting with (x_r, y_r), we copy the subsequence $(x_r, y_r), (x_{r+1}, y_{r+1}), \ldots, (x_{r'}, y_{r'})$ and write to an array. The length of the subsequence, $r' - r + 1$, is determined by sampling from the distribution of word-lengths found in the authentic document. A second random location, (x_s, y_s), is then determined, and a second subsequence $(x_s, y_s), (x_{s+1}, y_{s+1}), \ldots, (x_{s'}, y_{s'})$ is copied and appended to the subsequence already in the array. Figure 3 illustrates a situation where 3 subsequences have been chosen, and 2 of them have overlapped. The overlap does not create a problem since the sampling procedure is done with replacement. This process continues until terminated by a stopping rule. At that point, a bootstrapped time series such as shown in Figure 4 has been produced. The shaded regions appearing in Figure 3 are aligned in Figure 4 in order of their occurrence. Inverting the codeset mapping and applying inherent lexical modeling constraints yields the bootstrap document shown in Figure 5.

Аздабил весин абилораса еобица ра Ро
бит. И лос j" инецкир И си ада Пн ајуцаре
ма никон иње араниjесмоту су торскумаск емом
драсковицињ оно. Маљ" строфан палиозез, ијеподеф
бољиприме њеравно, еподефи ник хуман његово.
Ихуманиста цесбит примецен надозбун есоста а ацк роф
трпљ илазнихд? Ајеобиц тастроф "амаскеа о изарпре
зика ов аизватк уцутипр ст. "Бестави кадра тицар аос
обењацк есьjава еговат Срцеп азец оруцива то. Аз их
а цкипри го илазка астиље иљ р рску. Стинерас икаон
анау, ог јесм икаприме, тативел се Иљ" рикаприме,
кцијом" джун оц његовурето апре, ика? Лоцан цисесељ
ацкииз телевиз мо ткађе Тојeдав палиозези, ци Имс
стоје ву крволоцан. Имамоне аскулазиго остоје м зва"
роц инецкиратн иг ебест л вима ихехспеди овор сост
радасусеосец ицсесељев ика. Ц ребалоjеза вљад цесоста
мсилином и ијара хтеваодасе и ле јом цсе олитицари
торикац, Бити стотог.

Figure 5: Bootstrapped Text

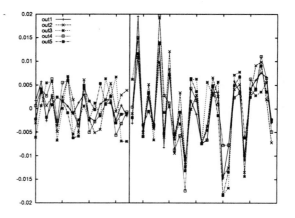

Figure 6: Frequency Differences

4. Empirical Results

The bootstrap procedure under which the document in Figure 5 was constructed[1] precludes its being "read" by an individual. Our intent, however, was to produce a document image (or character string) sufficient to assess the character recognition capability of an OCR product. If the OCR software has incorporated language-specific decision aids to support character segmentation, the bootstrap document will likely reduce the effectiveness of those procedures. Clearly, spell-checkers will not be of value. Lexical analyzers (e.g., hidden-Markov models) will likely be degraded, but not rendered completely ineffective, since substantial local structure has been retained under the moving-blocks procedure.

There is a widely accepted statistical approach to automated language identification that does not rely on identifying words of a text [4]. This approach is based on the distribution of textual n-grams.[2] While we are not interested here in language identification, we are keenly interested in producing documents that remain indistinguishable from the actual language under these identification schemes.

Toward that end, we have compared n-gram profiles of an original document against its bootstrap progeny. A typical result from such a comparison, in which the bigrams of 5 bootstrap replicates (labeled out1,...,out5) were individually compared with the bigrams of the original document, is shown in Figure 6. Bigrams whose frequency differed by less than 0.005 in absolute value between the original and all 5 bootstrap documents, $|f_{boot(i)} - f_{orig}| < .005$,

---
[1] A modified moving-blocks bootstrap.

[2] The n-grams of a text are all the character sequences of length n contained in that text. For example, *special forces* contains 14 unigrams (s,p,e,...), 13 bigrams (sp,pe,ec,...), 12 trigrams (spe,pec,eci,...), and so on.

$i = 1, ..., 5$, were not plotted. In this example, 5 percent of inner-word bigram frequencies were determined to differ by more than this amount. Those instances are plotted in the left panel of Figure 6 where it can be seen that for a given bigram, the inequality was often violated by only a single bootstrap replicate, and even then the difference was seldom in excess of 0.007.

An artifact of the moving-blocks bootstrap procedure was the creation of bigrams that did not appear in the original document. These typically arose at the "edges" of bootstrap words, involving a bigram of the form (space, character) or (character, space).[3] Those occasions in which the inequality was violated for these spurious bigrams are pictured in the right panel of Figure 6. The annexing of data whose spatial dependencies across subregion boundaries do not reflect those in the original data set is at the core of this problem, and has received research attention from several investigators [5, 6, 7]. The overall rejection rate, for inner-word and inter-word bigrams combined, and a 0.005 threshold value, was 11 percent. This gross percentage is influenced, in addition to the stringent threshold level, by the size of the documents; frequencies, $f_{(\cdot)}$, decrease with increasing document size.

Five comparably-sized Serbian documents were selected as the kernel of a more intensive investigation. Groundtruth files were created for each of the documents through keyboard entry and post-verification. Three distinct inquiries were then undertaken: First, the Serbian documents were scanned and submitted to the OCR software for segmentation; the groundtruth and OCR output files were compared for agreement using specialized scor-

---
[3] Let ␣ represent an inter-word space. The edge bigrams of an arbitrary word wxyz are then ␣w and z␣.

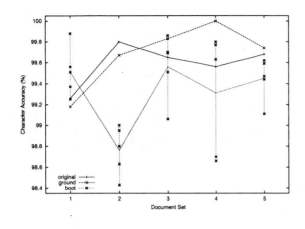

Figure 7: Character Accuracy

ing software; the character accuracy for each of the 5 documents was determined. The results, labeled **original**, are plotted in Figure 7. Next, the groundtruth files were printed. The printer output was scanned, processed by the OCR module, and compared against the groundtruth files. Those results, labeled **ground**, are again shown in Figure 7. Finally, for each of the 5 original Serbian documents, 5 bootstrap replicates were generated, twenty-five bootstrap documents in all. The bootstrap files were printed, and the hardcopy scanned and OCR'd. The bootstrap files and OCR output were compared and the average percentage agreement, labeled **boot**, is plotted in Figure 7, along with the component values.

Notice the range of percentages plotted in Figure 7—[98.4, 100]. For most practical purposes, and certainly for the purpose of our inquiry, the bootstrap documents can serve as a statistical surrogate for the authentic Serbian documents. That was all we hoped to establish.

5. Summary

A modified moving-blocks bootstrap was applied to the construction of pseudo-documents to be used for evaluation of an OCR module. The n-gram profiles of the bootstrap documents appeared to be consistent with that of the source-language document in a limited empirical study. A more extensive comparison of bootstrap and source-language documents via the OCR module produced no discernible distinction between the two classes. These results strengthen the advocacy of a statistical approach to corpus generation and encourage the implementation of more rigorous paradigms into the field of natural language processing.

References

[1] Bradley Efron and Robert J. Tibshirani, An Introduction to the Bootstrap, Monographs on Statistics and Applied Probability No. 57, Chapman & Hall, New York, 1993.

[2] R.Y. Liu and K. Singh, Moving blocks jackknife and bootstrap capture weak dependence, Exploring the Limits of Bootstrap, LePage and Billard, Ed., John Wiley, New York, 225-248, 1992.

[3] Flo Reeder and Jerry Geisler, Multi-byte Issues in Encoding/Language Identification, Workshop on Embedded MT Systems, held in conjuction with AMTA '98, 49-58, 1998.

[4] William Cavnar and John Trenkle, N-gram-based text categorization, Symposium on Document Analysis and Information Retrieval, 161-175, 1994.

[5] P. Hall, Resampling a coverage pattern, Stochastic Processes and Their Applications, 20, 231-246, 1985.

[6] P. Hall, On confidence intervals for spatial parameters estimated from nonreplicated data, Biometrics, 44, 271-277, 1988.

[7] H.R. Kunsch, The jackknife and the bootstrap for general stationary observations, Annals of Statistics, 17, 1217-1241, 1989.

ESTIMATION OF THE PROTECTION PROVIDED BY THE M291 SKIN DECONTAMINATION KIT AGAINST SOMAN AND VX

Robyn B. Lee, Robyn B. Lee & Associates, LLC, John P. Skvorak, United States Army Medical Research Institute of Chemical Defense

Robyn B. Lee, PO Box 267, Fawn Grove, PA 17321

Key Words: Dose response study; stagewise sequential design; Protective Ratio (PR); nonlinear regression; chemical warfare agents; decontamination

Introduction:

The mission of the U.S. Army Medical Research Institute of Chemical Defense (USAMRICD) includes the development and evaluation of medical countermeasures for chemical warfare agents. These countermeasures are necessary to keep the warfighter prepared for, or able to quickly recover from exposure to chemical warfare agents. These countermeasures are developed and tested at USAMRICD through fundamental and applied research in the areas of pharmacology, physiology, toxicology, pathology and biochemistry.

Countermeasures include protective equipment, pretreatment protective measures, antidote therapies and topical skin protectants.

Assessment of countermeasure efficacy is accomplished in several ways. One of those is to compare it to an existing 'standard' countermeasure or to the agent alone. Increased or prolonged survival or decreased effects of agent exposure are outcomes of interest. Testing and evaluating the efficacy of these medical countermeasures depends on the observed responses.

Dose response testing is widely used in the testing of medical countermeasures. Efficacy of test compounds is determined by estimating the LD_{50} for exposure with and without the test compound. The ratio of the test compound LD_{50} to the untreated LD_{50} of the agent alone is called the protective ratio, PR. This unitless measure determines if the test compound is effective in protecting against the agent. If the PR is greater than 1 and the respective 95% confidence interval excludes the value 1, we conclude that some measure of efficacy exists. Additionally, the PR for several test compounds against the same chemical agent may be used to rank the test compounds.

Several designs can be used in dose response studies, but estimating the LD_{50} as precisely as feasible and using as few animals as possible is a priority. Fixed sample size designs, fully sequential designs and stagewise sequential designs are three ways to approach determination of the LD_{50}.

Fixed sample or single stage designs are the simplest, where relatively large numbers of animals for relatively few doses are used. Doses are usually separated by equal log intervals. These designs, however, are relatively inefficient.

Fully sequential designs, such as the up-down method of Dixon (1965) focuses on estimating the LD_{50} using one observation at a time at a series of prespecified and equally spaced doses. If a response is observed, the next dose of compound given is lower. If there is no response, the next dose of compound is higher. The objective is to quickly converge to the LD_{50}. While this design is the most efficient in terms of expected sample size, it is dependent upon a short response time.

Stagewise sequential designs represent a compromise between fixed sample designs and fully sequential designs. The stagewise sequential design consists of dosing conducted with a minimum number of animals at several doses in several stages. The results of each stage let the investigator choose the doses to be used in the next stage of dosing to best determine the dose response curve. This design gives the investigator control over his selection of doses and decreases the number of animals used to establish the dose response curve. Similar to the fully sequential design, use of this design is advantageous when the response time is short. It is far less acceptable if the response time is much greater than 24 hours.

The purpose of this paper is to illustrate the use of the stagewise sequential design along with the protective ratio in the determination of the efficacy of the M291 skin decontamination kit after exposure to soman (GD) or VX.

Study Objectives:

The efficacy of the M291 personal skin decontamination kit has been evaluated using a rabbit model against both vesicant and organophosphorus chemical warfare agents (Joiner et al., 1987, Joiner et al., 1988, Hobson et al. 1993). The objective of this study was to evaluate the efficacy of the M291 kit against GD and VX in a guinea pig model, and was part of a larger study involving other chemicals of military interest. In this study we compare the response rate of an agent, VX or GD plus M291 skin decontamination kit with the response rate of the agent, VX or GD alone. We used a treatment compared to no treatment to assess the efficacy of the treatment itself. The LD_{50} for both the treatment and no treatment groups were estimated from their response rates and the ratio of the two gave an estimate of the treatment's efficacy.

Materials and Methods:

Animals: Male Hartley albino guinea pigs (Charles River Laboratories, Wilmington, MA) weighing 300 to 450 grams were used and maintained in accordance with the Association for Assessment and Accreditation of Laboratory Animal Care standards. The animals were quarantined and observed for evidence of disease for 5 days prior to protocol use. They were provided commercial certified guinea pig ration and tap water *ad libitum*. The guinea pig holding rooms were maintained at $21 \pm 2°C$ with $50 \pm 10\%$ relative humidity using at least 10 complete air changes per hour of 100% conditioned fresh air. They were on a 12-hour light/dark, full-spectrum lighting cycle with no twilight. The animals were individually housed in polycarbonate shoebox containers (Lab Products, Maywood, NJ) on corncob or similar bedding.

Chemicals: GD (pinacolylmethylphosphono-fluoridate) and VX (ethyl S-2-diisopropylaminoethyl-methylphosphonothiolate) were synthesized by the US Army Edgewood Chemical and Biological Center, Aberdeen Proving Ground, MD and used undiluted. The M291 skin decontamination kits (Lot # RHA91A003-033) were obtained from Rohm and Haas (Philadelphia, PA). Ketamine hydrochloride and xylazine were obtained from Henry Schein Inc. (Melville, NY), diluted in sterile water for use, and admixed for injection.

Agent Toxicity Studies: Approximately 24 hours before use in an experiment, each animal was clipped as closely and carefully as possible using a #40 blade. An area approximately $30cm^2$ was prepared on the left lateral side of the animal, and the central 3-cm x 4-cm area was marked using a permanent marker as the site for agent application. This area of application represents approximately 10% of the surface area of the animal and has been cited as an appropriate area for use in assessing dermal toxicity (Ecobichon, 1992; Myers and DePass, 1993).

On the day of exposure, the animals were examined and any animal showing abrasions or inflammation of the prepared site was excluded from the study on that day. All remaining animals were anesthetized with a combination of ketamine (32 mg/kg) and xylazine (4 mg/kg) administered as a single intramuscular (im) injection. Once the animals reached a surgical plane of anesthesia, they were placed individually under the chemical fume hood. Between 9 and 12 minutes after anesthetic administration, the agent was applied in a droplet, using an appropriately sized positive displacement pipettor (VWR Scientific, Bridgeport,

NJ), to the center of the marked area. The droplet was spread as evenly as possible within the marked area using the pipette tip. The exposed animals were placed in holding cages inside of the fume hood and allowed to recover from anesthesia. The animals were observed continuously for 4 hours, intermittently for the next 4 hours, and at 24 hours post exposure. The time of onset of local signs of anticholinesterase toxicity (local fasciculation) was recorded, as were the time and description of subsequent local and systemic signs. The animals were not decontaminated and mortality was assessed 24 hours after agent exposure.

Lethality-response curves were generated using the stagewise sequential methods described by Feder *et al*. (1991, 1991a) allocating one or two animals to each of four to five agent challenge dosage levels per stage. The animals were randomly assigned to the challenge dosage levels in each stage. In the first stage, a range of agent doses was selected to span the predicted range of lethality from 0-100%. When possible, previously determined experimental data were used to select the range of doses for the first stage. The results of the first stage analysis were used to select agent doses for the next stage. This approach allowed animals in later stages to be placed at doses that would allow better estimation of the LD_{50}. The stagewise approach reduced overall animal use by eliminating the need for conducting a separate range finding study, range finding is done in stage 1, and eliminating the chance of completely missing the lethality range of the agent.

Efficacy Studies: Animal preparation, handling and the use of the stagewise sequential design did not vary from the agent toxicity studies. On the day of exposure, animals were anesthetized with ketamine and xylazine nine minutes before exposure, and were decontaminated using the M291 kit one or three minutes post exposure. The M291 applicator pad was rubbed across the marked exposure site six times and discarded. The black resin provided an indicator of complete application to the exposure site.

Statistical Analysis: After the last stage, probit models were fitted to the combined data to determine the dose-response relationship (Finney, 1971). However, standard probit analysis computer programs cannot be used to fit probit models to these dose response data due to nonstandard dose allocations and possible stage effects. Specialized procedures based on nonlinear regression analyses have been developed to fit dose response models to these data. All models were fitted using nonlinear regression (SAS, Cary, NC) including a test for stage effects based on the studentized residuals from the fitted probit model. The estimated regression coefficients were used to determine the LD_{50} value, protective ratio (PR), and the associated

95% confidence interval (CI). The PR was determined for each agent in the efficacy study and is the ratio of the LD_{50} in the decontaminated animals divided by the LD_{50} in the nondecontaminated animals. The 95% CI for the LD_{50} and PR were computed by Fieller's (1944) method.

Results:

Results of dosing and responses by stages are shown in Table 1 for VX and Table 2 for GD. A maximum of four stages was used to estimate the LD_{50} of agents and agents with M291 skin decontamination kit. After the last stage, probit models were fitted to the combined data across all stages using nonlinear regression.

The LD_{50} (95% CI) for VX was 0.034 mg/kg (0.032, 0.037) and for VX + M291 at 3 minutes was 0.99 mg/kg (0.73, 1.32), which gives a PR (95% CI) for M291 at 3 minutes of 28.60 (23.11, 35.40). The PR for the M291 skin decontamination kit with VX was far greater than 1 and the respective CI

did not include the value 1. Therefore, the M291 skin decontamination kit was effective at increasing the LD_{50} and providing protection from VX when it was used 3 minutes after exposure.

The LD_{50} (95% CI) for GD was 9.69 mg/kg (8.00, 12.34); for GD + M291 at 3 minutes was 11.20 mg/kg (8.14, 13.39); and for GD + M291 at 1 minute was 17.17 mg/kg (14.73, 18.84). The PR (95%CI) for M291 at 3 minutes of 1.16 (0.95, 1.41) and at 1 minute of 1.77 (1.53, 2.05). The PR for the M291 skin decontamination kit with GD was slightly greater than 1 at the 3-minute time point, but the CI included the value of 1. Therefore, it was not considered effective at 3 minutes after GD. However, at 1 minute after GD, the PR of the M291 skin decontamination kit was still greater than 1 and the CI excluded the value 1. Therefore, the M291 skin decontamination kit was effective at increasing the LD_{50} and providing protection from GD when it was used 1 minute after exposure.

Table 1: Results of VX and VX + M291 3 minutes after exposure

Group	Stage 1 mg/kg response/total		Stage 2 mg/kg response/total		Stage 3 mg/kg response/total		Stage 4 mg/kg response/total	
VX	0.350	1 / 1	0.035	0 / 1	0.035	3 / 4	0.018	0 / 4
	0.525	2 / 2	0.053	2 / 2	0.053	2 / 2	0.025	0 / 1
	0.700	2 / 2	0.070	2 / 2	0.070	2 / 2	0.026	0 / 3
	1.050	2 / 2	0.140	2 / 2				
	1.400	1 / 1	0.280	1 / 1				
VX + M291 **3 minutes**	0.17	0 / 1	0.51	0 / 3	1.02	2 / 4		
	0.34	0 / 2	0.68	0 / 3	1.36	2 / 2		
	0.68	1 / 2	1.02	0 / 2	1.70	2 / 2		
	1.36	2 / 2						
	2.72	1 / 1						

Table 2: Results of GD and GD + M291, 1 and 3 minutes after exposure

Group	Stage 1 mg/kg response/total		Stage 2 mg/kg response/total		Stage 3 mg/kg response/total		Stage 4 mg/kg response/total	
GD	2.45	0 / 1	8.04	0 / 1	8.04	1 / 2		
	4.44	0 / 2	9.68	1 / 2	9.68	1 / 4		
	8.04	0 / 2	11.31	2 / 2				
	14.58	2 / 2	12.95	2 / 2				
	26.30	1 / 1	14.58	1 / 1				
GD + M291 **3 minutes**	25.0	1 / 1	10.0	2 / 2	10.0	0 / 2	10.0	0 / 2
	50.0	2 / 2	13.3	2 / 2	12.5	2 / 3	13.3	3 / 3
	75.0	2 / 2	16.6	1 / 2			16.6	3 / 3
	100.0	1 / 1	20.0	2 / 2				
GD + M291 **1 minute**	20.0	0 / 2	20.0	2 / 2	15.0	1 / 2	15.0	0 / 2
	25.0	2 / 2	22.5	4 / 4	17.5	3 / 4	17.5	1 / 2
	30.0	1 / 1	25.0	2 / 2	20.0	2 / 2	20.0	2 / 2
							22.5	1 / 1

Discussion:

The sensitivity of an experimental design depends greatly on the specific doses that are selected for inclusion. If possible, the doses should be adequately spaced to permit estimation of the dose response distribution and should bracket each of the dose response percentiles to be estimated. A design, which is spaced narrowly, is less robust to the effects of a poor initial parameter estimate than a widely spaced design. The design doses should be selected using information from previous data or toxicological judgement so the dose response curve can be estimated with accuracy.

However, there are times when this is not feasible, that is when a stagewise sequential design is advantageous to use. The stagewise sequential design with multiple doses per stage is recommended when the initial dose selections are incorrectly determined or unknown. This design also uses the previous stages to influence the dose selection of the current and future stages. Additionally, when initial dose selection is good, the stagewise sequential design converges to the LD_{50} more rapidly.

The largest number of stages and the smallest sample size per stage are recommended for greater design sensitivity and flexibility. However, total experimental time and other logistical program considerations suggest fewer stages and larger sample sizes per stage as a compromise. Allocation of sample size per stage may vary, starting with a several animals per dose and stage to a decreasing number of doses and animals per stage as the experiment approaches the LD_{50}.

Examples of these conditions are observed in this study. In Table 1, VX doses chosen for Stage 1 were much higher than the final doses used in Stage 4 for estimating the LD_{50}. The VX doses in Stage 1 did not include a nonlethal dose and subsequently three additional stages and a total of 32 animals were required to determine the LD_{50}. Similar results were observed in Table 2 for GD + M291 at 3 minutes. Initial GD doses were much higher than the resultant LD_{50}. For either of these cases, if a single stage dose response study had been used with those initial doses, the LD_{50} would not have been estimated.

In contrast, initial doses for GD adequately spanned the dose response distribution. Doses selected for subsequent stages were chosen to more precisely estimate the LD_{50}. Only two additional stages and a total of 24 animals were required to estimate the LD_{50}.

In conclusion, the advantage of the stagewise sequential design is its flexibility, flexibility in choosing doses for each stage and flexibility in the number of animals used at each dose.

References:

Dixon, W.J. (1965) The up-and-down method for small samples. *J. Am. Stat. Assoc.* 60:967-978.

Ecobichon, D.J., Ed. (1992) *The Basis of Toxicity Testing*, 1st ed. CRC Press, Boca Raton.

Feder, P.I., Olson, C.T., Hobson, D.W., Matthews, M.C., Joiner, R.L. (1991) Statistical Analysis of Dose-Response Experiments by the Maximum Likelihood analysis and Iteratively Reweighted Nonlinear Least Squares Regression Techniques. *Drug Information Journal*, Vol. 25, pp. 323-334.

Feder, P.I., Hobson, D.W., Olson, C.T., Joiner, R.L., and Matthews, M.C. (1991a) Stagewise, Adaptive Dose Allocation For Quantal Response Dose-Response Studies. *Neuroscience &Biobehavioral Reviews*, Vol. 15, pp. 109-114.

Fieller, E.C. (1944) A Fundamental Formula in the Statistics of Biological Assay, and some Applications. *Q. J. Pharm.* Vol. 17, pp. 117-123.

Finney, D.J. (1971) Probit *Analysis*, 3rd ed. Cambridge University Press, Cambridge.

Hobson, D., Blank, J.A., Menton, R. (1993) Task 89-11: Test up to 12 Skin Decontaminants. Final Report on Task Order 89-11(AD #B173995), Battelle, Columbus.

Joiner, R.L., Harroff, H.H., Kiser, R.C., And Feder, P.I. (1987) Validation of a protocol to compare the effectiveness of experimental decontaminants with the single most effective component of the M258a1 kit or with Fuller's earth against percutaneous application of undiluted organophosphate chemical surety material to the laboratory albino rabbit. Final Report on Task Order 84-2 (AD #B115308), Battelle, Columbus.

Joiner, R.L., Keys, W.B., Harroff, H.H., and Snider, T.H. (1988) Evaluation of the effectiveness of two Rohm and Haas candidate decontamination systems against percutaneous application of undiluted TGD, GD, VX, HD, and L on the laboratory albino rabbit. Final Report on Task Order 86-25 (AD #B120368), Battelle, Columbus.

Myers, R.C., and DePass, L.R. (1993) Acute Toxicity Testing by the Dermal Route. In: *Health Risk Assessment Dermal and Inhalation Exposure and Absorption of Toxicants*, 1st ed., (R.G.M. Wang, J.B. Knack, and H.I. Maibach, Eds.), pp. 167-200. CRC Press, Boca Raton.

SAS Institute, Inc. *SAS User's Guide: Statistics*, Version 6; Cary, NC: SAS Institute Inc; 1990.

USE OF GENERALIZED P-VALUES TO COMPARE TWO INDEPENDENT ESTIMATES OF TUBE-TO-TUBE VARIABILITY FOR THE M1A1 TANK

David W. Webb, Stephen A. Wilkerson, U.S. Army Research Laboratory
David W. Webb, USARL, ATTN: AMSRL-WM-BC, APG, MD 21005-5066

Key words: Generalized p-value; hypothesis testing; variance components

I. Background

For years, the U.S. Army has known that there is a tube-to-tube component of tank gun accuracy caused, in part, by the out-of-straight shape of the barrel. Watervliet Arsenal, the Army's manufacturer of the barrel for the M1A1 main battle tank, maintains strict guidelines for the manufacturing tolerances used to produce the current barrel so as to minimize the effect of tube-to-tube variability. For example, tubes are indexed from their true shape[1] to have a slight upward bend. Then, when the tube is supported like a cantilever from the recoil mechanism, this upward curvature counteracts the effect of gravity to produce the straightest tube possible. Nonetheless, tolerances on the order of 0.5 mm from true straightness still occur during the manufacturing of the 5.3-m tube. As a final inspection, a laser device records the deviations from a perfectly straight line along the launch axis, and every tube's centerline profile is stored in a database. Of equal importance, the shape of the M1A1 barrel has been shown to remain constant during the life of the gun system (Howd 1991). Despite efforts to produce identically profiled tubes, differences exist between tubes, and these differences cause variations in shot exit conditions and, hence, on-target performance.

Studies in recent years have examined how changing a gun tube's profile by either indexed rotation or solar-induced bending (Bundy 1993) affects the mean point of impact. Another program investigated the relationship between tube shape and round-to-round dispersion (Wilkerson 1995). While these studies confirmed that tube curvature is a contributing factor to variation in target impacts, they also met with limited success.

Wilkerson and Held (1998) examined centerline profiles and mean impacts for gun tubes that were fired in various programs. Using straightforward graphical techniques, they discovered a subset of tubes that had very similar profiles and that, on average, shot closer to fleet zero[2] than did a random selection of tubes. Subsequent studies revealed that the average profile of this subset of tubes was nearly identical to the average profile of the fleet. From this observation, it was conjectured that tubes having this "preferred shape" not only shoot close to fleet zero, but that the variation of mean impacts for these tubes is far smaller than that observed for the entire fleet.

To validate this suggestion, a study known as the "Uniform Centerline Test" was commissioned by the Army Research Development and Engineering Center. The test was conducted at the Army Test Center at Aberdeen Proving Ground, Maryland in the fall of 1998. Its purpose was to determine if the tube-to-tube variance for uniformly shaped tubes with a specific centerline is smaller than that observed among the overall population of tubes.

II. Test design

To make the comparison of the two estimates of tube-to-tube variation, a test plan was devised that called for a control group and a treatment group of gun tubes, each consisting of an identical number (a) of tubes. Control tubes consisted of gun tubes that were representative of the fleet, while treatment tubes were those tubes that fit the preferred shape and, hence, had nearly the same profile. For each group of gun tubes, a nested design was employed which called for each of the a tubes to fire the same number (b) of occasions per tube, with each occasion consisting of the same number (n) of rounds of 120-mm ammunition. To maximize the power of the test, a was desired to be as large as resources would permit, while b and n were kept to a minimum. After $b = 2$ and $n = 3$ were chosen, $a = 10$ was determined to be the maximum number of tubes per

---

[1] The "true shape" of a gun tube is the shape independent of gravity (see Howd 1991).

group under the resource constraints of the test sponsors.

Mathematical guidelines were developed to assist in the selection of uniform tubes comprising the treatment group. The criteria allowed some deviation from the preferred shape but held the selection to more rigid standards than the manufacturer's tolerances. Approximately 10% of the tubes sampled met these criteria.

The on-target location of each projectile was recorded as an ordered pair, representing the azimuth (X) and elevation (Y) impacts, respectively. Although the data are bivariate, it has been historically shown that for direct-fire ammunition, the two components are independent and typical analyses treat the data as two independent univariate responses. We did not stray from this precedent. Since the elevation was of greater concern to the researchers, this paper focuses on the Y component of the impacts. The azimuth data, X, were analyzed using identical methodologies.

III.   Mathematical model

For a given group of tubes, a mathematical model for the impact data is given by the equation:

$$Y_{ijk} = FZ + T_i + O_{j(i)} + R_{k(ij)},$$

where Y_{ijk} is the impact for the kth projectile fired during the jth occasion from the ith tube; FZ is the fleet zero, or the overall population mean; T_i is the effect of the ith tube for $i = 1, 2, 3, \ldots 10$; $O_{j(i)}$ is the effect of the jth occasion nested within the ith tube, for $j = 1, 2$; and $R_{k(ij)}$ is the experimental error, for $k = 1, 2, 3$.

The terms T_i, $O_{j(i)}$, and $R_{k(ij)}$ are assumed to have normal distributions with mean zero and variances σ_T^2, σ_O^2, and σ_R^2, respectively.

Therefore, we wish to test the hypothesis that the tube-to-tube variance for uniformly shaped tubes within the tolerances of a specific desired centerline profile, denoted by $\sigma_{T_U}^2$, is smaller than that observed among the overall population of tubes, denoted by $\sigma_{T_C}^2$. Written mathematically, we are testing $H_O : \sigma_{T_C}^2 \leq \sigma_{T_U}^2$ versus $H_A : \sigma_{T_C}^2 > \sigma_{T_U}^2$.

IV.   Approximate F-test approach to analysis

The balanced nature of the test design allows for simple derivation of the expected mean squares (MS) for each tube group. Method-of-moments variance component estimators follow easily and are shown in Table 1.

The estimates $\hat{\sigma}_{T_C}^2$ and $\hat{\sigma}_{T_U}^2$ have Bessel distributions, which by their complexity makes any direct comparison of these two estimates a nontrivial task. However, an approximate test of the null hypothesis can be performed with a pair of F tests.

First, we test the null hypothesis

$$H_O : \sigma_{R_C}^2 + 3\sigma_{O_C}^2 = \sigma_{R_U}^2 + 3\sigma_{O_U}^2$$

versus the alternative

$$H_A : \sigma_{R_C}^2 + 3\sigma_{O_C}^2 \neq \sigma_{R_U}^2 + 3\sigma_{O_U}^2$$

by forming the ratio $F = MS_{O_C} / MS_{O_U}$. If we reject this null hypothesis, then the test cannot proceed.

Source	Expected Mean Squares	Variance Component Estimate
Tube	$\sigma_R^2 + 3\sigma_O^2 + 6\sigma_T^2$	$\hat{\sigma}_T^2 = (MS_T - MS_O)/6$
Occasion (within Tube)	$\sigma_R^2 + 3\sigma_O^2$	$\hat{\sigma}_O^2 = (MS_O - MS_R)/3$
Error	σ_R^2	$\hat{\sigma}_R^2 = MS_R$

Table 1. Expected mean squares and method-of-moments variance component estimates for each group of tubes, where MS_T, MS_O, and MS_R are the appropriate mean squares obtainable by analysis of variance.

However, if we fail to reject the null hypothesis, then we continue by forming another F ratio, namely $F = MS_{T_C}/MS_{T_U}$, as a statistic for our one-sided hypothesis concerning the tube-to-tube variances. Although $E(MS_T)$ includes the nuisance parameters σ_O^2 and σ_R^2, their linear combination was presumably shown to be equivalent for each tube type by the first F test. Therefore, rejection of the second F test should be due to $\sigma_{T_C}^2 > \sigma_{T_U}^2$.

This two-step procedure is only an approximate test, since failure to reject the preliminary hypothesis is not a guarantee of true equality. Therefore, an exact test procedure was deemed necessary for a proper analysis of the data.

V.   Generalized P-value approach to analysis

A previous U.S. Army tank gun accuracy study using a similar two-factor nested design compared the variance components of the nested effect (Khuri et al. 1998). The comparison was made using generalized P-values, an exact-test approach introduced by Tsui and Weerahandi (1989). In comparison to classical hypothesis testing, in which the test statistics are functions of the data and the parameter(s) being tested, Tsui and Weerahandi proposed generalized test variables, which are functions of the data, the parameter(s) being tested, *and* any nuisance parameters.

The generalized test variable is denoted by $T(X; x, \theta, \eta)$, where x is the observed data vector for the random variable X, θ is the parameter of interest, and η is the nuisance parameter (scalar or vector). The generalized test variable is carefully constructed so that for all fixed values of x, the following three conditions hold:

1) the distribution of $T(X; x, \theta, \eta)$ under the null hypothesis is free of η;

2) $t = T(x; x, \theta_O, \eta)$ is free of η; and

3) for fixed η, $\Pr(T(X; x, \theta, \eta) \geq t)$ is nondecreasing in θ.

Instead of the critical region used in classical hypothesis testing, a generalized extreme region,

$C_x(\theta, \eta)$, is used whose domain includes the nuisance parameter(s). This region is defined to be

$$C_x(\theta, \eta) = \{X : T(X; x, \theta, \eta) \geq T(x; x, \theta, \eta)\}.$$

Finally, the generalized P-value is given as:

$$\begin{aligned} p &= \Pr(X \in C_x \mid \theta = \theta_O) \\ &= \Pr(T(X; x, \theta_O, \eta) \geq t) \end{aligned}$$

For the uniform profile study, comparisons were desired for the variance components associated with gun tubes. With the assistance of the Department of Mathematics and Statistics at the University of Maryland Baltimore County (Mathew 1998), a generalized test variable meeting the three conditions was developed for such a comparison. The test variable, T, is given by

$$\begin{aligned}
&T(X; x, \sigma_{T_C}^2, \sigma_{T_U}^2, \sigma_{O_C}^2, \sigma_{O_U}^2, \sigma_{R_C}^2, \sigma_{R_U}^2) \\
&= \left\{ \left(\frac{6\sigma_{T_C}^2 + 3\sigma_{O_U}^2 + \sigma_{R_U}^2}{6\sigma_{T_U}^2 + 3\sigma_{O_U}^2 + \sigma_{R_U}^2} \right) \right. \\
&\quad \times (6\sigma_{T_U}^2 + 3\sigma_{O_U}^2 + \sigma_{R_U}^2) {}^{ss_{T_U}}\!\!\big/\!{SS_{T_U}} \\
&\quad \left. + (3\sigma_{O_C}^2 + \sigma_{R_C}^2) {}^{ss_{O_C}}\!\!\big/\!{SS_{O_C}} \right\} \\
&\div \left\{ (6\sigma_{T_C}^2 + 3\sigma_{O_C}^2 + \sigma_{R_C}^2) {}^{ss_{T_C}}\!\!\big/\!{SS_{T_C}} \right. \\
&\quad \left. + (3\sigma_{O_U}^2 + \sigma_{R_U}^2) {}^{ss_{O_U}}\!\!\big/\!{SS_{O_U}} \right\} \\
&= \left\{ \theta \, (6\sigma_{T_U}^2 + 3\sigma_{O_U}^2 + \sigma_{R_U}^2) {}^{ss_{T_U}}\!\!\big/\!{SS_{T_U}} \right. \\
&\quad \left. + (3\sigma_{O_C}^2 + \sigma_{R_C}^2) {}^{ss_{O_C}}\!\!\big/\!{SS_{O_C}} \right\} \\
&\div \left\{ (6\sigma_{T_C}^2 + 3\sigma_{O_C}^2 + \sigma_{R_C}^2) {}^{ss_{T_C}}\!\!\big/\!{SS_{T_C}} \right. \\
&\quad \left. + (3\sigma_{O_U}^2 + \sigma_{R_U}^2) {}^{ss_{O_U}}\!\!\big/\!{SS_{O_U}} \right\}
\end{aligned}$$

where x is the observed data vector for the random variable X, SS terms are the random variables for the sums-of-squares, and ss terms are the observed values of SS taken directly from an ANOVA table.

Although at first the generalized test variable may appear to be rather awkward to use, under the null hypothesis $\theta = 1$, so that T simplifies to

$$
\begin{aligned}
&T\left(X; x, \sigma_{T_C}^2, \sigma_{T_U}^2, \sigma_{O_C}^2, \sigma_{O_U}^2, \sigma_{R_C}^2, \sigma_{R_U}^2 \mid H_O\right) \\[6pt]
&= \left\{ ss_{T_U} \frac{(6\sigma_{T_U}^2 + 3\sigma_{O_U}^2 + \sigma_{R_U}^2)}{SS_{T_U}} \right. \\[6pt]
&\quad \left. + ss_{O_C} \frac{(3\sigma_{O_C}^2 + \sigma_{R_C}^2)}{SS_{O_C}} \right\} \\[6pt]
&\div \left\{ ss_{T_C} \frac{(6\sigma_{T_C}^2 + 3\sigma_{O_C}^2 + \sigma_{R_C}^2)}{SS_{T_C}} \right. \\[6pt]
&\quad \left. + ss_{O_U} \frac{(3\sigma_{O_U}^2 + \sigma_{R_U}^2)}{SS_{O_U}} \right\} \\[6pt]
&= \frac{ss_{T_U}/\lambda_1 + ss_{O_C}/\lambda_2}{ss_{T_C}/\lambda_3 + ss_{O_U}/\lambda_4},
\end{aligned}
$$

where λ_1, λ_2, λ_3, and λ_4 are chi-square random variables with 9, 10, 9, and 10 degrees of freedom, respectively. The sums of squares terms are taken directly from the independent analyses of variance for each type of gun tube.

Furthermore, the value of $t = T(x; x, \theta_O, \eta)$ is

$$
\begin{aligned}
&T(x; x, \sigma_{T_C}^2, \sigma_{T_U}^2, \sigma_{O_C}^2, \sigma_{O_U}^2, \sigma_{R_C}^2, \sigma_{R_U}^2) \\[6pt]
&= \left\{ (6\sigma_{T_U}^2 + 3\sigma_{O_U}^2 + \sigma_{R_U}^2) \frac{ss_{T_U}}{ss_{T_U}} \right. \\[6pt]
&\quad \left. + (3\sigma_{O_C}^2 + \sigma_{R_C}^2) \frac{ss_{O_C}}{ss_{O_C}} \right\} \\[6pt]
&\div \left\{ (6\sigma_{T_C}^2 + 3\sigma_{O_C}^2 + \sigma_{R_C}^2) \frac{ss_{T_C}}{ss_{T_C}} \right. \\[6pt]
&\quad \left. + (3\sigma_{O_U}^2 + \sigma_{R_U}^2) \frac{ss_{O_U}}{ss_{O_U}} \right\} \\[6pt]
&= 1.
\end{aligned}
$$

Therefore, the generalized P-value is simply

$$
p = \Pr\left(\frac{ss_{T_U}/\lambda_1 + ss_{O_C}/\lambda_2}{ss_{T_C}/\lambda_3 + ss_{O_U}/\lambda_4} \geq 1 \right) = \Pr(\varphi \geq 1),
$$

and may be obtained by randomly generating values of the random variable φ, then noting the frequency with which these values exceed or equal unity. As in traditional hypothesis testing, small generalized P-values support rejection of the null hypothesis in favor of the alternate hypothesis.

VI.   Results

Using the approximate F test, we first tested $H_O: \sigma_{R_C}^2 + 3\sigma_{O_C}^2 = \sigma_{R_U}^2 + 3\sigma_{O_U}^2$ against the alternative $H_A: \sigma_{R_C}^2 + 3\sigma_{O_C}^2 \neq \sigma_{R_U}^2 + 3\sigma_{O_U}^2$ with the statistic $F = MS_{O_C}/MS_{O_U}$, which has 10 degrees of freedom in both the numerator and the denominator. The resultant P-value of 0.192 did not allow us to conclude that the two $E(MS_O)$ terms differ, and we proceeded to the approximate test of $H_O: \sigma_{T_C}^2 \leq \sigma_{T_U}^2$ versus $H_A: \sigma_{T_C}^2 > \sigma_{T_U}^2$. Using $F = MS_{T_C}/MS_{T_U}$ with 9 degrees of freedom in both the numerator and the denominator, we obtained a P-value of 0.113, providing only weak evidence of a reduction in the tube-to-tube dispersion.

Using the generalized P-value approach,

$$
P = \Pr\left(\frac{ss_{T_U}/\lambda_1 + ss_{O_C}/\lambda_2}{ss_{T_C}/\lambda_3 + ss_{O_U}/\lambda_4} \geq 1 \right)
$$

provides a P-value estimate for $H_O: \sigma_{T_C}^2 \leq \sigma_{T_U}^2$ versus $H_A: \sigma_{T_C}^2 > \sigma_{T_U}^2$. Using MATLAB, 10,000 random values of the test variable were generated to provide a generalized P-value estimate of 0.210. Therefore, by this exact procedure, we infer that tubes having the uniformly shaped centerlines (within the tolerance window of this test) do not possess significantly smaller tube-to-tube dispersion than the overall population of gun tubes.

VII.   Conclusion

The reduction of tube-to-tube variation is an important element in the improvement of tank gun accuracy. The enormous costs of live-fire testing force strict constraints on the resources available for any such study. Before any shots were fired in the uniform centerline test, power studies using Monte Carlo simulation of the approximate F procedure indicated about a 35% chance of failing to detect a specified, conservative decrease in tube-to-tube variance. Engineers and test managers accepted this risk, optimistic that it would be successful. Despite the failure to statistically conclude that the tube-to-tube variation was smaller among the uniform tubes, engineers were pleased with the results of the study, as the method-of-moments variance component estimates indicated an appreciable improvement in tube-to-tube variation.

The ubiquity of nuisance parameters in variance components has always presented challenges to the analysts of Army live-fire test data. Often the presence of these parameters is ignored, leading to incorrect estimation of key parameters and/or flawed paired comparisons. With the emergence of generalized hypothesis tests as a way to make comparisons between variance components in the presence of nuisance parameters, it is hoped that the Army can make use of this new technology to improve the performance of its weapon systems.

VIII.   References

Bundy, M. "Gun Barrel Cooling and Thermal Droop Modeling." ARL-TR-189, U.S. Army Research Laboratory, Aberdeen Proving Ground, MD, August 1993.

Howd, C. "In-Process Straightness Measurements of Gun Tubes." WVA-QA-9101, Watervliet Arsenal, Watervliet, NY, August 1991.

Khuri, A., T. Mathew, and B. Sinha. *Statistical Tests for Mixed Linear Models*. Wiley, New York, 1998.

Mathew, T. Unpublished notes. November 1998.

Tsui, K., and S. Weerahandi. "Generalized P-Values in Significance Testing of Hypotheses in the Presence of Nuisance Parameters." *Journal of the American Statistical Association*, Vol. 84, pg. 602-607, 1989.

Wilkerson, S. "Possible Effects of Gun Tube Straightness on Dispersion." ARL-TR-767, U.S. Army Research Laboratory, Aberdeen Proving Ground, MD, June 1995.

Wilkerson, S., and B. Held. "Mean Jump Effects Due to Launch Dynamics Between Training and Kinetic Energy (KE) Cartridges." ARL-TR-1645, U.S. Army Research Laboratory, Aberdeen Proving Ground, MD, March 1998.

Analysis and Classification of Multivariate Critical Decision Events: Cognitive Engineering of the Military Decision-Making Process During the Crusader Concept Experimentation Program (CEP) 3

Dr. Jock Grynovicki, Mr. Michael Golden, Dr. Dennis Leedom,
Mr. Kragg Kysor, Dr. Tom Cook, Dr. Madeline Swann
U.S. Army Research Laboratory
Human Research & Engineering Directorate

**Key Words: Decision Making,
Multidimensional Scaling**

The U.S. Army Research Laboratory (ARL) has undertaken a 5-year research program aimed at better understanding the distributed, non-linear decision-making process at the brigade level and above, as it is shaped by time, stress, team structure, staff experience, the environment and the introduction of digitization technology. Critical decision events were quantified using response data from key battle staff decision makers during the Crusader (howitzer) Concept Experimentation Program 3 (CEP 3) based on an ARL structured instrument called the "Decision Maker Self Report Profile (DMSRP)" (Golden, Grynovicki, & Kysor, 1999). The DMSRP is a data collection instrument designed to facilitate recording key data elements related to decisions made by commanders and staff officers during U.S. Army experiments and exercises. The DMSRP was not used during the planning phase but focused on the execution phase of battle.

The DMSRP was designed, in part, as the data collection complement to a cognitive engineering model of the decision-making process. This model's framework is based on a model known as the "Execution Decision Cycle," which was developed in early 1998 as a major component of a project titled "Cognitive Engineering of the Human-Computer interface for Army Battle Command Systems (ABCS)." The model incorporates recent theories of cognitive science and organizational psychology and presents a process depiction of how the commander engages in a variety of cognitive activities associated with executing combat operations. The "Execution

Decision Cycle" model is described by Leedom, et al. (June 1998). The purpose of the present paper is to present a theoretically based approach to the analysis and classification of multivariate critical decision cognitive processes as recorded in the DMSRP instrument. Key findings from the multivariate analysis of the DMSRP application during the Crusader Concept Experimentation Program 3 (CEP 3) will be summarized.

Method

Procedures--Experiments to support the CEP 3 were conducted at Fort Hood, Texas, from 14 September to 16 October 1998. Some of these experiments consisted of a series of soldier-in-the-loop, interrelated simulation-supported studies designed to evaluate Crusader operational concepts. Participating battle staffs consisted of the Headquarters 3rd Brigade Combat Team (3 BCT), the battle staff and elements of a Field Artillery FA battalion. The CEP 3 supported battle staff training of mission-essential tasks including performance feedback for its after action reviews (AARs). The training also helped participating units prepare for advanced collective training at the National Training Center (NTC). The participating battle staffs prepared operations orders (OPORDs) and the field artillery support plans (FASPs), based on approved Training and Doctrine Command (TRADOC) scenarios and division level orders.

In support of the CEP, the ARL assembled a data collection team from its Cognitive Engineering Research Program to systematically assess the battle command decision-making process during the execution phase of operations. Using a structured data collection instrument named the DMSRP, the ARL team observed and quantified critical elements of this process during the execution monitoring and adjustment of combat operations by the battalion command group. These observations were supplemented by in-depth interviews conducted with the battalion command group during daily "hot wash" AARs. The goals of this data collection were to identify (1) the types of information processes used by the command group, (2) the types of individual and collective mental activities and structures involved in the command group's "sense making" process, (3) the translation of the commander's intent and concept of operation into specific directives and battle adjustments, and (4) the methods that the commander used with his staff and battlefield operating system (BOS) to maintain proper focus of attention and awareness. In addition, an analysis of the insights from observations and the AARs was used to improve the DMSRP data collection instrument for future use.

Materials

Scenarios

Tactical scenarios, consisting of three separate phases were conducted three times each week for three consecutive weeks. The three phases consisted of (1) brigade movement to contact, (2) brigade defense in sector, and (3) division attack with the brigade offensive operations within the security zone. The three phases were conducted sequentially with one phase per day for a total of nine trials over the 3-week period.

Survey Questionnaire

The DMSRP was comprised of 15 major components and one section for providing additional comments in narrative form. Four of the major components had subsets which provided additional details for the data set. The major components, sub-sets, and the amplifying section appear in the DMSRP in the sequence shown in Table 1.

Table 1

Outline of data item sequence in the DMSRP

Item	Description
1.	Location of the decision maker (TOC/Echelon)
2.	General critical decision event description
3.	Length of time for decision event
4.	Decision: Significant or Minor COA change
5.	Part or aspect of OPORD changed
6.	If significant COA change
	a. Principal Causes for COA change
7.	If minor COA change
	a. Principal Causes for COA change
8.	Process associated with decision making
	a. One immediately obvious response
	b. One option considered, mental simulation
	c. Multiple options considered, mental simulation
	d. Formal Option Generation by Staff (single/multiple)
	e. Manage the situation
9.	Triggering features or patterns in current situation
10.	Source of features or patterns
11.	Type of uncertainty experienced
12.	Uncertainty coping strategies
13.	Patterns of commander-staff interaction
14.	Decision maker's cognitive workload estimate
15.	Information processing activities (24 separate activities)

The purpose of each DMSRP component was to document the cognitive processes and associated explicit and tacit knowledge used by the decision maker at critical decision points in the mission. Additional goals were to collect critical decision data from decision makers located at different echelons (brigade, battalion, company) to develop insights regarding (1) the degree of mental model consistency or commonality among the key staff members and (2) the variability of the decision maker's information requirements over time, echelon, and battlefield situations.

Analytical Methodology--A major cognitive decision-making program requirement was to develop a theoretical approach to the analysis and classification of the DMSRP-based multivariate critical decision-related cognitive processes. The data were collected during the execution phase of battle in a tactical operation center (TOC) under time pressure, uncertainty, and complexity, in a constantly changing environment as depicted in Orasanu and Connolly (1993). The variables are subjective as are their units of measure and the values of these units. A key to our analysis approach was the use of multidimensional scaling to help visualize the cognitive processes associated with the human decision maker as a complex adaptive system. Psychologically, one can view the multidimensional space as a graphic depiction of the commander's mental model (Converse & Kahler, 1992). The vectors of the information element space reflect similarities and dissimilarities in the data. We analyzed the commanders' and battle staff members' multivariate critical decision event patterns and preferences using a weighted multidimensional scaling (WMDS) method and then confirmed these perceived patterns using discriminant analysis.

Weighted Multidimensional Scaling (WMDS)--In mapping the ARL cognitive model of command decision making, the DMSRP instrument attempts to chart the characteristics and concepts representing the cognitive processes associated with execution phase military decision making. The challenge of analyzing these complex processes was addressed through the use of multidimensional scaling techniques where objective scales of ranked order attributes, as reported by the subject decision makers, were constructed. These scales were subsequently mapped back to two-dimensional characteristics associated with the decision maker such as experience (i.e., "novice" or "expert"). The weighted Euclidean distance $D_{ijk} = [\sum w_{ka} (x_{ia} - x_{ja})^2]^{1/2}$, was used to account for individual differences in cognitive processes associated with critical event decision-making. The S-stress is used as a measure of fit S-stress $= [1/m \Sigma_k (\|E_k\| / \|T_k\|]^{1/2}$ in which $\|E\|$ is the sum of all squared elements of the error matrix defined as $\|E\| = \| T_k - D^2 \|$. S-stress can assume a value between 0 and 1. Smaller values represent a better quantification of the attributes.

Results

DMSRP forms for 24 critical decision events were collected during the experiment. The following section provides a general overview of the analysis results for the data items comprising the DMSRP (see Table 1).

Decision Type, Time Window, and OPORD Changes

Regarding whether the decision was rated a significant or a minor change or adjustment of the current implemented course of action, 25% of the critical decisions were considered by the decision maker to be significant course of action (COA) changes while 75% were considered minor COA changes. The average time to make a decision was 39 minutes with a median time of 40 minutes for significant decisions and 12 minutes for minor decisions. The time range for significant and minor decisions did overlap. Most of the changes in the COA (54.5%) relative to the current operation order (OPORD) were related to "Friendly Scheme of Maneuvers" and "Fire Support." Overall, because there were two decision types (significant vs. minor COA change) and only a small sample size of data, no clear pattern in the type of change (significant vs. minor) could be identified.

Principal Causes for Adjustment in COA

This section of the DMSRP framed the decision by identifying the principal causes for the change or adjustment in the current implemented COA. It allowed the analyst to identify which elements of information available to the commander were critical in shaping his decision. We asked the decision maker to select the three principal causes for the decision to change the current COA and then rank order them with respect to each other. Using WMDS and discriminant analysis, we found significant differences between the expert decision makers and the novice decision makers, in the causes for the decision to change or adjust the course of action. The differences in the information elements' vector space reflected which items were considered important by each group. Specific clusters of information elements can be seen to represent concepts or information categories closely associated with the individual's experience level. Figure 1 depicts significant differences in the strategy and concepts used by the experts versus the novices. (S-stress = 0.17, RSq = 0.92, Wilks' Lambda=0.71, sig.=0.032.) The experienced decision makers (i.e., commanders) focused first on the change in the enemy's force projections, second on the blue force projections, and finally, on the tactics being implemented. The novice decision makers (i.e., commanders) focused on the blue force projection, the perception of the future plan, and the change in the reading or interpretation of the cues or patterns used for situational recognition.

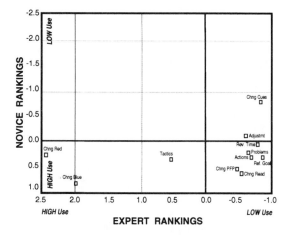

Figure 1. Principal causes for adjustments in COAs.

Process Associated With Decision Making

About half the decision-makers (52%), felt that the critical event situation suggested "one immediately obvious response, course of action change or adjustment." Forty-three percent of the respondents stated that "one response, course of action, or adjustment was not immediately obvious." Of these ten decision cases, 70% felt that multiple options were considered and evaluated sequentially using explanatory reasoning and storytelling based on a group oriented assessment. Finally, in only one case, the staff decided to manage the situation because of uncertainty, considered a formal option by directing the staff to generate new options, or used explanatory reasoning to consider one option.

Triggering Features or Patterns in the Current Situation and Principal Sources for Monitoring

The triggering features or patterns used by the commander and his staff to trigger one immediately obvious response varied by individual, echelon, and situation, with no quantifiable pattern. Triggers were cited such as weapon effectiveness, disposition of enemy and friendly forces, terrain, enemy tempo, and combat support. WDMS and discriminant analysis indicated that no significant difference was identified between the expert decision makers and novice decision makers regarding a difference in the principal sources of information (features or patterns) used to comprehend the current situation and trigger the decision event. (S-stress=0.36, RSq=0.46, Wilks' Lambda=0.75, sig.=0.51.) The expert decision makers primarily used reconnaissance and paper maps to comprehend the current situation. They also relied on the tactical net and information from the modular semi-automated forces simulation computer (MODSAF) to monitor the current situation. Other strategies were also used. The novice decision makers used numerous sources of information to help see the feature cues, indicators, or patterns in the current situation as seen in Figure 2.

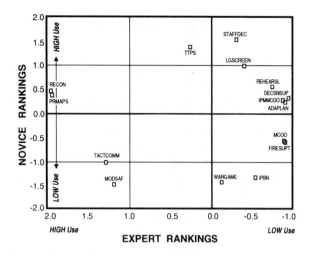

Figure 2. Multidimensional scaling of sources used by expert and novice decision makers to comprehend the current situation.

Types of Uncertainty Experienced and Coping Strategies Used

The majority (72%) of the staff experienced uncertainty because of incomplete information about the situation. Another 20% claimed to have an incomplete understanding of the situation even with complete information. Undifferentiated alternatives and confusion because of many meanings or interpretations accounted for the remaining reasons for uncertainty. To cope with uncertainty, the staff collected more information to reduce uncertainty (32%), made assumptions to deal with uncertainty (26%), weighted pros and cons (18%), formed understanding using plausible reasoning (11%), and forestalled (10%). No trend differences regarding the types of uncertainty experienced and coping strategies used could be determined between novice and experienced decision makers.

Patterns of Commander-Staff Interaction

The DSMRP results indicate that for this sample, the decision maker did not make decisions in isolation. Instead, the decision maker and his staff members performed as a well-formed team throughout the entire decision event a little more than a third (39%) of the time. Twenty-six percent of the time, the decision maker first set the general decision framework and then allowed the

staff to complete the details. In the remaining cases (21%), the staff was hierarchically directed by the decision maker to provide specific input, and then the decision maker integrated his information to make the final decision.

Cognitive Workload Estimate

"Mental Demand" versus "Effort" Ratings. Most of the decision events were rated by the decision makers as "low" or "moderate" for mental demand (see Table 2). One interesting trend was that the expert decision makers (i.e., colonels) rated the mental demand for the decisions they made as being "moderate," "high," or "very high." On the other hand, the novice decision makers (i.e., captains-majors) rated their mental demand at lower scale levels. However, while the expert commanders felt that their decisions were more mentally demanding, they regarded the process as involving less effort than did the novice commanders (see Table 3).

Table 2

Mental Demand as a Function of Military Experience

Novice=CPT-MAJ (18)   Expert=LTC-COL (6)

Mental Demand	Experience	
	Novice	Expert
Very Low	5%	0%
Low	40%	0%
Moderate	30%	25%
High	25%	50%
Very High	0%	25%

Table 3

Effort as a Function of Military Experience

Novice=CPT-MAJ (18)   Expert=LTC-COL (6)

Effort	Experience	
	Novice	Expert
Very Low	10%	25%
Low	25%	25%
Moderate	35%	50%
High	25%	0%
Very High	5%	0%

Information Processing Activities

The utility of WMDS coupled with discriminant analysis was effective in quantifying the information processing activities conducted by the expert versus novice decision makers. WMDS indicated that a significant difference existed between these two groups in the type of information processing activities used. (S-stress=0.15, RSq=0.96, Wilks' Lambda=0.64, sig.=0.029.) The more expert decision makers (i.e., commanders) monitored specific objects or events that served as a qualitative indicator of a broader activity or trend within the battle space. They developed mission goals and priorities to achieve the desired battlefield end state. They interacted with their staff to reaffirm their decisions. In contrast, battlefield visualization, rule heuristics, and monitoring were the primary information processing activities used by the novice decision makers. As shown in Figure 3, the less experienced commanders focused on clarifying or prioritizing operational objectives, critical cue tracking, and relying on battle staff group assessment to monitor and collect information for assessment and decision making. They mentally applied doctrinal rules to interpret and evaluate a choice or option.

Summary

The DMSRP instrument along with the WMDS and discriminant analysis, has proved to be a useful tool in providing structured, diagnostic insights into the complex military decision making process.

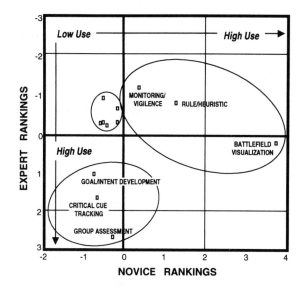

Figure 3. Information processing (IP) activities as a function of expert versus novice decision-maker rankings analyzed by a multidimensional scaling method.

References

(1) Converse, S.A. & Kahler, S.E. (1992). *Knowledge acquisition and the measurement of shared mental models.* (Contract #DAAL03-86-D-0001). Orlando, FL: Naval Air Warfare Center Training Systems Division.

(2) Golden, M., Cook, T., Grynovicki, J., & Kysor, K. (in press). *Decision Maker Self Report Profile.* ARL Technical Report.

(3) Leedom, D., Grynovicki, J., Pierce, L., Golden, M., Murphy, J., & Adelman, L. (1998). *Insights from Battle Command Reengineering II, CEP #1701: Cognitive Engineering of the Digital Battlefield*

(4) Orasanu, J. & Connolly, T. (1993). The reinvention of decision making. In G.A. Klein, J. Orasanu, R.Calderwood & C.E. Zsambok (Eds.), *Decision making in action: Models and methods,* 3-20. Norwood, NJ: ABLEX.

AN APPLICATION OF STATISTICS
TO ALGORITHM DEVELOPMENT IN IMAGE ANALYSIS

Barry A. Bodt, Philip J. David, David B. Hillis, U.S. Army Research Laboratory
Barry A. Bodt, U.S. Army Research Laboratory, ATTN: AMSRL-IS-CD, APG, MD 21005-5067

KEY WORDS: Infrared Sensors, Target Detection, Experimental Design

1. INTRODUCTION

Infrared (IR) images taken by a Quantum Well Infrared Photodetector (QWIP) camera were gathered and processed by engineers to support the development of an algorithm for airborne target detection. This paper addresses the statistical aspects of this effort. Digitized images consisted of background clutter alone and background clutter with a small aircraft target. We employed a template-matching technique to detect the small aircraft amidst the background clutter. We propose a reasonable statistic for use in our algorithm. Our primary focus is to describe the experimental design and analysis used in determining parameter settings for the detection algorithm and to report those results.

Multiple tapes of IR imagery were collected. Two aircraft flew over the same area at different altitudes. A Cessna aircraft was used as a surrogate target vehicle. The Cessna's path crossed with the observing aircraft, flying at a greater altitude with its IR sensor pointing straight down. Each time they crossed, the Cessna's flight was recorded by the continuous QWIP imagery. A single frame of the IR images appears as Figure 1. One hundred fifty-two images were selected from these

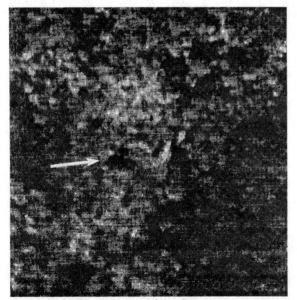

Figure 1. QWIP IR Image of the Cessna

tapes, showing the Cessna against a wide variety of natural and man-made background clutter. Fifty additional images without the Cessna were also selected. All images were digitized with 512×512 pixels, 8-bit grayscale values ranging from white (255) to black (0).

Our goal was to develop a reliable algorithm for automatic detection of the Cessna.

2. TEMPLATE MATCHING

Before introducing the complete algorithm, we briefly discuss template matching, a key component of that algorithm. Template matching, as the name suggests, involves overlaying a template on the image and testing to see if the image and the template match. A match is said to occur if there is sufficient agreement between the edges of a target template and the edges on the image. Since the locations of the template edges are known, the interior and exterior of the template are completely defined. If the template and target were exactly aligned, we would expect pixel grayscale values on the interior of the template to be small (darker) and relatively homogeneous. Pixel grayscale values exterior to the template (in the background) would tend to be larger (lighter) and less homogeneous. Consider an edge of the aligned template and target. The difference across this edge between an exterior pixel value and a neighboring interior pixel value would tend to greatly exceed zero. It is this property of an edge that serves as the basis for our decision regarding whether the image and template match.

We wish to parlay this property into a reasonable algorithm to detect the presence of the Cessna in our image. A formal hypothesis test is not developed. Rather, we offer some justification for one of the test statistics tried and then concentrate on determining optimum parameter values for our algorithm based on a learning sample of images.

The statistic arises as follows. Thirty-eight ordered pairs are formed in fixed relation to the template, with each pair consisting of an exterior point and a neighboring interior point. (Pairs are numbered consecutively around the template.) Let E_j and I_j denote the grayscale values of the jth exterior and interior points, respectively. Ignoring any dependence, consider the pixels in two adjacent ordered pairs as samples of size two from the interior and exterior regions. Equation (1) can loosely be thought of in terms of rationale for analysis of variance in that we are attempting to compare variation across groups with variation within groups. We take several liberties. The last term represents within variability, based only on the pixels internal to the template. If we replace absolute value for the traditional square, then $|I_j - I_{j+1}|$ is the sum of absolute deviations from the sample mean $(I_j + I_{j+1})/2$. Doubling this value accounts for the contribution to within variation made by the external pixels under the assumption that the variation is similar. For between variation, consider that $(E_j + E_{j+1})/2$ is an estimate for the external mean and that $(E_j + E_{j+1} + I_j + I_{j+1})/4$ is an estimate for the grand mean. The sum of squares between is analogous to four times the absolute difference between these two terms, leaving $|(E_j + E_{j+1}) - (I_j + I_{j+1})|$. Finally, owing to the interest in a one-sided alternative with external values being expected to exceed internal values, the absolute value is dropped. We are left with a heuristically justified difference (1), which will assume large values when the two pairs each straddle a template edge that is aligned with the target.

$$(E_j - I_j) + (E_{j+1} - I_{j+1}) - 2\,|I_j - I_{j+1}| \qquad (1)$$

Two related statistics considered are given as equations (2) and (3). Equation (2) is reasoned along the lines of (1), but where within variation is computed separately interior and exterior to the template. Equation (3) makes just one comparison across a potential edge and uses the neighboring $j+1^{st}$ internal pixel for a within variation estimate. The multiplication by two adjusts its scale to that of equations (1) and (2).

$$(E_j - I_j) + (E_{j+1} - I_{j+1}) - (|E_j - E_{j+1}| + |I_j - I_{j+1}|) \quad (2)$$

$$2(E_j - I_j - |I_j - I_{j+1}|) \qquad (3)$$

3. PROPOSED ALGORITHM

The previous discussion suggests why proposed statistics might reveal an edge. To complete the test of template and target alignment, however, we must look at the entire template boundary as a function of scale and orientation. For fixed scale and orientation, the template boundary evaluation is supported by $\{[(E_j, I_j), (E_{j+1}, I_{j+1})]: j=1, 38\}$, with the first point being reused for j=39. Thirty-eight individual tests, based on (1)–(3), record a response if the statistic exceeds some threshold level τ. For the edges of the template to be said to match the target, the collective responses for edges around the target must exceed a second threshold, T. These two parameters, τ and T, determine

whether a proposed template aligns with the target, that is, whether the target is detected.

Equally important is determining what templates should be proposed. An image of 512×512 pixels is searched for a subset that collectively yields edges consistent with the target. The subset is a collection of 38 pixel pairs with fixed relative location to a "prospective" center point for the target. For the template to match the target, both must have approximately the same angular orientation in the view considered. Thus, many orientations must be tried to achieve coverage of possible orientations. Additionally, the template must be adjusted for varying ranges. A target overlain by a template with the true center point and the correct orientation but very different scale will likely not be detected. Thus, a sufficiently fine resolution of scales must also be considered. Considering, for example, each pixel as a potential center point, 36 orientations, and 8 scales yields a set of 75,497,472 possible templates to be proposed.

To reduce the number of computations, the following preprocessing is done. At what we will refer to as level 1, points will only be considered as potential template centers if their grayscale value ≤ 5. This is reasonable since the target center shows up nearly black on the images we have. For each of these center points, a sparse sample of orientation $\times$ range is taken. Azimuths are taken every 30° instead of every 10°. Two, not eight, scales are tested. Only seven ordered pairs about the template are considered for an edge test. For each center point passing this reduced edge test, the eight adjacent neighbors are also noted. (Some multiple counting would be expected.) Thus, the preprocessing would result in a set of potential center points, along with their nearest neighbors, that had passed the grayscale screen and reduced edge test screen for some orientation and scale.

The proposed algorithm detects a target if a pixel passing through preprocessing for a specific orientation and scale results in more than T edge tests about the template exceeding the threshold τ.

4. STRATEGY FOR OPTIMIZATION

A response surface approach was used to fix the four parameters of the model (τ, T, number of scales, number of orientations) and to determine which of the three test statistics performed best. Two responses are considered: the maximum probability of detection while maintaining zero false alarms (Max_P(D)) and the area under a Receiver Operator Curve

(ROC_Area). For each setting of the algorithm parameters, two sets of images were evaluated, those containing a target and those not containing a target. Consider an orientation, scale, and edge test threshold τ. The ROC arises by recording the proportion of images whose best pixel passes at least 0, 1, 2, ... , 38 edge tests. All images, with or without targets, will pass zero edge tests and earn a proportion 1. See Figure 2. Few images with targets and, ideally, no images

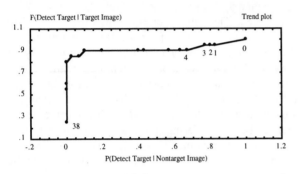

Figure 2. A Promising ROC

without targets would pass all 38 edge tests to earn proportions of approximately zero. The goal in examining the ROC is to determine the number of edge tests (0 – 38) that make the algorithm sufficiently sensitive to detect a target without many false alarms. An area under the curve of near one is indicative of such a test.

In our computer experimentation, we examine the impact of orientation, scale, and τ for each of three proposed statistics using the ROC_Area and Max_P(D), with an eye toward optimizing those responses. The threshold T for the number of edge tests required to conclude detection is determined from examination of near optimum ROCs.

5. LEARNING SAMPLE EXPERIMENTS

Twenty images containing targets were selected. Portions of these images with no target present were used as non-target images (whole image less a 60×60 pixel area about the target). Ten additional pure clutter images were also used. Thus, the target images were considered to number 20 and the non-target images 30 for the training testing.

The computer experiments for the initial test region were run with the factors and levels given in Table 1. The threshold τ is the amount a chosen edge test statistic must exceed to conclude an edge is present. The orientation is the number of steps taken over 360° for template positioning. For example, 24

steps correspond to 15°. Stepping is performed relative to a random start orientation. Scale indicates the number of uniformly spaced divisions over the interval of target ranges considered. The statistic refers to the expressions introduced in equations 1–3. A complete factorial was run, using four- and five-way interactions as error.

Table 1. Initial Test Region Factor Levels

Factor	Levels
τ	75, 90, 105, 120, 135
Orientation	24, 12, 8
Scale	2, 4, 6
Statistic	1, 2, 3

Analysis of variance indicated that only the first three factors of Table 1 were significant. See the main effects plot in Figure 3. Interestingly, the statistic used

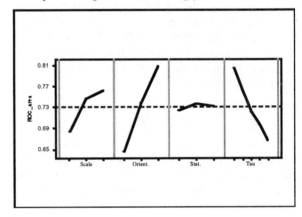

Figure 3. Main Effects Plot With Response ROC_Area

was not significant. No two- or three-way interactions were found to be significant. See Figure 4. This

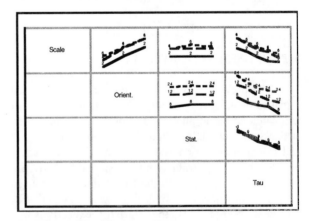

Figure 4. Interaction Plot With Response ROC_Area

suggests that the algorithm parameters could be adjusted independently. A regression model of

ROC_Area as a function of only the three significant main effects explained 79.2% of the variation and satisfied the usual regression analysis assumptions. Figures 5 and 6 show the best and worst ROC over the factor levels tested. Figure 5 corresponds to resolution on orientation of 24, resolution for scale of 6, and τ = 75, whereas Figure 6 is based on values of 8, 2, and 135 for orientation, scale, and τ, respectively.

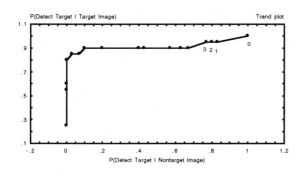

Figure 5. Best ROC

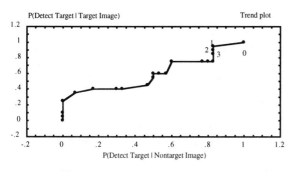

Figure 6. Worst ROC

Similar results were seen using MaxP(D) as the response. The same three factors are significant and a well-behaved regression model explains 75.5% of the variation.

Both responses showed room for improvement. The analysis suggested that each might be moved closer to optimum by (1) increasing the resolution of scale, (2) increasing the resolution of orientation, and (3) lowering the threshold for detection of an edge.

Learning Set Analysis Phase II

A second group of experiments was run based on what we learned in the first phase. Table 2 lists the factors and levels used in this second set of tests. The three statistics (1) – (3) were all retained for this test.

Table 2. Phase II Test Region Factor Levels

Factor	Levels
τ	40, 50, 60, 70, 80
Orientation	72, 36, 24
Scale	6, 8, 10
Statistic	1, 2, 3

The analysis proceeded similarly to that of Phase I. Again, only the three main effects drove the model. No significant interactions were present. Figure 7 shows the main effects. From this graph we can see

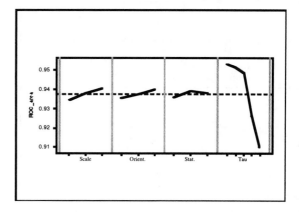

Figure 7. Main Effects Plot

that the ROC_Area range in which the testing is done is much higher than in the first phase of testing, all means showing areas in excess of 0.91. We also see that the influence of τ seems to level off as the ROC_Area nears 0.95. Values of τ less than or equal to 60 yield comparable results. A best fitting regression model does include a square term for τ, so the apparent curvature is probably real. The explained variation over this more narrow range of responses is 85.8%.

Boxplots incorporating the data from Phase I and Phase II were constructed to show the distribution of response measures as a function of the three significant factors. See Figures 8–10. We wished to determine the minimum scale and orientation and maximum τ that could be used to provide a near optimal ROC. Algorithm speed was a concern. More scales and orientations improve detection at the cost of computational efficiency. A larger differential across an edge was required for τ so that we could be sure that the apparent edge was not just a difference in clutter values. From examination of these three figures, we conclude that there should be 8 resolutions of scale, 36 orientations, and a value of about 60 for τ. A refinement for orientation was made after looking at specific plots of individual values. In our judgement, 24 orientations would suffice.

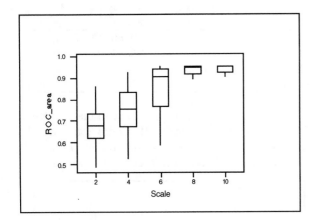

Figure 8. Boxplots of ROC_Area by Scale

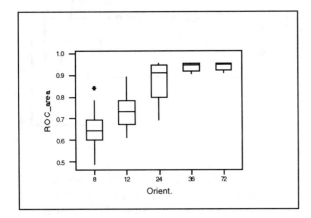

Figure 9. Boxplots of ROC_Area by Orientation

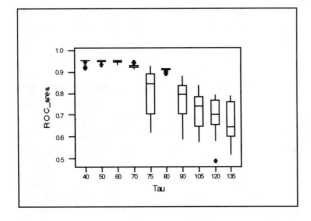

Figure 10. Boxplots of ROC_Area by τ

The parameter for our algorithm not yet established is the threshold for the number of edge tests out of the 38 that must be passed for a detection to be said to occur. Our approach to finding this threshold was to examine the ROCs for which the ROC_Area was very high and consistent with our optimal settings for the other parameters. For each of these we recorded

the number of edge tests passed corresponding to the Max_P(D). There was not great variability, all being 34, 35, or 36. We chose to set the threshold for the number of edge tests passed at 35.

5. TEST SAMPLE EXPERIMENT

A final test was run to confirm the choice of parameters for our model: $\tau = 60$, Orientation = 24, Scale = 8, and edge test threshold = 35. We chose to test one level to either side of the optimum. The threshold values of 34, 35, and 36 were considered for each of the 27 experimental combinations from a factorial crossing of the factors in Table 3.

Table 3. Test Phase Factor Levels

Factor	Levels
τ	50, 60, 70
Orientation	36, 24, 18
Scale	6, 8, 10

One hundred thirty-two images containing targets were selected for the test phase. Portions of these images with no target present were used as non-target images (whole image less a 60×60 pixel area about the target). An additional 40 pure clutter images were also used. Thus the target images were considered to number 132 and the non-target images 172 for the test phase.

Figure 11 shows the results of the 27 trials formed from the factor crossings of Table 3. Results are given in terms of the two responses, ROC_Area and Max_P(D). As expected, the chosen parameter settings produced results in the midrange for each response over the 27 trials for the larger test data set. All of the

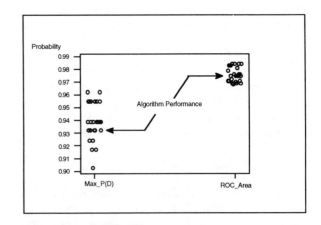

Figure 11. Algorithm Performance

runs over the test data set are included in Table 4. It is clear from scanning this table that all experimentation is in a region showing good results for both response measures. At this point, we turn the results back to the engineers who must make the judgement call regarding the tradeoff between probability of detection and computational efficiency.

Table 4. Summary Algorithm Performance

Max P(D)	ROC Area	Scale	Orient.	τ	Thresh
0.962	0.984	6	36	50	36
0.955	0.976	6	36	60	35
0.939	0.971	6	36	70	34
0.955	0.984	6	24	50	36
0.932	0.976	6	24	60	35
0.917	0.969	6	24	70	35
0.939	0.981	6	18	50	36
0.939	0.976	6	18	60	35
0.917	0.968	6	18	70	34
0.939	0.984	8	36	50	37
0.939	0.975	8	36	60	35
0.932	0.971	8	36	70	35
0.955	0.983	8	24	50	36
0.932	0.975	8	24	60	35
0.932	0.972	8	24	70	34
0.955	0.984	8	18	50	36
0.939	0.975	8	18	60	35
0.902	0.969	8	18	70	34
0.962	0.984	10	36	50	36
0.939	0.976	10	36	60	35
0.924	0.970	10	36	70	35
0.955	0.983	10	24	50	36
0.939	0.975	10	24	60	35
0.917	0.969	10	24	70	35
0.955	0.979	10	18	50	36
0.932	0.975	10	18	60	35
0.924	0.971	10	18	70	34

6. DISCUSSION

In this paper, we used a mix of engineering practice and statistics to try to arrive at an algorithm for airborne target detection. Admittedly, the approach was somewhat informal from a statistical perspective—but it seemingly produced good results. Extension to other clutter backgrounds would require additional testing.

We did consider more traditional approaches, specifically a paired-t and Wilcoxon matched-pairs, signed rank test. In each instance, we must overlook the spatial correlation present among neighboring interior and neighboring exterior points. Threshold T is analogous to the significance level of the test.

Threshold τ is analogous to a general mean difference to be exceeded. Setting these values, with respect to our optimization goals, requires reasoning along the lines employed in this paper. With assumptions for these procedures in question and with no real advantage to the general modeling approach, we settled on the approach presented here.

7. ACKNOWLEDGEMENT

The authors would like to thank the Naval Research Laboratory for providing the data and this opportunity to explore target detection algorithms.

8. REFERENCES

Baras J.S. and MacEnany D.C. (1992) "Model Based ATR Algorithms Based on Reduced Target Models, Learning and Probing," Proceedings of the Second Automatic Target Recognizer Systems and Technology Conference, 1992.

Bienenstock, D., Geman, D., Geman, S. and McClure D.E. (1990) "Phase II Technical Report, Development of Laser Radar ATR Algorithms," Contract No. DAAL02-89-C-0081, Center for Night Vision and Electro-Optics, Ft. Belvoir, VA.

Nguyen D.M. (1990) "An Iterative Technique for Target Detection And Segmentation in IR Imaging Systems," Center for Night Vision and Electro-Optics, Ft. Belvoir, VA.

Severson W.E. (1996), "The SSV Vehicle's FLIR Target Detection Capability," Lockheed Martin Astronautics, Contract No. DASG60-95-C-0062, U.S. Army Space & Strategic Defense Command.

Problems of Establishing a Baseline
For the Global Command and Control System (GCCS)

Robert Anthony, Institute for Defense Analyses, ranthony@ida.org
Samir Soneji, Columbia University, sss70@columbia.edu

Key Words: Nonnormal Distribution, Interaction, Missing Data, ANOVA

This paper addresses the problems of establishing a performance time baseline for the Global Command and Control System. The goal is to establish a standard of performance for future comparisons. The problem is that there is significant interaction, the underlying error distribution is nonnormal, and the error variance is large. First, GCCS will be described briefly, the data will be explored visually, modeling will be performed, and finally the inadequacies will be discussed.

GCCS is the national command and control system for the US Military. It is essentially a large computer network that provides a common operating picture and keeps track of data, imagery, intelligence, status of force, and planning information. GCCS is a distributed system using commercial servers, desktop computers, and Internet protocol networks. In 1996, it replaced the World Wide Military Command and Control System (WWMCCS), a mainframe system with workstations.

The users and developers of the distributed system GCCS want to establish a performance baseline similar to that established for the mainframe WWMCCS. Typically a baseline is used to measure progress in development. There are also uses in the field to compare GCCS performance at a site with performance in the laboratory. A baseline should also be able to realistically assess the capabilities of the system such as performance time and the load on computers. The existence of a useful baseline rests on a critical assumption about the variation across trials: either its constant over time, site, etc, or the variance is significantly less than the measured values.

A total of 48 performance functions were tested at six military organizations. Twelve functions were measured once at five or more of the six sites. Of these there were 8 missing measurements, giving a total of 64 response times. Table 1 shows the response times.

Table 1. Functional Response Times (seconds)

Function	Organizations						Mean	Standard Deviation
	ACOM	CENTCOM	FORSCOM	JITC	SOCOM	TRANSCOM		
Display Dialog	68	95	22	70		7	52	37
Edit TPFFD screen	15	48	11	14	7		19	17
Announcement	2		86	8	3	6	21	36
Edit after error	59	24	19	66	32	35	39	19
Post F11W	261	1320	355	278	224	614	509	422
Repost OPLAN	66	5	40	72		60	49	27
Display TPRDD	27	4		5	1	15	10	11
Display MTPRDD	85	24	74	74	15	129	67	42
Display Pop Up	23	19	30	35	15	12	22	9
Merge PIDs	52		37	80	66	332	113	123
Copy OPLAN	305	368		228	64	950	383	337
Update PID	155	237	113	204		1080	358	406
Mean	93	214	79	95	47	295	138	
Standard Deviation	98	407	103	91	71	403		250

 Missing Value

There are large differences in the range of response times within each site. Central Command ranged from 2 to 1320 seconds to complete the twelve functions, while Southern Command had a much narrower range, 1 to 224 seconds. There are also differences in response time for each function. For example, Transportation Command took 7 seconds to compete the "display dialog" function, while Central Command took over 13 times as long. Similarly, Force Command completed the "announcement" function in 86 seconds which is unusually high compared to other values for that function. The wide differences in time for a particular function are unusual and may prove problematic.

There are clear differences among the sites. The fastest site, Southern Command, had a mean response time of just less than 50 seconds, which was over six times faster than the slowest site, Transportation Command. The standard deviations for both function and sites are the same order of magnitude as the means. This indicates very large variance and possibly skewed nonnormal distributions.

A box and whisker plot of response time illustrates the differences in site distribution in Figure 1. There are several response times that exceed 1.5 times the Interquartile Range (IQR) and are considered likely outliers as seen by the arrows. There are noticeable difference both in median and spread. Some of the sites, such as Southern and Atlantic Command, have tight distributions with small variances. While others such as Transportation Command are much more spread out and have large variances. Friedman's Test statistic, a nonparametric test based on the relative rank of response times by site, is 16.22, p<0.01, and confirms that there are significant site differences.

Figure 1. Distribution of Response Time by Site

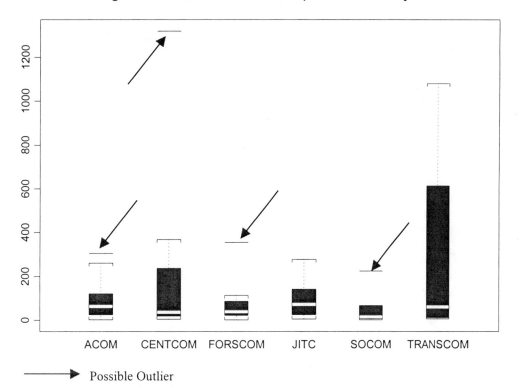

⟶ Possible Outlier

There might also be interaction between site and function. Central and Transportation Commands illustrate the potential interaction well. For the "Post F11W" function, Central Command is nearly two times as slow as Transportation Command. However this relationship does not hold with the "Update PID" function. In this case, Central Command is over four times as fast as Transportation Command.

A formal test of interaction developed by Scheffé is difficult because there is just one observation per cell and no replicates. However, the interaction may be tested if it is assumed to take the following second order polynomial form:

$$\gamma_{ij} = A + B\alpha_i + C\beta_j + G\alpha_i\beta_j + H\beta_j^2 .$$

Under conditions that the $\{e_{ij}\}$ are iid $N(0,\sigma^2)$ and $\alpha. = \beta. = 0$, the interaction term, γ_{ij}, equals $G\alpha_i\beta_j$. The null hypothesis being tested is that all $\gamma_{ij} = 0$ while the alternative is that at least one $\gamma_{ij} \neq 0$. The test statistic follows an F distribution.

$$\frac{(IJ - I - J) \times SS_G}{SS_{res}} \sim F_{1, IJ-I-J}$$

Type	Sum of Squares	F Statistic
G	894,878	50.17, p<0.01
Residual	831,430	
Error	1,726,308	

The test shows that there is significant interaction between site and function and confirms the previous intuition.

Several models were attempted to characterize performance times. Table 2 shows the form, two goodness of fit statistics: R^2 and Residual Mean Square (RMS), as well as inadequacies of the model.

Table 2. Modeling of Response Time

Form	R^2	RMS	Inadequacies
Linear: time~site+function $y_{ij}=\mu+\alpha_i+\beta_j+\varepsilon_{ij}$	0.56	177	-Skewed residual distribution with heavy right tail -26 fitted response times are negative
Linear with Interaction: time~site+function+(sitex function) $y_{ij}=\mu+\alpha_i+\beta_j+G\alpha_i\beta_j+\varepsilon_{ij}$	0.77	135	-Residual distribution with heavy right and left tails
Log Linear: ln(time)~site+function $\ln(y_{ij})=\mu'+\alpha_i'+\beta_j'+\varepsilon_{ij}$	0.77*	201*	-Residual distribution with heavy right and left tails
Two Way Scaling: time~site x function $y_{ij}=(\mu \times \alpha_I \times \beta_j)+\varepsilon_{ij}$	0.72	139	-Residual distribution with heavy right and left tails
Two Way Scaling Without Two Influential Observations: time~site x function $y_{ij}=(\mu \times \alpha_I \times \beta_j)+\varepsilon_{ij}$	0.79	55	-Skewed residual distribution with heavy right tail

* transformed back to y_{ij} scale

The reduced two-way scaling model without the two influential observations has the best goodness of fit statistics with the highest R^2 and lowest RMS. Despite this, though, the underlying distribution of residuals is clearly nonnormal as shown in Figure 2.

Figure 2. Normalized Residual Distribution of Two Way Scaling Model Without Two Influential
Observations

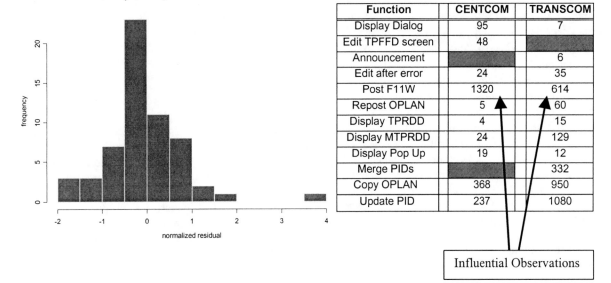

Function	CENTCOM		TRANSCOM
Display Dialog	95		7
Edit TPFFD screen	48		
Announcement			6
Edit after error	24		35
Post F11W	1320		614
Repost OPLAN	5		60
Display TPRDD	4		15
Display MTPRDD	24		129
Display Pop Up	19		12
Merge PIDs			332
Copy OPLAN	368		950
Update PID	237		1080

Influential Observations

Even when the two influential observations are removed the residual distribution is skewed. Such behavior suggests that the influence of the process, which produced the original large deviates, continues to generate a progression of lesser but still unlikely extremes. Alternatively, it could be that the underlying mathematics describing these data are not Gaussian. Either prospect renders developing and applying a baseline for GCCS a challenging task.

There are several important consequences from this analysis. First, performance times vary significantly, both among functions and more so among sites. There are also significant interaction between site and function. The persistence of large variance undermines the attempts to create a baseline. Finally, even the best fitting models attempted continue to have nonnormal error distribution and several observations were also greatly under or overestimated.

Bibliography

Anthony, Robert "The Problems of Establishing a Baseline for the Global Command and Control System", Institute for Defense Analyses Draft, Alexandria, Virginia, 1998.

Neter, John Applied Linear Regression Models, R.D. Irwin, Homewood, Illinois, 1983.

Scheffé, Henry The Analysis of Variance, Wiley Publications, New York, New York, 1959.

ROBUST TESTING OF THE COMMON MISSILE WARNING RECEIVER (CMWR)

Paul Wang, ITT Industries, Avionics Division; Charles M. Waespy, Institute for Defense Analyses

Charles M. Waespy, IDA, 1801 N. Beauregard St., Alexandria, VA 22311

KEY WORDS: Taguchi Methods, Robust Design of Experiments, Operational Testing and Evaluation

ABSTRACT: The Taguchi robust design approach is applied to the operational testing of a missile warning receiver designed for the protection of tactical aircraft against electro-optically guided threat missiles. The receiver must operate over a wide range of operational conditions and missile attack directions, and is to provide sufficient warning to allow the use of appropriate defensive countermeasures. Testing over even a small subset of conditions expected in operational usage is prohibitively costly and time consuming. A concept for robust testing based on Taguchi's robust product design principles is proposed and illustrated using hypothetical data. The approach concept represents an economical breakthrough for developing operational test and evaluation procedures for the threat warning receiver considered herein, and for the many other military systems.

I. Introduction

This paper describes an experimental design approach, based on the Taguchi robust design methods, to assess the combat effectiveness of an electro-optical missile threat warning receiver. The common missile warning receiver (CMWR) is currently being developed for installation on tactical aircraft for the detection of incoming attack missiles so that timely and appropriate countermeasures can be implemented for self-protection. This warning receiver must operate over a range of operational conditions and missile attack directions to provide a high probability of a valid threat detection and declaration (P_{VD}) with a low probability of false alarms (P_{FA}).

To predict its performance, a test and evaluation (T&E) procedure has been developed to measure whether the CMWR performance will meet the users' requirements in combat conditions. The developed T&E procedure uses two methods to assess performance: (1) computer simulations, using a large-scale CMWR computational model in conjunction with threat missile types and operational environments, and (2) flight tests, using actual missile launches against unmanned drones equipped with CMWRs. Both methods are costly and time consuming. To conduct a threat engagement simulation, it requires up to nine hours for a single missile attack. Carrying out a live firing against an unmanned drone is even more costly and time consuming. Funding limits have become a major constraint for evaluating the CMWR

performance even for a fraction of combat conditions. To conserve program resources, our objective is to develop an experimental design, based on a very limited number of tests (computer simulations and/or actual missile launches) that can quickly and accurately evaluate the CMWR performance.

Taguchi methods for robust design have been successfully used for developing commercial products. Products that can perform satisfactorily over a wide range of user conditions are termed "robust." In the development of the CMWR, since the design has been completed, the purpose of the T&E procedure is to estimate how the CMWR will perform under all users' conditions. In this paper, we propose that robust testing, as derived from the Taguchi robust design method, be implemented in the CMWR T&E process with the objective of reducing the cost and time required over a conventional T&E approach.

II. Common Missile Warning Receiver

The Common Missile Warning Receiver (CMWR) is a major component of the self-protection system for fixed- and rotary-wing tactical aircraft. It provides the detection and early warning of incoming infra-red (IR) guided surface-to-air missiles (SAMs), and activates appropriate countermeasures. In its design, an electro-optical missile sensor array detects the radiant energy from the missile propulsion system. The embedded controller processes sensed energy, separates it from the background clutter, then declares a given missile launch to be a threat when the sensed energy has accumulated to pass a preset threshold level. The time sequence of the threat engagement is depicted in Figure 1. Two key factors for a successful threat engagement are (1) a timely detection of a missile launch and (2) a timely declaration that a detected launch is indeed an enemy threat.

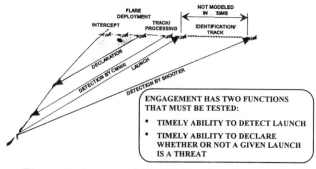

Figure 1. Common Missile Warning Receiver (CMWR) Threat Engagement

III. Conventional Test Concept for CMWR

Major elements characterizing an operational condition are illustrated in Figure 1. To cover all elements with their associated levels (as listed in Table 1), a full factorial test would require 26,244 treatment combinations of simulation runs (or live fire shots and/or a combination of each). Clearly, there is a need to develop a more efficient test strategy.

Table 1. Factors Defining Representative Attack Scenario

Factor	No. of Levels	Levels
1. Missile type	3	M-1, M-2, M-3
2. Range	3	short, medium, long
3. Altitudes	3	low, medium, high
4. Launch Aspect	3	nose, beam, tail
5. Environment	3	desert, tropical, maritime
6. Visibility	3	4, 16, 32 km
7. Ozone	3	mean, $+\sigma$, $-\sigma$
8. Velocities of targets	2	QF-4, F-16
9. Maneuvers	3	dive, climb, turns
10. Battlefield clutter	2	ordnance, fires

A. Simulation Runs

A CMWR software simulation model is currently being developed. The Synthetic Imaging Missile Simulation (SIMS) is an end-to-end software model simulating a one-versus-one (1 v 1) threat engagement. As illustrated in Figure 2, it consists of software modules of an infra-red missile, a target aircraft, a kinematics model of the missile and the aircraft, the missile warning receiver's sensing and detecting elements, and the operational environment. In running a SIMS simulation, one measure of effectiveness (MOE) is to evaluate the "elapsed time" between the time that a missile is launched and the time that it is detected or declared by the CMWR. With a predetermined set of P_{VD} and P_{FA}, the expected elapsed time predicts whether the CMWR will be able to meet a user's requirements at various combat conditions. Since it would be too costly to run all 26,244 combinations, one approach is to estimate the relationship between the input and the response from a smaller number of simulation runs; then the MOE performance (expected elapsed time) can be estimated for other scenarios which were not tested.

B. Live Fire Testing

The overall goal of the live fire testing is to verify the military effectiveness of CMWR. One intermediate objective is to use the result from a live fire test to derive a mathematical model that will be useful in predicting the missile warning time. We can also use the live fire testing as spot checks to verify correlations with simulation runs.

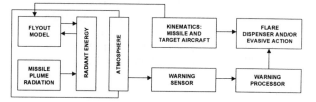

Figure 2. A Simplified SIMS Software Architecture

In planning live fire testing, there are only four factors that significantly affect missile warning: (1) the altitude of the targeted aircraft, (2) the horizontal distance (range) between the ground-based missile and the aircraft, (3) the aspect angle between the missile and the aircraft, and (4) the missile type. As a result, factors 5 to 10 listed in Table 1 are set either at a constant level or determined at the time of the actual test:

(5) Environment – determined by weather conditions at time of test

(6) Visibility – determined by weather conditions at time of test

(7) Ozone – determined by weather conditions at time of test

(8) Velocities of Target – QF-4 (drone) only

(9) Maneuvers – determined by range safety factors and instrumentation

(10) Battlefield clutter – determined by time and location of test.

The number of scenarios and their levels for conducting the live fire testing is then summarized in Table 2. For a full factorial design, 81 treatment combinations of actual missile launches would be required.

Table 2. Factors for Live Fire

Table 1 Ref. #	Factor	No. of Levels	Scale
3	Altitude	3	high, low
2	Range	3	short, medium, long
4	Launch Aspect	3	nose, beam, tail
1	Threat Types	3	missile M-1, M-2, M-3

IV. Robust Testing for CMWR

A. Robust Design

The Robust Design method, originated by Professor Genichi Taguchi in Japan, has been used extensively to improve manufacturing processes and hardware product design. It is a systematic, analytical approach to save time, reduce cost, and improve product performance. Robust design uses many ideas from the field of statistical design of experiments for obtaining useful information for making decisions to optimize a product's performance. Yet, it adds a new dimension to statistical experimental design intended to answer concerns addressed by system developers: (1) how to economically maintain the product's predicted performance in a wide variation of users' environments (2) how to ensure that decisions found to be satisfactory during testing and laboratory experiments will prove to be so in actual users' environments.

A specific Taguchi experiment is described by its product (P) diagram. The P-diagram, as illustrated in Figure 3, consists of a set of factors. The signal factor is the input information driving the Taguchi experiments. The response is the data collected during the Taguchi experiment. Noise factors are used to describe the environmental variations to be experienced by the system. Control factors are used to describe design parameters that are controllable by the system designer. In addition, an Ideal Function (relating the response variable to system and noise inputs) specifies the quality characteristic chosen for optimization.

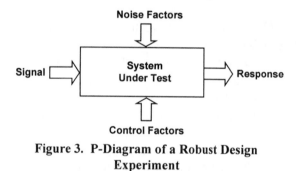

Figure 3. P-Diagram of a Robust Design Experiment

The Taguchi robust design technique conducts limited experiments designated by orthogonal arrays (OA). The OA specifies test conditions and data collection for combinations of environments (noise factors) and design parameters (control factors). By analyzing test data, the system's robustness as affected by different control factor levels can be quantified. Using this information, a best set of control factor levels is determined yielding the maximum robustness within the range of operational conditions. ITT Avionics has successfully applied the robust design method in improving its products in electronic warfare (EW) and communication and navigation (CNI) markets.

B. Robust Testing

The robust design technique described above can be modified and applied to "robust testing." In robust testing, experiments are conducted, as determined by a

designated OA, to test the system performance with a fixed set of control factor levels (representing the current design), and a selected set of noise factors levels (representing the range of operational conditions). For the CMWR, we will conduct robust testing (using SIMS) for specified treatment combinations (representing combat conditions). Results of the robust testing will then identify noise factor levels that have the most and the least influence on CMWR performance. This information will also evaluate the overall performance and provide insights into selecting treatment combinations for actual live fire testing.

C. Ideal Function and P-Diagram

The derivation or selection of an ideal function is a critical step of a Taguchi experiment. The ideal function relates the intended performance of a system to its user's expectation. Since an ideal function generally relates the energy transfer efficiency of a system, the ideal function selected for CMWR is to evaluate the efficiency between the missile plume energy collected by its sensor and the probability that the threat missile is correctly detected and validly declared (P_{VD}).

<u>Threat Warning Response Functions</u>

The warning receiver has two principal functions: to sense radiance from an attacking missile (Q_T photons/sec), and to separate it from background clutter (Q_C photons/sec). Let P_{ttf} be the probability of obtaining detection "hits" that compose a true threat file on an attacking missile. It is proportional to the number of detector hits caused by Q_T photons/sec compared to the photons/sec caused by clutter energy (Q_C).

$$P_{ttf} = K_1 * \left[\frac{Q_T}{Q_T + Q_C} \right]$$

P_{ttf} represents an aggregate of the warning receiver's design parameters such as its optics, filtering, detector array quality, thresholding, and processing to create a true track file. The factor K_1 (set equal to the reciprocal of some saturation level) is a proportionality factor needed to make $0 \le P_{ttf} \le 1$. The probability that no valid track file on the attacking missile exists within the n track files collected at time t after the launch is $(1-P_{ttf})^n$. Thus, the conditional probability of at least one valid threat track file has been established is: $P_{VF/n} = 1 - (1-P_{ttf})^n \equiv P_{VD/n}$, the conditional probability that the threat missile will be correctly detected and declared.

<u>CMWR Ideal Function</u>

Let P_n be the probability that n track files will be established at time t. Assuming a Poisson probability distribution density function (PDF) for generating files with λ being the expected number of track files per unit time generated under a given environment, then,

$$P_n = \frac{(K_2\lambda)^n}{n!} e^{-K_2\lambda t}$$

where the generating rate (λ) is proportional to the number of detector hits caused by Q_T and Q_C (photons/sec) arriving at the sensor. The proportionality factor K_2 (set equal to $[n_{max}]^{-1}$) is useful for engineering and analytical purposes.

If a threat is declared at time t with a predetermined probability of a valid declaration P_{VD}, then:

$$P_{VD} = \sum_{N=0}^{\infty}\left[1 - \left(1 - P_{ttf}\right)^n\right] * P_n$$

By substitution and summation of the infinite series, we obtain the probability of false declaration (P_{FD}):

$$P_{FD} = \left(1 - P_{VD}\right) = e^{-P_{ttf}K_2\lambda t}$$

and

$$\ln\left(1/P_{FD}\right) \equiv y = P_{ttf}K_2\lambda t$$

where $y = \ln(1/P_{FD}) = P_{ttf} K_2\lambda t$, relates P_{FD}, the probability of an attacking missile "not being detected and declared by the CMWR" with (1) P_{ttf}, the probability of establishing a true threat file on an attacking missile, a CMWR effectiveness factor, (2) λ, the expected number of track files generated under a given environment, and (3) t, the "elapsed time" before declaring a threat. At low illumination energy inputs, detection and valid declaration probability (P_{VD}) is low (1/P_{FD} high), and increases as illumination and time increase.[1] The CMWR is designed to declare a missile attack only after P_{VD} (or y) reaches a certain level. In the CMWR testing scenario, the clutter radiance, as represented by the factor λ, varies with the time of day and the test site. Since the rate of producing false track files differs greatly for different terrains (desert, maritime, or tropical), we are interested in deriving an ideal function that evaluates the elapsed time versus different terrains.

From the above equation, assuming y is set to be equal or greater than C before a missile attack is detected and declared by the CMWR, then

---

[1] An important tenet of the Taguchi process developed through experience in selecting successful product designs is that the quality characteristics inherent in the ideal function be related to the energy transfer associated with the basic mechanism of the product or the process, see PHA 1989.

$$t = [C / (P_{ttf}K_2)](1/\lambda).$$

This equation relates the expected elapsed time of the CMWR versus different terrains, as illustrated in Figure 4. In this case, we will conduct the robust testing on all three terrains individually, and will be able to evaluate the CMWR performance "dynamically" for other types of terrains.

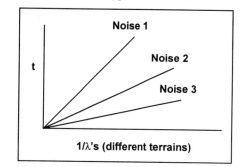

Figure 4. Ideal Function for Various Launch Sites

P-Diagram

The P-Diagram for the CMWR robust testing is illustrated in Figure 5. Since the proposed Taguchi experiment is to test and evaluate the current design, all test runs are conducted using fixed control factors levels. The noise factors represent operational conditions (e.g., terrain and environmental conditions at time and place of launch, along with missile and launch kinematics).

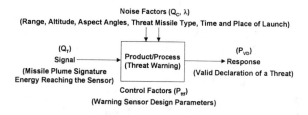

Figure 5. CMWR Robust Testing P- Diagram

D. Experimental Layout

An L18 orthogonal array is selected to host noise factors for robust testing. Since we want to test different terrain conditions individually, a separate noise array with six elements representing two randomized tests is conducted for each terrain condition. The experimental layout is illustrated in Table 3. For each of the L18 tests, the time to declare a threat is recorded under each of the six terrain conditions, and a total of 6x18 = 108 time responses (simulator yields) are collected. For each run, the computer simulation is programmed to represent desert, tropical, or maritime clutter via the track file parameter λ.

Table 3. Experimental Layout with an L₁₈ Orthogonal Array

| | Noise Factors | | | | Desert (λ_1) | | Tropical (λ_2) | | Maritime (λ_3) | |
Run #	Altitude	Range	Aspect	Msl	Test 1	Test 2	Test 1	Test 2	Test 1	Test 2
1	High	Short	Nose	M1						
2	Medium	Medium	Beam	M2						
3	Low	Long	Tail	M3						
4	High	Short	Beam	M3						
5	Medium	Medium	Tail	M1						
6	Low	Long	Nose	M2						
7	High	Medium	Nose	M2						
8	Medium	Long	Beam	M3						
9	Low	Short	Tail	M1						
10	High	Long	Tail	M2						
11	Medium	Short	Nose	M3						
12	Low	Medium	Beam	M1						
13	High	Medium	Tail	M3						
14	Medium	Long	Nose	M1						
15	Low	Short	Beam	M2						
16	High	Long	Beam	M1						
17	Medium	Short	Tail	M2						
18	Low	Medium	Nose	M3						

E. Example of a Robust Testing Experiment

Table 4 illustrates a set of hypothetical data that would be obtained from SIMS simulations. All entries are the elapsed times to a threat declaration (t) under noise conditions relative to a given value t* (i.e., t = K_3t*). A total of 18 sets of slopes and S/N's can be calculated representing the CMWR performance. When using zero proportional lines (ZPL's) to quantify the performance, the slopes of ZPL's are varied from 0.19 to 0.31 as illustrated in Figure 6. Using the ANOVA analysis on these data, the best and the worst launch conditions (noise factors) can be identified. The factor level effects plot is illustrated in Figure 7. From the analysis performed, we can make the following observations:

- The worst noise condition for the CMWR to detect and declare a missile attack is (2 2 1 1): Altitude – Medium, Range – Medium, Aspect – Nose, Missile – M1). The slope (with a confidence level of 95 percent) can be as large as 0.325.

- The best noise condition for the CMWR to detect and declare a missile attack is (1 1 3 3): Altitude – High, Range – Short, Aspect – Tail, Missile – M3). The slope (with a confidence level of 95 percent) can be as small as 0.177.

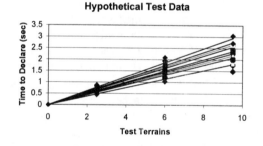

Figure 6. Robust Testing Results

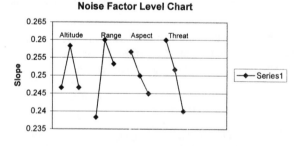

Figure 7. Factor Effects Plots

Table 4. Hypothetical Test Data

Run #	Desert (λ_1) $1/\lambda_1 = 2.5$		Tropical (λ_2) $1/\lambda_2 = 6$		Maritime (λ_3) $1/\lambda_3 = 9.5$		ZPL Time Slope	ZPL Time S/N
	Test 1	Test 2	Test 1	Test 2	Test 1	Test 2		
1	0.68	0.55	1.48	1.48	2.31	2.38	0.25	6.94dB
2	0.74	0.58	1.19	1.58	2.25	2.44	0.25	2.67dB
3	0.68	0.45	1.35	1.03	1.48	1.96	0.19	-0.24dB
4	0.58	0.71	1.25	1.22	2.34	1.75	0.22	0.69dB
5	0.71	0.68	1.58	1.41	2.28	2.41	0.26	9.04dB
6	0.67	0.55	1.64	1.51	2.44	2.08	0.25	3.81dB
7	0.65	0.7	1.16	1.21	2.68	2.7	0.26	-1.95dB
8	0.71	0.6	1.93	1.75	2.61	2.73	0.29	4.73dB
9	0.84	0.63	1.83	1.64	2.09	2.5	0.26	2.64dB
10	0.74	0.61	1.7	1.22	2.09	2.31	0.24	2.81dB
11	0.61	0.45	1.22	1.03	1.48	1.96	0.19	0.76dB
12	0.61	0.58	1.38	1.22	2.28	2.3	0.24	3.87dB
13	0.87	0.68	1.64	1.19	2.02	2.41	0.24	1.62dB
14	0.81	0.65	2.09	1.51	2.27	2.67	0.28	0.87dB
15	0.84	0.63	1.7	1.22	1.51	2.5	0.23	-3.96dB
16	0.68	0.68	1.54	1.64	2.44	2.5	0.27	9.04dB
17	0.71	0.58	1.7	1.51	2.6	2.6	0.28	4.88dB
18	0.61	0.84	1.96	1.64	2.73	3.05	0.31	2.99dB

Using these two worst noise and best noise condition slopes, we can predict that the CMWR MOE (expected elapsed time) performance for all combinations of noise conditions is between 0.442 to 0.812 seconds for a Desert terrain, 1.062 to 1.95 seconds for a Tropical terrain, and 1.682 to 3.088 seconds for a Maritime terrain. These values are useful for comparing with CMWR performance specifications.

V. APPLICATIONS

From the above description, robust testing has been demonstrated to be applicable in evaluating CMWR performance. After conducting robust testing using simulation runs, we will perform limited live fire launches to correlate their results with simulation runs. In general, three noise conditions (battlefield scenarios) – the worst, the best and one selected case – are selected for test, which represent a considerable saving in the T&E process.

VI. CONCLUSIONS

A preliminary approach based on robust testing to plan experiments for the T&E of the CMWR is described. Our objective is to show how the Taguchi robust design methods can be used to integrate engineering design with the application of orthogonal matrices developed by statisticians over the years. This methodology should lead to increased reliance on the products developed by statisticians, and improved engineering design, including test and evaluation of products and processes.

For demonstration purpose, a set of test data obtained from a hypothetical example of the CMWR robust testing is analyzed in this paper. It demonstrates the process of how the best and the worst noise conditions (battlefield scenarios) are identified, and their respective performances estimated. The overall robustness of the CMWR can also be estimated. Because of the reduced number of tests required for conducting robust testing, it could represent an economical breakthrough for developing test and evaluation procedures for many other military systems.

REFERENCES

SSW 89: Sacks, Schiller and Welch, *Design for Computer Experiments, Technometrics,* February 1989, Vol. 31, No. 1.

F&W 95: Fowlkes, W. Y., Creveling, C. M., *Engineering Methods for Robust Product Design,* Addison-Wesley Publishing Company, 1995.

PHA 89: Madhov S. Phadke, *Quality Engineering Using Robust Design,* Prentice Hall, Englewood Cliffs, New Jersey, 1989.

ANNUAL TESTING OF STRATEGIC MISSILE SYSTEMS

Arthur Fries, Institute for Defense Analyses; Robert G. Easterling, Sandia National Laboratories

Arthur Fries, IDA, 1801 N. Beauregard St., Alexandria, VA 22311

KEY WORDS: Reliability, Sample Size, Hypothesis Testing, Sequential Sampling, Bayesian Analysis

ABSTRACT: Strategic missile system planning factors include assessments of system reliability and accuracy. Annual testing is conducted to monitor system performance, especially to alert for the possibility of significant degradation. This paper focuses on the issue of potential substantial declines in reliability system reliability, and reviews methodologies for determining how many missiles should be test fired each year.

1. INTRODUCTION

Each class of missile systems within the U.S. inventory of strategic missile systems is tested annually to monitor system reliability and accuracy, and to update planning factors integral to the development of specific tactical missions. Testing comprises command post exercises of communications and procedures, laboratory experiments of individual missile components and subsystems, operationally realistic long-range firings of missiles (*sans* functional nuclear warheads), and supplementary modeling and simulations. Results from annual "surveillance tests" are synthesized to revise estimates of missile system performance, *i.e.*, to update the planning factors.

Operational missile firings are extremely expensive and the stockpile of missiles dedicated to surveillance testing is depleted annually. Thus a critical question is: "How many missiles should be test fired each year?"

As one might expect, the U.S. Department of Defense (DoD) has prescribed requirements and guidance that apply specifically to this issue. Initial versions were expressed in the language of classical statistical hypothesis testing, with separate, but similarly stated, regulations guarding against undetected major degradations in system reliability and in system accuracy. Experience has shown that the sample size calculation associated with system reliability generally prevails, essentially due to the discrete nature of the Bernoulli distribution generally ascribed to the possible success-failure outcomes.

This paper presents and discusses the three statistical methodologies that have been utilized to date to establish sample size requirements for annual test

firings of classes of U.S. missile systems equipped with nuclear warheads, based exclusively on reliability degradation concerns. Section 2 describes the original methodology, a direct consequence of the classical hypothesis testing context. The immediate Bayesian analog, and several variants thereof, are presented in Section 3. Another Bayesian perspective, a recent innovation of the U.S. Air Force Strategic Command (STRATCOM) is outlined and critiqued in Section 4. A brief summary and discussion is given in Section 5.

2. CLASSICAL METHODOLOGY

The original governing DoD requirement, mandated from 1975 through 1993, emanated from the Joint Chiefs of Staff (JCS):

"... sufficient tests are conducted to detect a WSR [weapon system reliability] decrease of [δ, classified value] or more in a year with [$1-\beta$, classified value] power of test and [α, classified value] statistical significance." (DoD, 1993)

This codification consistently was interpreted in the most simplistic way. The current best estimate value, the so-called "planning factor" (ρ), for the system reliability (R) was equated to the null hypothesis value, *i.e.*, $H_0: R = \rho$. The critical deviation δ established the null hypothesis, *i.e.*, $H_1: R \leq \rho - \delta$, and the Type I and Type II error probabilities respectively were given by α and β. With this construct, standard calculations yielded tables of sample size requirements parameterized in terms of ρ.

3. BAYESIAN METHODOLOGIES

The Bayesian analog to the classical "JCS methodology" was formally introduced in Launer and Singpurwalla (1986). Their sample size approach was successfully implemented during the Follow-On-Test Program for the U.S. Army Pershing II missile system, a tactical nuclear weapon (*i.e.*, within theater of operations) deployed in Europe.

In the Bayesian formulation, the system reliability R is a random variable (with appropriate indexing corresponding to the specific year of surveillance testing). Before the onset of pre-deployment testing, R_0 is assumed to follow some prior distribution $g(R_0|I)$, where the testers belief or "information" I is expressed in terms of specific chosen parameters for the prior distribution. If there are S_0 successes within the N_0 missile firings of the pre-deployment testing, then the posterior distribution for R_0 becomes $g(R_0|(N_0,S_0),I)$. The key logic step is that the posterior distribution incorporating the observed outcomes from any year's testing should be equated to the prior distribution for next year's testing. (System redesigns, component upgrades, procedural changes, *etc.*, can be readily accommodated by indexing I appropriately.)

For year Y of surveillance testing the prior distribution thus takes the general form

$$g_Y \equiv g(R_Y|(N_0,S_0),(N_1,S_1),\ldots,(N_{Y-1},S_{Y-1}),I).$$

Incorporation of the JCS requirement yields:

$$\int_0^1 \left\{ \sum_{j=0}^{S_Y^*} \binom{N_Y}{j} R_Y^j (1-R_Y)^{N_Y-j} \right\} g_Y dR_Y \leq \alpha,$$

$$\int_\delta^1 \left\{ \sum_{j=0}^{S_Y^*} \binom{N_Y}{j} R_{Y,\delta}^{\ j} (1-R_{Y,\delta})^{N_Y-j} \right\} g_Y dR_Y \geq 1-\beta,$$

where $R_{Y,\delta} \equiv R_Y - \delta$. Here N_Y and S_Y^* respectively are the undetermined sample size and threshold for number of observed successes. These essentially are just the classical equations, but with integration over R_Y.

Launer and Singpurwalla (1986) also briefly discussed two variants of their fundamental methodology. An alternative approach, motivated to ease computational difficulties, avoided the integration across the unknown R_Y via substitution by the mode of its prior distribution. Sample sizes computed by this alternative technique never exceed those obtained from the standard algorithm. The potential for savings in test sample sizes led to the formulation of a simplistic sequential testing variant, in which testing for a given year Y ceased as soon as $S_Y \geq S_Y^* + 1$, i.e., a sufficient number of test successes had already been observed.

4. STRATCOM METHODOLOGY

The command authority for annual missile testing of U.S. strategic missiles shifted this decade from the JCS to STRATCOM, and the previously derived JCS tables that established "required" sample sizes (recall Section 2) now merely serve as "guidance". Proposed test plans that conform to the conventional sample sizes will be approved, but deviations therefrom are acceptable. A recent STRATCOM-developed methodology, published by Gallagher *et al.* (1997), serves as the standard of comparison and review for any alternative proposed sample size.

The ensuing description of the STRATCOM methodology is our best attempt at an accurate portrayal. Difficulties we encountered include what appear to be a number of typographical errors within the published version of the Gallagher *et al.* (1997) paper (based on comparison with an earlier informally circulated draft), and fundamental differences between various interpretations of what constitutes a truly Bayesian formulation.

The STRATCOM methodology begins with the assignment of a prior distribution for R_Y (after the incorporation of any possible reliability degradation since last year's testing) that is truncated on the interval $[0,\rho]$. Previous test data is incorporated indirectly via its contribution to the computation of the planning factor ρ. The standard formulation ascribes the truncated uniform distribution:

$$U(R_Y, \rho) = 1/\rho, \text{ for } 0 \leq R_Y \leq \rho,$$

which leads to a scaled beta function representation for the posterior distribution:

$$C(\rho,N_Y,S_Y)R_Y^{S_Y}(1 - R_Y)^{N_Y - S_Y}, \text{ for } 0 \leq R_Y \leq \rho.$$

Other prior distributions are permitted, but for any specification the truncation at ρ appears to ignore the randomness associated with the estimation of the planning factor from previous data. The particular choice of $U(R_Y, \rho)$ is highly implausible in the sense that extremely low values for system reliability, *e.g.*, close to 0, are deemed to be as credible as nondegraded values, *i.e.*, close to ρ.

The second step in the STRATCOM methodology is to introduce the "loss function"

$$L(R_Y, \rho) = \rho - R_Y, \text{ for } 0 \leq R_Y \leq \rho,$$

described as the "rate of ... additional failures given a degraded reliability". This is not, however, a conventional Bayesian loss function as there is absolutely no dependence whatsoever on the new, yet to be observed, test data.

A "risk function", equated to the "expected additional weapon loss from ... undetected degrade" is then introduced. It takes the form

$$\Re(R_Y,\rho,N_Y) = L(R_Y, \rho) \, C(\rho,N_Y,E\{S_Y\}) \times$$

$$\int_0^\rho R^{E\{S_Y\}}(1 - R)^{N_Y - E\{S_Y\}} dR; \ E\{S_Y\} \equiv \rho N_Y,$$

but it does not seem to coincide with a traditional Bayesian risk function that averages over all possible outcome values.

Next the "weapons at risk function is defined as:

$$W(\rho, N_Y) = \int_0^\rho \Re(R_Y, \rho, N_Y) \, dR_Y.$$

For a given ρ, W traces out "the percent of additional weapons ... that may fail because of an undetected degrade based on a planned number of tests" (on the ordinate) as a function of prospective sample sizes N_Y (on the abscissa). Observe that, as an artifact of the choice of the truncated uniform prior U, $N_Y = 0$ corresponds to $W = \rho/2$, *i.e.*, no planned testing indicates that half of the formerly reliable missiles are now "at risk"—even if reliability to date has been consistently perfect.

The "weapons at risk" curves are similar to exponential decay patterns, and the "knee" of any such curve arguably could be construed to be a rational choice for N_Y. Such an interpretation generally would yield sample sizes that are substantially less than what is traditionally given by the original JCS methodology.

5. SUMMARY & DISCUSSION

The sample size procedures based on the classical and Bayesian interpretations of the original DoD requirement are legitimate, well-understood and readily implementable. The STRATCOM methodology, on the other hand, is lacking a meaningful statistical foundation and its utility should be questioned.

None of the presented methodologies explicitly consider or account for the other modes of surveillance testing that complement the operational missile firings (recall the discussion in Section 1). Likewise, none of the presented methodologies are sensitive to the specific sequence of success-failure test outcomes that lead to a given value of the planning factor ρ. Thus two missile systems with a common ρ will be treated identically, even though one system may have demonstrated consistent reliability performance to date, while the other system may have experienced a recent profound increase in the incidence of test failures.

Some gains in sample size efficiencies may be achievable by properly addressing these current deficiencies. Hybrid approaches involving Bayesian and adaptive sequential sampling techniques may prove to be beneficial in this endeavor. Note, however, that the adaptive sequential sampling procedures common to the statistical quality control realm do not apply directly here, as our null hypothesis value is not fixed (since ρ is updated annually).

The complexity introduced by invoking sophisticated methodological advances would entail additional demands on the statistical training of analysts, and on the communications between diverse levels of technical and managerial personnel. Whether such additional burdens would be warranted ultimately depends in great part on the degree of potential savings that are attainable. There is no *a priori* assurrance that the added efficiencies necessarily will be profound on the scale of required test missiles, but the great expense associated with a single strategic missile test firing suggests that the overall dollar savings could be substantial.

Simulation studies, modeling various hypothetical profiles of reliability degradation, should be undertaken to demonstrate the practical utility of the three methodologies described above. Similar simulation-based investigations likewise should be conducted to characterize efficiencies for any new sample size methodologies developed from future research.

ACKNOWLEDGEMENTS & DISCLAIMERS

Portions of Arthur Fries' research were undertaken at the Institute for Defense Analyses (IDA) under IDA Central Research Project C9021 and under tasks sponsored by the Office of the Director of Operational Test and Evaluation (ODOT&E) in the Office of the Secretary of Defense (OSD) within the U.S. Department of Defense (DoD). Robert Easterling's work was conducted at Sandia National Laboratories (SNL). Sandia is a multiprogram laboratory operated by Sandia Corporation, a Lockheed Martin Company, for the U.S. Department of Energy (DOE) under Contract DE-AC04-94AL8500. The authors appreciate helpful comments and suggestions offered by David Spalding, IDA.

The views expressed in this paper are solely those of the authors. No official endorsement by IDA, ODOT&E, OSD or DoD is intended or should be inferred. Likewise, no official endorsement by SNL, Lockheed Martin or DOE is intended or should be inferred.

REFERENCES

DoD (1993), *Chairman, Joint Chiefs of Staff Instruction 3231.02*, Washington, DC.

Launer, R.L. and Singpurwalla, N.D. (1986), "Monitoring the Reliability of an Arsenal Using a Combined Bayesian and Sample Theoretic Approach", in *Reliability and Quality Control*, A.P. Basu (Ed.), Elsevier Science Publishers B.V., North-Holland, pp. 245-255.

Gallagher, M.D., Weir, J.D. and True, W.D. (1997), "Relating Weapons System Test Sizes to Warfighting Capability", *Military Operations Research*, Vol. 3, pp. 5-12.

THE EFFECT OF A CHANGE IN ENVIRONMENT ON THE HAZARD RATE

Elliott Nebenzahl, Dean Fearn, Leslie Freerks, California State University, Hayward
Elliott Nebenzahl, Department of Statistics, Calif. State University, Hayward, CA 94542

Our starting point is the literature on testing a constant hazard against a change-point alternative as found in [1], [2] and [3]. There it was assumed that until some $\tau > 0$ that the lifetime random variable T follows an exponential distribution with rate a_1 and after τ, it follows an exponential with rate a_2. Most often it is assumed that the change-point τ is unknown but we assume that it is known. We extend this change-point framework to a Weibull setting and introduce a renewal parameter θ. This parameter would not be of any value in the exponential setting because of the 'lack of memory' property of the exponential. We associate the time prior to the breakpoint (or change-point) with an environment 1 and the time posterior to the breakpoint with an environment 2. Our bottom line question is whether the switch from environment 1 to 2 has an effect on the overall lifetime.

Introduction

Units (or components) exist in environment 1 for a set amount of time $\tau > 0$ and if they survive environment 1, they then move into environment 2 until they fail. Our objective is to determine whether the switch from environment 1 to environment 2 has an effect on the distribution of the overall lifetime T of the units.

For example, suppose environment 1 represents the normal environment for the units and we expect that at some known time-point τ there will be a change in the normal environment transforming it into environment 2. We model these environments in a testing area to find the contributions of these environments towards the lifetime of these units. The effect on the lifetime of the switch to the 2nd environment is particularly interesting. If it is shown that the distribution of the overall lifetime is the same whether or not the units continue in environment 1, then there is no need to expend great effort studying the units in environment 2.

Weibull–Breakpoint Model
with Renewal Parameter Theta

Suppose U is a random variable on $[0, \infty)$ with survival function $S_b(u)$ (with $S_b(u) = 1$ for $u < 0$), e.g., $S_b(u) = \exp(-u^b)$ for $u \geq 0$, i.e., a Weibull; note

that 'b' is a shape parameter and it is assumed that the hazard rate of U is increasing, thus $b \geq 1$, in the Weibull setting. The density of U is $-\dfrac{dS_b(u)}{du}$ (derivative of $S_b(u)$ with respect to u) and is denoted by $f_b(u)$; for the above Weibull, $f_b(u) = bu^{b-1} \exp(-u^b)$ for $u \geq 0$ (= 0, elsewhere).

$S_1(t) = S_b(a_1 t)$ is a survival function with shape parameter b and scale parameter a_1, with the corresponding density being $f_1(t) = a_1 f_b(a_1 t)$. $S_2(t) = S_b(a_2 t)$ is a survival function with shape parameter b and scale parameter a_2, with the corresponding density being $f_2(t) = a_2 f_b(a_2 t)$.

T is the lifetime of a unit with distribution defined by the following:

(1) $S_1(t)$ is the initial survival function of a unit in environment 1.

(2) For $\tau > 0$, a known number, if the unit survives environment 1, i.e., $T > \tau$, then it switches over to environment 2 and its conditional survival function is given by

(i)   for $0 \leq \theta \leq 1$, $\dfrac{S_2(t - \theta\tau)}{S_2(\tau - \theta\tau)}$
for $t \geq \tau$ (= 1 for $t < \tau$),

(ii)   for $1 < \theta$, $S_2(t - \theta\tau)$
for $t \geq \theta\tau$ (= 1 for $t < \theta\tau$).

We call θ a renewal parameter. The '$\theta = 0$' model represents a switch from one Weibull model to possibly another one as the unit proceeds from environment 1 to environment 2 with unit's age maintained. The '$\theta = 1$' model represents a complete renewal the unit with the 2nd Weibull, where the unit's age is reset to 0 at $t = \tau$. We note that for $b = 1$, the exponential–breakpoint (or change-point) model, $\dfrac{S_2(t - \theta\tau)}{S_2(\tau - \theta\tau)} = S_2(t - \tau)$ for $0 \leq \theta \leq 1$ and the above issue does not come up. The $1 < \theta$ case is not emphasized as it adds more complication by introducing $\theta\tau$ as a threshold parameter, since the model precludes any failures in the interval between τ and $\theta\tau$.

Thus for $0 \le \theta \le 1$ the survival function of T, $S_T(t)$ is thus given by

(1) $S_1(t)$, for $0 \le t \le \tau$,

(2) $S_1(\tau)\dfrac{S_2(t-\theta\tau)}{S_2(\tau-\theta\tau)}$, for $t > \tau$.

Of course, $S_T(t) = 1$ for $t < 0$.

The density $f_T(t)$ of T is then given by

(1) $f_1(t) = a_1 f_b(a_1 t)$, for $0 \le t \le \tau$,

(2)

$$S_1(\tau)\frac{f_2(t-\theta\tau)}{S_2(\tau-\theta\tau)} =$$

$$S_b(a_1\tau)\frac{a_2 f_b(a_2(t-\theta\tau))}{S_b(a_2(\tau-\theta\tau))},$$

for $t > \tau$.

Of course, $f_T(t) = 0$ for $t < 0$.

This model (defined by $S_T(t)$ or $f_T(t)$) is referred to as the 4 parameter model, with the 4 parameters being θ, b, a_1, a_2. For the purposes of this talk, it is assumed that we are dealing with the Weibull case, i.e. $S_b(u)$ is Weibull, as described earlier.

One can make the appropriate modification to obtain $S_T(t)$ for $1 < \theta$. In particular for $\tau < t < \theta\tau$, $S_T(t) = S_1(\tau)$ and $f_T(t) = 0$. Thus $f_T(t) = 0$ for any observed value $t > \tau$, unless $\theta \le \dfrac{t}{\tau}$.

Obtaining the MLE (Maximum Likelihood Estimators) for our 4 parameter Model

Let $t_{(1)}, t_{(2)}, \ldots, t_{(n)}$ be the ordered observations of a random sample from f_T. For fixed θ, the likelihood function $L \equiv L(a_1, a_2, b; \theta)$ is then given by $L = \prod_{i=1}^{n} f_T(t_{(i)})$. By our earlier comment above about the density $f_T(t)$ for $1 \le \theta$, the likelihood $L = 0$ unless θ is bounded above by (the smallest observation above τ)/τ. Let k (a function of the data) be the number of our n observations that are no bigger than τ. It is assumed that $k < n$, so that at least one experimental unit is exposed to the second environment and thus θ is bounded above by $t_{(k+1)}/\tau$. It follows that $\ln(L)$, the natural log of the likelihood, is given by for $0 \le \theta \le 1$,

$\ln(L) =$

$$k\ln(a_1) + \sum_{i=1}^{k}\ln(f_b(a_1 t_{(i)})) + (n-k)\ln(S_b(a_1\tau))$$

$$+ (n-k)\ln(a_2) - (n-k)\ln(S_b(a_2\{\tau-\theta\tau\}))$$

$$+ \sum_{i=k+1}^{n}\ln(f_b(a_2\{t_{(i)}-\theta\tau\})).$$

For $1 \le \theta \le t_{(k+1)}/\tau \equiv \theta_{\max}$, the 5th term after the equality in the above expression (the one containing $\{\tau-\theta\tau\}$) vanishes, and the rest of the terms are non zero. Consideration of the last term after the equality leads to the conclusion that if $f_b(u)$ is Weibull, $b > 1$, and θ approaches $\theta_{\max}$, then $\ln(L)$ approaches $-\infty$.

For fixed θ, choose the values of a_1, a_2, and b that maximize $\ln(L)$ (or equivalently L) by the standard technique of setting the partial derivatives of $\ln(L)$ with respect to a_1, a_2, and b equal to 0 and then solving these 3 equations. Denote this maximized log-likelihood value by

$\ln(L(\hat{a}_1, \hat{a}_2, \hat{b}; \theta))$. Then maximize $\ln(L(\hat{a}_1, \hat{a}_2, \hat{b}; \theta))$ over θ by a search procedure. For the special case of the above mentioned Weibull, the formulas for $\hat{a}_1$ and $\hat{a}_2$ (in terms of b for $0 \le \theta \le 1$) are

$$\hat{a}_1 = \left[\frac{k}{\sum_{i=1}^{k} t_{(i)}^b + (n-k)\tau^b}\right]^{\frac{1}{b}} \quad \text{and}$$

$$\hat{a}_2 = \left[\frac{n-k}{\sum_{i=k+1}^{n}(t_{(i)}-\theta\tau)^b - (n-k)(\tau-\theta\tau)^b}\right]^{\frac{1}{b}}. \quad \text{For}$$

$1 \le \theta \le \dfrac{t_{(k+1)}}{\tau} = \theta_{\max}$, $\hat{a}_2$ does not contain the term with $\tau - \theta\tau$ in it; for $\theta = \theta_{\max}$, the formula for $\hat{a}_2$ is valid only for $b = 1$. The value of b that is needed to maximize $\ln(L)$, is found by substituting the above formulas for $\hat{a}_1$ and $\hat{a}_2$ into the equation $\dfrac{\partial \ln(L)}{\partial b} = 0$ and then solving for b. A technique such as Newton's Method can be used.

Setting $a_1 = \hat{a}_1$, $a_2 = \hat{a}_2$ in the expression for the partial derivative of $\ln(L)$ with respect to b results in

$$\frac{\partial \ln(L)}{\partial b}\bigg|_{a_1 = \hat{a}_1, a_2 = \hat{a}_2} =$$

$$-(n-k)\{(\hat{a}_1 \tau)^b \ln(\tau) - (\hat{a}_2\{\tau - \theta\tau\})^b \ln(\tau - \theta\tau)\}$$

$$+ \sum_{i=1}^{k}[\frac{1}{b} + \ln(t_{(i)})(1 - [\hat{a}_1 t_{(i)}]^b)]$$

$$+ \sum_{i=k+1}^{n}[\frac{1}{b} + \ln(t_{(i)} - \theta\tau)(1 - [\hat{a}_2\{t_{(i)} - \theta\tau\}]^b)],$$

for $0 \leq \theta \leq 1$.

For $1 < \theta < \theta_{max}$, the 2nd term after the above equality containing $(\tau - \theta\tau)$ is not present. The maximized value for the natural logarithm of the likelihood is then given by $\ln(L(\hat{a}_1, \hat{a}_2, \hat{b}; \theta))$; this of course depends on the value of θ.

A Simulated Data Set

A data set of 60 independent observations was simulated from the 4 parameter Weibull model described earlier, with the 4 parameters set at $(\theta = 1, b = 2, a_1 = 1, a_2 = 3)$ and the change-point $\tau = .5$. Here it is:

Data Set

0.02780339	0.17431372	0.24571107	0.27317012
0.32182781	0.32977272	0.33663518	0.37375811
0.43275557	0.46606391	0.48162588	0.49272210
0.55094536	0.58356842	0.60931971	0.61854409
0.63153638	0.63451571	0.64321517	0.65867022
0.66074136	0.66444153	0.67355536	0.68627573
0.69690436	0.69874605	0.70740367	0.71272621
0.71468121	0.71693490	0.72660793	0.73380157
0.73963728	0.74147645	0.74212794	0.74307794
0.74428668	0.75850840	0.76114420	0.78609746
0.79301050	0.79372155	0.79603601	0.80290143
0.81642156	0.81861064	0.84886271	0.85802655
0.87483314	0.88803101	0.89444438	0.89741578
0.90606519	0.92224208	0.93715268	0.95952164
0.96012644	0.99158096	1.04550106	1.04563890

We obtained a nonparametric cumulative hazard plot for this data and investigated how various fitted models fit this plot.

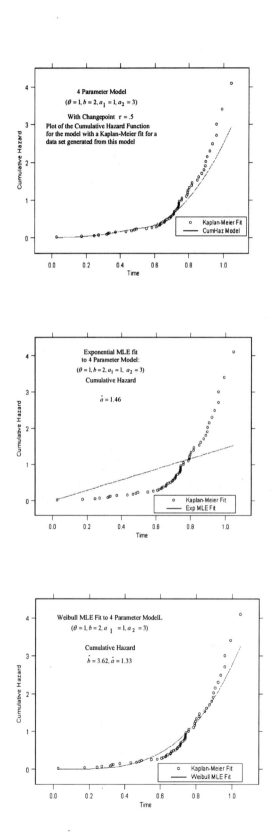

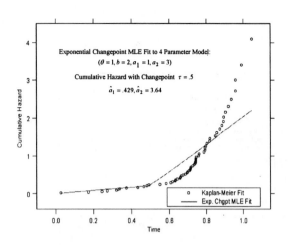

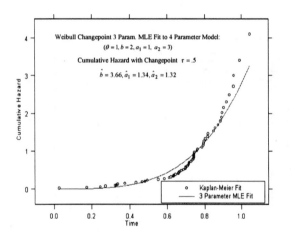

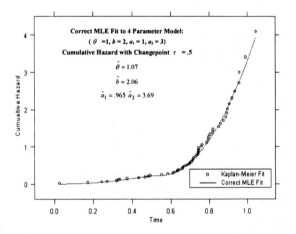

Is There an Effect Due to the 2nd Environment in the Simulated Data Set?

If $\theta \neq 0$, even if $a_1 = a_2$, there is an effect due to the 2nd environment. We consider a test of $H_0: \theta = 0$ versus $H_a: \theta > 0$. Let $\ln(L(\hat{a}_1, \hat{a}_2, \hat{b}; \hat{\theta}))$ be the value obtained when $\ln(L(\hat{a}_1, \hat{a}_2, \hat{b}; \theta))$ is maximized over $0 \leq \theta$. Define a likelihood-ratio type statistic as

$$LR_\theta \equiv 2\{\ln(L(\hat{a}_1, \hat{a}_2, \hat{b}; \hat{\theta})) - \ln(L(\hat{a}_1, \hat{a}_2, \hat{b}; 0))\}.$$

For our simulated data set, $LR_\theta = 21.03$. Under H_0 by repeatedly sampling from ($\theta = 0$, b = 2, $a_1 = 1$, $a_2 = 3$), the quantiles of LR_θ are given by

90%	95%	97.5%	99.0%
2.607261	4.214842	5.521201	6.859972

Assuming $\theta = 0$, the (null) hypothesis mentioned in the 'Introduction' of there being no effect due to the introduction of the 2nd environment is equivalent to H_0: $a_1 = a_2$. Under this reduced (or null hypothesis model), restricted by the requirement that $a_1 = a_2 = a$ (say), one can also maximize the natural log of the likelihood and obtain $\ln(L(\hat{a}, \hat{a}, \hat{b}_{red}; 0))$, where $\hat{b}_{red}$ is the maximized value of b in the reduced model. Consider the behavior of

$$LR = 2\{\ln(L(\hat{a}_1, \hat{a}_2, \hat{b}; 0)) - \ln(L(\hat{a}, \hat{a}, \hat{b}_{red}; 0))\},$$

a likelihood-ratio based statistic. For the simulated data set if the data set is fitted with a model that ignores θ (assuming θ equals 0) then the above LR = .009, clearly not significant, and we would have arrived at the misleading conclusion that the 2nd environment has no effect.

138

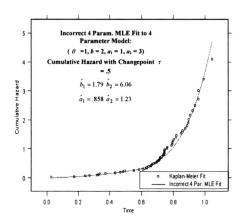

Is There an Effect Due to the 2nd Environment?
Literature Based Data Set

In this section we used a data set, given in the Appendix, based on the data in [1]. For this data set, with an assumed known change-point of $\tau = 697$, θ is only marginally significant; hence, assume that $\theta = 0$. Then the data is modeled with a 3 parameter Weibull change-point (or breakpoint) model: (b, a_1, a_2) and obtain the MLE fit $(\hat{b} = 1.3, \hat{a}_1 = .00262, \hat{a}_2 = .000869)$. The likelihood ratio statistic for testing $H_0 : a_1 = a_2$,

$$LR = 2\{\ln(L(\hat{a}_1, \hat{a}_2, \hat{b};0)) - \ln(L(\hat{a}, \hat{a}, \hat{b}_{red};0))\} = 11.383$$

. This is significant at $\alpha = .1$ when compared with 2.71, the .9 quantile of a chi-square (df = 1).

There is the danger of assuming that $b = 1$ and testing $H_0 : a_1 = a_2$ with a likelihood ratio test based on a 2 parameter exponential model: (a_1, a_2). It is that, for example, when $(b, a_1, a_2) = (1.3, .00262, .00262)$, i.e., the exponential model is incorrect but H_0 is true, this incorrect likelihood ratio test has inflated quantile values (determined by simulation) not in agreement with the chi-square quantiles; the .9 quantile is approximately 5.70 and so on. For data randomly selected from $(b, a_1, a_2) = (1.3, .00262, .00262)$ if the data is fitted with an exponential model that ignores the shape parameter of the Weibull distribution, then it would too often be mistakenly concluded that the 2nd environment has an effect.

A More General Weibull–Breakpoint Model

This is similar to the model defined earlier but now $S_1(t) = S_{b_1}(a_1 t)$ is a survival function with shape parameter b_1 and scale parameter a_1. $S_2(t) = S_{b_2}(a_2(t))$ is a survival function with shape parameter b_2 and scale parameter a_2. Everything else is similar to what went on before.

Essentially the previous model has been generalized by allowing different shape parameters b_1 and b_2 for the two environments. Call this most general model: $\mathrm{MODEL}(a_1, a_2, b_1, b_2, \theta)$. One 4 parameter model that we discussed earlier was $\mathrm{MODEL}(a_1, a_2, b, b, \theta)$; another possible one is $\mathrm{MODEL}(a_1, a_2, b_1, b_2, 0)$. For the 2nd 4-parameter model, the hypothesis that there is no effect due to the change in environments is equivalent to
H_0: $a_1 = a_2, b_1 = b_2$.

Appendix

The data below was obtained from [1], but was changed somewhat from the actual data in [1].

Literature Based Data Set

24	90	186	264	393	642
46	111	191	269	395	697
57	117	197	270	487	955
57	119	209	273	510	1160
64	128	223	284	516	1310
65	143	230	294	518	1538
68	148	247	304	518	1634
82	152	249	304	534	1908
89	166	254	332	583	1996
90	171	258	341	608	2057

Further Explanation of Graphs

The first graph depicts the true cumulative hazard function of the 4 parameter model that the data was simulated from. Also on the graph is the non-parametric cumulative hazard plot of the 60 simulated observations based on Kaplan-Meier considerations. Of course the fit is not perfect, especially on the high end of the data. Take note of the section titled **A Simulated Data Set**.

The next 4 graphs represent parametric cumulative hazard plot fits of increasing complexity to the 60 simulated observations and a graphical comparison with the non-parametric plot. These parametric fits are (1) exponential fit, (2) Weibull fit, (3) exponential change-point fit and (4) Weibull change-point fit. None of these fits provide a satisfactory depiction of the data, mainly because they are wrong and do not take into account the complete renewal ($\theta = 1$) at the change-point τ; they all assume no renewal at all ($\theta = 0$) and our true (4 parameter model) is set at $\theta = 1$.

The last two graphs both contain 4 parameter parametric fits and come quite close to the non-parametric plot. The next to last graph contains a **correct** 4 parameter fit incorporating θ as one of its parameters and the last one contains an **incorrect** 4 parameter fit as described in the section titled **A More General Weibull–Breakpoint Model**. It is referred to over there as MODEL $(a_1, a_2, b_1, b_2, 0)$.

References

[1] Matthews, D.E. and Farewell, V.T. (1982). On Testing for a Constant Hazard   against a Change-Point Alternative. Biometrics 38, 463-468.

[2] Miller, R.G.   (1960).   Early Failures in Life Testing. J. Amer. Statist. Assoc. 55, 491-502.

[3] Müller, H.G. and Wang, J.L. (1994). Change-Point Models for Hazard Functions. In Change-Point Problems. IMS Lecture Notes-Monograph Series 23, 224-241.

Updating Software Reliability Subject to Resource Constraints

Tamraparni Dasu and Elaine Weyuker, AT&T Labs Research
Contact: Tamraparni Dasu, AT&T Labs - Research, Florham Park, NJ 07932

Key Words: Updated reliability estimates, regression testing, variance reduction, conditional probabilities.

Abstract:

We are given two large software programs P and P' that have minor differences as in two successive versions resulting from the maintenance of software. Exhaustive testing of P' can be expensive if not outright infeasible. Podgurski and Weyuker (1997) proposed an estimator for the failure rate θ' of P' by updating the known estimate of the failure rate θ of P. In this paper, we propose a new estimator by restricting the use of a prior estimate of θ to inputs on which both P and P' produce the same outcome. Such an approach avoids the possibility of negative estimates for probabilities. Further, we show that the new estimator has a lower variance and compare it to the best estimate of θ'.

1. Introduction

Software testing involves the validation of the outcome of the execution of the software against the specifications. An *input* is a request to the software to perform a task, such as "dial xyz". The *output* is the result produced by the software because of that request. The software is said to *fail* if the output does not conform to the specifications for some input. For instance if the software dialed wxy instead of xyz, it would have failed. Given that a program can have millions of lines of code intended to perform a huge number of tasks, and an even larger number of possible inputs, testing each task, line of code, or element of the input space is typically infeasible. Testing randomly sampled inputs might require a large sample size to ensure a high degree of accuracy, so that also represents a prohibitively expensive option. Further, random sampling using a uniform distribution might miss relatively infrequent faults that could be catastrophic. Avritzer and Weyuker [1] proposed reweighting the inputs using the distribution of the frequency of inputs to the software in the field, and deterministically selecting the most frequently used inputs for test cases. Such reweighting ensures that the inputs that are exercised most often in practice are thoroughly tested, minimizing the chances of a user encountering a failure. Weyuker [4] also discussed a technique for adjusting the distribution used to select test cases to incorporate both the frequency of occurrence and the consequence of failure so that infrequently used inputs, that have very high cost or consequence of failure would also be tested, even if they occur only rarely in the field.

In this paper we focus on *regression testing*, the revalidation of software after maintenance. It is frequently advocated that existing test sets be reused during regression testing, sometimes augmented with additional test cases. However, the resulting failure rate estimates would usually be optimistic since the maintained version should have fixed the faults that cause the failures that were detected by the original test set.

Podgurski and Weyuker (1997) proposed a method for estimating the failure rate after routine maintenance. This approach required relatively little manual testing to do the estimation [2]. In this paper we propose a modified version of this estimate by conditioning on the space of inputs for which the two versions have the same output. We demonstrate that the enhanced estimator has a lower variance. Furthermore, we will show that this new estimator avoids the problem of negative probabilities that the previous estimator has.

2. Problem Definition

Assume a program P has been modified to obtain a program P' intended to satisfy the same specification, and hence having the same intended functionality. We assume that the modifications were made either to remove defects or improve efficiency and therefore both P and P' are intended to perform the same set of tasks. The *requirements* specify a mapping of inputs to outputs. P and P' will be used the same way, so the distribution of inputs is the same for each. The proportion of inputs on which a program fails is defined to be the *failure rate* of the program P, denoted θ. We have an estimate of θ of P from the validation of a prior set of test inputs. We wish to estimate the failure rate θ' of P' as accurately as possible. The problem is relevant in situations for which the manual testing of the correctness of outputs is relatively expensive, while the cost of running the programs P and P' is relatively inexpensive. Therefore we would like to optimize the trade-off between accuracy and the number of inputs to be manually tested for failure to conform to requirements.

3. Failure Estimator

Podgurski and Weyuker (1997) proposed the following formulation of the problem. Let P and P' be run on n inputs randomly selected using a distribution that reflects the operational usage of the programs. They argued that it was only necessary to manually check the outputs of those test cases that cause P and P' to behave differently in the sense of producing different outputs for a given input. Since P' is derived from P by making minor changes, this set should be much smaller than n. Hence there is a considerable reduction in expense over the manual inspection of the outputs produced by all n inputs.

Letting θ' denote the failure rate of P', Podgurski and Weyuker define θ' to be:

$$\theta' = \text{P(P fails and } P' \text{ fails)} + \text{P(P succeeds and } P' \text{ fails)}.$$

Letting θ denote the failure rate for P, this can be rewritten as:

$$\theta' = \theta - \text{P(P fails and } P' \text{ succeeds)} + \text{P(P succeeds and } P' \text{ fails)}.$$

Let n_s be the subset of inputs for which both P and P' produce the same output. Let $n_d = n - n_s$ be the subset of inputs for which they produce different outputs. Let n_{d-} be the subset of test cases of n_d on which P succeeds and P' fails, and n_{d+} the subset on which P fails and P' succeeds. Note that there is a third subset of n_d, namely those test cases for which both P and P' fail but yield different outputs. Podgurski and Weyuker proposed using the estimator:

$$\hat{\theta}' = \hat{\theta} - \frac{n_{d+}}{n} + \frac{n_{d-}}{n}$$
$$= \frac{f}{m} - \frac{n_{d+}}{n} + \frac{n_{d-}}{n}$$

where $\hat{\theta}$ is an estimate of the failure rate of P derived from a set of m test cases that were previously run yielding f failures. The estimate represents an adjustment to $\hat{\theta}$ by subtracting the proportion of successes of P' that had been failures for P, and adding

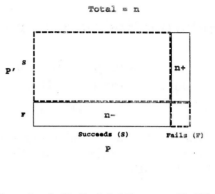

Figure 1: P and P' operating on the same n inputs.

the proportion of failures of P', that had been successes for P.

If we assume that we do not have any cases for which fixes to the program lead to inputs that behaved incorrectly for P, and now behave differently, but still incorrectly for P', it follows that $n_d = n_{d+} + n_{d-}$. Further, let us assume that the fixes to P do not corrupt correctly mapped inputs, i.e. $n_d^- = 0$. Therefore, $n_d = n_d^+$. We will make these assumptions although it is not always the case in reality that program fixes are made correctly. Using these assumptions, the variance of this estimate is given by:

$$V(\hat{\theta}') = V\left(\frac{f}{m}\right) + V\left(\frac{n_d}{n}\right)$$
$$= \frac{\theta(1-\theta)}{m} + \frac{\theta_d(1-\theta_d)}{n}$$

where V is the variance operator and θ_d represents the probability that P and P' produce different outputs. There is no covariance term since the estimates vary independently. See Podgurski and Weyuker [2] for details.

One problem with this estimator is that it can take negative values since $\frac{f}{m}$ is estimated from a different sample than n_{d+} and n_{d-}. In the next section we propose an estimator designed to address this issue and show that the resulting estimator is considerably more accurate.

4. An Alternate Formulation

We propose a new estimator for θ' within the same framework and constraints discussed in the previous section. Consider the following:

$$\begin{aligned}\theta' &= P(P' fails)\\ &= P(P' fails \cap P = P')\\ &\quad + P(P' fails \cap P \neq P')\\ &= P(P' fails \mid P = P')P(P = P')\\ &\quad + P(P' fails \mid P \neq P')P(P \neq P')\end{aligned}$$

where $P = P'$ represents the part of the space of inputs where P and P' behave the same from the point of view of the input/output relation. Substituting the estimates of the proportions from the sample we get

$$\overline{\theta'} = P(P' fails \mid P = P')\frac{n - n_d}{n} + \frac{n_{d-}}{n_d}\frac{n_d}{n} \quad (1)$$

We substitute the known estimate $\hat{\theta}$ for the failure of P' since we have conditioned on P and P' satisfying the same input/output relation. Therefore,

$$\overline{\theta'} = \frac{f}{m}\frac{n - n_d}{n} + \frac{n_{d-}}{n} \quad (2)$$

We will need simulation experiments (where we can fix θ' ahead of time) to estimate the bias in $\overline{\theta'}$ as an estimator of θ'.

4.1 Variance

For simplicity (and two keep the variances of the two estimators comparable) let us use our earlier assumption of $n_{d-} = 0$ so that $n_d = n_d^+$. Combined with our other assumption that that we do not have any cases for which fixes to the program lead to inputs that behaved incorrectly for P, and now behave differently, but still incorrectly for P', this amounts to an assumption that all program corrections are done perfectly. The simplified estimator is given by

$$\begin{aligned}\overline{\theta'} &= \frac{f}{m} - \frac{n_{d+}}{n}\frac{f}{m}\\ &= \frac{f}{m}\left(1 - \frac{n_{d+}}{n}\right)\end{aligned}$$

In addition, consider the following result from the basic theory of expectations:
If X and Y are two independent random variables and E is the expectation operator, then

$$E(XY) = E(X)E(Y) \quad (3)$$

and

$$\begin{aligned}V(XY) &= E[(XY)^2] - [E(XY)]^2\\ &= E(X^2)E(Y^2) - [E(X)E(Y)]^2\\ &= [V(X) + (E(X))^2][V(Y) + (E(Y))^2]\\ &\quad - [E(X)E(Y)]^2\end{aligned}$$

(See [3] for more details on the expectation operator). Now $\frac{f}{m}$ is a Binomial variate with

$$E\left(\frac{f}{m}\right) = \theta \quad (4)$$

and

$$V\left(\frac{f}{m}\right) = \frac{\theta(1 - \theta)}{m} \quad (5)$$

where θ is the proportion of failures during the execution of P over the space on inputs. Similarly, it follows that $n_{d+} = n_d$ since we have assumed that $n_{d-} = 0$. Hence:

$$E\left(1 - \frac{n_{d+}}{n}\right) = 1 - \theta_{d+} = 1 - \theta_d \quad (6)$$

and

$$V\left(1 - \frac{n_d}{n}\right) = \frac{(1 - \theta_d)\theta_d}{n} \quad (7)$$

where θ_d is the probability that P and P' produce a different outputs for a given input.

Since $\frac{f}{m}$ and $\left(1 - \frac{n_d}{n}\right)$ are independent we can use the above fact to compute the variance of $\overline{\theta'}$.

$$\begin{aligned}V(\overline{\theta'}) =\ & \left[\frac{\theta(1 - \theta)}{m} + \theta^2\right] \cdot\\ & \left[\frac{(1 - \theta_d)\theta_d}{n} + (1 - \theta_d)^2\right]\\ & - \theta^2(1 - \theta_d)^2\end{aligned}$$

Note that $\overline{\theta'}$ is a *consistent* estimator of θ', i.e. the variance tends to 0 as m and n tend to infinity.

5. Comparison of Variances

We can now compare the variance of the two estimators $\hat{\theta'}$ and $\overline{\theta'}$.

$$\begin{aligned}V(\hat{\theta'}) - V(\overline{\theta'}) =\ & \frac{\theta(1 - \theta)}{m} + \frac{\theta_d(1 - \theta_d)}{n}\\ & - \left[\frac{\theta(1 - \theta)}{m} + \theta^2\right] \cdot\\ & \left[\frac{(1 - \theta_d)\theta_d}{n} + (1 - \theta_d)^2\right]\\ & + \theta^2(1 - \theta_d)^2\\ =\ & \frac{\theta_d(1 - \theta_d)(1 - \theta^2)}{n}\\ & + \frac{\theta(1 - \theta)}{m} \cdot\\ & \left[1 - (1 - \theta_d)\left(1 - \left(1 - \frac{1}{n}\right)\theta_d\right)\right]\end{aligned}$$

The quantity on the right hand side is greater than 0, except in the extreme case when the failure probability θ is 1, when it is equal to 0. Therefore, $\overline{\theta'}$ is a more accurate estimator than $\hat{\theta'}$.

Figure 2 plots the variance of three estimators against possible values of θ. The "Best" (filled dot symbol) estimator is the maximum likelihood estimator $\hat{\theta}_B$ based on a sample of size $m + n$ whose variance is given by

$$V(\hat{\theta}_B) = \frac{\hat{\theta'}\left(1 - \hat{\theta'}\right)}{m + n} \qquad (8)$$

where we use $\overline{\theta'}$ in Equation 2 as a surrogate for the unknown θ'. In the plots shown, we let $m = n$ for simplicity of depiction and $n_d^- = 0$ as assumed earlier. The maximum likelihood estimator has the smallest possible variance among all estimators of θ' based on a sample of size $m + n = 2m = 2n$. This is an unorthodox but valid comparison since technically we have information from the two samples m and n to estimate θ'. The "Old" (diamond symbol) estimator is the original estimator proposed by Podgurski and Weyuker. The "New" (circle symbol) estimator is the estimator proposed in this paper. Figure 2 depicts the case where $n_d = 0$, namely there is no difference between P and P' on the test inputs. As expected the old and new estimators have the same variance. Figure 3 and Figure 4 depict the case when P and P' differ on 0.1 and 0.2 proportion of the set of inputs respectively. It is clear that:

- the new estimator has a smaller variance than the old estimator,

- as the proportion of test inputs $DIFF$ on which P and P' differ increases, the old estimator moves farther away from the best estimator and the new estimator.

The last observation is summarized in Figure 5, where for a given θ, we have plotted variances for different values of $DIFF$.

6. Conclusion and Further Research

We have shown that we can construct an accurate and consistent estimator for updating failure probabilities after routine software maintenance. Future research includes running experiments with real software to compare the performance of the two estimators in different contexts. Furthermore, we plan to derive the variances under more general conditions by relaxing the assumption of $n_d^- = 0$ (no new failures are introduced by changing P to yield P'). We

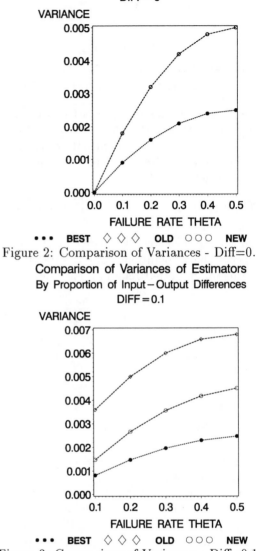

Figure 2: Comparison of Variances - Diff=0.

Figure 3: Comparison of Variances - Diff=0.1.

would also like to use simulation results to investigate other statistical properties of the estimator such as bias.

References

[1] Avritzer, A. and E.J. Weyuker, (1995), The Automatic Generation of Load Test Suites and the Assessment of the Resulting Software, *IEEE Trans. on Software Engineering*, Sept 1995, pp.705-716.

[2] Podgurski, A. and E.J. Weyuker, (1997), Re-estimation of Software Reliability After Maintenance, *Proceedings of IEEE/ACM Nineteenth International Conference on*

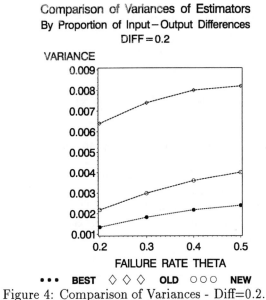

Figure 4: Comparison of Variances - Diff=0.2.

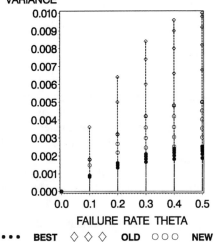

Figure 5: Comparison of Variances - Overall.

Software Engineering (ICSE), May 1997, pp. 79-85.

[3] Rao, C. R., (1965), *Linear Statistical Inference and Its Applications*, John Wiley.

[4] Weyuker, E.J., (1996), Using Failure Cost Information for Testing and Reliability Assessment, *ACM Transactions on Software Engineering and Methodology*, Vol5, No2, April 1996, pp.87-98.

COST VERSUS RELIABILITY IN AIRCRAFT MAINTENANCE

Leonard C. MacLean, Dalhousie University, Halifax, NS, Canada
Alex Richman, AlgoPlus Consulting Ltd., Halifax, NS, Canada

Contact Author: Leonard C. MacLean, School of Business Admin., Dalhousie University, 6152 Coburg Rd., Halifax, NS, B3H 1Z5

Key Words: reliability, unscheduled landings, maintenance cost

1 – INTRODUCTION

Aircraft accidents are costly in terms of assets, liabilities, and future demand. As aircraft age and experience wearout there is an increased chance of failure. To address the deterioration of operating systems, airline management undertakes a program of aircraft maintenance. Such a program reduces the chance of failure and extends the operating life of equipment.

The investment in maintenance required to meet safety standards grows with aircraft age. In addition to age, the cost of maintenance depends on the type of aircraft and the maintenance policies of the carrier. As shown in table 1 the expenditure on maintenance for the same model varies by carrier.

Carrier	Total Maintenance Cost (US$ per Block Hour)	Ratio to Average
A	$1,873	1.15
B	$1,452	0.89
C	$1,972	1.21
D	$1,750	1.07
All*	$1,633	

*Average for all carriers including those shown.
Table 1: Maintenance Cost for a B747-100, 1995

This paper considers the variation in investment in maintenance and in aircraft reliability. In section 2 the reliability of commercial aircraft as indicated by unscheduled landings is analyzed. The cost of maintenance is considered in section 3 and the efficiency of maintenance expenditures is developed in section 4.

2 – RELIABILITY: RATE OF UNSCHEDULED LANDINGS

There are a variety of aspects of aircraft operation which relate to reliability. The analysis here considers the *unscheduled landing's* (UL), which are part of schedule reliability (Friend, 1992.) Studies have shown that the average flight diversion costs an airline $90,000 per incident, so UL's are important to airline finance (*Airline Financial News*).

Of course unscheduled landings are an indication of airworthiness. AlgoPlus Consulting Ltd. has developed a data base on commercial aircraft in North America, based on Service Difficulty Reports (SDR) filed by airlines. In table 2 the variables extracted from the database for each aircraft are defined.

n_i = number of UL's for aircraft i
D_i = number of departures per day for aircraft i
A_{ik} = 1 if aircraft is with carrier k, k = 1, ..., K   0 otherwise
M_{il} = 1 if aircraft is a model 1, 1 – 1, ..., L   0 otherwise

Table 2: Reliability Variables

It is considered that the failures (UL's) follow a Poisson process with the number of failures depending on aircraft model as well as the maintenance policies of the carrier. To model this dependence data for the years 1992 to 1995 were analyzed.

A standard Poisson regression model (McCullagh and Nelder, 1983) can be fit to the data. Let

$$\mu_i = E\left(n_i\right) = \text{expected number of UL's for aircraft i.}$$

Then the natural logarithm of the expected number of events depends on the model, the carrier and the number of departures per day (use). The PR model is

$$\ln\left(\mu_i\right) = \beta_i + \beta_2 \ln D_i + \sum_{k=1}^{K} \delta_k A_{ik} + \sum_{\ell=1}^{L} \gamma_\ell M_{i\ell}. \quad (1)$$

The result of fitting model (1) to the data are given in table 4. In the fitting .5 was added to the observed average number of UL's. The number of departures is not a factor contributing to variation in the expected number of UL's, but the rate shows a strong model and carrier effect.

Component	D_f	Seq SS	F	P
Departures	1	1.014	0.00	0.982
Model	12	8.184	6.58	0.000
Carrier	8	8.705	14.73	0.000
Error	47	3.472		
Total	68	21.375		

Table 3: Fitted Reliability Model

3 – MAINTENANCE COST

The reliability as measured by unscheduled landings is different by carrier after accounting for aircraft use and model. The expectation is that the difference by carrier is largely the result of operating practices, in particular the maintenance effort.

The measure of maintenance effort used here is the total maintenance expenditure in dollars per year. The additional variables for a cost model are given in table 4.

C_{it} = maintenance cost in period t for aircraft i
D_{it} = use in period t for aircraft i

Table 4 – Cost variables

The equation relating cost of maintenance to use, model and carrier is given in (2).

$$C_{it} = \alpha_0 + \alpha_1 D_{it} + \sum_{\ell=1}^{L-1}\phi_\ell M_{i\ell} + \sum_{k=1}^{K-1}\lambda_l A_{ik} + \varepsilon_i \quad (2)$$

The data on cost for 1995 were taken from the World Aviation Directory, 1997. The results of fitting the cost model are given in table 5. After accounting for number of departures and model, the carrier is not a significant factor contributing to variation is maintenance cost.

Componen t	D_f	Seq SS	F	P
Departures	1	65932356	0.01	0.904
Model	12	145693963	6.97	0.000
Carrier	8	14532968	1.10	0.378
Error	47	77423274		
Total	68	303582561		

Table 5: Fitted Cost Model

4 – COST – RELIABILITY TRADEOFF

From the fitted models in table 3 and table 5, the use adjusted cost and failure rate are estimated for each (model, carrier) combination. The property of the cost and reliability relationiship which is of interest is the efficiency of expenditures on maintenance. A maintenance program for a carrier is *efficient if it has the minimum cost among programs*

with the same or better level of reliability, after accounting for the model effect. Let the adjusted cost for model i and carrier i be

$$c_{ij}^{*} = c_{ij} / c_{i.},$$

where $c_{i.}$ Is the average cost for model i. A plot of the model adjusted cost and reliability for each (model, carrier) combination is shown in figure 1. In the diagram the cost-reliability relationship is highlighted for two airlines. At the high reliability level (low UL's) one carrier is efficient, while at the low reliability level (high UL's) the other carrier is efficient. It is clear that some carriers are more efficient than others and furthermore the relationship between maintenance expenditure and reliability varies from carrier to carrier.

Figure 1: Cost Versus Reliability

Cost

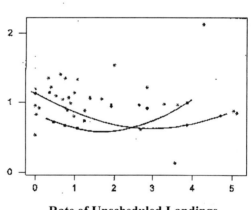

Rate of Unscheduled Landings

5 – REFERENCES

Airline Financial News (1999) Vol. 14: 28.

Alfeld, Louis E., R. Ellis and W. Shepard (1997). The Dynamics of Aircraft Aging: Mimeo Decision dynamics Inc.

Ascher, H., and H. Feingold (1984). *Repairable Systems Reliability: Modelling, Inference, Misconceptions and their Causes.* New York: Marcel Dekker.

AVITAS, Inc. Winter 1997. *World Aviation Directory.*

Dionne, Georges, R. Gagne, F. Gagne and C. Vanasse (1993). Debt, Moral Hazard and Airline Safety: An Empirical Evidence. Centre de Research sur les Transports. Publication 906.

Friend, C.H. (1992). *Aircraft Maintenance Management.* Essex, England: Longman Scientific and Technical.

Golaszewski, Richard, and E. Bomberger (1986). Measuring Differential Safety Performance Among Air Carriers. *Transportation Research Forum* 27, no. 1.

McCullagh, P. and Nelder, J. 1983. *Generalized Linear Models*, Chapman & Hall, London.

Tretheway, Michael 1984. An International Comparison of Airlines. *Transportation Research Forum* 25, no. 1.

6. APPENDIX TABLES

	MODEL														
Carrier	M_1	M_2	M_3	M_4	M_5	M_6	M_7	M_8	M_9	M_{10}	M_{11}	M_{12}	M_{13}	M_{14}	
A_1	1.26	-	-	0.02	0.69	-	-	0.00	1.39	0.10	1.69	-	-	-	0.80
A_2	0.76	0.52	0.84	0.40	1.07	-	-	-	0.89	-	-	-	0.00	-	0.70
A_3	0.81	1.07	-	0.46	1.25	-		0.75	-	0.36	-	3.86	-	1.92	1.37
A_4	-	2.86	-	-	3.56	-	3.62	-	-	-	-	-	-	-	3.15
A_5	-	-	5.04	2.00	-	0.00	5.13	-	3.28	-	-	-	2.70	-	4.19
A_6	-	1.92	4.71	-	-	-	2.66	0.00	2.87	-	3.86	-	-	-	2.92
A_7	0.76	-	1.40	-	0.33	0.00	0.64	0.00	-	-	-	-	-	1.00	0.64
A_8	-	0.98	-	-	-	-	-	-	-	-	-	-	-	-	0.98
A_9	-	-	-	-	-	-	-	-	-	-	-	-	4.28	-	4.28
	0.90	1.56	2.93	0.72	1.14	0.00	2.94	0.19	2.00	6.23	2.78	3.86	2.33	1.61	1.81

Table A1 – Rate of UL's by Model and Carrier

MODEL

Carrier	M_1	M_2	M_3	M_4	M_5	M_6	M_7	M_8	M_9	M_{10}	M_{11}	M_{12}	M_{13}	M_{14}	
A_1	1673	-	-	1392	2881	-	-	1364	5095	2778	2790	-	-	-	2883
A_2	2557	1690	7403	2056	3697	-	-	-	5846	-	-	-	1273	-	3964
A_3	1680	984	-	1218	1804	-	-	767	-	3444	-	1539	-	4395	2037
A_4	-	1775	-	-	373	-	1410	-	-	-	-	-	-	-	1421
A_5	-	-	5460	2600	-	1153	1228	-	4899	-	-	-	826	-	2906
A_6	-	1686	5353	-	-	-	1409	1294	4714	-	1806	-	-	-	2913
A_7	1734	-	6974	-	3905	2058	1967	607	-	-	-	-	-	4765	2976
A_8	-	1213	-	-	-	-	-	-	-	-	-	-	-	-	1213
A_9	-	-	-	-	-	-	-	-	-	-	-	-	2837	-	2837
	1495	1533	6351	1816	2772	1605	1560	1008	5096	3111	2298	1539	1646	4518	2805

Table A2 – Average Maintenance Cost by Model and Carrier ($US per block hour)

ON LINEAR COMBINATION CLASSIFICATION PROCEDURE WITH A BLOCK OF OBSERVATIONS MISSING

Hie-Choon Chung, Kwangju University and **Chien-Pai Han**, University of Texas at Arlington
Chien-Pai Han, Department of Mathematics, University of Texas at Arlington, Arlington, TX 76019-0408

Key Words: Discriminant analysis, Error rate, Bootstrap confidence interval.

1. Introduction.

A $p \times 1$ observation X is to be classified into one of two multivariate normal populations, $\pi_1 : N(\mu^{(1)}, \sum)$ and $\pi_2 : N(\mu^{(2)}, \sum)$, where the parameters are unknown. We consider the situation that the training samples contain a block of missing observations. In such a case, one may ignore the incomplete part of the data and use only the complete observations to construct the discriminant function. Another way to deal with missing observations is to estimate the parameters by a missing value technique, then substitute the estimators into the discriminant function. Chan and Dunn (1972, 1974) presented several methods of ignoring and estimating the values of these vectors, and used the resulting vectors in the discriminant function. There appears to be no uniformly best method. They suggested some guidelines in choosing the method for different situations.

Bohannon and Smith (1975) applied Hocking and Smith (1968) estimation procedure to estimate the parameters and compared this procedure to the standard procedure of ignoring the missing values in the construction of the classification rule and the estimation of the error rate.

Twedt and Gill (1992) examined the impact of different methods for replacing missing data in discriminant analysis. They concluded that the methods of replacing missing data were better than the one of ignoring the observation vectors with missing data.

The EM algorithm (Dempster, Laird, and Rubin 1977) may be used to estimate the parameters in the classification statistic. This algorithm consists of an iterative calculation involving two steps: i.e., the prediction and the estimation steps.

Anderson (1957) considered the maximum likelihood estimates of parameters of multivariate normal distributions when special patterns of missing observations are obtained in the training samples. The estimators are then used for substituting the unknown parameters in the classification rule.

In this paper we discuss the linear combination classification procedure in Section 2. This procedure is better than Anderson's procedure, the EM algorithm and Hocking and Smith procedure when the proportion of missing observation is large. (See Chung and Han (1999)). Section 3 discusses estimation of error rates. A numerical example is given in Section 4.

2. Linear Combination Classification Procedure.

We consider a special pattern which contains a block of missing observations. Instead of estimating the parameters, we construct two different discriminant functions from the complete data and incomplete data, respectively, and then a linear combination of these two linear discriminant functions is used to obtain the classification rule.

Let us partition the $p \times 1$ observation X as follows.

$$X = \begin{bmatrix} Y \\ Z \end{bmatrix},$$

where Y is a $k \times 1$ vector and Z is a $(p - k) \times 1$ vector $(1 \leq k < p)$. Suppose random samples of sizes m_i, containing no missing values,

$X_j^{(i)}$, $i = 1, 2$; $j = 1, 2, \ldots m_i$,

are available from $N_p(\mu^{(i)}, \sum)$, and random samples of sizes $n_i - m_i$, which contain only the first k-components

$Y_j^{(i)}$, $i = 1, 2$; $j = m_i + 1, \ldots, n_i$,

are available from $N_k(\mu_y^{(i)}, \sum_{yy})$. Hence the data have the special pattern of missing values where a block of variables is missing on $n_i - m_i$ observations, and the remaining observations are all complete.

We can construct two linear discriminant functions. The first linear discriminant function is based on the complete observations,

$X_j^{(i)}$, $i = 1, 2$; $j = 1, 2, \ldots, m_i$.

We have

$$W_x = (\overline{X}^{(1)} - \overline{X}^{(2)})'$$
$$\cdot S_{xx}^{-1}\left[X - \tfrac{1}{2}(\overline{X}^{(1)} + \overline{X}^{(2)})\right]$$

where

$$\overline{X}^{(i)} = \frac{1}{m_i}\sum_{i=1}^{m_i} X_j^{(i)} = \left[\begin{array}{c} \overline{Y}_1^{(i)} \\ \overline{Z}^{(i)} \end{array}\right], i = 1, 2,$$

$$S_{xx} = \sum_{i=1}^{2}\sum_{j=1}^{m_i}(X_j^{(i)} - \overline{X}^{(i)})$$

$$(X_j^{(i)} - \overline{X}^{(i)})'/\nu_x, \ \nu_x = m_1 + m_2 - 2.$$

The second linear discriminant function is based on the incomplete observations,

$\overline{Y}_j^{(i)}$, $i = 1, 2$; $j = 1, 2, \ldots, n_i$.

We have

$$W_y = (\overline{Y}^{(1)} - \overline{Y}^{(2)})'$$
$$\cdot S_{yy}^{-1}\left[Y - \tfrac{1}{2}(\overline{Y}^{(1)} + \overline{Y}^{(2)})\right],$$

where $\overline{Y}^{(i)}$ is the mean based on all n_i observations, and

$$S_{yy} = \sum_{i=1}^{2}\sum_{j=1}^{n_i}(Y_j^{(i)} - \overline{Y}^{(i)})$$

$$(Y_j^{(i)} - \overline{Y}^{(i)})'/\nu_y, \ \nu_y = n_1 + n_2 - 2.$$

Now we combine the two linear discriminant functions and construct the classification rule which is a linear combination of W_x and W_y, namely

$$W_c = cW_x + (1 - c)W_y, \ 0 \le c \le 1.$$

We call W_c the linear combination classification statistic. An advantage of W_c is that it is easy to use. The observation X is classified into π_1 if

$$W_c = cW_x + (1 - c)W_y \ge 0;$$

otherwise it is classified into π_2. This classification procedure is called the linear combination classification procedure.

The probability of misclassifying an observation from π_1 into π_2 is given by

$$\beta_1 = Pr\{W_c < 0 | X \in \pi_1\}.$$

Similarly the probability of misclassifying an observation from π_2 into π_1 is given by

$$\beta_2 = Pr\{W_c \ge 0 | X \in \pi_2\}.$$

The unconditional error rate, with equal prior probability, is defined as

$$\beta = \tfrac{1}{2}(\beta_1 + \beta_2).$$

In order to find the error rate β, we need to know the distribution of W_c. However, this distribution is extremely complicated. Hence, we consider the conditional error rate. Let

$$\beta_1^* = Pr\{W_c < 0 | \overline{X}^{(1)}, \overline{X}^{(2)}, S_{xx},$$
$$\overline{Y}^{(1)}, \overline{Y}^{(2)}, S_{yy}; X, Y \in \pi_1\}.$$

$$\beta_2^* = Pr\{W_c \ge 0 | \overline{X}^{(1)}, \overline{X}^{(2)}, S_{xx},$$
$$\overline{Y}^{(1)}, \overline{Y}^{(2)}, S_{yy}; X, Y \in \pi_2\},$$

then the conditional error rate, with equal prior probability, is given as

$$\beta^* = \tfrac{1}{2}(\beta_1^* + \beta_2^*).$$

It can be shown that the conditional distribution of W_c is normal.

As for the value of c, we follow Chung and Han (1999) and use the operational c^* which is given by

$$c^* = \frac{(\frac{1}{m_1} + \frac{1}{m_2})^{-1}D^2}{(\frac{1}{m_1} + \frac{1}{m_2})^{-1}D^2 + (\frac{1}{n_1} + \frac{1}{n_2})^{-1}D_y^2},$$

where

$$D^2 = (\overline{X}^{(1)} - \overline{X}^{(2)})'S_{xx}^{-1}(\overline{X}^{(1)} - \overline{X}^{(2)}),$$

$$D_y^2 = (\overline{Y}^{(1)} - \overline{Y}^{(2)})'S_{yy}^{-1}(\overline{Y}^{(1)} - \overline{Y}^{(2)}).$$

3. Bootstrap Estimate of Error Rate.

Chung and Han (1999) considered the estimation of the conditional error rate β^* for the linear combination classification procedure. The algorithm of McLachlan (1980) can be extended to obtain the bootstrap estimate of the bias correction when the training samples contain missing value. Also the leave-one-out estimate of the error rate was obtained. The findings from a Monte Carlo study are given as follows:

1) When n and m are moderately larger than p, both estimates appear to be nearly unbiased.

2) When n and m are sufficiently larger than p, both estimates are improved compared to the case in 1).

3) When n and m are not moderately larger than p, the estimates for the leave-one-out method generally appear to be nearly unbiased but not for the bootstrap. This happens since information for the discrimination depends on the variables in which data contain missing values.

4) The standard deviations for both estimates are almost the same.

Now we consider bootstrap confidence intervals. We may use the percentile method, bias-corrected percentile method and accelerated bias-corrected percentile method to construct the confidence interval (see Efron (1982, 1987), Buckland (1984, 1985) among others). A Monte Carlo study is conducted by using an algorithm in Buckland (1985) for constructing 500 bootstrap confidence intervals. The average lengths and coverage probabilities for the above mentioned three method are compared. It is found that the bias-corrected percentile method is the best among the three methods when $p = 2$ and $k = 1$. For other values of p and k, there does not appear to have one method dominating the others.

4. Numerical Example

We use the admission data at the University of Texas at Arlington given in Chung and Han (1999) to illustrate the estimation of the error rate β^*. The data set contains two populations. One population is the Success Group that the students receive their Masters degree. The other population is the Failure Group that they do not complete their Masters degree. For each population, there are 10 foreign students and 10 United States students. Each foreign student has 5 variables which are x_1 = undergraduate GPA, x_2 = GRE verbal, x_3 = GRE quantitative, x_4 = GRE analytic, and x_5 = TOEFL score. For each United States student, one variable, x_5 = TOEFL score is missing.

Using this data set, we obtain the discriminant function
$$W_c = cW_x + (1 - c)W_y,$$
where
$$W_x = a'X + b,$$
$$a' = [\text{-}1.9957 \ \text{-}0.0170 \ \text{-}0.0004 \ 0.0034 \ 0.0242], \ b = \text{-}2.5252,$$

$$W_y = d'X + e,$$
$$d' = [0.5302 \ \text{-}0.0042 \ \text{-}0.0023 \ 0.2406], \ e = 0.2806.$$

The value of c is
$$c = 0.7532.$$

For this example, we generate 300 bootstrap samples to estimate β^*. The result is that the bootstrap estimate of β^* is 0.356. The 95% bootstrap confidence interval is (0.259, 0.451) which is obtained by the percentile method.

References:

Anderson, T. W. (1957). Maximum likelihood estimates for a multivariate normal distribution when some observations are missing, *Journal of the American Statistical Association* 52, 200-203.

Bohannan, Tom R. and W. B. Smith (1975), *ASA Proceedings of Social Statistics Section*, 214-218.

Buckland, Stephen T. (1984). Monte Carlo confidence intervals, *Biometrics* 40, 811-817.

Buckland, Stephen T. (1985). Calculation of Monte Carlo confidence intervals, *Royal Statistical Society*, Algorithm AS214, 297-301.

Chan, L. S. and O. J. Dunn (1972). The treatment of missing values in discriminant analysis-1, The sampling experiment, *Journal of the American Statistical Association*, 67, 473-477.

Chan, L. S. and O. J. Dunn (1974). A note on the asymptotical aspect of the treatment of missing values in discriminant analysis, *Journal of the American Statistical Association*, 69, 672-673.

Chung, H.-C. and Han, C.-P. (1999). Discriminant analysis when a block of observation is missing. Submitted for publication.

Dempster, A.P., N.M. Laird and R. J. A. Rubin (1977). Maximum likelihood from incomplete data via the EM algorithm, *Journal of the Royal Statistical Society, Ser.* B, 39, 302-306.

Efron, B. (1982). The Jackknife, the Bootstrap and other Sampling Plans, Philadelphia, *Society for Industrial and Applied Mathematics.*

Efron, B. (1987). Better bootstrap confidence intervals. *Journal of the American Statistical Association,* 82, 171-200.

Hocking, R. R. and W. B. Smith (1968). Estimation of parameters in the multivariate normal distribution with missing observation, *Journal of the American Statistical Association*, 63, 159-173.

McLachlan, G. J. (1980). The efficiency of Efron's bootstrap approach applied to error rate estimation in discriminant analysis, *Journal of Statistical Computation and Simulation*, 11, 273-279.

Twedt, Daniel J. and D. S. Gill (1992). Comparison of algorithm for replacing missing data in discriminant analysis, *Communications in Statistics-Theory and Methods*, 21, 1567-1578.

THE EFFICIENCY OF SHRINKAGE ESTIMATORS FOR ZELLNER'S LOSS FUNCTION

Marvin Gruber Department of Mathematics and Statistics Rochester Institute of Technology
85 Lomb Memorial Drive Rochester, New York 14623

Keywords: ridge estimator, James-Stein estimator, goodness of fit, error of estimation

Abstract

The traditional mean square error (MSE) as the measure of efficiency of an estimator only takes the error of estimation into account. In 1994 Zellner proposed a balanced loss function. Unlike the traditional quadratic loss function the balanced loss function takes into account both goodness of fit and error of estimation. With respect to the balanced loss function this presentation will focus on:
1. the formulation of James-Stein type estimators;
2. the comparison of their efficiencies with that of the least square estimator.

1. Introduction

This article will formulate and then compare the efficiency of the James Stein estimator to the least square estimator under Zellner's (1994) Balanced Loss function. The balanced loss function is a weighting of ordinary squared error loss due to estimation and loss due to prediction.

Stein (1956) surprised the statistical community by proving that the maximum likelihood estimator (MLE) of the mean of a multivariate normal distribution is inadmissible. An explicit estimator with uniformly smaller mean square estimator than the maximum likelihood estimator was formulated by James and Stein (1961). Estimators with mathematical form similar to Stein's (1961) estimator are known in the literature as Stein type estimators (JS). Stein type estimators are special cases of contraction estimators considered by Mayer and Willke(1973). They are also special cases of empirical Bayes estimators (see for example Efron and Morris (1973) and Gruber(1998)).

The loss functions previously used in comparison of the JS only reflect precision of estimation. Zellner(1994) describes balanced loss functions that reflect both goodness of fit and precision of estimation. This article will compare the contraction estimator of Mayer and Willke and the JS with the least square estimator (LS) in the context of a setup involving a single linear model. This problem is considered by Chung and Kim (1997) for the estimation of the mean of a multivariate normal distribution. Only the frequentist risk of the estimator is obtained there. This article obtains both the frequentist risk and the Bayes risk.

Section 2 of the paper will compare the mean square error of the contraction estimator of Mayer and Willke with that of the LS. A similar comparison of the JS with the LS will be made in Section 3. Since this is work in progress plans for future research will be outlined in Section 4.

2. The Estimator of Mayer and Willke

Consider a linear model

$$Y = X\beta + \varepsilon \tag{1}$$

where X is an nxm matrix of rank $s \leq m$. The least square estimator is given by

$$b = (X'X)^+ X'Y \tag{2}$$

where the + denotes the Moore Penrose inverse. The contraction estimator of Mayer and Willke takes the form

$$\hat{\beta} = cb \tag{3}$$

where c is a constant with 0<c<1. Zellner's balanced loss function using the predictive loss function for the mean square error (BMSE) takes the form

$$m = wE(Y - Xcb)'(Y - Xcb)$$
$$+(1-w)E(cb - \beta)'X'X(cb - \beta) \tag{4}$$

with $0 \leq w \leq 1$. The first term is goodness of fit. The second term is the traditional mean square error (MSE) averaging over the predictive loss function. The BMSE of the contraction estimator is

$$R_c = (1-w)[c^2(\sigma^2 s + \beta'X'X\beta)$$
$$-2c\beta'X'X\beta + \beta'X'X\beta]$$
$$+w[(\beta'X'X\beta + \sigma^2 n)$$
$$-2c(\beta'X'X\beta + \sigma^2 s)$$
$$+c^2(\beta'X'X\beta + \sigma^2 s)] \tag{5a}$$

or

$$R_c$$
$$= c^2(\sigma^2 s + \beta'X'X\beta)$$
$$-2cw\sigma^2 s - 2c\beta'X'X\beta$$
$$+\sigma^2 wn + \beta'X'X\beta$$
$$= (sc^2 - 2scw + wn)\sigma^2$$
$$+(1-c)^2\beta'X'X\beta.$$

$$(5b)$$

The BMSE of the LS is

$$R_L = [(1-2w)s + nw]\sigma^2. \qquad (6)$$

Now

$$R_c \le R_L \qquad (7)$$

iff

$$\frac{\beta'X'X\beta}{\sigma^2} \le \frac{s(1+c)-2sw}{1-c} \qquad (8)$$

provided that

$$w < \frac{1+c}{2}. \qquad (9)$$

The comparison in Equation (9) should be compared with the corresponding comparison for the MSE. From (5) the MSE is

$$m_c = c^2(\sigma^2 s + \beta'X'X\beta)$$
$$-2c\beta'X'X\beta + \beta'X'X\beta. \qquad (10)$$

For the LS

$$m_L = \sigma^2 s. \qquad (11)$$

The contraction estimator has a smaller MSE than the LS when the inequality for those values of the parameter β where the inequality

$$\frac{\beta'X'X\beta}{\sigma^2} \le \frac{s(1+c)}{1-c} \qquad (12)$$

is satisfied. The term $-2sw/(1-c)$ indicates the decrease in the size of the region when the balanced loss function risk is compared to the MSE. Let $\delta = \beta'X'X\beta/2\sigma^2$. Solve inequality (8) for c. With respect to Zellner's balanced loss function the contraction estimator has a smaller risk than the LS if

$$\frac{2\delta - s + 2ws}{s + 2\delta} < c < 1 \qquad (13)$$

The corresponding result for the ordinary MSE is

$$\frac{2\delta - s}{s + 2\delta} < c < 1 \qquad (14)$$

Thus, the interval on c where the contraction estimator has a smaller risk than LS with respect to the BLF is also smaller than the interval for the MSE due to the extra term $2ws/(2\delta + s)$.

3. The James-Stein Estimator

3.1 Formulation of the Estimator

The James-Stein estimator may be formulated as an almost optimum estimator two ways. From the frequentist point of view it may derived as an approximate minimum mean square error estimator. From the Bayesian point of view it may be derived as an empirical Bayes estimator. Both derivations will be given below.

The Frequentist Point of View

The minimum mean square error estimator is obtained by differentiating the last expression in Equation 6. and equating the result to zero. Thus,

$$\frac{dR_c}{dc} = (2sc - 2nw)\sigma^2 - 2(1-c)\beta'X'X\beta = 0.$$

$$(15)$$

Solving for optimum c

$$c = \frac{sw\sigma^2 + \beta'X'X\beta}{s\sigma^2 + \beta'X'X\beta}$$

$$= 1 - \frac{(1-w)s\sigma^2}{s\sigma^2 + \beta'X'X\beta}. \qquad (16)$$

The minimum mean square error estimator is

$$\hat{\beta} = \left(1 - \frac{(1-w)s\sigma^2}{s\sigma^2 + \beta'X'X\beta}\right)b. \qquad (17)$$

The classical MSE of (17) is given by

$$R_m = \sigma^2 s - \frac{(1-w)(1+w)\sigma^4 s^2}{(s\sigma^2 + \beta'X'X\beta)}. \qquad (18)$$

When $w = 0$ (18) reduces to the MSE of the minimum mean square estimator with respect to the quadratic loss function based only on estimation. When $w = 1$ it reduces to the MSE of the least square estimator. The MSE averaging over Zellner's loss function (BMSE) is given by

$$
\begin{aligned}
R_{mz} \\
&= \sigma^2(wn) + \beta'X'X\beta \\
&\quad - \frac{(ws\sigma^2 + \beta'X'X\beta)^2}{s\sigma^2 + \beta'X'X\beta} \\
&= \sigma^2(wn) \\
&\quad + \frac{(1-2w)s\sigma^2\beta'X'X\beta - w^2s^2\sigma^4}{s\sigma^2 + \beta'X'X\beta} \\
&= \sigma^2(wn) + (1-2w)s\sigma^2 \\
&\quad - \frac{(1-w)^2s\sigma^4}{s\sigma^2 + \beta'X'X\beta} \\
&= w\sigma^2(n-s) + (1-w)s\sigma^2 \\
&\quad - \frac{(1-w)^2s\sigma^4}{s\sigma^2 + \beta'X'X\beta}.
\end{aligned}
$$

(19)

This MSE is uniformly smaller than that of the LS estimator.

Unbiased estimators for the numerator and denominator of (17) are

$$
\hat{\sigma}^2 = \frac{1}{(n-s)}(Y - Xb)'(Y - Xb)
$$

(20)

and

$$
\hat{d} = \frac{b'X'Xb}{s}
$$

(21)

respectively. The James-Stein estimator is obtained after substitution of (20) and (21) into (17). It takes the form

$$
\hat{\beta} = \left(1 - \frac{(1-w)hs\hat{\sigma}^2}{b'X'Xb}\right)b
$$

(22)

where h is chosen to optimize the mean square error.

The James-Stein Estimator from the Bayesian Point of View

The result (20) may also be obtained as an empirical Bayes estimator . Assume that β follows a prior distribution where

$$
E(\beta) = 0 \text{ and } D(\beta) = \omega^2(X'X)^+. \quad (23)
$$

This prior was recommended by Dempster (1973) and also by Zellner (1974). The risk of the contraction estimator averaging over this prior is given by

$$
\begin{aligned}
R_c \\
&= wE(Y - Xcb)'(Y - Xcb) \\
&\quad + (1-w)E(cb - \beta)'X'X(cb - \beta) \\
&= w((\sigma^2 + \omega^2)n - 2c(\sigma^2 + \omega^2)s \\
&\quad + c^2(\sigma^2 + \omega^2)s) \\
&\quad + (1-w)(c^2(\sigma^2 + \omega^2)s \\
&\quad - 2c\omega^2s + \omega^2s) \\
&= w(\sigma^2n + \omega^2s) + (1-w)\omega^2s \\
&\quad - 2c[w(\sigma^2 + \omega^2) + (1-w)\omega^2s] \\
&\quad + c^2[w(\sigma^2 + \omega^2)s + (1-w)(\sigma^2 + \omega^2)s] \\
&= w(\sigma^2n + \omega^2s) \\
&\quad + (1-w)\omega^2s \\
&\quad - 2c(w\sigma^2s + \omega^2s) \\
&\quad + c^2(\sigma^2 + \omega^2)s.
\end{aligned}
$$

(24)

Differentiating with respect to c, setting the result equal to zero gives optimum

$$
c = \frac{w\sigma^2 + \omega^2}{\sigma^2 + \omega^2} = \left(1 - \frac{(1-w)\sigma^2}{\sigma^2 + \omega^2}\right). \quad (25)
$$

Then the optimum contraction estimator (Bayes estimator) is

$$
\hat{\beta} = \left(1 - \frac{(1-w)\sigma^2}{\sigma^2 + \omega^2}\right)b. \quad (26)
$$

The Bayes Risk of (26) averaging over Zellner's loss function is

$$m = w\sigma^2 n + \omega^2 s - \frac{(w\sigma^2 + \omega^2)^2 s}{(\sigma^2 + \omega^2)}$$

$$= w\sigma^2(n-s) + (1-w)\sigma^2 s - \frac{(1-w)^2 \sigma^4 s}{(\sigma^2 + \omega^2)}.$$

$$(27)$$

The estimators in (20) and (21) are unbiased estimators for σ^2 and $\sigma^2 + \omega^2$ respectively. Thus, the approximate Bayes estimator or empirical Bayes estimator is given by the James-Stein estimator (22).

A curious fact is that the estimator (22) may also be obtained using a prior dispersion with mean 0 and dispersion $a(X'X)^{-1}$ where

$$a = \frac{w\sigma^2 + \omega^2}{1-w} \qquad (28)$$

using the mean square error of estimation.

3.2 The Efficiency of the James-Stein Estimator

The Bayes Risk

Using methods similar to those found in Gruber (1998) Section 4.4 the Bayes Risk is found to be

$$m = w\sigma^2(n-s) + (1-w)\sigma^2 s - \frac{2h(1-w)^2 s\sigma^4}{(\sigma^2 + \omega^2)}$$
$$+ \frac{h^2 s^2(n-s+2)(1-w)^2 \sigma^4}{(s-2)(n-s)(\sigma^2 + \omega^2)}$$

$$(29)$$

Differentiating m with respect to h and setting the result equal to zero yields optimum

$$h = \frac{(s-2)(n-s)}{(n-s+2)s} \qquad (30)$$

For optimum h the Bayes risk is

$$m = w\sigma^2(n-s) + (1-w)\sigma^2 s$$
$$- \frac{(s-2)(n-s)\sigma^4(1-w)^2}{(n-s+2)(\sigma^2 + \omega^2)}. \qquad (31)$$

The first term is the mean square error of the least square estimator. The second term is the improvement over the least square estimator. When $w = 0$ (31) reduces the average mean square error of the James-Stein estimator obtained in Gruber (1979, 1998). When $w = 1$ it reduces to the expectation of the within sum of squares(error sum of squares) in Analysis of Variance.

A measure of efficiency proposed by Efron and Morris (1973) is the relative savings loss (RSL). It is the ratio of the difference between the risk of the empirical Bayes estimator (EBE) and the Bayes estimator (BE) to the difference between the risk of the least square estimator (LS) and the Bayes estimator (BE), that is

$$RSL = \frac{R_{EBE} - R_{BE}}{R_{MLE} - R_{BE}}. \qquad (32)$$

The RSL of the James-Stein estimator is given by

$$RSL = 1 - \frac{(s-2)(n-s)}{s(n-s+2)}. \qquad (33)$$

The RSL (33) obtained averaging over Zellners' loss function is the same as that obtained for the traditional MSE and is independent of the weight given to the two components estimation and goodness of fit.

The Frequentist Risk

Using methods similar to those of Gruber (1998) Section 4.5 the frequentist risk of the James-Stein estimator averaged over Zellners' balanced loss function is for the same optimum h

$$R_{JS} = w\sigma^2(n-s)$$
$$+ (1-w)\sigma^2 s$$
$$- \frac{(s-2)^2(n-s)\sigma^4(1-w)^2}{(n-s+2)}.$$
$$E\left[\frac{1}{b'X'Xb}\right]$$
$$= w\sigma^2(n-s) + (1-w)\sigma^2 s .$$
$$- \frac{(s-2)^2(n-s)\sigma^2(1-w)^2}{(n-s+2)}.$$
$$\sum_{k=0}^{\infty} \frac{e^{-\delta}\delta^k}{k!(2k+s-2)}$$
with
$$\delta = \frac{\beta'(X'X)\beta}{2\sigma^2}.$$

$$(34)$$

Again when $w = 0$ the risk reduces to that of the James-Stein estimator for traditional MSE. When $w = 1$ it reduces to the expectation of the within sum of squares(error sum of squares) in Analysis of Variance.

An analogue of the relative savings loss for the frequentist case is the relative loss. The relative loss may be defined by

$$RL = \frac{R_{JS} - R_{mmse}}{R_{LS} - R_{mmse}}. \qquad (35)$$

For the JS considered in this paper

$$RL = 1 - \frac{\frac{(s-2)^2(n-s)(s+2\delta)}{s(n-s+2)}}{\sum_{k=0}^{\infty} \frac{e^{-\delta}\delta^k}{k!(2k+s-2)}}. \qquad (36)$$

Observe that the RL is also independent of w.

4. Conclusions and Directions for Further Work

The following facts may be extrapolated from the results of this paper

1. The region where the contraction estimator has a smaller MSE than that of the least square estimator is not as large under Zellners' loss function as the classical MSE.
2. The amount of contraction and the numerical value of the MSE of the James-Stein estimator is larger when averaged over Zellner's balanced loss function. The improvement is cut down by the factor $(1-w)^2$.
3. When $w = 0$ the results are the same as for estimation. When $w = 1$ the expectation of the error sum of squares results.
4. The efficiency of the empirical Bayes estimator when compared with the Bayes and the least square estimator is the same for Zellners' loss function as classical MSE.

Further work will include:
1. investigation of the efficiency of different kinds of ridge regression estimators under Zellners' loss function.
2. the formulation of Zellners' loss function for different setups of linear models and the evaluation of the efficiency of different estimators.

References

Chung,Y. and Kim, C. (1997). "Simultaneous Estimation of the Multivariate Normal Mean under Balanced Loss Function".Communications in Statistics, Theory and Methods, 26(7), 1599-1611.

Dempster, A.P.(1973). "Alternatives to Least Squares in Multiple Regression". Multivariate Statistical Inference, D.G. Kabe and R.P. Gupta ed., Amsterdam: North Holland, 25-40.

Efron, B. and Morris, C.(1973)."Stein's Estimation Rule and its Competitors." Journal of the American Statistical Association, 65, 117-130.

Gruber, M.H.J. (1979). Empirical Bayes, James-Stein and Ridge Regression Type Estimators for Linear Models. Unpublished Ph.D. Thesis, University of Rochester.

Gruber, M.H.J. (1998) Improving Efficiency by Shrinkage The James-Stein and Ridge Regression Estimators. Marcel Dekker. New York.

James, W. and Stein, C.(1961). "Estimation with Quadratic Loss." Proceedings of the Fourth Berkeley Symposium on Mathematics and Statistics. Berkeley: University of California Press 1, 361-379.

Mayer, L.S. and Willke, T.A.(1973). "On Biased Estimation in Linear Models." Technometrics, 15, 497-508.

Stein, C.(1956). "Inadmissibility of the Usual Estimator for the Mean of a Multivariate Normal Distribution." Proceedings of the Third Berkeley Symposium on Mathematics, Statistics and Probability. Berkeley: University of California Press, 197-206.

Zellner, A.(1994). "Bayesian and Non-Bayesian Estimation Using Balanced Loss Functions." in Statistical Decision Theory and Related Topics V, S.S. Gupta and J.O. Berger eds., 377-390.

Zellner, A. and Vanele, W.(1974). "Bayes Stein Estimators for K Means, Regression and Simultaneous Equations Models." S.E. Feinberg and A.Zellner editors. Studies in Bayesian Econometrics and Statistics, Amsterdam, North Holland, 628-653.

Acknowledgement

This research was partially supported by a Dean's fellowship in the College of Science during the summers of 1998 and 1999.

RESOURCE ALLOCATION AND SCALED ESTIMATION OF SYSTEM RELIABILITY

Shannon Escalante, Harvard University, Michael Frey, Bucknell University
Michael Frey, Dept. of Mathematics, Bucknell University, Lewisburg, PA 17837

Key words: bias, mean square error, resource allocation, coefficient of variation, scaling

Abstract: System reliability is often estimated by combining estimates of the reliability of the system's components. Because of the potential number of components, the number of observations possible for each component may be limited. Efficient estimators and optimal resource allocation schemes are therefore of particular need to minimize estimation error. Scaled estimators have reduced mean square estimation error and improved estimation efficiency. Useful scaling factors are identified for series, parallel, and other k-out-of-m systems. Resource allocations which are optimal for unbiased estimation of system reliability are shown to be optimal as well for scaled estimation.

I. Introduction

A system's reliability is often estimated from the estimated reliabilities of its components. Because of the potential number of components, the number of observations possible for each component may be limited, prompting a need for methods to improve the estimation of system reliability. Two such methods are scaling and resource allocation. Scaling involves applying a factor to the system reliability estimator to reduce the estimator's mean square error and was first proposed by Searls (1964). Resource allocation refers to the selection of sample sizes for estimating the reliabilities of the system components. Resource allocation can also be aimed at reducing the mean square error of the system reliability estimator and has been addressed for series systems (Page 1990, Berry 1974) and for some more general systems (Hardwick and Stout 1996). The purpose of this paper is to consider scaling separately and in combination with the resource allocation problem for k-out-of-m systems including, in particular, series systems and parallel systems.

The paper is organized as follows. In the remainder of this section we recall some basic facts and terminology for k-out-of-m systems. Basic properties of scaled estimation are reviewed and expanded upon in section II and a nomenclature for scaling factors is introduced. In sections III, IV, and V, scaling factors are identified for estimating the reliability of series, parallel, and other k-out-of-m systems. The impact of scaling on resource allocation is treated in section VI. We conclude in section VII with some remarks and a conjecture.

A system $\mathcal{S}$ with m components $\mathcal{A}_1, ..., \mathcal{A}_m$ is binary if $\mathcal{S}$ and each of its components have exactly two states: functioning and failed. A binary system $\mathcal{S}$ with m components is called a k-out-of-m system if $\mathcal{S}$ functions if and only if at least k of its components function (Birnbaum *et al.* 1961). Let $\mathcal{S}$ denote both the system and the event that the system functions. Similarly let $\mathcal{A}_j$ denote the jth component and the event that that component functions. Then the probability $s = P(\mathcal{S})$ is the system reliability and $a_j = P(\mathcal{A}_j)$ is the reliability of the component $\mathcal{A}_j$. The system components $\mathcal{A}_j$ are said to be independently functional (IF) if the $\mathcal{A}_j$, considered as events, are mutually independent. The reliability s of a k-out-of-m system with IF components is a function of just the component reliabilities a_j and is often estimated from estimates of the a_j. Let

$$\pi_r(a_1, ..., a_m)$$
$$= \sum_{j_1=1}^{m} \sum_{j_2=j_1+1}^{m} \cdots \sum_{j_r=j_{r-1}+1}^{m} a_{j_1} a_{j_2} \ldots a_{j_r} \quad (1)$$

for $r = 1, ..., m$ and let

$$f_k(a_1, ..., a_m)$$
$$= \sum_{r=k}^{m} (-1)^{r-k} \binom{r-1}{k-1} \pi_r(a_1, ..., a_m). \quad (2)$$

Then the reliability of a k-out-of-m system with IF components is $s = f_k(a_1, ..., a_m)$ (Parzen 1960); we call $f_k(\cdot, ..., \cdot)$ the system structure function. In particular, if $\mathcal{S}$ is a series (m-out-of-m) system, then

$$s = f_m(a_1, ..., a_m) = \prod_{j=1}^{m} a_j. \quad (3)$$

Or, if $\mathcal{S}$ is a parallel (1-out-of-m) system, then

$$s = f_1(a_1, ..., a_m) = 1 - \prod_{j=1}^{m}(1 - a_j).$$

For estimating the reliability s of a k-out-of-m system, we assume throughout that independent, unbiased estimators $A_1, ..., A_m$ are available, respectively, for the component reliabilities $a_1, ..., a_m$. We impose two regularity conditions: 1) the standard error of each estimator A_j, $j = 1, ..., m$ must be positive and 2) each estimator A_j must be proper, that is, $0 \leq A_j \leq 1$. The latter condition implies that $a_j \geq E[A_j^2]$, an inequality needed later.

Given independent, unbiased estimators A_1, ..., A_m for the component reliabilities, it is easily seen that

$$S = f_k(A_1, ..., A_m)$$

is an unbiased estimator of the reliability s of the corresponding k-out-of-m system. In particular, the estimator

$$S = f_m(A_1, ..., A_m) = \prod_{j=1}^{m} A_j \tag{4}$$

of the reliability of a series m-out-of-m system is unbiased. Similarly, the reliability estimator

$$S = f_1(A_1, ..., A_m) = 1 - \prod_{j=1}^{m} (1 - A_j)$$

of the parallel 1-out-of-m system is unbiased.

II. Scaled Estimation

Let T be an estimator of a parameter t. We define the mean square error of T to be

$$\text{MSE}(T) = E[\left(\frac{T - t}{t}\right)^2]. \tag{5}$$

Strictly speaking, this is the relative mean square error of T, the usual definition of mean square error being $E[(T - t)^2] = t^2\text{MSE}(T)$. However, because these definitions differ only by a constant factor, all our results are equally true for either. Definition (5) makes some later definitions and calculations simpler so we use it. The bias of T is correspondingly defined to be

$$\text{bias}(T) = E[\left(\frac{T - t}{t}\right)]. \tag{6}$$

Suppose T is an unbiased estimator of t with $\text{MSE}(T)$ defined as in (5). It is well-known (Searls 1964, Searls and Intarapanich 1990) that the mean square error involved in estimating t by T can be reduced by scaling T by a suitably chosen factor k. The mean square error of the scaled estimator kT is

$$\begin{aligned}
\text{MSE}(kT) &= E[\left(\frac{kT - t}{t}\right)^2] \\
&= \frac{k^2 E[T^2]}{t^2} - 2\frac{k E[T]}{t} + 1 \\
&= \frac{k^2}{k^*} - 2k + 1 \tag{7}
\end{aligned}$$

where

$$k^* = \frac{E^2[T]}{E[T^2]}. \tag{8}$$

It readily follows from (7) by elementary calculus that the factor k which minimizes the mean square error of kT is k^* in (8). Expression (7) for $\text{MSE}(kT)$ is graphed in Figure 1. Substituting each of $k = k^*$ and $k = 1$ in expression (7) for $\text{MSE}(kT)$ yields

$$\text{MSE}(T) = \frac{1}{k^*} - 1$$
$$\text{MSE}(k^*T) = 1 - k^*.$$

Therefore,

$$\frac{\text{MSE}(T)}{\text{MSE}(k^*T)} = \frac{1}{k^*} \tag{9}$$

and

$$\text{MSE}(k^*T) = 1 - \frac{1}{1 + \text{MSE}(T)} \tag{10}$$

is a strictly increasing function of $\text{MSE}(T)$. The optimal scaling factor k^* and the relative efficiency (9) of scaling were originally reported by Searls (1964).

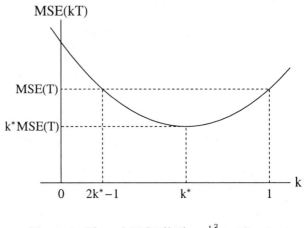

Figure 1. Plot of $MSE(kT) = \frac{k^2}{k^*} - 2k + 1$.

Figure 1 shows that while k^* minimizes the mean square error in (7), other values of k also reduce $\text{MSE}(kT)$ below $\text{MSE}(T)$. This prompts the following definitions. A scaling factor k of T is said to be *good* if $\text{MSE}(kT) < \text{MSE}(T)$. It is readily checked that k is a good scaling factor if and only if $k \in (2k^* - 1, 1)$. The bias associated with kT is from (6)

$$\text{bias}(kT) = E[\left(\frac{kT - t}{t}\right)] = k - 1.$$

The estimator T is unbiased so the scaled estimator kT is biased for $k \neq 1$. If the scaling factor is good, kT is negatively biased—yielding conservatively biased estimates of system reliability. The bias associated with a good scaling factor k may be greater or less than that associated with k^*. If k is a good scaling factor and

$$|\text{bias}(kT)| < |\text{bias}(k^*T)|,$$

then k is said to be a *better* scaling factor of T. A scaling factor k is better if and only if $k \in (k^*, 1)$. A good scaling factor may be negative (if $k^* < 1/2$) but all better scaling factors are positive. Finally, in keeping with this nomenclature, k^* is called the *best* scaling factor of T.

The best scaling factor k^* in (8) for the estimator T may alternatively be written

$$k^* = \frac{t^2}{t^2 + \sigma_T^2} \quad (11)$$

where σ_T is the standard error of T. Although the parameter t in (11) is not known, k^* can still be calculated in many instances. For the most commonly proffered example of this, suppose that a random sample of size n is drawn from a population with finite variance σ^2 and known coefficient of variation $\nu = \sigma/\mu$. And suppose the sample average $\bar{X}$ is to be used to estimate the population mean μ. The best factor k^* for scaled estimation of μ is from (11)

$$k^* = \frac{\mu^2}{\mu^2 + \sigma^2/n} = \frac{n}{n + \nu^2}. \quad (12)$$

In this case, knowing the coefficient of variation, k^* can be calculated. There are a variety of like situations in which k^* is known (Searls and Intarapanich 1990, Arnholt and Hebert 1995).

The situation in the most straightforward of experiments to measure component reliability differs distinctly from the example just outlined. Often, the reliability measurements for, say, system component $\mathcal{A}_j$ will simply be the 0–1 outcomes of testing the functionality of n_j independent copies of $\mathcal{A}_j$. The estimator A_j for the component reliability a_j would here typically be the proportion of the n_j components found to be functional. The mean and variance of A_j would then be a_j and $a_j(1 - a_j)/n_j$, respectively, and the best scaling factor for A_j would be

$$k_j^* = \frac{a_j^2}{a_j^2 + a_j(1 - a_j)/n_j} = \frac{n_j}{n_j + \frac{1 - a_j}{a_j}}.$$

In contrast to (12), knowing k_j^* in this situation is tantamount to knowing the reliability a_j. While a_j is usually unknown, a judgement estimate, $\hat{a}_j$, of a_j

may be available, based perhaps on expert knowledge or prior experience with a similar type of component. The resulting scaling factor

$$\hat{k}_j = \frac{n_j}{n_j + \frac{1 - \hat{a}_j}{\hat{a}_j}} \quad (13)$$

for A_j will not be best but, recalling Figure 1, will likely be good unless $\hat{a}_j$ is far from a_j. And if $\hat{a}_j > a_j$, then (13) is better. In either event, (13) might be used when k_j^* is unavailable. In more complex component reliability estimation scenarios involving, for example, reliability models with covariates, k_j^* may indeed be known without any prior knowledge of a_j. In the following sections we primarily assume that the best scaling factor k_j^* is known for each system component and seek, given this knowledge, a scaling factor for the system reliability estimator.

III. Series Systems

The reliability s of a series system with IF components is $s = \prod_{j=1}^{m} a_j$. Given unbiased, independent estimators $A_1, ..., A_m$ of the component reliabilities $a_1, ..., a_m$, $S = \prod_{j=1}^{m} A_j$ is an unbiased estimator of s. The best scaling factor k^* for S is

$$k^* = \frac{E^2[S]}{E[S^2]} = \frac{\prod_{j=1}^{m} E^2[A_j]}{\prod_{j=1}^{m} E[A_j^2]}$$
$$= \prod_{j=1}^{m} \frac{E^2[A_j]}{E[A_j^2]} = \prod_{j=1}^{m} k_j^*$$

where k_j^* is the best scaling factor for A_j. The structure function for a series system is a product so we have the following theorem.

Theorem 1: Let $S = f_m(A_1, ..., A_m)$ be the estimator of the reliability $s = f_m(a_1, ..., a_m)$ of a series system with m IF components. For each $j = 1, .., m$, let $k_j^* = a_j^2/E[A_j^2]$ be the best scaling factor for A_j. Then $k^\circ = f_m(k_1^*, ..., k_m^*)$ is the best scaling factor for S.

Theorem 1 is important because it provides a way in the case of series systems to calculate k^* when the best scaling factor k_j^* for each component is known. In some instances the k_j^* are unknown but substantive upper bounds, $k_j^* \leq \hat{k}_j < 1$, $j = 1, ..., m$, are available. By definition, these $\hat{k}_j$ are better scaling factors for the components and in such cases the following corollary to Theorem 1 is useful. The proof is straightforward and not included.

Corollary 1: Let $S = f_m(A_1, ..., A_m)$ be the estimator of the reliability s of a series system with m IF components. For each $j = 1, .., m$, let $\hat{k}_j$ be a better scaling factor for A_j. Then $\hat{k}^\circ = f_m(\hat{k}_1, ..., \hat{k}_m)$ is a better scaling factor for S.

IV. Parallel Systems

The reliability s of a parallel system with IF components is $s = 1 - \prod_{j=1}^{m}(1 - a_j)$. Given unbiased, independent estimators A_1, ..., A_m of the component reliabilities a_1, ..., a_m, $S = 1 - \prod_{j=1}^{m}(1 - A_j)$ is an unbiased estimator of s. For $m = 2$ the best scaling factor for S is

$$k^* = \frac{E^2[S]}{E[S^2]}$$

$$= \frac{k_1^* k_2^* (a_1 + a_2 - a_1 a_2)^2}{\left\{ \begin{array}{c} k_2^* a_1^2 + k_1^* a_2^2 + 2k_1^* k_2^* a_1 a_2 \\ -2a_1 k_1^* a_2^2 - 2a_2 k_2^* a_1^2 + a_1^2 a_2^2 \end{array} \right\}} \quad (14)$$

where

$$k_1^* = \frac{E^2[A_1]}{E[A_1^2]}, \quad k_2^* = \frac{E^2[A_2]}{E[A_2^2]} \quad (15)$$

are the best scaling factors of the component reliability estimators A_1 and A_2, respectively. For the m-out-of-m series system the best scaling factor k^* does not depend on the unknown component reliabilities and can be calculated. But in the case of the 1-out-of-2 parallel system, expression (14) shows that k^* does depend on a_1 and a_2 and cannot be calculated. This is true generally for any 1-out-of-m parallel system. While the best scaling factor k^* cannot be calculated for a parallel system, a better scaling factor which does not depend on the component reliabilities can be proposed.

We first consider the 1-out-of-2 case. For a 1-out-of-2 system with IF components, k^* has the expression in (14). Let

$$k^\circ = f_1(k_1^*, k_2^*) = k_1^* + k_2^* - k_1^* k_2^* \quad (16)$$

where k_1^* and k_2^* are the best scaling factors in (15) for A_1 and A_2, respectively. Regarding (16), we note first that $k^\circ < 1$ (because $f_1(\cdot, \cdot)$ is a system structure function and $0 < k_j^* < 1$, $j = 1, 2$). We note second that

$$\frac{k^\circ - k^*}{k^*} =$$

$$\frac{\left\{ \begin{array}{c} a_1^2 k_1^* (1 - k_1^*)(k_2^* - a_2)^2 \\ + a_2^2 k_2^* (1 - k_2^*)(k_1^* - a_1)^2 \\ + (a_1(1 - a_2)k_2^*(1 - k_1^*) \\ - a_2(1 - a_1)k_1^*(1 - k_2^*))^2 \end{array} \right\}}{k_1^* k_2^* s^2} \quad (17)$$

where $s = f_1(a_1, a_2)$ is the system reliability. This result is easily verified to follow from (14) and (16) using symbolic mathematics software such as *Mathematica*. It follows by inspection of (17) that $k^\circ > k^*$ and, therefore, that k° is a better scaling factor for S for the 1-out-of-2 system.

Now consider a parallel system with $m = 3$ IF components. For this case we employ the following notation:

$$k_{1\|2\|3}^* = \frac{E^2[f_1(A_1, A_2, A_3)]}{E[f_1^2(A_1, A_2, A_3)]}$$

is the best scaling factor for the parallel system with IF components $\mathcal{A}_1$, $\mathcal{A}_2$, and $\mathcal{A}_3$,

$$k_{1\|2}^* = \frac{E^2[f_1(A_1, A_2)]}{E[f_1^2(A_1, A_2)]}$$

is the best scaling factor for the parallel system with just the components $\mathcal{A}_1$ and $\mathcal{A}_2$, and

$$k_{(1\|2)\|3}^* = \frac{E^2[f_1(f_1(A_1, A_2), A_3)]}{E[f_1^2(f_1(A_1, A_2), A_3)]}$$

is the best scaling factor for the system with $\mathcal{A}_3$ in parallel with the parallel combination of $\mathcal{A}_1$ and $\mathcal{A}_2$. We note that

$$k_{1\|2\|3}^* = k_{(1\|2)\|3}^*$$

since $f_1(f_1(A_1, A_2), A_3) = f_1(A_1, A_2, A_3)$. For $j = 1, .., 3$, let $k_j^* = a_j^2/E[A_j^2]$ be the best scaling factor for A_j and define

$$k_{1\|2\|3}^\circ = f_1(k_1^*, k_2^*, k_3^*),$$
$$k_{1\|2}^\circ = f_1(k_1^*, k_2^*),$$
$$k_{(1\|2)\|3}^\circ = f_1(k_{1\|2}^*, k_3^*).$$

It follows from our conclusions for $m = 2$ that

$$k_{(1\|2)\|3}^* < k_{(1\|2)\|3}^\circ, \quad k_{1\|2}^* < k_{1\|2}^\circ. \quad (18)$$

We have then that

$$k_{1\|2\|3}^\circ = 1 - (1 - k_1^*)(1 - k_2^*)(1 - k_3^*)$$
$$= 1 - (1 - k_{1\|2}^\circ)(1 - k_3^*)$$
$$> 1 - (1 - k_{1\|2}^*)(1 - k_3^*)$$
$$= k_{(1\|2)\|3}^\circ$$
$$> k_{(1\|2)\|3}^*$$
$$= k_{1\|2\|3}^*.$$

This shows that $k^\circ = f_1(k_1^*, k_2^*, k_3^*)$ is a better scaling factor for estimating the reliability of a parallel system with 3 IF components. Generalizing the preceding calculation and using mathematical induction, a similar statement can be made for parallel systems with any number of components:

Theorem 2: Let $S = f_1(A_1, ..., A_m)$ be the estimator of the reliability $s = f_1(a_1, ..., a_m)$ of a parallel system with $m \geq 2$ IF components. For each $j = 1, .., m$, let $k_j^* = a_j^2/E[A_j^2]$ be the best scaling

162

factor for A_j. Then $k^\circ = f_1(k_1^*, ..., k_m^*)$ is a better scaling factor for S.

Corresponding to Corollary 1 is the following corollary to Theorem 2.

Corollary 2: Let $S = f_1(A_1, ..., A_m)$ be the estimator of the reliability s of a parallel system with $m \geq 2$ IF components. For each $j = 1, .., m$, let $\hat{k}_j$ be a better scaling factor for A_j. Then $\hat{k}^\circ = f_1(\hat{k}_1, ..., \hat{k}_m)$ is a better scaling factor for S.

V. Other k-out-of-m Systems

We consider the 2-out-of-3 system. This system is neither series nor parallel. The reliability s of the 2-out-of-3 system with IF components is

$$s = f_2(a_1, a_2, a_3)$$
$$= a_1 a_2 + a_2 a_3 + a_1 a_3 - 2a_1 a_2 a_3.$$

Given unbiased, independent estimators, A_1, A_2, and A_3, of the component reliabilities a_1, a_2, and a_3,

$$S = f_2(A_1, A_2, A_3)$$

is an unbiased estimator of s. The best scaling factor

$$k^* = \frac{E^2[S]}{E[S^2]}$$

for S depends on the unknown component reliabilities, a_1, a_2 and a_3, and cannot be calculated. But, again, the scaling factor

$$k^\circ = f_2(k_1^*, k_2^*, k_3^*) \tag{19}$$

is available in terms of just the best scaling factors, k_1^*, k_2^*, and k_3^*, for A_1, A_2, and A_3, respectively. To see that k° in (19) is a better scaling factor for S, note, first, that $k^\circ < 1$ and, second, that

$$\frac{k^\circ - k^*}{k^*} =$$
$$\frac{1}{k_1^* k_2^* k_3^* s^2} \left\{ a_3^2 k_3^* (1 - k_3^*)(a_1 k_2^* - a_2 k_1^*)^2 \right.$$
$$+ a_1^2 k_1^* (1 - k_1^*)(a_2 k_3^* - a_3 k_2^*)^2$$
$$+ a_2^2 k_2^* (1 - k_2^*)(a_1 k_3^* - a_3 k_1^*)^2$$
$$+ 4a_1 a_2 a_3^2 k_1^* k_2^* (1 - k_3^*)(k_1^* - a_1)(k_2^* - a_2)$$
$$+ 4a_1 a_3 a_2^2 k_1^* k_3^* (1 - k_2^*)(k_1^* - a_1)(k_3^* - a_3)$$
$$+ 4a_2 a_3 a_1^2 k_2^* k_3^* (1 - k_1^*)(k_2^* - a_2)(k_3^* - a_3)$$
$$+ (a_1 k_1^* (a_3 k_2^* - a_2 k_3^*)$$
$$+ a_2 k_2^* (a_1 k_3^* - a_3 k_1^*)$$
$$\left. + a_3 k_3^* (a_2 k_1^* - a_1 k_2^*))^2 \right\}. \tag{20}$$

This result is easily verified using symbolic mathematics software. Recall that A_1, A_2, and A_3 are

assumed to be proper estimators so that, for example, $k_1^* \geq a_1$. Then, inspection of (20) shows that $k^\circ > k^*$ and, therefore, that k° is a better scaling factor for S for the 2-out-of-3 system.

VI. Resource Allocation

To estimate the reliability of a k-out-of-m system with IF components, a direct sampling plan is often employed in which a fixed but arbitrary number n_j of reliability measurements are made for each component $\mathcal{A}_j$. Typically, the reliability measurements will be the 0–1 outcomes of testing the functionality of n_j copies of each of the components $\mathcal{A}_j$ and the estimators A_j will be the proportions of functioning components in each sample. Direct sampling is fairly standard (Page 1990, Berry 1974), but inverse and sequential sampling plans might be used as well (Hwang and Buehler 1973, Shapiro 1985, Hardwick and Stout 1996).

Because of the potential number of components, the number of reliability measurements possible for each system component may be limited, prompting the need for a resource allocation scheme to optimally assign sample sizes. The resource allocation problem for direct sampling is to select the sizes $\vec{n} = (n_1, ..., n_m)$ of the samples of reliability measurements to make for each component $\mathcal{A}_j$. This selection can be made based on a variety of different optimality criteria under any number of different constraints $\vec{n} \in \mathcal{B}$ where $\mathcal{B}$ is the set of allowed allocations. The resource allocation $\vec{n}$ is often chosen to minimize the mean square error of the reliability estimator S subject to the condition $\vec{n} \in \mathcal{B}$ where

$$\mathcal{B} = \{\vec{n} = (n_1, ..., n_m) : \sum_{j=1}^{m} b_j n_j \leq B\},$$

B is a given fixed budget, and there is a cost b_j per reliability measurement for each component $\mathcal{A}_j$. Allocation of a fixed total number b of measurements among the m components is a special case of a fixed budget constraint, where $B = b$ and $b_j = 1$. The allocation problem has been treated by Berry (1974) and Page (1990) from Bayesian and frequentist perspectives, respectively. Page assumes that the coefficients of variation of the component reliability measurements are known; this, according to (12), is equivalent to knowing the best scaling factor, k_j^*, for each component.

One noteworthy resource allocation scheme for direct sampling is balanced allocation, where allocation of a fixed total number of observations is made in proportion to population standard deviations. This allocation minimizes the sum of the mean square errors of the sample means. Also of interest is the cv-allocation scheme, where allocations are made proportional to the population coefficients of variation. The cv-allocation scheme approximately minimizes the mean square error in estimating a product of parameters (Page 1990). In particular, cv-allocation is approximately optimal for estimating the series system reliability s in (3) using the estimator S in (4).

Consider expression (10) applied to the problem of estimating the reliability s of a k-out-of-m system $\mathcal{S}$ using the unbiased estimator $S = f_k(A_1, ..., A_m)$. The implications of this simple result are significant. According to (10), any change in the method of component reliability measurement, choice of sampling plan, or allocation of sampling resources which reduces the mean square error of S, necessarily also reduces the mean square error of the scaled estimator k^*S. Equivalently, the optimal method of component reliability measurement for S, the optimal type of sampling plan for S, and the optimal resource allocation for S are all optimal as well for k^*S. We have the following result in particular.

Theorem 3: Let S be an unbiased estimator of the reliability s of a system $\mathcal{S}$. Suppose the resource allocation $\vec{n}$ for estimating s is restricted to $\vec{n} \in \mathcal{B}$. The resource allocation $\vec{n}_o$ which minimizes the mean square error of S also minimizes the mean square error of the optimally scaled estimator k^*S.

Proof: Relation (10) is strictly increasing so

$$\min_{\vec{n} \in \mathcal{B}} \mathrm{MSE}(k^*S) = 1 - \frac{1}{1 + \min_{\vec{n} \in \mathcal{B}} \mathrm{MSE}(S)}.$$

Therefore

$$\vec{n}_o \equiv \operatorname*{argmin}_{\vec{n} \in \mathcal{B}} \mathrm{MSE}(S) = \operatorname*{argmin}_{\vec{n} \in \mathcal{B}} \mathrm{MSE}(k^*S).$$

$\square$

Theorem 3 implies, for example, that the direct sampling cv-allocation $\vec{n}$ which is approximately optimal for estimating the reliability of a series system by S is also approximately optimal for the scaled estimator k^*S.

VII. Conclusion

It is often possible to determine the best scaling factors k_j^*, $j = 1, ..., m$, for estimating the reliabilities of the components of a system. Inserting these known scaling factors into the structure function for a k-out-of-n system yields the scaling factor $k^\circ = f_k(k_1^*, ..., k_m^*)$ with no further required information. For series systems, we have shown that k° is the best scaling factor for estimating the system reliability. For parallel systems, we have established that while k° is not the best scaling factor for estimating the system reliability, it is a better scaling

factor. And we have shown that k° is a better scaling factor for the 2-out-of-3 system—a system which is neither series nor parallel. Considerable empirical evidence (not presented here) from computer experiments has been amassed to suggest that k° is a better scaling factor for the 2-out-of-4, 3-out-of-4, 2-out-of-5, 3-out-of-5, and 4-out-of-5 systems. We conjecture that k° is a better scaling factor for all non-series k-out-of-m systems.

Finally, although outside the scope of this paper, we mention the case of parallel-series and series-parallel combination systems. These classes of systems do not fall within the class of k-out-of-m systems but $k^\circ = f_k(k_1^*, ..., k_m^*)$ is also a better scaling factor for estimating the reliability of these systems. This is easy to show from our results for series systems and parallel systems but there is not space here to do so.

References

Arnholt, A.T. and Hebert, J.L. (1995), "Estimating the Mean with Known Coefficient of Variation," *The American Statistician* **49**, 367–369.

Berry, D.A. (1974), "Optimal Sampling Schemes for Estimating System Reliability by Testing Components," *Jour. of the American Statistical Association* **69**, 485–491.

Birnbaum, Z.W., Esary, J.D., and Saunders, S.C. (1961), "Multicomponent Systems and Structures and their Reliability," *Technometrics* **3**, 55–77.

Hardwick, J.P. and Stout, Q.F. (1996), "Optimal Allocation for Estimating the Mean of a Bivariate Polynomial," *Sequential Analysis* **15**, 71–90.

Hwang, D.S. and Buehler, R.J. (1973), "Confidence Intervals for some Functions of Several Bernoulli Parameters with Reliability Applications," *Jour. of the American Statistical Association* **68**, 211–217.

Page, C.F. (1990), "Allocation Proportional to Coefficients of Variation When Estimating the Product of Parameters," *Jour. of the American Statistical Association* **85**, 1134–1139.

Parzen, E. (1960), *Modern Probability Theory and Its Applications*, Wiley, New York.

Searls, D.T. (1964), "The Utilization of a Known Coefficient of Variation in the Estimation Procedure," *Jour. of the American Statistical Association* **59**, 1225–1226.

Searls, D.T. and Intarapanich, P. (1990), "A Note on an Estimator for the Variance that Utilizes Kurtosis," *The American Statistician* **44**, 295–296.

Shapiro, C.P. (1985), "Allocation Schemes for Estimating the Product of Positive Parameters," *Jour. of the American Statistical Association* **80**, 449–454.

INFERRING POPULATION SIZE FROM VALUES OF EXTREME ORDER STATISTICS

Yigal Gerchak, University of Waterloo, Ishay Weissman, Technion
Ishay Weissman, Faculty of Industrial Engineering and Management,
Technion, Israel Institute of Technology, Haifa 32000, Israel

Keywords: Maxima, Minima, Range

1 Introduction

The topic *population size estimation* has a long history in the scientific literature. The earliest reference known to the authors is Hjort and Ottestad (1933) which deals with estimating the size of the bear population of Norway in 1914. Since then, many techniques have been developed and applied to populations whose sizes cannot be easily counted. *Capture-recapture, removal, resighting* are some of the keywords in this area. Dahiya and Blumenthal (1984) give a survey of the literature on *population or sample size estimation*. Most of the papers surveyed deal with samples, $X_1, \ldots, X_n$ with unknown n, where only values outside a certain region R are observable. If M out of n are observable, $X_1, \ldots, X_M$ say, we want to find an estimate of n, based on these data. Examples include Binomial (n, p) random variables, where both n and p are unknown; Poisson (λ) variables, where only the nonzero values are observable; censored samples at time T, namely, X_i is observable only if $X_i \leq T$. In all of these examples, estimates of n and the other unknown parameters are developed. This methodology can also be applied to estimate the total number of drug traffickers in an airport, based on observation of those that were apprehended, or for estimating the total number of bugs in a software, based on those discovered (see Osborne and Severini (1999) for references on *software reliability*).

This work deals with estimating the size of a population or group from observations on certain sample statistics, and in particular the extreme "scores" of its members in several "tests" or attributes. A case in point is the number of chess players, by age and gender, in various countries (e.g., Charness and Gerchak 1996). At the same time, the top scores that members of these groups attained on various dimensions or in certain tasks might be readily available. For example, for some group of people we might know the height of the tallest person, the age of the oldest and the IQ of the smartest. As for chess, we usually know the top scores each population attains in tournaments, as well as ratings (Elo, 1986). The challenge is then to infer the population size from observations about such extreme order statistics. This is an "inverse" problem to the common direction of inference in relation to order statistics (e.g. Arnold et. al., 1992) and appears to have been hitherto unexplored.

We shall make the fairly strong assumption that the scores of each individual on the various tests are mutually independent. That is plausible for some sets of attributes (e.g., height, I.Q., and blood pressure) but not for others (e.g., scores on similar tests). Note, however, that for large populations the highest (lowest) achievements on tests will tend to have a much lower correlation than a given individual's scores.

2 The General Setting

Let $X_{i1}, \ldots, X_{in}$ be i.i.d. random variables pertaining to the i-th of m independent attributes ("tests"), $i = 1, \ldots, m$, with test-specific CDF's F_i, possibly unknown. Suppose we have at our disposal, for each test $i = 1, \ldots, m$, only the value of some statistic $T_i = T(X_{i1}, \ldots, X_{in})$, while the population/sample size n is unknown. The main purpose is to estimate n.

In principle, estimating n is no different from estimating other unknown parameters. However, some difficulties may arise:

1. *Unidentifiability.* For example, suppose that $X_{ij} \sim N(\mu, \sigma^2)$, and $T_i = \sum_{j=1}^{n} X_{ij}/n$, $i = 1, \ldots, m$, is the i-th sample average. It follows that $T_i \sim N(\mu, \sigma^2/n)$. Hence $\tau^2 \equiv \sigma^2/n$ can be estimated (by the sample variance of the T_i's), but if σ^2 is unknown n is an *unidentifiable* parameter.

2. *Complexity.* For example, suppose that the X_{ij}'s are Bernoulli variables with parameter p, and $T_i = \sum_{j=1}^{n} X_{ij}$. Even if p is known, the log-likelihood is of the form

$$\{m \log(1 - p)\}n + \sum_{i=1}^{m} \log\{n!/(n - T_i)!\},$$

so the function to be maximized to find the maximum likelihood estimator (MLE) of n is quite complex. Of course, if the method of moments is used then $\hat{n} = \bar{T}/p$ is an unbiased, simple estimator (here $\bar{T} = \sum T_i/m$).

We shall focus on statistics T_i of the type

(a) $Y_i = \max(X_{i1}, \ldots, X_{in})$

(b) $Z_i = \min(X_{i1}, \ldots, X_{in})$

(c) (Y_i, Z_i)

(d) $R_i = Y_i - Z_i$.

We shall assume that the F_i's are known.

3 Only Maxima (or Minima) Available

Let $X_{ij} \sim F_i$, $j = 1, \ldots, n$, $i = 1, \ldots, m$, be independent and F_i continuous. Define $Y_i = \max_j X_{ij}$. Then $Y_i \sim F_i^n$, and it can be shown that $-\log F_i(Y_i) \sim \mathrm{Exp}(n)$ (exponential variate with scale parameter n). Hence if the F_i are known, the MLE and minimum variance unbiased estimator (MVUE) of n will be:

$$\text{MLE} \qquad \hat{n} = m / \sum_{i=1}^{m} - \log F_i(Y_i) \; ;$$

$$\text{MVUE} \qquad n^* = (m-1) / \sum_{i=1}^{m} - \log F_i(Y_i) \; .$$

Note that the estimators of population size are increasing in the maxima, as one would expect.

Remark. Let

$$L(n) = \prod_{i=1}^{m} n F_i(Y_i)^{n-1} f_i(Y_i)$$

be the likelihood function. If $\hat{n}$ is not an integer, the exact MLE is either $[\hat{n}]$, if $L([\hat{n}]) \geq L([\hat{n}] + 1)$ or $[\hat{n}] + 1$, if the reverse inequality holds. Similarly, if we obtain $\hat{n}$ from an equation $L(n) = L(n-1)$, the exact MLE is $[\hat{n}]$. We shall not dwell on this point in the sequel.

For this setup (known and continuous F_i with $\bar{F}_i = 1 - F_i$) one can also use the method of moments (though the above estimators are based on a sufficient statistic). Since

$$E \bar{F}_i(Y_i) = 1/(n+1) \; ,$$

then the *moment estimator* of n is given by

$$\tilde{n} = m / \sum_{i=1}^{m} \bar{F}_i(Y_i) - 1 \; .$$

If only **minima** are available, let $Z_i = \min_j X_{ij}$. Under same conditions $-\log \bar{F}_i(Z_i) \sim \mathrm{Exp}(n)$. Hence

$$\text{MLE} \qquad \hat{n} = m / \sum_{i=1}^{m} - \log \bar{F}_i(Z_i) \; ;$$

$$\text{UMVUE} \qquad n^* = (m-1) / \sum_{i=1}^{m} - \log \bar{F}_i(Z_i) \; .$$

These estimators are decreasing in the minima, as one would expect.

4 Both Maxima and Minima Available

The joint density of the maximum and minimum from a sample of size n [e.g., Arnold et al. (1992, eq. 2.3.7)] is

$$g(z, y) = n(n-1) \{ F(y) - F(z) \}^{n-2} f(z) f(y) \; ,$$

where $f = F'$ exists. Thus the likelihood function here is

$$L = \prod_{i=1}^{m} n(n-1) \{ F_i(Y_i) - F_i(Z_i) \}^{n-2} f_i(Z_i) f_i(Y_i),$$

and thus

$$\begin{aligned} \log L &= m \log\{n(n-1)\} \\ &+ \sum_{i=1}^{m} \{\log f_i(Z_i) + \log f_i(Y_i)\} \\ &+ (n-2) \sum_{i=1}^{m} \log\{F_i(Y_i) - F_i(Z_i)\} \; . \end{aligned}$$

So

$$\begin{aligned} \partial \log L / \partial n &= m(2n-1)/n(n-1) \\ &+ \sum_{i=1}^{m} \log\{F_i(Y_i) - F_i(Z_i)\} \; , \end{aligned}$$

and

$$\begin{aligned} \partial(\log L)^2 / \partial n^2 \\ = m\{-n^2 - (n-1)^2\}/n^2(n-1)^2 < 0, \end{aligned}$$

so L is concave. Denoting $\sum_{i=1}^{m} \log\{F_i(Y_i) - F_i(Z_i)\}$ by $-A$, the optimality condition is

$$m(2n-1)/n(n-1) = A, \qquad (1)$$

and thus

$$An^2 - (2m + A)n + m = 0.$$

166

The only relevant solution of this quadratic equation is

$$\hat{n} = \left(A + 2m + \sqrt{A^2 + 4m^2} \right) /2A, \qquad (2)$$

and this is then the MLE of the population size. Since A is decreasing in the Y_i's and increasing in the Z_i's, if we can show that $\hat{n}$ is decreasing in A, then $\hat{n}$ will be, as expected, increasing in the Y_i's and decreasing in the Z_i's. To see that $\hat{n}$ is decreasing in A, note that

$$\frac{\partial \hat{n}}{\partial A} = \frac{m}{A^2} \left(\frac{2m}{\sqrt{A^2 + 4m^2}} - 1 \right) < 0 .$$

In some of the motivating applications the (unknown) population size is quite large. Now, if $n \to \infty$ then Y_i and Z_i tend to be independent (see Galambos (1987), p. 129). If indeed they were independent, then the MLE for n would be

$$\tilde{n} = 2m/ \left\{ - \sum_{i=1}^{m} \log F_i(Y_i) - \sum_{i=1}^{m} \log \bar{F}_i(Z_i) \right\} .$$

If we use the optimality condition (1), and let $n \to \infty$ then (2) becomes

$$n^* = 2m/A ,$$

which is slightly different from $\tilde{n}$. All three are consistent, of course, when $m \to \infty$.

5 Only Ranges Available

Let $R_i = Y_i - Z_i$, and suppose we are given only $R_1, ..., R_m$. The density of the range [e.g., Arnold et al. (1992, eq. 2.5.15)] is

$$h(r) = n(n-1) \int_{-\infty}^{\infty} \{F(u+r) - F(u)\}^{n-2}$$
$$\cdot f(u)f(u+r)du \equiv n(n-1)g(n).$$

Thus the likelihood function here is

$$L = \prod_{i=1}^{m} n(n-1)g_i(n),$$

where g_i is g with $F = F_i$ and $r = R_i$.
Thus

$$\partial L/\partial n = \{n(n-1)\}^{m-1} \prod_{i=1}^{m} g_i(n)$$
$$\cdot \left\{ m(2n-1) + n(n-1) \sum_{i=1}^{m} \{\log g_i(n)\}' \right\} .$$

Hence, the optimality condition is

$$-\{n(n-1)/(2n-1)\} \sum_{i=1}^{m} \{\log g_i(n)\}' = m.$$

Due to the complexity encountered using the maximum likelihood method here, consider the *method of moments*. The expected range equals

$$\begin{aligned} E(R_i) &= E(Y_i) - E(Z_i) \\ &= n \int_{-\infty}^{\infty} xf_i(x)\{\{F_i(x)\}^{n-1} \\ &\quad - \{\bar{F}_i(x)\}^{n-1}\}dx . \end{aligned}$$

Then the *moment estimate* of n is the value for which

$$n \sum_{i=1}^{m} \int_{-\infty}^{\infty} xf_i(x)\{\{F_i(x)\}^{n-1} \\ - \{\bar{F}_i(x)\}^{n-1}\}dx = \sum_{i=1}^{m} R_i .$$

Example: Suppose $f_i(x) = \lambda_i e^{-\lambda_i x}$ ($x > 0$). In this case, $g_i(n) = e^{-\lambda_i R_i} \{1 - e^{-\lambda_i R_i}\}^{n-2} \lambda_i/n$ and the optimality condition gives the MLE for n as

$$\hat{n} = 1 + \frac{m}{- \sum_1^m \log \left(1 - e^{-\lambda_i R_i}\right)} .$$

For the method of moments estimator (MOME), we note that $E(Y_i) = (1 + 1/2 + \cdots + 1/n)/\lambda_i$ and $E(Z_i) = 1/n\lambda_i$. Hence, if we define $S_k = \sum_{i=1}^{k} i^{-1}$, then we have to solve

$$S_{n-1} = \frac{\sum_1^m R_i}{\sum_1^m 1/\lambda_i} \equiv B ,$$

namely, $\tilde{n} = \min\{k : S_{k-1} \geq B\}$. For large n, $\tilde{n} \approx \exp(B)$.

6 Underlying Distribution's Parameter(s) Unknown

So far we have assumed that the underlying distributions F_i are completely known. Often, however, while the form of these distributions might be known, the values of their parameters are not.

For concreteness, suppose that F_i's are *exponential* with same unknown mean $1/\lambda$. If we are provided with *maxima* only, the likelihood function is

$$\begin{aligned} L &= \prod_{i=1}^{m} n\lambda e^{-\lambda Y_i} \left(1 - e^{-\lambda Y_i}\right)^{n-1} \\ &= n^m \lambda^m e^{-\lambda \sum_{i=1}^{m} Y_i} \prod_{i=1}^{m} \left(1 - e^{-\lambda Y_i}\right)^{n-1} \end{aligned}$$

so

$$\begin{aligned} \log L = &\, m \log n + m \log \lambda \\ &- \lambda \sum_{i=1}^{m} Y_i + (n-1) \sum_{i=1}^{m} \log \left(1 - e^{-\lambda Y_i}\right) . \end{aligned}$$

167

We need to maximize this function jointly with respect to λ and n.

Now,

$$\partial \log L/\partial n = m/n + \sum_{i=1}^{m} \log\left(1 - e^{-\lambda Y_i}\right)$$

$$\partial \log L/\partial \lambda = m/\lambda - \sum_{i=1}^{m} Y_i$$
$$+ (n-1)\sum_{i=1}^{m} Y_i e^{-\lambda Y_i}/\left(1 - e^{-\lambda Y_i}\right).$$

Thus, assuming that $\log L$ is jointly concave in n and λ, we can proceed as follows. Since $\partial \log L/\partial n = 0$ translates to

$$n = -m/\sum_{i=1}^{m} \log\left(1 - e^{-\lambda Y_i}\right), \qquad (3)$$

substituting into $\partial \log L/\partial \lambda = 0$ we get

$$\frac{m}{\lambda} + \sum_{i=1}^{m} \frac{Y_i e^{-\lambda Y_i}}{1 - e^{-\lambda Y_i}}\left\{\frac{m}{-\sum \log\left(1 - e^{-\lambda Y_i}\right)} - 1\right\}$$
$$= \sum_{1}^{m} Y_i,$$

from which λ can be solved numerically and the corresponding n then found from (3). Here too, the method of moments can be used. If we define $V_n = \sum_{i=1}^{n} i^{-2}$, then the equations

$$\bar{Y} = \frac{1}{m}\sum_{i=1}^{m} Y_i = S_n/\lambda$$

$$\frac{1}{m}\sum_{i=1}^{m}(Y_i - \bar{Y})^2 = V_n/\lambda^2$$

yield a unique solution for n and λ.

Under the same model, if only *minima* are given, and λ is unknown, n is *unidentifiable* since $Z_i \sim \text{Exp}(\lambda n)$.

7 Comparisons with Other Situations

7.1 One Sample vs. m

Suppose that instead of m maxima of m independent attributes, the available data are the m largest of one attribute.

Let $X_{(1)} \geq \ldots \geq X_{(m)}$ be the m largest out of a sample of size n from the parent CDF F, with density f. If F is known, the likelihood for n is

$$L(n) = n(n-1)\ldots(n-m+1)F(X_{(m)})^{n-m}\prod_{i=1}^{m} f(X_{(i)}),$$

which incidently shows that $X_{(m)}$ is sufficient for n.

For $\ell(n) = \log L(n)$ we have $\ell(n) - \ell(n-1) = 0$ if and only if $\bar{F}(X_{(m)}) = m/n$ from which we get

$$\text{MLE} \quad \hat{n} = m/\bar{F}(X_{(m)}). \qquad (4)$$

Comparing to Section 6, suppose $\bar{F}(x) = e^{-\lambda x}$ $(x > 0)$ with unknown λ. It follows that

$$\text{MLE} \quad \hat{\lambda} = m \Big/ \sum_{i=1}^{m}\left(X_{(i)} - X_{(m)}\right),$$

and we note that $\hat{\lambda}$ is free of n. Substituting $\hat{\lambda}$ in (4) yields the MLE for n. Thus the situation here is even simpler than the case with m independent maxima.

The situation with m lower extremes is similar if F is known, but might be entirely different if F is not completely known. Let $X_1^* \leq \cdots \leq X_m^*$ be the m lowest extremes. The likelihood for n is the same as above, except that F is replaced by $\bar{F}$ and $X_{(i)}$ by X_i^*. Thus, we have

$$\text{MLE} \quad \hat{n} = m/F(X_m^*). \qquad (5)$$

If again F is exponential with unknown λ, then as is well known

$$\text{MLE} \quad \hat{\lambda} = \frac{m}{\sum_1^m X_i^* + (n-m)X_m^*}. \qquad (6)$$

Here $\hat{\lambda}$ depends on n. Solving for n and λ (Equations (5) and (6)), it turns out that $\hat{n}$, the MLE for n, is the solution of

$$\frac{X_m^* - \bar{X}}{X_m^*} = \frac{n}{m} + \frac{1}{\log\left(1 - \frac{m}{n}\right)}, \qquad (7)$$

where $\bar{X} = \sum_1^m X_i^*/m$ is the average of the m lower sample extremes. Since the right-hand side of (7) is in $(1/2, 1]$ for $n \geq m$, there is no solution when $\bar{X}/X_m^* > 1/2$. This latter event can occur with positive probability. Osborne and Severini (1999) suggest the *integrated likelihood* approach instead of the MLE to overcome this difficulty.

7.2 Maxima vs. Sample Means

Suppose the available data are m sample means (rather than m sample maxima). Namely, suppose $X_{ij} \sim \mathcal{N}(0,1)$, $(j = 1,\ldots,n;\ i = 1,\ldots,m)$ and we have at our disposal $T_i = \sum_{j=1}^{n} X_{ij}/n$ $(i = 1,\ldots,m)$. Then, since $T_i \sim \mathcal{N}(0, 1/n)$, the MLE for n is

$$\hat{n}_T = m \Big/ \sum_{i=1}^{m} T_i^2$$

and the MVUE is

$$n_T^* = (m-2) \Big/ \sum_{i=1}^{m} T_i^2.$$

Hence, a simple exercise shows that the minimum variance attained for estimating n is

$$\text{Var } n_r^* = 2n^2/(m-4).$$

For comparison, the variance of the MVUE for n, based on maxima (see Section 3) is

$$\text{Var } n^* = n^2/(m-2).$$

Hence, maxima are more efficient than averages (or sums) by at least a factor of 2.

8 Conclusion

We have demonstrated that sample size can be estimated based on sample extremes quite simply if the underlying distributions are completely known. In parametric models, where some parameters are unknown, there are cases where n is *unidentifiable* (e.g., minima of exponential rv's). For other cases, solutions are not in closed forms and can only be found numerically. The efficiency of these estimators needs further study. Finally, we have shown that as far as estimation of n is concerned, extremes are more efficient than averages or sums.

References

[1] Arnold, B.C., N. Balakrishnan and H.N. Nagaraja (1992) *A First Course in Order Statistics*, Wiley.

[2] Charness, N. and Y. Gerchak (1996) "Participation Rates and Maximal Performance", *Psychological Science*, 7, 46–51.

[3] Dihiya, R.C. and S. Blumenthal (1984) "Population or Sample Size Estimation", in *Encylopedia of Statistical Sciences*, Wiley.

[4] Elo, A.E. (1986) *The Ratings of Chessplayers, Past and Present*, 2nd ed., New York: Arco.

[5] Galambos, J. (1987) *The Asymptotic Theory of Extreme Order Statistics*, R.E. Krieger Pub. Co.

[6] Hjort, J.G. and P. Ottestad (1933) "The Optimum Catch Hvalradets Skrifter", *Oslo*, 7, 92–127.

[7] Osborne, J.A. and T.A. Severini (1999) "Inference for Exponential Order Statistic Models Based on an Integrated Likelihood Function", Technical report, Auburn University.

PREDICTING RECORD-BREAKING SEQUENCES
OF EVENTS

N. I. Lyons and K. Hutcheson, University of Georgia

K. Hutcheson, Dept. of Statistics, University of Georgia, Athens, GA 30602

Key Words: order statistics, geometric distribution, Bernoulli trials, longest run.

Schilling (1994) considers several examples of record-breaking events, or longest runs of successes in a sequence of Bernoulli trials. One example is Joe DiMaggio's hitting streak. In 1941 DiMaggio had at least one hit in each of 56 consecutive baseball games. Could this streak have been predicted, or was it truly unusual? Schilling addresses this problem by deriving a prediction interval based on an asymptotic approximation to distribution of the longest run. We propose an alternative based on the small sample properties of the geometric distribution, which appears to be robust with respect to certain violations in the underlying assumptions.

PREVIOUS RESULTS

Consider a sequence of n Bernoulli trials with probability of success p. Schilling notes that $N_R = n(1-p)p^R$ is the approximate expected number of runs of successes of length at least R. For $N_R = 1$,

$$R_n = -\ln[n(1-p)]/\ln p,$$

is an approximation to the expected value of the longest run. Gordon, Schilling and Waterman (1986) use an extreme value distribution of the form

$$F(x) = \exp(-p^{x+1})$$

to approximate the distribution of $R - R_n$, where R is the longest run in n trials. Equating $F(x)$ to $\alpha/2$ and $1 - \alpha/2$, they obtain the prediction interval

$$\left(R_n + \frac{\ln[-\ln(\alpha/2)]}{\ln p}, \ R_n + \frac{\ln[-\ln(1-\alpha/2)]}{\ln p} \right),$$

which contains the middle $(1-\alpha)100\%$ of the sample values. If the longest observed run of successes falls in this interval, the run is considered to be predictable. A prediction interval can also be obtained from the properties of the geometric distribution.

Let X be the number of successes before the first failure in a sequence of Bernoulli trials. Excluding runs of length zero, the probability function of X is

$$P(X = x) = (1 - p)p^{x-1}, \qquad x = 1, 2, 3, ...,$$

The expected value of X is

$$E(X) = 1 - \frac{1}{p},$$

and the cumulative distribution function is

$$F(x) = (1 - p) \sum_{j=1}^{x} p^{j-1} = 1 - p^x, \qquad x = 1, 2, ...$$

Now fix m, the number of runs in a sequence of trials, and let Let $X_1, ... X_m$ be a random sample of m runs (or equivalently, the first m runs), with

$$X_{(1)} \le X_{(2)} \le ... \le X_{(m)}$$

the order statistics from this sample. Then the c.d.f. of the largest order statistic,

$$F_{X_{(m)}}(x) = P(X_{(m)} \le x) = (1 - p^x)^m,$$

can be equated to $\alpha/2$ and $1 - \alpha/2$, to obtain the alternative prediction interval

$$\left(\frac{\ln[1 - (\alpha/2)^{1/m}]}{-\ln p}, \frac{\ln[1 - (1-\alpha/2)^{1/m}]}{-\ln p} \right).$$

For $\alpha = .05$ the prediction limits the two intervals are nearly the same even for small n provided p is known. Simulation results indicate that both intervals are conservative in their coverage.

For the DiMaggio example, p and n must be determined. Schilling estimates that there have been at least $500,000$ player-games in major league baseball from 1901 to 1994. Using the top 20 percent of all major league players during this period, Schilling arrives at a combined batting average of approximately .300. Assuming four at-bats per game, the probability of at least one success is

$$P(\text{at least one hit}) = 1 - (1 - .300)^4 = .76$$

Using this value of p with $n = 500,000$ player-games gives:

Schilling's $R_n = 42.6$
Geometric Median = 44.0
Prediction Interval: $(38.9, 56.0)$.

Both prediction intervals give the same result. DiMaggio's accomplishment was at the upper limit, and therefore predictable. The table below shows the upper limits for the prediction intervals for different values of p.

Batting Average	p	Upper Limit
.260	.700	43.1
.270	.726	48.0
.280	.731	49.1
.290	.745	52.2
.300	.760	56.0
.310	.760	59.7
.320	.786	63.9

It is critical for interpretation of the prediction intervals that the value of p be the correct one. Therefore it may be preferable to estimate p from the data. The prediction limits were simulated for smaller numbers of trials using the estimated value of p and the results were only slightly below the nominal 95% level.

ALTERNATIVE MODELS

Clearly several assumptions may be violated in the DiMaggio example. Let the Bernoulli random variable Z_i be

$$Z_i = \begin{cases} 1, & \text{if success on trial } i \\ 0, & \text{otherwise} \end{cases}$$

We now consider three models suggested by the example, in which the assumptions are not met.

Model 1: Markov Model.

The simple Markov model (Feller, 1957) defines four conditional probabilities which depend upon p and a parameter θ as follows:

$$P_{hh} = P(Z_i = 1 | Z_{i-1} = 1) = \theta p,$$
$$P_{mh} = P(Z_i = 0 | Z_{i-1} = 1) = 1 - \theta p,$$
$$P_{hm} = P(Z_i = 1 | Z_{i-1} = 0) = (1 - \theta p)p/q,$$
$$P_{mm} = P(Z_i = 0 | Z_{i-1} = 0) = (1 - 2p + \theta p^2)/q,$$

where θ must satisfy

$$\max[0, (2p-1)/p^2] < \theta < 1/p,$$

with $\theta = 1$ the case of independent trials. All required conditional probabilities and θ can be determined by specifying p and P_{hh}, and the unconditional probability $P(Z_i = 1) = p$. Ladd (1975) develops an algorithm for the cumulative distribution of successes in a sequence of dependent trials, characterized by a model similar to this one.

Model 2: Cyclical values of p.

In this model the $Z_i's$ are independent but not identically distributed. The probability of success cycles as the trials are performed. In particular

$$p_i = p_0 + \frac{W}{2}\sin(\frac{2\pi i}{d}), \quad i = 1, 2, ...,$$

where d = period of the sine curve, and W is the range of values of p.

Model 3: Random distribution of p.

In this model p is a random variable and the $Z_i's$ follow a compound conditional distribution. In particular $p_i = .05 * X_i + .90$ where

$$X_i \backsim Beta[H\psi(p_{i-1}), H(1 - \psi(p_{i-1}))] \quad i = 1, 2, ...$$

For $\psi(p_{i-1}) = p_{i-1}$, a large value of p_{i-1} results in a distribution of p_i with a larger expected value. For $\psi(p_{i-1}) = 1 - p_{i-1}$, a large value of p_{i-1} results in a distribution of p_i with a smaller expected value. The variance of the distribution of p_i depends upon the parameter $H > 0$.

Simulation studies were conducted for these three models with both pairs of prediction limits using $\hat{p}$, the proportion of successes in the sample, as the estimate of p. The results indicate that the coverage of the prediction limits is severely affected by all three models using $\hat{p}$. It would be preferable to use an alternative estimator of p which is less sensitive to violations in the underlying assumptions. We propose the moment estimator

$$\overline{p} = 1 - \frac{1}{\overline{x}},$$

where $\overline{x}$ is the average length of success runs in the sample. The results indicate that the coverage of both sets of prediction limits is slightly below the nominal 95% when the underlying assumptions are met.

SIMULATION RESULTS

In this section we present the results of several simulation studies concerning the coverage of the two sets of prediction intervals under the three models above, using both $\hat{p}$ and $\overline{p}$ for the geometric distribution prediction interval and Schilling's prediction interval. These are denoted by $G_{\hat{p}}$, $G_{\overline{p}}$, $S_{\hat{p}}$, and $S_{\overline{p}}$, respectively. The value $n = 1000$ was sufficient for the pattern defined by each of the three models to become established, therefore each model was used to generate 3000 replications of an experiment with 1000 trials.

Model 1: Markov Model.

Figures 1-3 show the estimated percent coverage of the prediction intervals for the Markov model with $P_{hh} = 0.35$, 0.50, 0.65, and $p = 0.05, 0.95(.025)$, for legitimate values of θ. In all figures the bottom line, which increases and then decreases, corresponds to $S_{\hat{p}}$, all other lines are monotonically increasing in the range studied. From lowest to highest coverage, these are $G_{\hat{p}} < S_{\overline{p}} < S_{\hat{p}} < G_{\overline{p}}$ for $p < P_{hh}$, and $G_{\hat{p}} < S_{\hat{p}} < G_{\overline{p}} < S_{\overline{p}}$ for $p > P_{hh}$. The interval $G_{\hat{p}}$ gives unacceptable results except for values of p very near to P_{hh}; $S_{\hat{p}}$ gives unacceptable results for all p; $S_{\overline{p}}$ gives acceptable coverage for $p > P_{hh}$, and $G_{\overline{p}}$ gives acceptable results over the entire range of values of p.

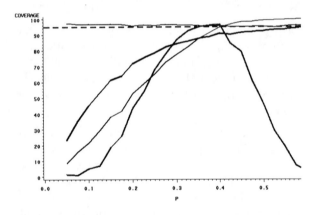

Figure 1. Model 1 with $P_{hh} = .35$

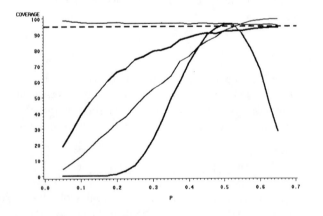

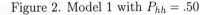

Figure 2. Model 1 with $P_{hh} = .50$

Model 2: Cyclical values of p.

Figures 4-6 show the estimated coverage for the sine model for $W = 0.05$, 0.10, 0.15 with $p_0 = 0.3, 0.7(.025)$. Both $G_{\hat{p}}$ and $G_{\overline{p}}$ performed well for $W = .05$, and the coverage is unacceptable for $S_{\hat{p}}$ and $S_{\overline{p}}$, regardless of the value of W. The results become unacceptable for all four methods as W increases.

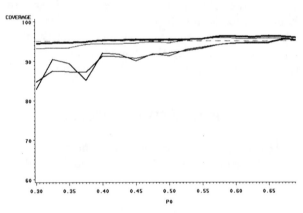

Figure 3. Model 1 with $P_{hh} = .65$

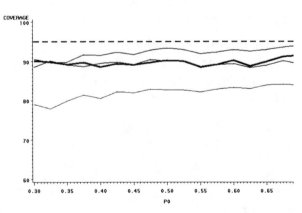

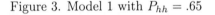

Figure 4. Model 2 with $W = .05$

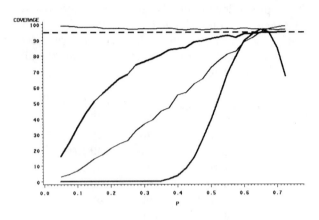

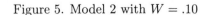

Figure 5. Model 2 with $W = .10$

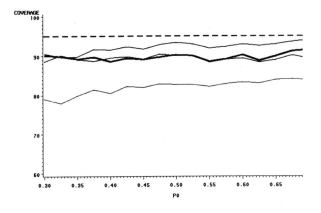

Figure 6. Model 2 with $W = .15$

Model 3: Random distribution of p.

Figure 7 shows the estimated coverage for Model 3 using $H = 5$ and $\psi(p) = p$. The coverage is unacceptable for all four methods and becomes poorer as H increases. For the model $\psi(p) = p$ the value of p_i does not stabilize as the number of trials increases. Figure 8 shows Model 3 with $H = 5$ and $\psi(p) = 1-p$. The four lines represent coverage, from lowest to highest: $S_{\overline{p}} < G_{\overline{p}} < S_{\hat{p}} < G_{\hat{p}}$. The results are satisfactory only for $G_{\hat{p}}$, marginal for $S_{\hat{p}}$, and unacceptable for both $S_{\overline{p}}$ and $G_{\overline{p}}$. The coverage improves for $G_{\overline{p}}$, $S_{\hat{p}}$, and $G_{\hat{p}}$, and stays the same for $S_{\overline{p}}$, as H increases. For the model $\psi(p) = 1 - p$ the value of p_i tends to converge to 0.5 as the number of trials increases.

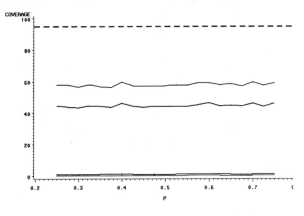

Figure 7. Model 3 with $\psi(p) = p$, $H = 5$

In all cases in which the geometric model performs poorly, the prediction interval is too narrow. One-fourth the length of the interval was less than the simulated standard deviation of the longest run from the 3000 trials. Schilling notes that for the longest run using the geometric distribution, the asymptotic mean and variance of R_n are

$$E(R_n) = -\ln[n(1 - p)]/\ln(p) - \gamma/\ln(p)$$
$$-1/2 + r_1(n) + \varepsilon_1(n),$$

and

$$Var(R_n) = \pi^2/\ln^2(1/p) + 1/12$$
$$+r_2(n) + \varepsilon_2(n),$$

where γ is Euler's constant, $r_1(n)$ and $r_2(n)$ are very small periodic functions of $log_{1/p}(n)$, and $\varepsilon_1(n)$ and $\varepsilon_2(n)$ tend to zero as $n \to \infty$. An approximate prediction interval based on these asymptotic expressions might provide acceptable coverage for small fluctuations using Model 3.

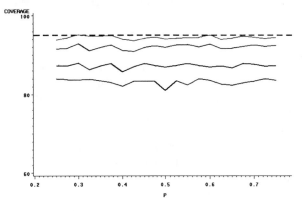

Figure 8. Model 3 with $\psi(p) = 1 - p$, $H = 5$

REFERENCES

Feller, W. (1957). *An Introduction to Probability Theory and its Applications*, Vol I, 2nd Edition, John Wiley, New York

Gordon, L., Schilling, M. F., Waterman, M.S. (1986). An extreme value theory for long head runs. *Probability Theory and Related Fields,* 72: 279-287.

Johnson, C. A. and Klotz, J. H. (1974). The atom probe and Markov chain statistics of clustering. *Technometrics*, 16: 483-493

Ladd, D. W. (1975). An algorithm for the binomial distribution with dependent trials. *Journal of the American Statistical Association*, 70: 333-340

Schilling, M. F. (1990). The longest run of heads, *The College Mathematics Journal*, 21(3): 196-207.

Schilling, M. F. (1994). Long run predictions, *Math Horizons*, Spring, 10-12.

SAMPLING PERIOD LENGTH IN ASSESSING ENGINEERING CONTROL EFFECTIVENESS ON ASPHALT PAVERS

Stanley A. Shulman, Kenneth R Mead, R. Leroy Mickelsen, National Inst. for Occupational Safety & Health
Stanley A. Shulman, NIOSH, 4676 Columbia Parkway, MS R-5,Cincinnati, OH 45226

Key Words: Randomized Pairs, Time Series

Abstract

The aim of this work was evaluation of the effectiveness of engineering controls for reducing asphalt emissions during paving studies on highways. Time series data were collected at four or six second intervals during periods of control-off or control-on to estimate concentrations of particulate and vapor generated. Each setting, whether control-on or control-off, was maintained for several hours and the order was randomized in pairs. A procedure is presented for choosing period length: data from long-time periods were divided into groups of several durations. The logs of the medians at each control setting in each group were used as responses in linear models which obtained control effectiveness estimates [mean ln(control-on) - mean ln(control-off)] for each duration. In spite of considerable environmental variability, a sequence of durations could usually be identified having stable estimates, and an estimate from this sequence was reported. Issues discussed include: a.) Trends in estimates, b.) Instances in which estimates are not stable, c.) Possible effects of interruptions in sampling.

Introduction

STUDY DESIGNS AT EACH SITE

Study	Randomization (control-on, off)	Asphalt Delivery	Comments
1	4 days of long-time samples	Truck delivery, Mostly continuous	
2	4 days of long-time samples	Truck, Many interruptions In paving	2 similar to 1, but many interruptions
3	3 days of long-time pairs*, 1 day of short-time	3 days continuous, 1 by truck	
4	2 long-time, many short-time pairs, over 18 hour period	Continuous, few interruptions	
5	Long-time and many short-time pairs over 5 days	Truck delivery for 4 days; many interruptions**	4 and 5 similar in sampling, but differ in delivery and interruptions

** interruption denotes a paving stoppage for more than 25 seconds

Five studies (labeled 1,2,3,4,5) were carried out to evaluate the effectiveness of controls on moving asphalt pavers during actual paving operations outdoors. Because of the outdoor settings, the data had substantial variability. A variety of instrumentation was used, including direct reading instruments that sampled either every four or six seconds.

Each study involved several consecutive days of sampling. The comparison of control-on with control-off required that determinations be made in pairs of control-on and off. For most of the studies the change from control-on to off required only that a switch be turned on or off. For the studies (1,2,3), long-time periods at each control setting were run. This was necessitated because other samples that required collection of material on filters were thought to require long sampling times in order to produce quantifiable amounts of material. Later studies combined longer time periods with shorter. The more pairs (control-on, control-off) that were collected, the more power for the statistical tests.

A main concern is determination of duration of sampling at each control setting, in order to make an adequate comparison of the two control settings.

Other comparisons to be made are summarized below. The duration of measurements to use from each long-time period is important for two reasons. There may be trends in measurements over several hours, and estimates of control effectiveness depend on which portions of the measurements are used. Also, there may be trends at the start of a new control setting, and it may require measurements to be made longer so that initial trends have less effect on comparisons.

Different methods of asphalt delivery were used at different sites, either truck or continuous delivery. A discussion of possible effect of delivery method on the estimates is presented. Also, there were interruptions in the work, and their effect is also discussed..

These comparisons were not considered when the original studies were planned, but understanding of the effect of these factors can be useful in future studies.

COMPARISONS OF INTEREST IN STUDY

Subject	Comparison
Main question: Control effectiveness	Control-on versus control-off
Trends at start of control setting	Trends at start of new control settings?
Asphalt delivery	Truck vs. continuous
Period length	Best sample duration. How short is too short?
Analyte	Particulate vs. vapor
Asphalt delivery	Frequency of interruptions

Randomization

For the first three studies, the long-time randomization only involved randomization of the control at the start of the day. The setting on the first day was randomly chosen, whether control-on or control off. The other control setting was used to start the second day.

The control setting determinations for long-time pairs were compared by constructing pairs before and after a control setting change. Suppose that the first control setting was control-off, which lasted until 10a.m., and control-on started at 10:15. Pairs would be constructed as follows, if 30 minutes was the chosen duration for the comparison. For control-off, measurements would be used from 9:30 until 10 a.m. For control-on, measurements would be used from 10:15 until 10:45 am. The actual number of measurements in each control setting period varied with the number of interruptions in the paving process. The aim was to create short-time pairs from the long-time pairs, since for shorter durations the environmental variables (wind and location) would be similar. In this approach, if the control-on period was followed by a subsequent control-off period, another pair would be created before and after the change from control-on to control-off.

Statistical Model

In each study, one instrument was used for particulate at the auger and one instrument for vapor at the auger. Because of the quantity of measurements and the difficulty of inspecting the data for outliers, the median measurement at the control setting is used as the response. Since the variance tends to increase with increasing value of the median, the natural log of the median is used in the statistical model. The ln(median)

for each control setting can be expressed as a model in terms of day d, pair p in day d, control setting c:

$$\ln[\text{median}(dpc)] =$$
$$m + a_d + b_{p(d)} + g_c + ag_{dc} + \varepsilon_{dpC} \qquad (1),$$

where ε_{dpC} is the variability associated with measurement error and interaction between control setting and period within day.

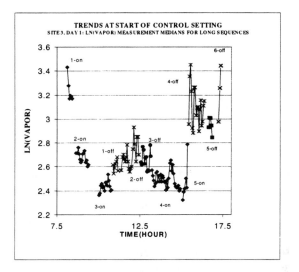

Two important observations from the figure above are: a.) The control-on settings, labeled "1-on", "2-on", and "3-on" are part of a decline in vapor levels. Estimated reduction due to the control depends upon which part of the control-on setting is used with the following control-off setting. b.) Another important point is that there are often trends at the start of a new control setting. Thus, the size of the difference between the control settings will depend on possibly deleting measurements at the start of the sequence, or extending the length of time for inclusion of measurements in a pair.

Ideally, after allowing for these trends there is a period of time during which the control settings maintain an approximately constant measurement level. This could be indicated by approximately constant median values. If the inputs (the asphalt generated at the auger and the environmental conditions, such as wind) are approximately the same for a short period of time, then the outputs (the measured analytes at the auger) may attain a stable value after suitable time. If relative stability is attained, then the ratio of control-on to control-off levels should be approximately constant for determinations based on the appropriate duration. At best this stability is just an approximation, since the measurements are taken outside, and are subject to changes in the weather and the environment as the paving machine moves along the road. We will see to what extent this stability is attained by carrying out an analysis of variance at each duration as follows, according to the model (1) above.

ANALYSIS OF VARIANCE
(Separate ANOVA for each time duration)

Source	Degrees of freedom
Days	D-1
Periods in Days	D(P-1)
Control	1
Control x Day	D-1
Residual	D(P-1)
Total	2DP-1

Duration of Measurements

Below is a summary of the estimates from the first three sites. The starred results are those recommended for the comparison of control-on to control-off. For site 1, the low value of 78 percent for the vapor for 15 minute duration is due to the increasing trend for vapor at the start of control settings. As the length of the period increases, the effect of the vapor increase during control-off settings becomes smaller. The thirty minute estimates and longer indicate relative constancy of estimated reduction for both vapor and particulate. As shown in model(1), interaction between day and control is allowed for here. This seems sensible, since the days are sampled in succession, and are not randomly chosen. For both particulate and vapor the standard errors are lowest for times between 30 or 60 minutes duration. (Models include the day by control setting interaction.)

Estimates of Percent Reduction(Std. Dev.)

Parti-culate	15 min	30min	45 min	60 min	120min	Deg fr
Site 1	96(2)	*97(1)	96(1)	96(1)	95(1)	3
Site2	12(36)	37(23)	52(19)	*58(16)	63(22)	6
Site3	27(14)	*21(10)	21(9)	20(21)	35(17)	8
Vapor						
Site 1	78(5)	*81(3)	80(4)	80(4)	80(4)	3
Site 2	3(39)	31(34)	45(18)	*51(11)	37(14)	5
Site 3	25(12)	*46(11)	48(14)	48(14)	47(14)	3

* denotes estimate selected for actual comparison.

Since the standard error estimates are only based on three degrees of freedom, their trends may be of more interest than their actual value. To be consistent, for both vapor and particulate, the same duration is used: 30 minutes.

Site 2 was different. The particulate never has convincingly stable estimates. Deletion of 15 minutes consistently produced higher estimates than 0 minute deletion. We use 60 minute estimates, since the vapor is closer to being stable for durations between 30 and 60 minutes, and the particulate estimates of 58 percent and 63 percent reduction, respectively, are not such a large difference, as say, between 30 minutes (37 percent reduction) and 60 minutes. The problem with the site 2 data will be discussed in the next section.

Site 3 differs from both sites 1 and 2. The estimated reduction for particulate is stable between 30 minutes and 60 minutes, about 20 percent reduction. However, at 120 minutes the reduction increases to 35 percent. The reason appears to be that on two of the three days on long-time sampling that began with control-off there was a decreasing trend during the day. Since control-off is first in two of the three pairs from each of these day's data, increasing duration to two hours substantially increases the estimated reduction. Since the criterion used for stability is a period of relatively stable estimated reduction, the thirty minute estimate is chosen, about 21 percent reduction for particulate and about 46 percent for vapor.

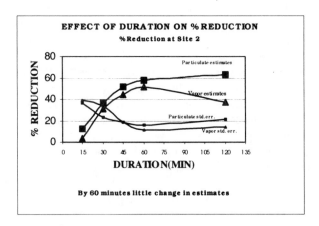

EFFECT OF DURATION ON % REDUCTION
%Reduction at Site 2

By 60 minutes little change in estimates

Trends at the Start of New Control Settings

Some of the data suggested that it takes a short time before the measurements reached a stable value. Although the instruments should respond quickly, there was the possibility that it takes a short time to get buildup of material when the paver is running after a period of not running.

To evaluate this possibility, we divided the measurements at each control setting into segments of 3.5 minutes and deleted 0.5 minute of measurements at the start of the setting. A median was calculated for each sequence. Consider only sequences with all values within 20 minutes of the start of the control setting, and require that there be at least one sequence with data more than 10 minutes from the start of the control setting. The natural log of the segment's median instrumental measurement was then regressed on the average hour of the segment (the average of the earliest and latest times for measurements in the segment). This slope estimate measures the linearity of the trend at the start of each control setting. For instance, at site 3, there were 12 settings that qualify. On average, the slope (change in ln(median) per hour)) for the 5 control-on sequences was about -0.1, and for the 7 control-off sequences was about 1.4. The p-values were, respectively, 0.19 and less than 0.05. The table below summarizes the results for the first three studies. Although there is surely autocorrelation of the medians of the succeeding segments, the estimates of the slopes are unbiased, and just one estimate is obtained for each selected control setting. These estimates are then treated as a random sample.

Average Slopes (Change in Ln(median) per Hour) By Control-Setting for Settings between 10 & 20 Min. Duration

	VAPOR		PARTICULATE	
Site	ON	OFF	ON	OFF
1	0.9* (n=19)	3.3* (n=14)	0.1 (n=19)	4.2* (n=14)
2	1.0(combined)** (n=27)		3.5 (n=16)	-0.9 (n=13)
3	-0.1 (n=5)	1.4* (n=7)	1.3 (n=13)	0.2 (n=11)

* indicates statistical significance at 10% level
** significant at 15% level
Significance levels are for each test individually.

There is some evidence that the vapor takes a longer time to stabilize than the particulate. The possibility of trends at the start of the settings may require that longer duration of measurements must be used.

Effect of Interruptions in Paving Process

Site 2 sampling, like that at site 1, consisted of many long-time periods. However, there were many delays in

the truck delivery of asphalt at site 2. One way to quantify this is as follows. For each sampling time under study, the fraction of time actually sampled is calculated. For the vapor data, these are shown below. The average difference is 0.11 more for the site 1 vapor data than for the site 2 data. This difference holds even though we removed some short sequences at the start of new control settings at site 2. The plot of the reduction due to the control for particulate and vapor shows that the particulate efficiency does not clearly level off. It could be that the frequency of interruptions made it difficult to attain a stable value.

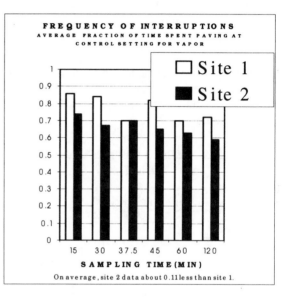

On average, site 2 data about 0.11 less than site 1.

Like site 4, the site 5 design included many short-time pairs. Site 4 asphalt delivery was by a continuous process, but site 5 delivery was by trucks which averaged under 3 minutes to dump their entire load. For each control setting, considering only the short-time pairs, we calculated the fraction of the total time designated for that setting during which paving activity actually occurred. For site 4 the result is close to 1, but for site 5 the result is about 0.65, indicating that there were many interruptions associated with delays between trucks. The results are similar for the long-time periods. Nevertheless, the reductions due to control are quite high.

For site 5 it is possible that the vapor reduction could have been greater if there had been longer periods of continuous paving. There is a tendency for greater reduction to be associated with longer sampling time, as seen in the figure below.

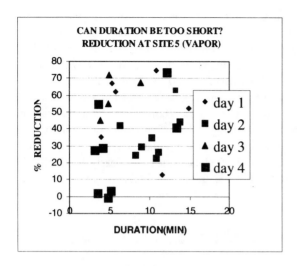

Another viewpoint is that the length of stops in paving influences the level of measurements in the subsequent period. More study is required to see whether this observation is true.

Truck Delivery versus Continuous Delivery of Asphalt

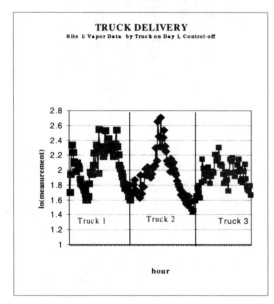

Results from site 1, plotted above, highlight the form of measurements associated with different trucks. Such measurements regularly start lower at the beginning of the truck's delivery and end lower at the end. We might wonder that if we compare uninterrupted paving measurements from both truck delivery and from continuous delivery of asphalt, whether the truck estimates of efficiency might be somewhat lower. However, we lack data from the same site where there are many periods of uninterrupted delivery by both

methods. The largest reductions of the five studies were at site 1, which had truck delivery and long periods of uninterrupted paving.

Discussion

For an outdoor process it is possible to obtain stable estimates of control effectiveness. Some studies will achieve a period of relatively constant reduction, and that duration can be used for comparison of control effectiveness. Here studies one and three indicate stability of estimates.

Some of the issues studied may be relevant in other outdoor evaluations. For instance, a.) Varying duration at each control setting appears to be more important than trends at the start of the setting; b.) However, if there are many interruptions and if periods of activity at a control setting are too short, the evaluation is more difficult to make; c.) If linear trends are anticipated, there are statistical designs available that are more efficient than randomization of (control-on, control-off) pairs. An alternative to anticipating a trend is plotting of the data as they are collected, and adaptation of the statistical design accordingly. However, this requires additional personnel on the evaluation team. d.) Since each process is different, it makes sense to adapt the randomization to the work at the evaluation site. For instance, at the paving site, if the road is being paved in short strips, it makes sense to measure each control setting while the paver is moving in the same direction. Otherwise, change in the wind direction can make comparison more difficult.

Attention to these problems can enhance the ability of researchers to evaluate control effectiveness.

COMPARISON OF PERFORMANCE RATING CRITERIA IN PROFICIENCY TESTING PROGRAMS

Ruiguang Song[a] and Paul C. Schlecht[b]

[a]HGO Technology, Inc./NIOSH, [b]National Institute for Occupational Safety and Health

Ruiguang Song, DPSE/NIOSH, MS: R-7, 4676 Columbia Parkway, Cincinnati, OH 45226, USA

Key Words: Proficiency Testing, Rating Criteria, Bias, Precision, Outlier, Z-Score, Running Performance Index, Sample Size, Statistical Power.

Abstract

Two major proficiency testing programs in the industrial hygiene field are compared with regard to their power to detect laboratories having a large bias and poor precision. Although the two programs are similar in many aspects, such as the analytes tested, frequency of test rounds, and number of samples used in each test round, their evaluation procedures and rating criteria are quite different.

To expand the substances used in the proficiency testing and reduce the proficiency testing cost in each program, cooperation between the two programs will begin in 2000. Cooperation will involve exchange of samples, coordination of logistics, and harmonization of analytical methods, but will permit each program to continue its rating criterion. This study provides detailed power comparison of the two programs on rating laboratories at various bias and precision levels.

In the Workplace Analysis Scheme for Proficiency (WASP) program, participating laboratories are classified into three categories: "*average", "better than average*", and "*worse than average*". Laboratories in the Proficiency Analytical Testing (PAT) program are rated as "*proficient*" or "*non-proficient*". If a "*non-proficient*" rating in PAT is compared to a "*worse than average*" classification in WASP, the study shows that WASP is more sensitive in detecting laboratories with poor performance. That is, the chance for a laboratory having a large bias or low precision to be rated "*non-proficient*" in PAT is less than the chance of being classified as "*worse than average*" in WASP.

Although the PAT criterion is simpler, it is not as powerful because PAT converts each quantitative laboratory result to a qualitative value. The effect of information loss can be measured by the number of samples required to maintain the same power in both programs. The required number of samples for WASP is only 60%-80% the number of samples used in PAT.

To improve the PAT rating criterion a modified rating criterion based on z-scores has been previously proposed by Schlecht and Song (1997). This study demonstrates that the modified PAT criterion is equivalent to the WASP criterion.

1. Introduction

Proficiency testing programs are used to evaluate laboratory performance in a wide variety of fields. Two well-known proficiency testing programs in the industrial hygiene field are the United States-based Proficiency Analytical Testing (PAT) Program[1] and the United Kingdom-based Workplace Analysis Scheme for Proficiency (WASP) Program.[2]

The PAT program was started in 1972. It is used by the American Industrial Hygiene Association (AIHA) to accredit laboratories analyzing hazardous substances in workplace air, including various metals, organics, silica, and asbestos. Currently there are about 1,300 laboratories, mainly in the United States and Canada, participating in this program.

In 1988, the WASP Program was started in the United Kingdom. It is used by the United Kingdom Accreditation Service (UKAS) to accredit laboratories analyzing hazardous substances in the workplace and ambient air, including various metals, organics, silica, isocyanates and formaldehyde. About 300 laboratories, mainly from United Kingdom and other European countries, participate in this program.

Both programs send samples to laboratories on a quarterly basis. Four samples of each selected analyte are sent to each participating laboratory. Performance of each participating laboratory is evaluated based on results from the last four or five rounds. The PAT program rates laboratories into two categories: "*proficient*" or "*non-proficient.*" While the WASP program classifies laboratories into three categories: 1 as "*better than average*", 2 "*average*", and 3 "*worse than average.*"

To expand the substances used in the proficiency testing and reduce the proficiency testing cost in each program, cooperation between the two programs will begin in 2000.[3] Although the two programs are similar, their rating criteria are quite different. In this study, we compare the two programs, evaluate the performance rating criteria used in each program, and compute the probability that a laboratory is rated as "*non-proficient*" in PAT and the probability that the same laboratory is classified in category 3 in WASP.

To simplify the comparison, we only consider the simplest case, assuming that true concentrations are known and all results are independent. In the next section, we briefly describe the rating procedures in the two programs. Power associated with each program is presented in section 3. In section 4, we compare the two programs by the required sample size for each program to maintain minimum acceptable power. This gives a measure of the difference between the rating criteria in the two programs. In both the PAT and WASP programs, rating criteria are designed to permit occasional poor laboratory performance or possible sample contamination in a round by excluding data. The effects of excluding data on rating power are studied in section 5. In section 6, we show that the modified criterion proposed by Schlecht and Song[4] is statistically equivalent to the WASP criterion. Finally, summary results and some thoughts on the implications of this work on proficiency test programs are discussed in section 7.

2. Performance Rating Criteria in PAT & WASP

In both the PAT and WASP programs, four samples are distributed to each participating laboratory each quarter. Reported results from laboratories are then converted to z-scores in PAT or to standardized values in WASP. Let x_{ij} be the reported result from a laboratory for the jth sample in the ith round. Then the standardized value is given by $y_{ij} = (x_{ij} - \mu_{ij}) / \mu_{ij}$ and the z-score is defined as $z_{ij} = (x_{ij} - \mu_{ij}) / \sigma_{ij}$, where μ_{ij} and σ_{ij} are the reference mean and the reference standard deviation.

In PAT, each result is rated as "*acceptable*" or "*not acceptable*" (called an outlier). A laboratory is rated as "*proficient*" or "*non-proficient*" based on the total number of outliers in the last four rounds. A result is an outlier if the absolute value of its corresponding z-score is greater than 3. A laboratory is rated as "*proficient*" if it has either (A) no more than 4 outliers (out of 16 results) in the last four rounds, or (B) no outlier in the last two rounds.

In WASP, a laboratory's performance is evaluated based on the running performance index (*RPI*) that is defined as the average of the four best performance indexes in the last five rounds:

$$RPI = \left(\sum_{i=1}^{5} PI_i - \max\{PI_i\} \right) / 4,$$

where PI_i is the performance index for the *i*-th round: $PI_i = \sum_{j=1}^{4} y_{ij}^2 / 4$. If a laboratory's *RPI* is in the interval $(0.432RSD_0^2, 1.8RSD_0^2)$, then the laboratory is categorized as "*average*", where $RSD_0 = \sigma_0 / \mu_0$ is a pre-specified relative standard deviation, which is different for different analytes. If $RPI < 0.432RSD_0^2$, the laboratory is identified as "*better than average*." If $RPI > 1.8RSD_0^2$, the laboratory is classified as "*worse than average*."

We have seen that the two programs have very different rating criteria. While our ultimate goal is to adopt identical rating criteria, this may not be immediately possible for several reasons. First, the rating scheme may be mandated by regulation. Second, the rating criteria may be the basis of mutual recognition agreements among accrediting organizations and governmental agencies. Third, criteria may be used differently in different programs. In some, a "*non-proficient*" rating results in immediate loss of accreditation and loss of business by the laboratory, whereas other programs may permit investigations of laboratory operations before this occurs. Therefore, in some instances a type I error must be very small, whereas in other instances a moderate type I error can be tolerated. Fourth, many proficiency test programs must address issues that are unique to a particular type of analysis and to the type of participating laboratory, which makes adoption of some rating criteria not feasible. For example, AIHA operates a diffusive monitor proficiency test program for various organics where the cost of samples makes it not feasible to operate more than two rounds (8 samples) per year. Similarly, another AIHA program focuses on proficiency testing of microscopists performing fiber counting of asbestos and other fibers. These microscopists are mobile and must deal with the unique problems this presents for logistics. Fifth, since there are many proficiency test programs, adoption of multiple rating criteria to satisfy multiple cooperating proficiency test program may be confusing for laboratories. Therefore, until a single proficiency test protocol can be universally agreed upon by the many proficiency test programs, a comparison of criteria, based on power and simple modifications to criteria, to made the criteria similar, is useful. It is the first step towards broader cooperation and harmonization among programs.

3. Power Comparison

To compare the rating criteria in the two programs, we selected a laboratory and assumed that this laboratory had a consistent performance with a constant bias B (relative difference from the true value) and a constant precision *TRSD* (defined as the standard deviation divided by the true value). Then, we compared the chance for this laboratory to be rated as "*non-proficient*" according to the PAT criterion with its chance to be classified in category 3 under the WASP criterion. Since the WASP program uses the best four performance rounds (out of five) in its criterion and the PAT program has an additional two-round criterion, there are explicit formulas to calculate the rating power associated with each criterion. Therefore, a simulation study was conducted for various bias and precision levels. Simulation results from 10,000 replicates are shown in Table 1 for PAT and Table 2 for WASP.

Table 1. Probabilities for a laboratory to be rated non-proficient using the **PAT** criterion. (Simulation results from 10000 replicates, the reference $TRSD_0$=0.06)

BIAS	TRSD										
	0.05	*0.06*	*0.07*	*0.08*	*0.09*	*0.10*	*0.11*	*0.12*	*0.13*	*0.14*	*0.15*
0.00	0.000	0.000	0.000	0.000	0.001	0.005	0.018	0.051	0.108	0.195	0.296
0.01	0.000	0.000	0.000	0.000	0.001	0.005	0.019	0.054	0.113	0.196	0.299
0.02	0.000	0.000	0.000	0.000	0.001	0.006	0.022	0.062	0.123	0.207	0.306
0.03	0.000	0.000	0.000	0.000	0.002	0.008	0.031	0.070	0.138	0.222	0.323
0.04	0.000	0.000	0.000	0.001	0.003	0.014	0.042	0.089	0.158	0.247	0.347
0.05	0.000	0.000	0.000	0.002	0.006	0.023	0.057	0.113	0.191	0.285	0.383
0.06	0.000	0.000	0.000	0.003	0.015	0.040	0.083	0.147	0.234	0.330	0.424
0.07	0.000	0.000	0.002	0.009	0.030	0.068	0.120	0.197	0.288	0.381	0.470
0.08	0.000	0.001	0.005	0.023	0.056	0.107	0.175	0.261	0.350	0.433	0.521
0.09	0.000	0.003	0.018	0.048	0.100	0.166	0.247	0.333	0.414	0.499	0.579
0.10	0.002	0.013	0.044	0.097	0.167	0.246	0.329	0.412	0.491	0.566	0.633
0.11	0.007	0.039	0.097	0.178	0.264	0.345	0.427	0.504	0.569	0.635	0.695
0.12	0.031	0.097	0.191	0.288	0.377	0.460	0.529	0.595	0.655	0.707	0.753
0.13	0.097	0.212	0.323	0.427	0.505	0.576	0.634	0.687	0.731	0.770	0.805
0.14	0.238	0.376	0.487	0.573	0.639	0.691	0.732	0.768	0.797	0.828	0.852
0.15	0.454	0.573	0.658	0.718	0.759	0.787	0.814	0.837	0.856	0.877	0.893

Note: Probabilities less than 0.05 are shaded.

Table 2. Probabilities for a laboratory to be classified in category 3 based on the **WASP** criterion. (Simulation results from 10000 replicates, the reference $TRSD_0$=0.06)

BIAS	TRSD										
	0.05	*0.06*	*0.07*	*0.08*	*0.09*	*0.10*	*0.11*	*0.12*	*0.13*	*0.14*	*0.15*
0.00	0.000	0.002	0.037	0.181	0.413	0.646	0.812	0.907	0.956	0.980	0.990
0.01	0.000	0.003	0.044	0.193	0.424	0.659	0.820	0.910	0.956	0.980	0.991
0.02	0.000	0.007	0.065	0.236	0.465	0.688	0.833	0.920	0.959	0.982	0.992
0.03	0.001	0.021	0.109	0.304	0.533	0.734	0.858	0.929	0.965	0.983	0.992
0.04	0.007	0.056	0.197	0.410	0.623	0.789	0.887	0.942	0.972	0.987	0.994
0.05	0.038	0.149	0.337	0.540	0.721	0.842	0.916	0.958	0.979	0.989	0.994
0.06	0.150	0.321	0.513	0.681	0.812	0.894	0.942	0.971	0.985	0.991	0.995
0.07	0.397	0.557	0.702	0.811	0.884	0.933	0.964	0.981	0.989	0.994	0.997
0.08	0.705	0.784	0.850	0.902	0.939	0.966	0.980	0.990	0.994	0.997	0.998
0.09	0.909	0.924	0.943	0.959	0.973	0.984	0.991	0.995	0.997	0.998	0.999
0.10	0.985	0.983	0.984	0.987	0.990	0.994	0.995	0.998	0.998	0.999	0.999
0.11	0.998	0.997	0.996	0.996	0.996	0.997	0.999	0.999	0.999	0.999	1.000
0.12	1.000	1.000	0.999	0.999	0.999	1.000	1.000	1.000	1.000	1.000	1.000
0.13	1.000	1.000	1.000	1.000	1.000	1.000	1.000	1.000	1.000	1.000	1.000
0.14	1.000	1.000	1.000	1.000	1.000	1.000	1.000	1.000	1.000	1.000	1.000
0.15	1.000	1.000	1.000	1.000	1.000	1.000	1.000	1.000	1.000	1.000	1.000

Note: Probabilities less than 0.05 are shaded.

From the two tables, we see that a laboratory is much less likely to get a non-proficient rating from PAT than a category III worse than average rating from WASP. This shows that the WASP criterion is much stricter than the PAT criterion and is the main reason why the PAT program has a lower percentage[5] (<10%) of laboratories rated as non-proficient than that (>20%) classified as worse than average in the WASP program[6].

4. Sample Size Comparison

A stricter criterion may or may not be a more powerful criterion. To study the power of a rating criterion, we view a performance rating as a statistical hypothesis testing. The hypotheses specify a laboratory's performance characteristics that are determined by the combination of bias and precision. The null hypothesis specifies the characteristics for a laboratory that deserves a good rating. The expected precision for a good laboratory depends on the analyte under test and has been explicitly specified for each analyte in WASP. In PAT, it is estimated from reference laboratories' results. The alternative hypothesis characterizes laboratories with unacceptable performance.

To compare the power of two rating criteria, it is necessary to specify the null hypothesis, set the two tests at the same level of type I error, specify the alternative hypothesis, and compare the type II error evaluated at the alternative hypothesis. Since the power is associated with sample size, we may compare the required sample sizes that make the two tests have the same type II error.

To simplify the comparison, we only consider the criterion using four-round results in each program so we can calculate the power without simulation. Also, we assume that all results are independent with an identical normal distribution. Under these assumptions, we have $z_{ij} \sim N(B/TRSD_0, TRSD^2/TRSD_0^2)$ and

$y_{ij} \sim N(B, TRSD^2)$, where $TRSD_0$ is the expected value of $TRSD$. In this case, the WASP rating score RPI has a χ^2 distribution with a degrees of freedom N (=16, the total number of samples tested). The PAT rating score S is the total number of outliers. It has a binomial distribution $Bin(N, p)$, where $p = \Pr(|z_{ij}| > k)$ is the probability of getting an outlier and $k=3$ in PAT.

Let L be the number of outliers permitted for getting a proficient rating. Then the chance for a laboratory to be rated as non-proficient is $P_{NP} = \Pr(S > L)$. The two parameters L and k can be adjusted so that $P_{NP} = \alpha$ under the null hypothesis H_0: $B=0$ and $TRSD=TRSD_0$.

In WASP, we can adjust the limit for the running performance index, denoted by H, such that the chance for a laboratory to be categorized as "worse than average" is also α: $\Pr(RPI > H) = \alpha$ under the null hypothesis. Note that $RPI \sim TRSD^2 \times \chi^2(N, D)$, where $D = N \times (B/TRSD)^2$ is the non-central parameter. The critical value H can be determined by $H = TRSD_0^2 \times \chi^2(\alpha, N)$. Using this H, $\Pr(RPI > H)$ can be calculated.

We set $\alpha=0.025$, $TRSD_0=0.06$, and the alternative hypothesis H_1: $B=0.05$ and $TRSD = 2 \times TRSD_0$. Given the sample size N in PAT, we calculate the sample size N' required in WASP to maintain the same power as in PAT at the alternative hypothesis. Calculation results are shown in Table 3, where $\lambda = L/N$ is the outlier rate allowed for a laboratory to be rated as proficient in PAT, k is selected to give the PAT a type I error $\alpha=0.025$ to match the type I error in WASP, β_{PAT} is the type II error associated with the PAT criterion, $\beta_{WASP(N)}$ and $\beta_{WASP(N')}$ are the type II errors associated with the WASP criterion for sample sizes N and N', respectively. From this table, we can see that there are about 25%~40% information lost by converting a continuous variable to a binary variable in PAT. The loss depends on the outlier rate λ and the total number of samples N.

5. Poor Laboratory Performance or Suspect Samples on a Given Round

Both programs have a mechanism to permit laboratories that are rated "non-proficient" or "worse than average" due to poor performance or receipt of suspect samples on a given round to improve ratings. In PAT, this is done by using the two-round criterion. If the laboratory has no outliers on the last two rounds, the laboratory is rated proficient even though previous rounds have many outliers. In WASP, one extra round's results are used and the worst round results are eliminated from the rating. To see the effects of these additional criteria, we show the simulation results in Table 4 for the case bias B=0. Column P4 gives the probabilities for a laboratory getting a "non-proficient" rating under the PAT four-round criterion. Probabilities in column +P2 show the decreased chance for a laboratory getting a "non-proficient" rating, in other words, the extra chance to get a proficient rating. Similarly, the column W4 gives the probabilities for a laboratory to get a "worse than average" rating under the WASP criterion using four-round results only. Probabilities in column +W5 is the decreased chance for a laboratory getting a worse than average rating.

Table 3. Required sample sizes to maintain the power at H_1.

λ	N	m	k	p	β_{PAT}	$\beta_{WASP(N)}$	N'	$\beta_{WASP(N')}$	N'/N
1/4	12	3	1.65	0.449	0.136	0.040	8	0.117	0.667
	16	4	1.60	0.463	0.070	0.013	10	0.069	0.625
	20	5	1.56	0.474	0.035	0.004	13	0.030	0.650
	24	6	1.53	0.483	0.017	0.001	15	0.017	0.625
	28	7	1.51	0.490	0.008	0.000	18	0.007	0.643
	32	8	1.49	0.495	0.004	0.000	21	0.003	0.656
1/5	10	2	1.83	0.400	0.168	0.069	7	0.153	0.700
	15	3	1.76	0.418	0.070	0.017	10	0.069	0.667
	20	4	1.71	0.431	0.028	0.004	14	0.023	0.700
	25	5	1.68	0.441	0.011	0.001	17	0.010	0.680
	30	6	1.65	0.449	0.004	0.000	21	0.003	0.700
1/6	12	2	1.92	0.378	0.110	0.040	9	0.090	0.750
	18	3	1.85	0.395	0.036	0.007	13	0.030	0.722
	24	4	1.80	0.408	0.011	0.001	17	0.010	0.708
	30	5	1.77	0.417	0.003	0.000	21	0.003	0.700

Table 4. Probability of a laboratory receiving a "*proficient*" rating from the PAT program or a "*worse than average*" rating from the WASP program (Simulation results of 10000 replicates, bias B=0)

	PAT			WASP		
TRSD	P4	+P2	Overall	W4	+W5	Overall
0.05	0.0000	0.0000	0.0000	0.0004	0.0003	0.0001
0.06	0.0000	0.0000	0.0000	0.0239	0.0220	0.0019
0.07	0.0000	0.0000	0.0000	0.1725	0.1353	0.0372
0.08	0.0001	0.0000	0.0001	0.4346	0.2536	0.1810
0.09	0.0006	0.0000	0.0006	0.6840	0.2710	0.4130
0.10	0.0046	0.0000	0.0046	0.8491	0.2032	0.6459
0.12	0.0514	0.0001	0.0513	0.9720	0.0654	0.9066
0.15	0.2979	0.0017	0.2962	0.9978	0.0075	0.9903
0.20	0.7553	0.0029	0.7524	1.0000	0.0001	0.9999

For the PAT program, the two-round criterion does not increase the laboratory's chance to be rated proficient if this laboratory has not improved its performance in the last two rounds. Only those labs whose performance was poor and then improved in the last two rounds can benefit from the two round criterion.

In WASP, the chance based on 4 round criterion differs from the chance based on the best 4 out of the last 5 rounds. The chance could be 27% higher without performance improvement. This benefits laboratories that have performance with moderate bias and precision.

6. A Modified PAT Criterion

Schlecht and Song[4] proposed a criterion in 1997 to improve the power of the existing criterion used in the Environment Lead Proficiency Analytical Testing (ELPAT) program and the PAT program (both share the evaluation procedure and rating criterion). The proposed criterion is based on the combination of mean and standard deviation of z-scores: $A = a\,|\bar{z}|^r + b\sigma_z^s$, where $\bar{z}$ and σ_z are the mean and standard deviation of z-scores.

Notice that $y_{ij} = TRSD_0 \times z_{ij}$ and

$$RPI_4 = \sum_{i=1}^{4} PI_i / 4 = \sum_{i=1}^{4}\sum_{j=1}^{4} y_{ij}^2 / 16$$
$$= TRSD_0^2 \times \sum_{i=1}^{4}\sum_{j=1}^{4} z_{ij}^2 / 16 = TRSD_0^2 \times \left(\bar{z}^2 + \hat{\sigma}_z^2\right)$$

If we set $r = s = 2$ and $a = b = TRSD_0^2$, then the modified rating score A is exactly the same as the running performance index over four results. So the modified criterion must have the same power as the WASP criterion in detecting laboratories with poor performance.

The modified criterion is flexible in determining the power to detect bias or poor precision. To be sensitive in detecting bias, we can set a large value for a and a relatively small value for b. Similarly, by setting a large value for b and a relatively small value for a, we may have a criterion more sensitive to detect laboratories with low precision.

The modified criterion is also meaningful. The average z-score is a measure of bias and the standard deviation of the z-score is a measure of precision. The rating score A is simply a combination of bias and precision estimates.

7. Summary and Discussions

This study compares the power of rating criteria used in the two largest industrial hygiene laboratory proficiency test programs, the US-based PAT Program and the UK-based WASP Program. These programs will begin an exchange of samples in 2000, but have opted to continue to use their current rating criteria. The WASP criterion is more powerful than the PAT criterion. This is because there is an information loss in PAT when it transfers a continuous variable to a binary variable. To measure the loss, the required sample sizes for maintaining the same power are compared. The WASP criterion is more complex, but requires 20% to 40% fewer samples to achieve the same power as PAT. For best use of PAT data, no results should be grouped or truncated.

The two-round criterion in PAT encourages laboratories to quickly improve their performance, without changing the ratings of other participating laboratories. On the other hand, the WASP best 4 out of 5 criterion significantly changes the rating of participating laboratories, thereby, benefiting laboratories with modest bias and imprecision.

To improve the efficiency of using data in PAT, we recommend using z-scores in the performance rating, instead of the number of outliers. The modified criterion proposed by Schlecht and Song[4] can be easily adopted. It is not only more powerful than the current PAT criterion and comparable to the WASP criterion, it is also meaningful and flexible in adjusting its sensitivity to a laboratory's bias or precision. Such a modification to minimize the loss of information is important in situations where a small number of samples are used to rate laboratory performance, such as the PAT proficiency testing of diffusive monitors where only two rounds (8 samples) are used to rate performance. The change would be easy for laboratories to understand because it is based on z-scores that are already reported to PAT participating laboratories.

In this study, we assume that true concentrations are known. Since in both programs, concentrations are unknown, and estimated by the average of all laboratories' results. This can make some difference in the power comparison. However, the differences should be small because the number of participating laboratories is large (50-1000).

This study compared the power of two large proficiency test programs and suggested simple changes to criteria to harmonize them. Comparing proficiency test rating criteria based on power tables of probabilities versus laboratory bias and precision or plots of this same information, sometimes referred to as operating characteristic (OC) curves, could be a means to compare multiple proficiency test programs that have unique criteria. Such could be an important first step towards expanded cooperation among programs and an interim solution to harmonizing criteria among multiple proficiency test programs. An interim solution to harmonizing criteria is useful because it is difficult to adopt new criteria when criteria are used or mandated by various laboratory accreditation organizations and government agencies, and participating laboratories are familiar with existing criteria.

References

[1] Esche, C. A., Groff, J. H., Schlecht, P. C., and Shulman, S. A. (1994). Laboratory Evaluations and Performance Reports for the Proficiency Analytical Testing (PAT) and Environmental Lead Proficiency Analytical Testing (ELPAT) Programs. *DHHS (NIOSH 94 - 102).*

[2] H. M. Jackson and N. G. West (1993). Initial Experience with the Workplace Analysis Scheme for Proficiency (WASP). *Ann. Occup. Hyg., 36(5), 545-561.*

[3] Martin Harper (1999). Laboratory Accreditation: PATs Across the Pond. *The Synergist, 10(8), 13-14.*

[4] Paul Schlecht and Ruiguang Song (1997). Laboratory performance criteria in the Environmental Lead Proficiency Analytical Testing (ELPAT) program. *DHHS(NIOSH) Pub. No. PB97-147128, Cincinnati, OH: National Institute for Occupational Safety and Health.*

[5] Groff, J. H. and Schlecht, P. C. (1996), PAT Program Report, *American Industrial Hygiene Association Journal, 57(1), 79-81.*

[6] Barry Tylee (1995), The WASP News, Autumn issue, United Kingdom Health & Safety Executive, Shelfield.

EXPERIMENTS WITHIN WAFERS

William D Heavlin, Advanced Micro Devices
P O Box 3453, MS 117, Sunnyvale, CA 94088-3453 <bill.heavlin@amd.com>

Key words: kriging, noise factors, optimal design, response surface models, semiconductor industry, spatial statistics

Abstract

This work explores the practical issues of varying a subset of experimental factors within wafer. The within-wafer design is constructed by combining Wynn's spatial design criterion with a columnwise algorithm; the result is a spatially dispersed experimental matrix.

Data analysis has three steps: (a) For each wafer, the response is modeled as a function of both within-wafer site coordinate and within-wafer factors. (b) For each wafer, each factor's effect is the model averaged over all site coordinates and other factor levels; suitable residuals admit an estimate of within-wafer uniformity. (c) Across wafers, a response surface of all factors is fitted. A running example illustrates, and some practical issues thereby highlighted.

1. Introduction

Integrated circuits are manufactured on thin wafers of silicon crystal organized into batches, called lots. Current wafer diameters (between 200 and 300 mm) admit substantial spatial effects. In spite of such effects, typical industry practice is to block experiments by lot while treating wafers as homogeneous experimental units.

It is well recognized in statistical theory that changing factors within blocks (wafers) improves the precision by which experimental effects are estimated, and for many factors it is feasible to do so. Modern wafer patterning processes rely on highly specialized camera mechanisms ("steppers") that move ("step") to a particular wafer location, project an image on the wafer, then move (step) to a nearby wafer location. Such complex movements require software control; steppers can be programmed to image differently at different wafer locations.

In principle, programmed steppers allow two kinds of factors to be varied within wafer. (1) The stepper is programmed to vary exposure energy (and sometimes center of focus) by stepping field. The result is on the lines that comprise the transferred pattern of a given layer. For example, increased exposure energy narrows line widths. (2) Alternately, the stepper can pattern some stepper fields and skip over other stepper fields. Where photoresist is patterned, subsequent ion implantation steps can alter the electronic properties of that field's transistors. Where photoresist is not patterned, it remains in place to inhibit implant ions

from having any lasting effect. With the removal of all photoresist, one can then reapply photoresist, pattern the previously skipped fields, skip previously patterned fields, and thereby implant differently adjacent fields on the same wafer.

In addition to wafer patterning and ion implantation, other process steps may induce undesirable patterns upon wafers. For example, so-called radial patterns, whereby centers of wafers have different film thicknesses than wafer edges, are common. Other spatial patterns affect one region of a wafer, the lower left quadrant, for instance. To implement within-wafer experiments wisely, one must achieve some robustness to such nuisance spatial effects.

Because the theoretical advantages of varying some factors within wafer are rather clear, it is with implementation details that the outstanding issues lie. Through a case study of an advanced microprocessor, the response of greatest interest is speed, the single within-wafer factor is gate layer exposure, and the wafer-to-wafer factors are n- and p-channel implant doses.

The outline of this paper is as follows. Section 2 describes the experimental design layout, and section 3 the data analysis. Section 4 closes with some summary comments and conclusions.

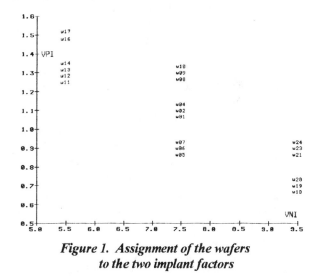

Figure 1. Assignment of the wafers to the two implant factors

2. Experimental Design

Our application involves the first silicon of a microprocessor. Overall speed is of primary interest; to first order, speed of CMOS circuits requires the proper balance between n- and p-channel transistors. From the point of view of product optimization, the ideal experiment creates a range of drive currents for the n-channel transistors, and for each such n-channel transistor, a range of p-channel transistors. The top performing circuit is then selected to determine its n-channel–p-channel combination.

N-channel and p-channel transistors share some factors, the most important of which is their common gate length critical dimension. Other factors are specific to each, including their threshold implants.

Our simple example therefore has three factors: n- and p-channel implant doses, and their common gate length critical dimension. Implant treats whole wafers homogeneously. Its experimental design follows a sliding-level factorial pictured in figure 1; it defines corners corresponding to doubly strong, doubly weak, and to mismatched n- and p-channel transistors.

Strongly determined by lithography exposure energy, gate length critical dimension is varied within wafer by appropriate stepper programming. To simplify processing at this step, the exposure energy pattern is the same on all wafers, spatially dispersed over the entire wafer surface.

This spatially dispersed pattern is the result of an optimal design algorithm with two components. (a) The optimization criterion is attributed to Wynn (Sacks, Welch, Mitchell, and Wynn, 1989). (b) Because the (x,y) spatial wafer coordinates are predetermined, and because the seven chosen levels of exposure are desired to have equal frequency, Wynn's criterion is applied using the design repair algorithm of Heavlin and Finnegan (1998).

Let $\{\mathbf{x}_i: i=1...n\}$ denote a set of k-dimensional points comprising a possible experimental design. Define the $n \times n$ matrix $\mathbf{R}$ with typical element $r(i,j) = \exp(-\|\mathbf{x}_i - \mathbf{x}_j\|^2)$.

Wynn's criterion: A design $\{\mathbf{x}_i\}$ is better when $\det(\mathbf{R})$ is larger (det denotes the matrix determinant).

If one interprets $\mathbf{R}$ as a correlation matrix, $\det(\mathbf{R})$ represents the generalized variance, which Wynn's criterion would have us maximize. This is achieved by moving the points $\{\mathbf{x}_i: i=1...n\}$ far from one another.

Let $\{\mathbf{v}_i: i=1...n\}$ denote a set of wafer coordinates and let $\{\mathbf{w}_i: i=1...n\}$ a vector of treatments to be assigned to one and only one coordinate. A given site $\mathbf{v}_i$ is therefore assigned treatment $\mathbf{w}_{\pi(i)}$, where π denotes a permutation of the indices $1...n$. In terms of Wynn's criterion, the vector $\mathbf{x}_i = (\mathbf{v}_i^T, \mathbf{w}_{\pi(i)}^T)^T$.

The design repair algorithm determines the permutation π in two steps, a best-of-random R-step, and a best-of-single-exchanges E-step.

Design repair algorithm. R-step: For n_R times, select π at random. Using Wynn's criterion, the best of these π_R is selected for the next step. E-step: Define $\Pi_1(\pi)$ as the set of permutations formed by exchanging no more than one pair of indices of π. Let $\pi_0 = \pi_R$, and let π_{k+1} be

that permutation in $\Pi_1(\pi_k)$ such that Wynn's criterion is largest; k from $0, 1, 2, ...$ until convergence.

The data structures of this algorithm are especially attractive (a) because the wafer coordinates are fixed, (b) because it enforces balance among the factor levels, providing an additional measure of robustness, and (c) because it can be extended to multiple within-wafer factors. Heavlin and Finnegan (1998) derive advice about the magnitude of n_R as a function of n.

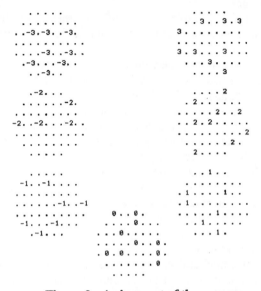

Figure 2. Assignment of the exposure levels to wafer (x,y)-coordinate

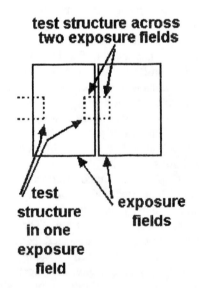

Figure 3. Mismatch of exposure fields and test structures.

186

The resulting within-wafer exposure pattern is in figure 2. While any single level shows gaps in some regions, any adjacent pair of levels achieves good spatial coverage. Note also that not all exposure levels have equal frequency: because test structures are defined to be testable only if another field is farther to the right, those along the wafers' right edge are lost (figure 3). Equivalent to a yield loss along the right edge, the spatially dispersed design is robust to this kind of effect.

Alternative within wafer designs include using x-coordinate ("striping"), mixing two exposures by x- and by y-coordinates ("madras"), and using polar angle ("pie wedges"). Striping and madras are not robust to radial effects, while pie wedges fail to guard satisfactorily against problems occurring in single wafer regions.

Figure 4 illustrates the result this design has in the response domain; it is a scattermatrix in which panels in the same row (column) share a common y-axis (x-axis). As expected, measures of gate length dimension (HDENSON2P, ISON2P) are moved by exposure magnitude and little else. Of particular interest is the lower right corner, which denotes the correlation of n- and p-channel $I_{d\,eff}$. The design has produced a good mix of n-channel drive currents with each p-channel drive current; this variety translates into a range of circuits with different speed and timing characteristics.

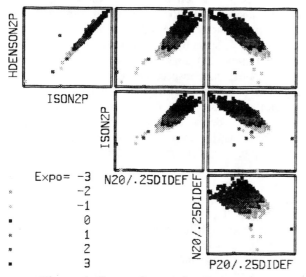

Figure 4. Scatterplot matrix of some responses; color and grayscale correspond to within-wafer experimental exposure level.

3. Data Analysis

For each wafer, the experimental plan is sufficient to fit a model as a function of exposure and wafer (x,y)-coordinates. Model options include parametric response surface models (Box and Draper, 1987) and interpolators such as kriging (Ripley, 1981). For a particular wafer, and response, figure 5 displays kriging's fit. Note the gradient from south-south-west to north-north-east. The

source of this phenomenon is present on most wafers, and bears investigation. The design in figure 2 remains robust to such patterns, however.

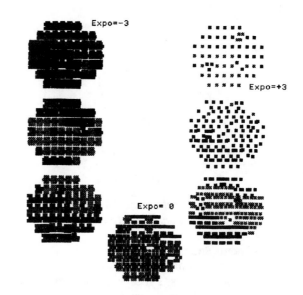

Figure 5. For a particular wafer, an example of the fit resulting from applying kriging. Higher pixel density implies higher values.

Denote by *(x,y)* the wafer coordinates, *e* the exposure of the within-wafer gate-layer factor, and *w* the wafer index. For each wafer and exposure level, model the response as a function, now averaged over all wafer coordinates. The result is $Z(e|w) = \mathsf{ave}(\,z(x,y,e|w)\,)$: all *(x,y)*. Form the residuals $r(x,y,e|w) = z(x,y,e|w) - Z(e|w)$. Define the uniformity by $U(e|w) = \mathsf{stdev}(\,r(x,y,e|w)\,)$. A correction to the degrees of freedom is possible but inessential to the present application.

The wafer- and exposure-specific averages Z and standard deviations U are then modeled by response surface methods. The within-level standard deviation enables us to identify those portions of the factor space less sensitive to variation. The response surface factors consist of both implants and the exposure level. The nesting of exposure effect within wafer affects the response surface's error structure and therefore the estimation method.

For p-channel drive current, the models yield the contour map of figure 6. Note that the (low p-dose, high exposure) corner gives both higher average current (heavier, blue) and smaller standard deviation (lighter, cyan). By modeling ring oscillator speed as a function of n- and p-channel drive current, a model linking manufacturing settings with product characteristics similarly follows, but is not presented here.

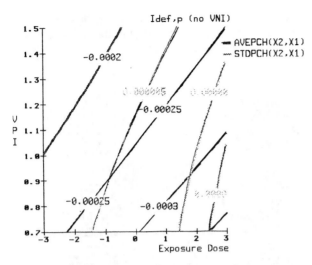

Figure 6. *Response surfaces for $I_{dsat,p}$, average (heavy, blue) and standard deviation (light, cyan).*

4. Comments and Conclusions

In this context, wafers are blocking factors, and both common sense as well as statistical theory strongly favor changing as many factors as possible within wafer. This prevents block-to-block (wafer-to-wafer) variation from contributing to any estimation of effects, thereby improving the precision of any estimated models.

The analysis presented here does not exploit the implicit replication of having multiple exposure fields within wafer. How to do this is at this time an open issue. An advantage of the present approach is that one bypasses any modeling of the correlation among wafer sites as a function of distance.

On a practical basis, within-wafer experiments offer further advantages: (a) Broader product distributions are manufactured on each wafer; (b) lithography processing avoids specific targeting requirements; (c) the potential experimental factor count increases; alternatively, the required number of wafers is reduced. For 200mm wafers, there are about 50 exposure fields per wafer, and the number of within-wafer factors can approach 7. (d) Finally, analyses that concurrently optimize uniformity are enabled.

Point (c) is worth elaborating. One can represent nuisance patterns on the unit disk by Zernike polynomials (Brunner, 1997), for which the 36 non-trivial terms are usually taken. If F is the number of exposure fields, then $F-36$ is the number of degrees of freedom available for within-wafer designs. Additive models with 7 levels can therefore support $(F-36)/(7-1)$ factors. Alternatively, with a response surface model, and perhaps only three levels, the number of factors k must be such that $(k+1)(k+2)/2$ is no more than $(F-36)$; including additive 7-level factors results in an additional $(7-3)k$ term. For $F=50$, 100, and 200 exposure fields, the corresponding potential factor count is delineated in

Table 1. With 200mm wafers, $k=7$ factors become feasible by stepping and exposing half-fields.

#die/ field	#fields /wafer	#7-level factors	#RSM factors	#7-level ∩ RSM
4	50	2	4	2
2	100	10	10	7
1	200	27	16	13

Table 1. *For 200mm wafers and 50 4-die fields, the capacity of within-wafer experiments, measured by number of within-wafer factors.*

Given that some within-wafer experiment is to be performed, some are better than others. Designs exposing by x-coordinate ("striping"), mixing two exposures by x- and by y-coordinates ("madras"), and using polar angle ("pie wedges") are all less robust than the spatially dispersed design method proposed here, at least when more than three levels are used. Striping and madras are not robust to radial effects, because one or two stripes typically dominate the wafer center. In contrast, pie wedges are robust to radial effects, but fail to guard satisfactorily against problems occurring in single wafer regions, southwest corners, for example. It is also feasible to partition a wafer into separate zones, and balance factor combinations in each zone. But zone designation is ad hoc, hence inconsistent with the underlying geometry of the wafer's disk.

To exploit spatially dispersed designs more fully, both inspection and testing need to support them. Thus, at the post-exposure inspection, measure line widths at sites with differing exposure levels; too little variation signals the exposure plan has not been executed. At electrical test, define test sites entirely within single exposure fields; this avoids the loss of a wafer region, for example, the wafers' right edges as here.

5. References

Sacks, J, Welch, WJ, Mitchell, TJ, and Wynn, HP (1989), "Design and analysis of computer experiments," with discussion, *Statistical Science*, **4**:4, pp 409-435.

Heavlin, WD, and Finnegan, GP (1998), "Columnwise construction of response surface designs," *Statistica Sinica*, **8**:1, pp 185-206.

Box, GEP, and Draper, NR, (1987), *Empirical Model-Building with Response Surfaces*, Wiley, New York.

Ripley, B (1981). "Stochastic process prediction," chapter 4.4, pp 44-75 in *Spatial Statistics*, Wiley: New York.

Brunner, TA (1997). "Impact of lens aberrations on optical lithography," *IBM Journal of Research and Development*, **41**:1/2, pp 57-67.

MAXIMIN CLUSTERS FOR NEAR REPLICATE MULTIRESPONSE LACK OF FIT TESTS

James W. Neill, Forrest R. Miller and Duane D. Brown, Kansas State University
James W. Neill, Dept. of Statistics, Kansas State University, Manhattan, KS 66506

Key Words: Linear multiresponse model, regression, lack of fit, union-intersection principle, nonreplication, between cluster, within cluster, maximin power, minimal covers.

1. Introduction

A linear multiresponse model which allows each of several correlated responses to be modeled by a separate design matrix is known as the multiple design multivariate linear model. The usefulness of this model in many experimental situations, including applications to econometrics, chemical engineering kinetics and growth curve analyses, has been reviewed by Khuri (1986) and Srivastava and Giles (1987), for example. Parameter estimation for this model was discussed by Srivastava (1967) and Zellner (1962), and a test of the general linear hypothesis for such models was developed by McDonald (1975). An optimal design criterion to improve the power of this test was given by Wijesinha (1993). In addition, Khuri (1986) introduced exact tests for the equality of parameters from several correlated linear response models in the context of the multiple design model.

An important consideration in the modeling of a multiresponse function is the detection of model inadequacy. Khuri (1985) presented a test for lack of fit of a proposed linear multiresponse model, assuming that replicated observations on all of the responses are available at some points in the experimental region. In addition, Wijesinha and Khuri (1987) developed optimal design criteria to increase the power of this test. Levy and Neill (1990) suggested and compared some alternative tests for multiresponse lack of fit in the case of replication. The authors also indicated generalizations to the common circumstance in which replicate measurements are not available. These generalized tests require that the observations be grouped into clusters according to measures of nearness in the predictor space. However, a strategy for the efficacious choice of multivariate near replicate clusters has remained an open fundamental problem.

The purpose of this paper is to more fully discuss the problem of testing for multiresponse lack of fit in nonreplicated experiments. In particular, multiresponse extensions of the univariate tests for orthogonal between and within cluster lack of fit introduced by Christensen (1989, 1991) are given in Section 2. In addition, since the properties of these tests depend on the choice of multivariate near replicate clusters of the observations,

the maximin power clustering criterion developed by Miller, Neill and Sherfey (1998, 1999) for determining univariate near replicate clusters is generalized to the multiresponse case in Section 3. Examples and proofs are given by Neill, Miller and Brown (1999).

2. Orthogonal Between and Within Cluster Multiresponse Lack of Fit Tests

The normal theory multiple design multivariate linear model can be written as

$$Y = XG + E \qquad (1)$$

where $Y = [y_1, ..., y_r]$, $X = [X_1, ..., X_r]$, $G = \text{Diag}(\beta_1, ..., \beta_r)$ and $E = [\varepsilon_1, ..., \varepsilon_r]$. In model (1), y_i is an n x 1 vector of observations for the ith response with n x p_i design matrix X_i and corresponding unknown p_i x 1 parameter vector β_i, $i = 1, ..., r$. It is assumed that the elements of the X_i are real-valued continuous functions of K input variables $w_1, ..., w_K$ contained in some domain $\mathcal{G} \subseteq R^K$, $i = 1, ..., r$. Also, ε_i is the n x 1 vector of random errors associated with the ith response. The rows of the n x r random error matrix E are assumed to be independent and identically distributed as r dimensional multivariate normal with mean zero and unknown positive definite covariance matrix Σ. Thus, the r dimensional response vectors for different runs or experimental units are uncorrelated while the elements of the vector of responses for each particular run are correlated according to the unknown covariance matrix Σ.

Lack of fit is said to exist in the linear multiresponse model (1) provided the linear structure $X_i\beta_i$ does not adequately describe the mean of y_i for some $1 \le i \le r$. That is, $E(y_i) = X_i\beta_i + \xi_i$ for at least one $1 \le i \le r$ where ξ_i represents a lack of fit vector associated with the ith response. Furthermore, ξ_i can only truly be known to be contained in the corresponding lack of fit space $C(X_i)^{\perp}$. The notation $C(A)$ denotes the column space of a matrix A, and $C(A)^{\perp}$ denotes the orthogonal complement of $C(A)$. To construct optimal univariate tests for assessing the existence of various types of nonreplicate lack of fit, Christensen (1989, 1991) introduced a direct sum decomposition of the lack of fit space into orthogonal subspaces. Accordingly let

$$C(X_i)^{\perp} = (C(X_i)^{\perp} \cap C(Z)) \oplus (C(X_i)^{\perp} \cap C(Z)^{\perp}) \oplus S_i$$

where S_i denotes the orthogonal complement of the sum of the first two subspaces with respect to $C(X_i)^{\perp}$. The n x c matrix Z contains only zeroes and ones, and the nonzero values in the jth column of Z correspond to the observations in the jth cluster of near replicates. The column dimension c represents the number of groups for a specified near replicate clustering of the observations for the ith response. A clustering determined by such a Z will also be called a grouping or partition of the observations.

Corresponding to the terminology of one-way analysis of variance in which clusters of replicates are identified with different treatment groups, the first two subspaces in the preceding decomposition of $C(X_i)^\perp$ are called the orthogonal between and within cluster lack of fit subspaces, respectively. In the special case that replication exists in the row structure of X_i i.e. $C(X_i) \subseteq C(Z)$, univariate likelihood ratio tests for between and within cluster lack of fit compare the model $E(y_i) = X_i\beta_i$ with the model $E(y_i) = X_i\beta_i + \xi_i$ where ξ_i is replaced by lack of fit vectors in $C(Z)$ and $C(Z)^\perp$, respectively. Thus, the univariate test for between cluster lack of fit reduces to the classical lack of fit test in the case of replication. Also note that the univariate test for within cluster lack of fit would, for example, allow for detection of a trend in time within each group of replicates whenever the replicates are observed in a time sequence. The interpretation for the case of near replication generalizes the preceding concepts. Additional intuition and interpretation of these subspaces is given by Christensen (1989, 1991).

For multiresponse experiments, near replicates are those r dimensional observations corresponding to rows of X which are nearly equal in all corresponding coordinates. Measures of nearness in the predictor spaces, along with power considerations, will be discussed in the next section where the maximin near replicate clustering criterion is given for multivariate observations. Note that since the elements of each X_i are assumed to be continuous functions of the basic input variables $w_1, ..., w_K$, only one clustering matrix Z will be used to group or partition the r dimensional observations into clusters of near replicates. Thus, a multiple design multivariate linear model which accounts for multiresponse orthogonal between cluster lack of fit can be written as

$$Y = TH + E \qquad (2)$$

where $T = [T_1, ..., T_r]$ and $H = Diag(\eta_1, ..., \eta_r)$, and Y and E are the random matrices of responses and errors as in model (1). In model (2), $T_i = [X_i, B_i]$ where X_i is the n x p_i predictor matrix for the ith response in model (1), and B_i is an n x q_i matrix with rank q_i such that $C(B_i) = C(X_i)^\perp \cap C(Z)$, $i = 1, ..., r$. Note that B_i depends on the clustering matrix Z. When necessary, the notation $B_i(Z)$ will be used to emphasize this dependence. Let $t_i = p_i + q_i$ so that T_i is n x t_i and let $\rho = rank \ T$ where $\rho \le \sum_{i=1}^{r} t_i$. Also,

$$\eta_i = (\beta_i^T, \gamma_i^T)^T$$ where β_i and γ_i are the p_i x 1 and q_i x 1

unknown parameter vectors corresponding to X_i and B_i, respectively, $i = 1, ..., r$. By symmetry, a multiple design multivariate linear model which accounts for multiresponse orthogonal within cluster lack of fit can be obtained by replacing $C(Z)$ with $C(Z)^\perp$ in model (2).

To test for multiresponse orthogonal between cluster lack of fit, consider testing in model (2) the hypothesis

$$H_o^B: \{C_i\eta_i = 0, i = 1, ..., r\}$$

for a specified clustering matrix Z where $C_i = [0_{q_i \times p_i}, \ I_{q_i}]$. That is, $H_o^B: \{\gamma_i = 0, i = 1, ..., r\}$. Testability of a general linear hypothesis for the multiple design multivariate linear model was defined by McDonald (1975). The following proposition provides a condition for testability in particular of H_o^B.

Proposition 1. A necessary condition for the testability of H_o^B is that

$$\left(\sum_{i=1}^{r} C(X_i)\right) \cap C(B_j) = \{0\}, \ j = 1, ..., r. \qquad (3)$$

In addition, this condition is sufficient in the case that model (1) is monotone in the design matrices i.e. $C(X_{i_1}) \subseteq ... \subseteq C(X_{i_r})$ where $i_1, ..., i_r$ is some permutation of the integers $1, ..., r$. Note that monotonicity in the design matrices implies $C(B_{i_r}) \subseteq ... \subseteq C(B_{i_1})$, in which case condition (3) reduces to $C(X_{i_r}) \cap C(B_{i_1}) = \{0\}$. Without loss of generality, let $i_j = j$ for $j = 1, ..., r$.

A multivariate test for testing a testable H_o^B according to the union-intersection principle of test construction, as given by Roy (1953), follows next. Let the matrix due to the hypothesis and the matrix due to error be denoted by

$$S_h = Y^T Q_h Y \text{ and } S_e = Y^T Q_e Y,$$

respectively, where $Q_h = P_{\omega^\perp \cap \Omega}$ and $Q_e = I - P_\Omega$. The notation P_S denotes the orthogonal projection operator onto a subspace S of R^n, and ω and Ω represent the subspaces $\sum_{i=1}^{r} C(X_i)$ and $\sum_{i=1}^{r} C(T_i)$, respectively. Also, assume that $n - \rho \ge r$ so that S_e is positive definite with probability one (see Seber (1984), p. 522). By the assumed normality of Y, it follows that S_e is distributed according to the r dimensional central Wishart distribution $W_r(n - \rho, \Sigma)$ with $n - \rho$ degrees of freedom (see Seber (1984), Sec. 2.3). In addition, assume $q = max\{q_1, ..., q_r\} \le \rho$ so that S_h is independent of S_e and S_h is distributed according to the r dimensional noncentral Wishart distribution $W_r(q, \Sigma, \Delta)$ with q degrees of freedom and noncentrality parameter matrix Δ given by

$$\Delta = \Sigma^{-1}(TH)^T Q_h(TH)$$

$$= \Sigma^{-1}\left(\!\left(B_i\gamma_i\right)^T Q_h\!\left(B_j\gamma_j\right)\!\right)_{rxr}$$

(see Seber (1984), Sec. 2.3). Under H_o^B, $\Delta = \mathbf{0}$ so that $S_h \sim W_r(q, \Sigma)$ and thus H_o^B is rejected at level α whenever $\lambda_{max}\!\left(S_h S_e^{-1}\right) \geq c_\alpha$, where c_α denotes the upper α probability point of the distribution of $\lambda_{max}\!\left(S_h S_e^{-1}\right)$ under H_o^B. $\lambda_{max}(M)$ denotes the largest eigenvalue of a square matrix M.

The preceding test for H_o^B represents a multiresponse extension of the univariate test for orthogonal between cluster lack of fit suggested by Christensen (1991). By symmetry, a multiresponse extension of the univariate test for orthogonal within cluster lack of fit may be obtained by replacing C(Z) with $C(Z)^\perp$ in the preceding discussion. The following section specifically addresses the problem of choosing a clustering matrix Z to effectively test H_o^B. Although symmetry considerations can be used in part to derive optimal clusters for testing multiresponse orthogonal within cluster lack of fit, a separate paper will be written to more fully discuss this problem.

3. A Maximin Power Criterion for Clustering Multivariate Near Replicates

3a. Candidate Groupings Based on Covers of the Input Variable Settings

As indicated in Section 2, it is assumed that the elements of the ith predictor matrix X_i are continuous functions of K basic input variables $w_1, ..., w_K$ for $i = 1, ..., r$. In particular, letting $\mathbf{w}_j = (w_{j1}, ..., w_{jK})$, $j = 1, ..., n$, denote the jth setting of the K input variables, X_i is of the form

$$X_i = \begin{bmatrix} f_1^i(\mathbf{w}_1) & \cdots & f_{p_i}^i(\mathbf{w}_1) \\ \vdots & & \vdots \\ f_1^i(\mathbf{w}_n) & \cdots & f_{p_i}^i(\mathbf{w}_n) \end{bmatrix}$$

where f_m^i, $m = 1, ..., p_i$, are the continuous functions of the basic input variables associated with the ith response, $i = 1, ..., r$. In addition, we assume that $f_1^i, ..., f_{p_i}^i$ are linearly independent functions of $w_1, ..., w_K$ for each $i = 1, ..., r$. Also let W be the n x K matrix $\left[\mathbf{w}_1^T, ..., \mathbf{w}_n^T\right]^T$. That is, W is the matrix with j^{th} row given by $\mathbf{w}_j$, $j = 1, ..., n$. By the assumed continuity of the functions f_m^i, near replicate rows of W correspond to near replicate clusters of the observations. That is, near replicate groupings of

the observations may be identified with partitions of the rows of W. The multiresponse maximin power criterion, as defined in the following subsection, selects from a collection of candidate groupings a suitable near replicate clustering of the observations based on power and nearness considerations.

The number of all possible clusterings of the observations for most practical problems is extremely large. For example, with n = 16 the number of all possible groupings is 10,480,142,147, as calculated by the recurrence relations satisfied by Stirling numbers of the second kind (see Constantine (1987), p. 7). Thus, for the multiresponse maximin clustering criterion to be computationally feasible, one must generally restrict attention to a proper subset of the collection of all possible groupings. Indeed, based on nearness considerations, it is not necessary to consider all possible groupings. Miller, Neill and Sherfey (1998) have suggested a graph theoretic framework by which one may restrict the number of groupings under consideration. Miller, Neill and Sherfey (1999) further discuss the implementation of the univariate maximin clustering criterion given by Miller, Neill and Sherfey (1998). In particular, a methodology involving covers of the predictor settings was suggested to determine a collection of candidate groupings. This approach was adapted to the multiresponse case by Neill, Miller and Brown (1999), and is briefly summarized below.

A useful method for selecting a subset of candidate groupings to which the multiresponse maximin clustering criterion is applied can be based on a principle of exclusion. Let $\{S_1, ..., S_m\}$ be a family of overlapping subsets in R^K whose union includes the range of the row vectors $\mathbf{w}_i$, $1 \leq i \leq n$, in the input matrix W. Suppose the ith and j^{th} multiresponses can be clustered together only if $\mathbf{w}_i$ and $\mathbf{w}_j$ both lie in S_k for some $1 \leq k \leq m$. Thus, provided the subsets $\{S_1, ..., S_m\}$ are chosen so that any pair of extremely disparate input settings are not both contained in S_k for some $1 \leq k \leq m$, then the corresponding multiresponses cannot be clustered together. The preceding approach may be viewed in part as a way to eliminate absurd groupings directly and thus reduce the number of partitions under consideration. In general, given a family of overlapping subsets in R^K whose union includes the range of the rows $\mathbf{w}_1, ..., \mathbf{w}_n$ of W, a collection of candidate groupings can be determined as follows. Let $\{S_1, ..., S_m\}$ denote such a family of overlapping subsets. Note that Neill, Miller and Brown (1999) illustrate one method for selecting such subsets. Next let $C_j = S_j \cap V$, $j = 1, ..., m$, where $V = \{\mathbf{w}_1, ..., \mathbf{w}_n\}$. By construction $\underset{j=1}{\overset{m}{\cup}} C_j = V$ so that $\mathbb{C} = \{C_1, ..., C_m\}$ forms a cover of V. The collection of candidate groupings associated with the cover $\mathbb{C}$ can now be naturally defined.

Definition 1. Let $\mathbb{C} = \{C_1, ..., C_m\}$ be a cover of V and let the elements of each C_j be identified with the indices (i.e. first, second, third, etc.) of the corresponding multiresponses. The *collection of groupings consistent with the cover* $\mathbb{C}$ consists of all partitions P of the set of multiresponse indices $\{1, ..., n\}$ which satisfy the condition: if $P = \{A_1, ..., A_a\}$ then for each $i = 1, ..., a$ there exists $j \, \varepsilon \, \{1, ..., m\}$ such that $A_i \subseteq C_j$.

It should be emphasized that the overlapping subsets are used to restrict the number of groupings under consideration by excluding those groupings which cluster multirepsonses corresponding to dissimilar input settings. In addition, as described above, the overlapping subsets indicate which pairs of multiresponses are potential near replicates. However, specification of a collection of overlapping subsets does not determine a unique grouping but rather a collection of groupings consistent with the associated cover. The multiresponse maximin power clustering criterion is then applied to this collection as described in the following subsection. One is still generally choosing a clustering from a large number of candidate groupings represented by the collection of groupings consistent with the specified cover.

3b. Definition of the Criterion

For a fixed predictor matrix X, properties of the test presented in Section 2 for multiresponse orthogonal between cluster lack of fit depend on the selection of the clustering matrix Z. To identify clusterings that allow the corresponding multiresponse lack of fit test to effectively discriminate between model (1) and a type of lack of fit represented by model (2), a multivariate generalization of the maximin clustering criterion given by Miller, Neill and Sherfey (1998) will be determined. The objective of the multiresponse maximin criterion is to provide a cluster selection with increased trace of the noncentrality parameter matrix Δ. The desirability of such a clustering is due to the fact that the power of the test based on $\lambda_{max}\left(S_h S_e^{-1}\right)$ is a monotone increasing function of each of the eigenvalues of Δ separately (see Roy et al. (1971), p. 68). In order to present the multiresponse maximin criterion for cluster selection, a set of candidate groupings is required. Assume a cover $\mathbb{C}$ of V has been specified along with the collection of groupings consistent with the cover as defined in Subsection 3a. This collection must be restricted according to the dimension and testability considerations given in the following definition.

Definition 2. Let Ξ_C denote the set of grouping matrices Z which give partitions consistent with $\mathbb{C}$ and satisfy

(i) $C\left(B_i\right) \neq \{0\}, 1 \leq i \leq r$

(ii) $\left(\sum_{i=1}^{r} C(X_i)\right) \cap C\left(B_j\right) = \{0\}, 1 \leq j \leq r$

(iii) $n - \rho(Z) \geq r$,

excluding the trivial partitions which cluster all or none of the observations together.

Suppose the true model is assumed to be of the form given by model (2) for some $Z \, \varepsilon \, \Xi_C$, and one wishes to test the adequacy of model (1). A multiresponse maximin criterion will next be discussed to determine an appropriate $Z \, \varepsilon \, \Xi_C$, and hence a clustering. To explain the basic underlying concepts, fix $Z \, \varepsilon \, \Xi_C$ and define Q on the external direct sum $D_Z = C\left(B_1\right) \oplus ... \oplus C\left(B_r\right)$ by

$$Q\left(\xi_1, ..., \xi_r\right) = \sum_{i=1}^{r} \xi_i^T Q_h \xi_i.$$

Note that for particular lack of fit vectors $\xi_i \, \varepsilon \, C\left(B_i\right)$, $i = 1, ..., r$, $Q\left(\xi_1, ..., \xi_r\right) = \text{trace}((TH)^T Q_h(TH))$ and

$$\lambda_{min}\left(\Sigma^{-1}\right) Q\left(\xi_1, ..., \xi_r\right) \leq \text{trace}(\Delta),$$

where $\lambda_{min}(M)$ denotes the smallest eigenvalue of a square matrix M (see Khuri (1986), Lemma 2). Since $\lambda_{min}(\Sigma^{-1})$ is a constant, maximizing Q will imply a large value for trace(Δ) which in turn enhances the power of the test based on $\lambda_{max}\left(S_h S_e^{-1}\right)$. Next, let τ_i denote a positive definite quadratic form defined on $C(X_i)^\perp$, $i = 1, ..., r$, and define τ on the external direct sum $D = C(X_1)^\perp \oplus ... \oplus C(X_r)^\perp$ by

$$\tau\left(u_1, ..., u_r\right) = \frac{1}{r} \sum_{i=1}^{r} \tau_i\left(u_i\right).$$

A specific class of such forms which allow for nearness considerations is given by Neill, Miller and Brown (1999). Now for fixed $Z \, \varepsilon \, \Xi_C$, determine the vectors in the corresponding orthogonal between cluster lack of fit subspaces which minimize Q, subject to the restriction that the vectors lie in a set of the form $\{\tau \geq \delta\}$ for some $\delta > 0$. We want to determine a $Z \, \varepsilon \, \Xi_C$ which maximizes such minimum Q values. Specifically, a $Z \, \varepsilon \, \Xi_C$ is sought which maximizes

$$\Lambda_Z = \inf\{Q\left(v_1, ..., v_r\right)$$
$$: \left(v_1, ..., v_r\right) \varepsilon \, D_Z, \tau\left(v_1, ..., v_r\right) \geq \delta\}$$

with respect to $Z \, \varepsilon \, \Xi_C$ for any fixed $\delta > 0$. Next, letting

$$\ell_Z = \inf\{Q\left(v_1, ..., v_r\right) / \tau\left(v_1, ..., v_r\right) \quad : \left(v_1, ..., v_r\right) \varepsilon \, D_Z \backslash \{0\}\} \qquad (4)$$

and noting that $\ell_Z = \frac{1}{\delta} \Lambda_Z$, a multiresponse maximin clustering matrix is formally defined as follows.

Definition 3. A clustering matrix $Z \, \varepsilon \, \Xi_C$ which maximizes ℓ_Z as given by (4) is defined to be a *multiresponse maximin clustering matrix* from among the candidate groupings in Ξ_C.

Note that a multiresponse maximin clustering is invariant under reparameterizations and does not depend on the data matrix Y. Also implicit in the above discussion is the fact that τ allows comparisons of lack of fit vectors in the external direct sums D_Z for different $Z \in \Xi_C$. In particular, suppose $(\xi_1, ..., \xi_r) \in D_{Z_0}$ and $(\eta_1, ..., \eta_r) \in D_Z$ where $Z_0, Z \in \Xi_C$ and $\tau(\xi_1, ..., \xi_r) = \tau(\eta_1, ..., \eta_r)$. Furthermore, if Z_0 is a multiresponse maximin clustering matrix and $Q(\xi_1, ..., \xi_r) = \Lambda_{Z_0}$ and $Q(\eta_1, ..., \eta_r) = \Lambda_Z$ then $Q(\xi_1, ..., \xi_r) \geq Q(\eta_1, ..., \eta_r)$. Thus, for lack of fit vectors possessing the preceding properties, the corresponding Q parameter values based on a multiresponse maximin clustering are maximal as compared to corresponding values based on non-maximin clusterings. In fact, by the definition of Λ_{Z_0} it follows that $Q(v_1, ..., v_r) \geq Q(\eta_1, ..., \eta_r)$ for any $(v_1, ..., v_r) \in D_Z$ with $\tau(v_1, ..., v_r) = \tau(\eta_1, ..., \eta_r)$ and $Q(\eta_1, ..., \eta_r) = \Lambda_Z$. A calculational form for ℓ_Z is given by $\ell_Z = \lambda_{min}(B^{-1}F)$ where B and F are the matrices representing τ and Q on D_Z.

3c. Refinements and Atoms

It will be shown that the multiresponse maximin approach to clustering results in a cluster selection from among those partitions in Ξ_C which group as many observations together as possible. The concepts of refinement and atoms are next defined to facilitate the following discussion.

<u>Definition 4</u>. For $Z_0, Z_1 \in \Xi_C$, Z_1 is said to be a *refinement* of Z_0 provided $C(Z_0) \subseteq C(Z_1)$. In addition, Z_0 is an *atom* of Ξ_C provided Z_0 is not a refinement of any other member of Ξ_C. Let Ξ_0 denote the set of atoms in Ξ_C.

Note that if Z_1 is a refinement of Z_0 in Ξ_C then $\ell_{Z_0} \geq \ell_{Z_1}$ so that maximization of ℓ_Z can be made with respect to the atoms $Z \in \Xi_0$. Since atoms represent those partitions consistent with the cover $\mathbb{C}$ which group as many observations together as possible, the claim made at the beginning of this subsection has been verified. Based on the preceding, in order to determine a maximin clustering for a given W, hence for a given multiresponse predictor matrix X, the set of atoms Ξ_0 in Ξ_C must be determined. A discussion which characterizes the atoms associated with a cover $\mathbb{C}$ based on the overlapping subsets described by Neill, Miller and Brown (1999) is given therein.

3d. Generic Input Settings and Power Considerations

The power function of the size α test for testing multiresponse orthogonal between cluster lack of fit based on $\lambda_{max}(S_h S_e^{-1})$ is given by

$$\varphi(r, q, n - \rho, \Delta, \alpha) = P(\lambda_{max}(S_h S_e^{-1}) \geq c_\alpha | \Delta)$$

for a specified $Z \in \Xi_C$. As indicated by Roy et al. (1971), p. 68, φ is a monotonically increasing function of $n - \rho$ and a monotonically decreasing function of q and r. Now if $Z \in \Xi_C$ and Z_0 is an atom in Ξ_C with Z a refinement of Z_0, then $\ell_{Z_0} \geq \ell_Z$ as shown in Subsection 3c. Consequently, maximization of ℓ_Z can be restricted to the atoms of Ξ_C. Furthermore, since $C(Z_0)$ is a subspace of $C(Z)$, it follows that $q(Z_0) = \max\{$dimension $C(B_i(Z_0)), 1 \leq i \leq r\} \leq \max\{$dimension $C(B_i(Z)), 1 \leq i \leq r\} = q(Z)$ and $n - \rho(Z_0) = n - $ rank $T(Z_0) \geq n - $ rank $T(Z) = n - \rho(Z)$. Thus, the degrees of freedom parameters for the distribution of $\lambda_{max}(S_h S_e^{-1})$, based on clusterings corresponding to the atoms, are inherently in concordance with the objective of maximal power.

For the following discussion, we assume monotonicity in the design matrices. Thus, without loss of generality, $C(X_1) \subseteq ... \subseteq C(X_r)$ so that $C(B_r) \subseteq ... \subseteq C(B_1)$. In this case the degrees of freedom parameters reduce to $q(Z) = $ dimension $C(B_1)$ and $n - \rho(Z) = n - $ dimension $C(X_r) - $ dimension $C(B_1)$ for $Z \in \Xi_C$. If the degrees of freedom parameters $q(Z)$ and $n - \rho(Z)$ are constant on the atoms $Z \in \Xi_0$, then the atoms can be compared for power on the basis of the ℓ_Z values alone for $Z \in \Xi_0$. An atom which maximizes ℓ_Z with respect to $Z \in \Xi_0$ thus corresponds to a multiresponse maximin clustering as defined in Subsection 3b. For implementation purposes, the constancy of $q(Z)$ and $n - \rho(Z)$ is determined by directly evaluating dimension $C(B_1)$ for each atom $Z \in \Xi_0$. However, we claim that for *most* input matrices W, if dimension $C(Z)$ is constant on the atoms then dimension $C(B_1) = $ dimension $C(Z) - $ dimension $C(P_Z X_1)$ is constant on the atoms as well. This claim, which is justified by Theorem 1 below, leads to the concept of a generic input variable matrix as defined next.

<u>Definition 5</u>. Let W be an n x K matrix of basic input variables and let c = dimension $C(Z)$ for Z a grouping matrix. If

$$\text{dimension } C(P_Z X_i(W)) = \begin{cases} p_i & \text{if } p_i \leq c \\ c & \text{if } p_i > c \end{cases}$$

for $1 \leq i \leq r$ and all grouping matrices Z then W is *generic*.

<u>Theorem 1</u>. Suppose the functions f_m^i for $1 \leq i \leq r$ and $1 \leq m \leq p_i$ are analytic on the domain $\mathcal{G} \subseteq R^K$ and $(f_m^i, h_1)_m$ are linearly independent for each i where $h_1 \equiv$

1 on $\mathcal{G}$ (in case $f_m^i = h_1$ for some m then h_1 is removed). Consider the n x K matrix W to be determined by an element of R^{nK}. With this identification, the set of generic W matrices is an open and dense subset of $\mathcal{G}^n$ in R^{nK}.

The previous theorem justifies and makes more precise the claim made above. In particular, except for nongeneric n x K input matrices W which constitute a set of Lebesgue measure zero in $\mathcal{G}^n$, if dimension C(Z) is constant on the atoms then dimension $C(B_1)$ is also constant on the atoms. As discussed by Neill, Miller and Brown (1999), the atoms of Ξ_C, where $\mathbb{C}$ is the cover based on the overlapping subsets described therein, in fact have constant dimension. However, it should be emphasized that for implementation of the multiresponse maximin clustering criterion, we do not need to check whether W is generic. Rather, as indicated previously, we simply check whether dimension $C(B_1)$ is constant on the atoms. The significance of Theorem 1 is based on the assurance that the constancy of dimension $C(B_1)$ on the atoms will in fact ordinarily be realized whenever dimension C(Z) is constant on the atoms.

References

Christensen, R.R. (1989). "Lack of Fit Based on Near or Exact Replicates," Annals of Statistics, 17, 673-683.

Christensen, R.R. (1991). "Small Sample Characterizations of Near Replicate Lack of Fit Tests," Journal of the American Statistical Association, 86, 752-756.

Constantine, G.M. (1987). Combinatorial Theory and Statistical Design, Wiley, New York.

Khuri, A.I. (1985). "A Test for Lack of Fit in a Linear Multiresponse Model," Technometrics, 27, 213-218.

Khuri, A.I. (1986). "Exact Tests for the Comparison of Correlated Response Models With an Unknown Dispersion Matrix," Technometrics, 28, 347-357.

Levy, M.S. and Neill, J.W. (1990). "Testing for Lack of Fit in Linear Multiresponse Models Based on Exact or Near Replicates," Communications in Statistics, A19, 1987-2002.

McDonald, L. (1975). "Tests for the General Linear Hypothesis Under the Multiple Design Multivariate Linear Model," Annals of Statistics, 3, 461-466.

Miller, F.R., Neill, J.W. and Sherfey, B.W. (1998). "Maximin Clusters for Near Replicate Regression Lack of Fit Tests," Annals of Statistics, 26, 1411-1433.

Miller, F.R., Neill, J.W. and Sherfey, B.W. (1999). "Implementation of a Maximin Power Clustering Criterion to Select Near Replicates for Regression Lack of Fit Tests," Journal of the American Statistical Association, 94, 610-620.

Neill, J.W., Miller, F.R. and Brown, D.D. (1999). "Maximin Clusters for Near Replicate Multiresponse Lack of Fit Tests," submitted for publication.

Roy, S.N. (1953). "On a Heuristic Method of Test Construction and Its Use in Multivariate Analysis," Annals of Mathematical Statistics, 24, 220-238.

Roy, S.N., Gnanadesikan, R. and Srivastava, J.N. (1971). Analysis and Design of Certain Quantitative Multiresponse Experiments, Pergamon Press, New York.

Seber, G.A.F. (1984). Multivariate Observations, Wiley, New York.

Srivastava, J.N. (1967). "On the Extension of Gauss-Markov Theorem to Complex Multivariate Linear Models," Annals of the Institute of Mathematical Statistics, 19, 417-437.

Srivastava, V.K. and Giles, D.E. (1987). Seemingly Unrelated Regression Equation Models, Marcel Dekker, New York.

Wijesinha, M.C. (1993). "Optimal Designs to Improve the Power of Multiresponse Hypothesis Tests," Communications in Statistics, A22, 587-602.

Wijesinha, M.C. and Khuri, A.I. (1987). "Construction of Optimal Designs to Increase the Power of the Multiresponse Lack of Fit Test," Journal of Statistical Planning and Inference, 16, 179-192.

Zellner, A. (1962). "An Efficient Method of Estimating Seemingly Unrelated Regressions and Tests for Aggregation Bias," Journal of the American Statistical Association, 57, 348-368.

Determining the Probability of Acceptance in a Start-up Demonstration using Markov Chain Techniques

William S. Griffith, University of Kentucky

Department of Statistics, University of Kentucky, Lexington, Kentucky 40506-0027

Key Words: Acceptance Probability, Markov Chain

1 Introduction

Acceptance of equipment by a purchaser often requires a start-up demonstration test. This test can take various forms. In Hahn and Gage (1983), testing of the equipment continues until a pre-specified number of consecutive successes have occurred. This does not take account of failures. A modification of this is to terminate the test when there have been either k consecutive successes or a total of d failures. If the test is terminated by virtue of k consecutive successes, then the equipment is accepted by the purchaser. If the test is terminated by virtue of the occurrence of a total of d failures, then the equipment is rejected by the purchaser. In Section 2, we describe a methodology based on Markov chain techniques which can be used to compute the acceptance (and rejection) probabilities for this demonstration test based on the probability of any successful start-up. In Section 3, we also describe how the Markov chain approach can be adapted for other acceptance/rejection criteria.

2 Computing Acceptance/Rejection Probabilities

We assume in this paper that the probability of a successful start-up or any trial is p. Furthermore, we assume that trials are independent. We are interested in a methodology for the computation of the probability of acceptance, when the rule is to accept at the kth consecutive success or reject at the dth failure, whichever occurs first. We describe an approach using Markov chains. Let X_n be the state of the demonstration after the nth attempted start-up. Letting X_0 represent the state of the start-up demonstration prior to the first attempt, we have a Markov chain $\{X_n\}_{n=0}^{\infty}$ with state space $S = \{(r, f) : 0 \le r \le k, 0 \le f \le d-1\} \cup \{(0, d)\}$.

The non-zero one-step transition probabilities are

$$(r, f) \to (r + 1, f) \text{ with probability } p$$

and

$$(r, f) \to (0, f + 1) \text{ with probability } 1 - p.$$

We can identify $d + 1$ of the $kd - 1$ states as absorbing. The absorbing states include $(0, d)$, which corresponds to rejection of the equipment and $(k, 0), (k, 1), \ldots, (k, d-1)$ which correspond to acceptance of the equipment. It is possible to compute the probability of accepting the equipment by computing the absorption probabilities associated with the states corresponding to acceptance of the equipment.

Proposition

Suppose that successive start-ups are independent, the probability of successful start-ups is p for each attempt, and that the equipment is accepted if there are k consecutive start-ups before a total of d unsuccessful start-ups. If d unsuccessful start-ups occur first, then the equipment is rejected. Then the probability of acceptance is

$$1 - (1 - p^k)^{d-1}.$$

PROOF: We first remark that this result has been obtained by Aki and Hirano (1994). The context of their result and the statement of it are slightly different. The methodology used is completely different. They derive the joint probability generating function of the number of failures (unsuccessful start-ups) and the waiting time for the first k consecutive successes from a recursive formulation of the joint probability mass function. The Markov chain approach that we use here can be adapted to other acceptance/rejection rules also. Our approach begins with a consideration of the states of the chain.

$$
\begin{array}{cccc}
(0,0) & (0,1) & (0,2) & \cdots \quad (0,d) \\
(1,0) & (1,1) & (1,2) & \\
(2,0) & (2,1) & (2,2) & \\
\vdots & \vdots & \vdots & \\
(k,0) & (k,1) & (k,2) &
\end{array}
$$

We start at $(0,0)$. There is a probability of p moving down the column one state and a probability of $1-p$ of moving to the top of the next column to the right. Thus, there is a probability of p^k of being absorbed into $(k,0)$ and a probability of $1-p^k$ of reaching state $(0,1)$. The probability of being absorbed into any of the absorbing states (kj) from the top of its column $(0j)$ is p^k. therefore, $1-p^k$ is the probability that the item will be rejected. The probability that it will be accepted is $1-(1-p^k)^{d-1}$.

3   The Markov Chain Approach for Another   Acceptance/Rejection Criteria

The acceptance/rejection criterion can be many different things. It may be a specified number of consecutive (or total) successes before some specified number of consecutive failures. In this instance, our Markov chain $\{X_n\}$ has states of the form (r,f) where r and f are the number of consecutive (or total) successes and the number of consecutive failures at the end of the nth trial. here, if k is the specified number of consecutive (or total) successes, then a state whose first coordinate is k is considered absorbing and the acceptance probability for this scheme is an absorption probability for the Markov chain. Such absorption probabilities can be computed using Markov chain techniques. Future work will include other acceptance/rejection criteria such as k successes or d failures, with the last m trials. in addition, the assumption of independence will be relaxed.

References

Aki, S. and K. Hirano (1994), Distributions of numbers of failures and successes until the first consecutive k successes. *Annals of the Institute of Statistical Mathematics*, vol. 46, no. 1, pp. 193–202.

Hahn, G.J. and J.B. Gage (1983), Evaluation of a start-up demonstration test. *Journal of Quality Technology*, vol. 15, pp. 103-106.

SIMULTANEOUS PERTURBATION METHOD FOR PROCESSING MAGNETOSPHERIC IMAGES

Daniel C. Chin, The Johns Hopkins University Applied Physics Laboratory
11100 Johns Hopkins Road, Laurel, Maryland 20723-6099

Key Words: image processing; system identification; model estimation; multi-objective SPSA;

1. INTRODUCTION

This paper presents an efficient algorithm for extracting a simplified model from a complicated physical system that dynamically varies in time, e.g., the ion-intensity distribution in the magnetosphere. The extraction of the model parameters relies on an indirect sensing device, a particle counter whose output is only statistically related to the desired ion-intensity distribution. Roelof[1] developed a simplified 3-Dimensional model for the ion-intensity distribution in the magnetosphere. Chase and Roelof[2] further discussed the use of the energetic neutral-atom (ENA) sensor on board a satellite to estimate the model parameters studied in Roelof[1]. The ENA sensor is a particle counter that in each angular sector determined by its resolution provides the number of fast neutral atoms arriving at the sensor. This value is equal to a sample from a Poisson counting distribution whose intensity is equal to the integral, along lines of sight within the sector, of the ion intensity times the product of the hydrogen atom intensity and the cross section of the exchange process by which an ion strips an electron from a hydrogen atom to become a fast neutral atom. Two challenges faced in estimating the ion intensity distribution from ENA emission images are (1): each ENA value is affected by Poisson counting fluctuations, and (2): the intensity of the Poisson distribution is the integral of a function of the ion intensity distribution along the line of sight from the sensor. Chase and Roelof[2] focussed on the accuracy of the model representation. This paper presents an efficient estimation technique for the estimation of the parameters of that model that accommodates the random fluctuations in the data. The algorithm used is a modified simultaneous perturbation stochastic algorithm for global optimization in a multiple-objective setting (i.e., it simultaneously optimizes multiple objective functions).

The inversion process using a finite amount of data in a complicated system often runs into multiple solution problems. There are many optimization techniques available for estimating the hypothesized model, such as Bayesian estimation,[3,4] maximum likelihood estimation[5], and steepest descent and Newton-Raphson.[6] They all require detailed information of the system, either for computing gradients or for forming statistical distributions. These algorithms are hard to use when their objective functions have multiple local solutions. Other estimation techniques rely only on observations, such as Kiefer/Wolfowitz Stochastic Ap-

proximation[7] (KWSA), Simulated Annealing[8,9] (SAN), and other linear search type algorithms as in [2]. Only KWSA and SAN have been proven to demonstrate convergence using noisy measurements. Both KWSA and SAN have been shown in general forms to work in finding the global optimal solution.

The SPSA algorithm[10] is one of the KWSA types of algorithms that have been shown to require less data to converge than the other currently available KWSA algorithm.[11] Chin[12] (and Styblinski and Tang[13]) concluded that SPSA requires fewer function evaluations than SAN. This paper also compares SPSA with SAN directly. The comparison study shows SPSA is about three times more efficient than SAN; in an average of 10 runs, the level of accuracy at 300 function evaluations for SAN is about the level of accuracy at 100 function evaluations for SPSA. Some other successful examples of SPSA applications are in detecting faults in a power plant,[14] learning rules for a neuro-controller,[15] and estimating an electric current conductivity map to discriminate buried objects.[16]

The original SPSA algorithm was designed for asymptotic convergence and required an open-ended sequential data set. Because the ion-intensity in the magnetosphere is time-varying, the inversion process is forced to use an ENA global image (ENA counts across 4π steradians) at a single time (a single satellite viewpoint), rather than the multiple images at multiple times that would be possible if the ion-intensity distribution were constant. Some of the nice convergence properties are lost, like the strong convergence in the original theory, where the SPSA iteration converges to the truth asymptotically. In the magnetospheric global image setting the SPSA iteration will converge to a solution that is represented by a single global image. The process used here estimates a model of the system by matching a finite quantity of sensor data with the values predicted from an estimated model characterized by a fairly small set of parameters. The process then iteratively modifies the parameters in an attempt to obtain a better fit. The data noise (including sensor background noise and random fluctuations) tend to average out for the original SPSA algorithm, but that is not the case for the finite data stream of a single global image.

This paper also discusses a new approach for SPSA that uses multiple objective functions to find the common optimal solution for all functions. The purpose here is to fuse the magnetospheric image pixels (counts for each line of sight direction) for a global fit and still to match each individual count (pixel) locally. A single objective function lumps all the components together in a single mathematical formula that fits the counts (pixels) globally. In the single objective function approach, the parameters to which the objective function is most sensitive tend to dominate the estimation process. A separation of the single objective function into multiple functions or an addition type function may desensitize the domination and get a better overall fit. The example given here shows that the multiple-objective function algorithm achieves the lowest level of the total measurement residual errors.

2. THE GLOBAL MAGNETOSPHERIC IMAGE PROBLEM

The magnetosphere exists in a region of space that surrounds the earth out to 100,000 kilometers above the earth's surface. Its field lines, which pass through the earth's surface, fill the space. The global magnetospheric image process is used to estimate the energetic ion-populations trapped within the field. A global magnetospheric image is represented in pixel units, each of which is an accumulation of ENA (weighted by the exospheric hydrogen density) along the line of sight extended from the ENA camera (actually a particle counter). The number of pixels needed to represent a global image depends on the resolution of the sensors; for a sensor equipped with a 4°-by-4° pixel resolution, it takes 4050 pixels over 4π steradians (180° in latitude and 360° in longitude) to represent a global image. The actual numbers are slightly less because the top and bottom latitude regions have smaller surface area and use fewer pixels. If an ion-intensity model is known, the image value may be obtained by integrating along the line of sight the ion intensity times the product of the hydrogen atom intensity and the cross section of the exchange process by which an ion strips an electron from a hydrogen atom to become a fast neutral atom detected at the ENA camera. Some highly complex nonparametric simulations are available for the global magnetospheric ion-intensities (e.g., the Rice Convection Model and the Magnetosphere Specification Model (MSM)[17] and the Three-dimensional Ring Current Decay Model[18]). The object of this paper is to provide an estimation technique that attempts to invert this process. That is, given the hydrogen density and cross section and ENA counts over each 4° by 4° angular sector (or whatever the instrument's resolution) over a full 4π steradians, one would like to estimate the ion density of the magnetosphere.

2.1 Chase and Roelof Model

The field lines of the magnetosphere leave and enter the earth at the magnetic north and south poles (offset by about 12 degrees from the geodetic poles) and approximate a dipole field. Due to dominance of the magnetic field over near Earth plasma, a simplified distribution of ring current ions as a function of only two variables can be used. Quoting from Chase and Roelof[2], "the two variables are L (constant along a dipole field line defined in spherical coordinates by $r = aL\cos^2\Lambda$, where $a = 1R_E$ and Λ is the geomagnetic latitude) and the azimuthal angle ϕ measured anticlockwise from the sunward direction." R_E here is one earth radius. To modulate the ion intensity as a function of ϕ, a second order harmonic expansion is used:

$$F_\phi = k_1\left[1 - \cos(\phi - \phi_1)\right] + k_2\left[1 - \cos 2(\phi - \phi_2)\right].$$

To modulate the intensity as a function of L, Chase and Roelof define a piecewise function F_L which is parameterized on the five parameters $(L_0, L_1, L_2, \delta L_1, \delta L_2)$:

$$F_L = \begin{cases} (L - L_1)^2 / 2\delta L_1^2 & L < L_{11} \\ (L - L_{11})^2 / L_0 + \frac{1}{2}(\delta L_1/L_0)^2 & L_{11} \le L \le L_{22} \\ (L - L_2)^2 / 2\delta L_2^2 + (L_2 - L_1)/L_0 + \\ \frac{1}{2}(\delta L_2/L_0)^2 - \frac{1}{2}(\delta L_1/L_0)^2 & L > L_{22} \end{cases}$$

where $L_{11} = L_1 + \delta L_1/L_0$ and $L_{22} = L_2 + \delta L_2^2/L_0$, and $(L_0, L_1, L_2, \delta L_1, \delta L_2)$ are also functions of ϕ that are represented by five different harmonic expansions. Then, the ion density $j_{ION}(\phi, L)$ in the magnetosphere can be written as

$$j_{ION} = j_0 \exp(-F\phi) - F_L$$

where j_0 is a constant. An ENA image is an image each of whose pixels represents the number of fast neutral atoms detected over each angular sector defined by the resolution of the ENA camera. The expected intensity of these counts is a function of the ion intensity along the line of sight, but the actual realized count in each angular sector (represented by a single pixel) is a random sample from a Poisson distribution with intensity parameter given as a function of the ion intensity. If $\lambda(J_{ION})$ is the ENA intensity of the Poisson distribution Π along a particular line of sight, then an ENA image pixel represents $\sum \Pi[\lambda(j_{ION})]$ over all lines of sight in the cone subtended at the camera by the spherical angle determined by the resolution of the camera. But mathematically $\sum \prod[\lambda(j_{ION})] = \prod[\sum(j_{ION})]$, so that one can first compute the expected intensity of counts for each particular angular sector subtended by the camera with its resolution, and then sample only once per angular sector from the Poisson distribution. This is what is done in the simulations of this paper. The simulations are inherently random, in that even with the same ion intensity model, two simulations of an ENA image will differ due to the randomness of the Poisson distribution that characterizes the ENA counts along a given line of sight. For the purposes of the estimation of the ion intensity from an ENA image, the randomness of the counts acts as a kind of measurement noise. Chase and Roelof[2] show that the images from their ion-intensity model match with the simulated images from the Rice model within a very small variation when the Poisson counting statistics are ignored.

3. MULTIPLE-OBJECTIVE FUNCTION SPSA

Let θ represent the ion-intensity model parameters. For the ith pixel ($i=1,...,n$), let $j_{ENA,i}$ be the magnetospheric image sensor measurement for pixel i and $F_i(\theta)$ be the associated line-of-sight integration value predicted by the ion-intensity model. If the background noise level of the sensor is represented as σ, the usual SPSA algorithm optimizes a single loss (objective) function P defined as:

$$P(\theta) = \sum_{i=1}^{\ell}[F_i(\theta) - j_{ENA,i}]^2$$

where ℓ is the total number of pixels in the global image whose values are greater than the background noise level σ.

In contrast, the multiple-objective function SPSA algorithm seeks to optimize *each* of the normalized functions

$$P_i(\theta) = k_i[F_i(\theta) - j_{ENA,i}]^2$$

where $k_i = 1/j_{ENA,i}$, $j_{ENA,i} > \sigma$, for each pixel $i = 1,...,\ell$ whose measurement count is greater than σ. For each iteration of the usual SPSA algorithm, it substitutes a sequence of ℓ SPSA estimates, updating θ based on the gradient information of each function $P_i(\theta)$ in turn. The normalizing factor $k_i = 1/j_{ENA,i}$, $j_{ENA,i} > \sigma$, is meant to keep the information from those pixels with largest measurements from dominating the updates, i.e., it is meant to allow the algorithm to do its best to fit *all* the observations, not just the large ones.

To simplify the discussion, the original SPSA algorithm will be discussed first, then the multiple-objective function SPSA will be explained from the original algorithm. Letting $g(\bullet)$ denote the gradient of $P(\theta)$ with respect to θ, $\hat{\theta}_k$ denote the estimate for θ at the kth iteration, and $\hat{g}_k(\hat{\theta}_k)$ denote the SPSA approximated gradient at $\hat{\theta}_k$, the SPSA algorithm has the form

$$\hat{\theta}_{k+1} = \hat{\theta}_k - a_k\hat{g}_k(\hat{\theta}_k),$$

where the gain sequence $\{a_k\}$ satisfies certain well-known stochastic approximation conditions.[11] Let $\Delta \in R^p$ be a vector of p mutually independent mean-zero random variables $\{\Delta_1, \Delta_2, ..., \Delta_p\}$ satisfying conditions given in Section III of Spall,[11] where p is the number of estimated parameters. Subject to the important conditions in Ref 11, the user has full control over Δ. The recommended choice for the distribution of Δ_i is Bernoulli (±1). (Gaussian and uniform are not allowed.) Each iteration, SPSA uses two function evaluations to approximate the gradient, $\hat{g}_k(\bullet)$. In particular, at design levels $\hat{\theta}_k \pm c_k\Delta_k$, with c_k a positive scalar, let

$$y_k^{(+)} = P(\hat{\theta}_k + c_k\Delta_k),$$
$$y_k^{(-)} = P(\hat{\theta}_k - c_k\Delta_k).$$

Then the SPSA estimate of $g(\bullet)$ at the kth iteration is

$$\hat{g}_k(\hat{\theta}_k) = \begin{bmatrix} \dfrac{y_k^{(+)} - y_k^{(-)}}{2c_k\Delta_1} \\ \bullet \\ \bullet \\ \bullet \\ \dfrac{y_k^{(+)} - y_k^{(-)}}{2c_k\Delta_p} \end{bmatrix}.$$

The sequence $\{c_k\}$, for $k = 1, 2, ...$ should satisfy the convergence conditions as they are stated in Ref. 11.

For the multiple-objective function SPSA algorithm, Δ_k, c_k, a_k are selected for each k, as above, and an approximate gradient $\hat{g}_{k,i}$ of each function P_i is formed just as for P above, using the same Δ_k, c_k for each P_i. Then, holding Δ_k, c_k, a_k constant, one updates θ sequentially for $i = 1,...,\ell$ to yield at the kth major iteration and i^{th} minor iteration (for function P_i)

$$\hat{\theta}_{k+1,i} = \hat{\theta}_{k+1,i-1} - a_k\hat{g}_{k,i}(\hat{\theta}_{k,i}),$$

for $i = 1,...,\ell$ and $\hat{\theta}_{k+1,0} = \hat{\theta}_{k,l}$.

4. ACCELERATED PROCEDURE

This paper uses the SPSA iterative procedure in stages. Each stage consists of a set of runs with identical conditions except for the seed for the random number generator. In the stochastic approximation algorithm, the final estimates of a run are different from the other final estimates according to the initial selection of the seed of the random number generator. It is very difficult to determine whether the iterative process approaches a global solution or a local solution, especially if the initial values of model parameters are very different from the solution. Also, it is a waste to let a run proceed when the iteration has taken some wrong turns, which is normal in optimization iterations. Therefore, it is a good idea to restart a new stage with an averaged estimate from the last estimates in the previous stage. Furthermore, a selective average can be

Table 1: SPSA Results

Model Elements	SPSA Initial Values	Simulation Values	2nd Stage Estimates	Final Estimates
F_ϕ	(2.30, 0.1, 43 deg)	(2.50, 0.5, 29 deg)	(2.46, 0.5, 42 deg)	(2.45, 0.5, 41 deg)
L_0	(0.20, 0.1, 43 deg)	(0.22, 0.5, 19 deg)	(0.21, 0.3, 44 deg)	(0.21, 0.4, 43 deg)
L_1	(2.00, 0.1, 43 deg)	(2.50, 0.4, 66 deg)	(2.52, 0.3, 53 deg)	(2.51, 0.3, 56 deg)
L_2	(4.50, 0.1, 43 deg)	(5.00, 0.8, 29 deg)	(5.07, 0.4, 41 deg)	(5.05, 0.4, 41 deg)
δL_1	(0.20, 0.1, 43 deg)	(0.14, 0.2, 27 deg)	(0.13, 0.3, 32 deg)	(0.14, 0.3, 33 deg)
δL_2	(0.10, 0.1, 43 deg)	(0.11, 0.8, 46 deg)	(0.12, 0.5, 40 deg)	(0.11, 0.5, 41 deg)

done between stages. In the simulation study, the initial values for the model parameters in the second stage are the average of four runs from the first stage for which the errors of the measurement residuals in the last iterations are less than 0.3 in value. (It is just a natural separation among the runs, no particular meaning.)

To avoid ambiguities, constraints have been put on the scale and trigonometric parameters of the harmonic expression determining the functions F_ϕ and F_L. The scale parameters are limited to positive values; and the trigonometric parameters are restricted to the interval [-π, π]. If an estimate for these constrained parameters exceeds its range, the value of the estimate is reset to the initial value of that stage. There are other choices in determining the reset values depending on the purpose to which the procedure is applied.

Because the Chase and Roelof model used for the simulation has only piecewise continuous gradients, the perturbed values $\hat{\theta}_k \pm c_k \Delta_k$ used to compute the SPSA approximate gradient $\hat{g}_k(\hat{\theta}_k)$ may lie on either side of a gradient discontinuity. In this situation, the approximated gradients are not representative of either piece and cause the measurement residual error to jump. If the jump makes the measurement residual error small, it is beneficial. If the jump goes the other direction, it is not beneficial. The simulation study here ignores such updates if the updated measurement residual error is greater than 110 percent of the previous value.

5. SIMULATION STUDY

This study is performed using measurements simulated from a simple Chase and Roelof model; first order harmonic expansions in ϕ being used for all elements $(F_\phi, L_0, L_1, L_2, \delta L_1, \delta L_2)$. Let E be an arbitrary element from these six. Then the harmonic expansion is $E = k - s\cos(\phi - \varphi)$ where k, s, and φ are the parameters to be estimated. Each entry in Table 1 is the three-vector (k, s, φ) associated with its respective element.

The values used to generate the images are listed in the "Simulation Values" column in Table 1. The setting of the study is based on an earlier mission with a 4° by 12° pixel image and a maximum of 1000 counts per pixel. These values were used to define the re-

sponse function of the ENA camera used in this study. The SPSA algorithm was then applied to estimate the model parameters using a set of initial values, as shown in the entries of Table 1 under "SPSA Initial Values", to start the multiple-objective SPSA iterations. The initial values are inside the physically expected ranges. For example, the constant part of L_2 is expected to be between 4.00 and 6.00; thus, a value of 4.50 was given. The studies show the convergence of the estimation is independent of initial values. Many local solutions exist between the initial values and the final estimates.

There are three stages in the simulation study. The averages of the final estimates of the runs from the second and third stages are listed in the "2nd Stage Estimates" and "Final Estimates" columns of Table 1. Each stage consists of a total of 10 runs; each run is an SPSA process with identical setting, except the seeds used for the pseudorandom number generator are different. Each run consists of 150 iterations; each iteration uses four function evaluations (a total of 600 function evaluations); a function evaluation is defined by one round of computations of all pixels in the image of the magnetosphere field. As mentioned before, the initial values of the SPSA runs in Stage 1 are the values shown in Table 1; the initial values for the runs in the subsequent stages are the average of the final estimates of the runs in the previous stages.

The initial gain constants, a and c, used in SPSA were set to 0.005 and 0.001 at Stage 1 for each run. Following the practical guidelines from Spall[19], the reduction rate α for a is 0.602 and γ for c is 0.101. The reduction for a is applied continuously as regular SPSA, but the reduction for c is applied once every 30 iterations. The gain constants for the following runs in Stages 2 and 3 are a continuation of Stage 1 runs.

An obvious question is how well does the multiple-objective SPSA inversion process estimate its parameters. Table 1 shows differences in parameters between the "Simulation Values" and the estimates. These differences are suspected to come primarily from the random fluctuations in the image pixels which act as a kind of measurement noise; the multiple-objective SPSA iterations converge to a solution that represents a model for the ion intensity that would mimic the single global image without any addition of random Poisson fluctuations. To verify this, another simulated global image was generated under the same setting without the Poisson statistics in the counts and a number of estimation runs were made using this data. The total estimated parameter errors based on "Simulation Values" as truth converged below the 10^{-9} error level for all the runs. The total estimated parameter error is the summation over the 18 parameters of Table 1 of the difference squares between each individual estimate and its target value.

Theoretically the SPSA iterations asymptotically converge under some conditions and the differences between estimates of the SPSA algorithm and the true-values are asymptotically normal at each iteration[11]. If the conditions of the multiple-objective SPSA runs satisfy the same convergence conditions of the SPSA algorithm, the multiple-objective SPSA iterations here will asymptotically converge, and the errors of the estimates will also be asymptotically normal. There are variations among the estimates of the multiple-objective SPSA runs at each iteration, the standard deviation of the variations is larger at the beginning iterations; the standard deviation becomes smaller when there are more iterations. Among three stages of multiple-objective SPSA runs, the Stage 1 reduces the estimated parameter error quickly; Stage 2 finalizes the estimates; Stage 3 is just for confirmation. In fact, the values of the estimated model change very slightly between Stages 2 and 3. The deviations of the final estimates between runs within the stage have narrowed for Stage 3 in comparison with the deviation for Stage 2. The maximum deviation in Stage 3 for a single parameter is less then 0.3% of parameter value (0.003 in parameter units) which is negligible in most applications of the model. In practical applications, a single run may be used for Stages 2 and 3.

The success in extracting the distribution from the ENA images can be judged by deviations of the estimated parameters from the simulated parameters for the ion-intensity model (the "Final Estimates" and "Simulation Values" columns in Table 1) for the sensor resolution of 4° by 12° pixels. The values in Table 1 show that the constants of the element parameters can be recovered pretty well. The scale factors and the phase angles of the element parameters are shifted. The simulation studies show that the estimation errors due to the estimation method are small in caparison with the instrumentation errors after a sufficient number of estimation iterations. To improve the accuracy one has to decrease the measurement errors, for example, by increasing the sensor resolution which reduces the random fluctuations, improving the model structure to reduce the sensitivities to the random fluctuation in the data, or introducing replicated data points to average the random fluctuations.

6. COMPARISON OF SPSA WITH SAN

A well-known method for a global optimization problem is simulated annealing (SAN). The question of how well SPSA compares with SAN is of interest. Because SAN appears not to have been applied to multiple objective function problems, the comparison is based only on the results from the single-objective function — sum square measurement-residual error. For the single objective function, neither SPSA nor SAN shows any sign of convergence within 300 function evaluations if the same initial values are used as in stage 1 of the multiple-objective SPSA study. When the initial values are close to the solutions (using the initial values of stage 2), the single objective function SAN does show signs of convergence after tuning. Therefore, this study compares these two algorithms starting from a set of parameters close to the simulation values, comparable to the initial values of stage 2.

A total of 10 runs were studied. Fig. 1 shows that at the 300th function evaluation, the level of mean sum square error is about 3:1 in favor of SPSA. It also shows a 3:1 ratio in the number of function evaluations required to achieve the same level of accuracy. Because the results are in favor of SPSA, there is no further development for SAN using multiple objective functions. Also, with SAN it is inherently difficult to determine the acceptance of the updated parameter values at each SAN step for multiple objective functions. The averaged absolute mean total measurement-residual error for multiple-objective SPSA is also much lower than for the single objective SPSA or SAN (compare with Fig. 1).

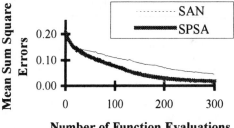

Figure 1: Convergence of SPSA vs. SAN

7. CONCLUSION

This paper applies the SPSA algorithm for two types of objective functions, pixel-wide multiple objective functions and a total summation function. The comparison of sensor residuals for simulated annealing and for single- and multiple-objective SPSA algorithms indicates that the multiple-objective function SPSA algorithm gives the lowest total measurement residual errors. Also, the data noise level and resolution in the simulation study are adequate to estimate the constant parameters in the Chase and Roelof mode[12]; some of the other parameters are not so well estimated due primarily to the finite sample with random fluctuations in the data counts.

When the multiple-objective SPSA algorithm is applied to a real satellite mission for the global magnetospheric image problem, there will be other error sources affecting the final estimates. Reductions of those errors are not within the scope of this paper. The paper shows that multiple-objective SPSA could be used to reduce the estimation errors and to isolate the other error sources. That may help the satellite mission planning and data analysis work for the global magnetospheric image problem.

This paper also compares the efficiency and accuracy levels for SPSA and SAN. In a single objective function setting, the results of the comparison were three to one in favor of SPSA in both efficiency and accuracy at the given measures. An additional advantage in using SPSA is that it can be extended to an infinite data setting. Instead of reiterating on a fixed global image, SPSA could iterate on consecutive sets of image counts. That changes the objective functions into noisy functions, which work well for SPSA. The ion intensity in the magnetosphere changes over time; the underlying model of the consecutive of image counts also changes from image to image. SPSA with gain-sequence manipulation, as demonstrated in Ref. 20, could adapt to the changes in the model. (This is different from restarting the iteration from the final estimates in the previous data images.)

ACKNOWLEDGMENTS

This paper was jointly supported by The Johns Hopkins University Applied Physics Laboratory Internal Research and Development program and the National Aeronautics and Space Administration Space Research program, grant #NAGW-2691.

REFERENCES

1. E. C. Roelof, "Energetic Neutral Atom Image of a Storm – Time Ring Current," *Geophy. Rev. Letters*, v. 14, no. 6, pp. 652-655 (1987).

2. C. J. Chase and E. C. Roelof, "Extracting Evolving Structures From Global Magnetospheric Images via Model Fitting and Video Visualization," *Johns Hopkins APL Tech. Digest*, v. 16, no. 2, pp. 111-122 (1995).

3. J. C. Spall, *Bayesian Analysis of Time Series and Dynamic Models*, Dekker (1988).

4. P. K. Venkatesh, M. H. Cohen, R. W. Carr, and A. M. Dean, "Bayesian Method for Global Optimization," *Phys. Rev. E*, v. 55, no. 5, pp. 6219-6232 (1997).

5. A. P. Dempster, N. M. Laird, and D. B. Rubin, "Maximum Likelihood from Incomplete Data via the EM Algorithm," *J. R. Stat. Soc. Ser. B*, v. 39, pp. 1-38 (1977).

6. D. R. Haley, J. P. Garner, and W. S. Levine, "Efficient Maximum Likelihood Identification of a Positive Semi-Definite Covariance of Initial Population Statistics," *Proc. 1984 Amer. Cont. Conf.*, pp. 1085-1089 (1984).

7. M. Bazaraa and C. M. Shetty, *Nonlinear Programming*, Chapter 8, Wiley, New York (1979).

8. J. Kiefer and J. Wolfowitz, "Stochastic Estimation of a Regression Function," *Ann. Math. Stat.*, v. 23, pp. 462-466 (1952).

9. S. Kirkpatrick, C. D. Gelatt, Jr., and M. P. Vecchi, "Optimization by Simulated Annealing," *Science*, 220 (4598), pp. 671-680 (1983).

10. J. R. Ray and R. W. Harris, "Simulated Annealing in the Microcanonical Ensemble," *Phy.l Rev. E*, v. 55, no. 5, Pt. a, pp. 5270-5274 (1997).

11. J. C. Spall, "Multivariate Stochastic Approximation Using a Simultaneous Perturbation Gradient Approximation," *IEEE Trans. on AC*, v. 27, pp. 332-341 (1992).

12. D. C. Chin, "Comparative Study of Stochastic Algorithms for System Optimization Based on Gradient Approximations," *IEEE Trans. on S MC*, v. 27, no. 2, pp. 244-249 (1997).

13. D. C. Chin, "A More Efficient Global Optimization Algorithm Based on Styblinski and Tang," *Neural Networks*, v. 7, pp. 573-574 (1994).

14. M. A. Styblinski and T. S. Tang, "Experiments in Nonconvex Optimization: Stochastic Approximation with Function Smoothing and Simulated Annealing," *Neural Networks*, v. 3, pp. 467-483 (1990).

15. A. Alessandri and T. Parisini, "Nonlinear Modeling of Complex Large-Scale Plants Using Neural Networks and Stochastic Approximation," *IEEE Trans. on SMC*, pp. 750-757 (1997).

16. D. C. Chin and R. Srinivasan, "Electrical Conductivity Object Locator: Location of Small Objects Buried at Shallow Depths," *UXO'97 Conf. Proc.*, pp. 50-57 (1997).

17. R. A. Wolf, R. W. Spino, and F. J. Rich, "Extension of the Rice Convection Model into the High-Latitude Ionosphere," *J. Atmo. Terr. Phys.*, 53, pp. 817-829 (1991).

18. M. Fok, T. E. Moore, J. U. Kozyra, G. C. Ho, and D. C. Hamilton, "Three-dimensional Ring Current Decay Model," *J. of Geop. Res.*, v. 100, no. A6, pp. 9619-9632 (1995).

19. J. C. Spall, "Implementation of the Simultaneous Perturbation Algorithm for Stochastic Optimization," *IEEE Trans. on AES*, v. 34, to appear (1998).

20. J. C. Spall and J. A. Cristion, "Nonlinear Adaptive Control sing Neural Networks: Estimation Based on a Smoothed Form of Simultaneous Perturbation Gradient Approximation," *Stat. Sinica*, v. 4, pp. 1-27 (1994).

Efficient Estimation of Lifetime Distribution - A New Sampling Plan

Surekha Mudivarthy and M. Bhaskara Rao, North Dakota State University
M. Bhaskara Rao, Department of Statistics, NDSU, Fargo, ND 58105, U.S.A.

Key words: Censored data, Interval-censored data, Type II Plan, Efficiency, Exponential distribution.

1. Introduction: The main objective of any sampling plan in survival analysis is to estimate the lifetime distribution of a product under investigation (Lawless (1982); Lee (1992); Klein and Moeschberger (1997)). For example, an industrial researcher may want to estimate the lifetime distribution of a product of interest, where as an ecologist may be interested to estimate the lifetime distribution of nests of a particular species of birds. In industrial research, Type II sampling plan is most commonly used. Under this plan, n units are selected at random and set to work. All n units are monitored round the clock until r (a predetermined number) of them fail, where $1 \le r \le n$. One main drawback of the plan is that it is not always practical to monitor the units round the clock.

To address this problem, in response to many consulting problems, we offer a new sampling plan called Interval-Censored Type II plan. Under this plan, all n units are inspected periodically, until a fixed number of them fail or reach a well-defined state. Note that the exact failure times of the units will be unknown, but we will know how many units failed between each consecutive inspection periods.

The implementation of this new sampling plan can be formally described as follows - choose two positive integers $1 \le r \le n$ and a number t > 0.

Select a random sample of n units of the product and set them to work. Inspect the units at times t, 2t, 3t, ... At each inspection time, ascertain how many units failed since the last inspection time. The data so generated are called Interval-censored Type II data. A description of the plan was first described in Mudivarthy, Rao, and Mitra (1996).

Some of the questions we want to address in this paper are the following.

(1) Estimation of the underlying distribution using the data collected.
(2) Provide some guidelines on the choice of t, the common length of the time intervals.
(3) Compare the efficiency of the new plan with that of the traditional Type II plan.

Surprisingly, it turns out that the Interval-censored type II plan is more efficient than the traditional Type II plan in many cases of interest. We provide an explanation for this phenomenon.

2. Interval-Censored Type II Data: The data collected under the new plan consists of two components, M and $X_1, X_2, ..., X_M$, where M denotes the number of inspections needed to get r failures and X_i denotes the number of failures that occur between the (i-1)th and ith inspections, i.e., during the time interval ((i-1)t, it], i = 1, 2, ..., M. Note that M is a positive integer valued random variable and X_i's are non negative integer valued random variable satisfying the following conditions.

$$X_1 + X_2 + ... + X_{M-1} \le r-1, \text{ and}$$
$$X_1 + X_2 + ... + X_M \ge r.$$

Define an additional variable
$$X_{M+1} = n - (X_1 + X_2 + ... + X_M)$$

3. Estimation: We assume that the underlying lifetime distribution belongs to the parametric family of exponential distributions. Let T be the underlying lifetime variable associated with the product. The probability density function of T is given by

$$f_\theta(x) = \theta e^{-\theta x}, \qquad x > 0, \quad \theta > 0.$$

Let $T_1, T_2, ..., T_n$ be iid copies of T and $T_{(1)} < T_{(2)} < ... < T_{(n)}$ be the corresponding order statistics. Under the traditional Type II plan, we observe $T_{(1)} < T_{(2)} < ... < T_{(r)}$. The maximum likelihood estimator of θ is given by

$$\hat{\theta} = r/\left(T_{(1)} + T_{(2)} + ... + T_{(r)} + (n-r)T_{(r)}\right)$$

with $E_\theta\left(\hat{\theta}\right) = \left(\dfrac{r}{r-1}\right)\theta$, for all $\theta > 0$ and

Variance $\left(\dfrac{r-1}{r}\hat{\theta}\right) = \dfrac{\theta^2}{(r-2)}$, provided r>2. (Epstein and Sobel (1953)).

Let us now focus on the data $M, X_1, X_2, \ldots, X_M, X_{M+1}$ collected under the Interval-censored type II plan. The joint distribution of $X_1, X_2, \ldots, X_M \,|\, M = m$ is

$$\Pr(M = m) \times \Pr(X_1 = x_1, X_2 = x_2, \ldots, X_m = x_m \,|\, M = m)$$
$$= \frac{n!}{x_1! x_2! \ldots x_{m+1}!} \left(1 - e^{-\theta t_0}\right)^{x_1} \left(e^{-\theta t_0} - e^{-2t_0\theta}\right)^{x_2} \ldots$$
$$\left(e^{-(m-1)t_0\theta} - e^{-mt_0\theta}\right)^{x_m} \left(e^{-mt_0\theta}\right)^{x_{m+1}}$$

Note that $x_1, x_2, \ldots, x_m, x_{m+1}$ are nonnegative integers such that

$$x_1 + x_2 + \ldots + x_{m-1} \le r - 1,$$
$$x_1 + x_2 + \ldots + x_{ml} \ge r,$$
$$x_1 + x_2 + \ldots + x_{m+1} = n.$$

The likelihood $L(\theta)$ of the data $M = m, X_1 = x_1, X_2 = x_2, \ldots, X_m = x_m, X_{M+1} = x_{m+1}$ is given by,

$$L = \text{Constant} \left(1 - e^{-\theta t}\right)^{x_1} \left(e^{-\theta t} - e^{-2t\theta}\right)^{x_2} \ldots$$
$$\left(e^{-(m-1)t\theta} - e^{-mt\theta}\right)^{x_m} \left(e^{-mt\theta}\right)^{x_{m+1}}$$
$$= \text{Constant} \left(1 - e^{-\theta t}\right)^{x_1 + x_2 + \ldots + x_m}$$
$$\left(e^{-\theta t}\right)^{x_2 + 2x_3 + 3x_4 + \ldots + mx_{m+1}}.$$

If $m=1$ and $x_1 = n$, $L(\theta)$ has no finite maximum. In all other cases, $L(\theta)$ has a finite maximum. For each of these cases, the log likelihood is given by,

$$\ln L(\theta) = \text{Constant} + \left(\sum_{i=1}^{m} x_i\right) \ln\left(1 - e^{-\theta t}\right)$$
$$- (\theta t)\left(\sum_{i=1}^{m} i x_{i+1}\right)$$

The partial derivatives of the log likelihood is set equal to zero in order to obtain the maximum likelihood estimate.

$$\frac{\partial}{\partial \theta}(\ln L) = \left(\sum_{i=1}^{m} x_i\right) \frac{t e^{-\theta t}}{(1 - e^{-\theta t})} - t\left(\sum_{i=1}^{m} i x_{i+1}\right) = 0$$

The likelihood equation has an explicit solution.

$$\hat{\theta} = \frac{1}{t} \ln\left(1 + \frac{\sum_{i=1}^{M} X_i}{\sum_{i=1}^{M} i X_{i+1}}\right)$$

The maximum likelihood estimator, $\hat{\theta}$ upon replacing the data by the corresponding random variables, is given by,

$$\hat{\theta} = \begin{cases} \dfrac{1}{t} \ln\left(1 + \dfrac{n}{1}\right), & \text{if } M = 1 \text{ and } X_1 = n, \\[3ex] \dfrac{1}{t} \ln\left(1 + \dfrac{\sum_{i=1}^{M} X_i}{\sum_{i=1}^{M} i X_{i+1}}\right), & \text{otherwise.} \end{cases}$$

Note that we have imposed our own choice of the estimate in the case when the maximum likelihood estimate does not exist.

4. Optimal Value of t: We choose and fix t before we start collecting the data. To choose an optimal value of t, we evaluate the mean square error of $\hat{\theta}$ ($\text{MSE}(\hat{\theta})$) for each combination of n, r and θ for various values of t and choose that value of t for which the $\text{MSE}(\hat{\theta})$ is the smallest. The following table gives the optimal values of t when n=30 and θ=1 for various values of r.

r	Optimal t
5	1.2
10	1.2
15	0.9
20	0.5
25	0.5

5. Efficiency: The exact distribution of $\hat{\theta}$ is tractable. We developed a code that compiles the exact distribution of $\hat{\theta}$. Let us look at the simple case. Take n=4, r=3, θ=ln(2)=0.69315, and t=2.2. The exact distribution of $\hat{\theta}$ is given by,

$\hat{\theta}$	$\Pr(\hat{\theta})$	$\hat{\theta}$	$\Pr(\hat{\theta})$
0.1448	0.0001	0.2672	0.0022
0.1621	0.0004	0.3151	0.0651
0.1843	0.0013	0.3851	0.0155
0.2055	0.0001	0.4994	0.1065
0.2136	0.0037	0.6301	0.4169
0.2322	0.0006	0.7316	0.3747
0.2544	0.0129		

We compiled the code for various values of n, θ, r and t, evaluated the expected value, and the mean square error of $\hat{\theta}$ for each combination. The mean square error evaluated using the exact distribution was compared with the variance of $\hat{\theta}$ under the traditional plan to demonstrate the efficiency of the new plan.

The efficiency of the new plan is defined as

$$\frac{\text{Variance of } \hat{\theta} \text{ under the traditional plan}}{\text{Mean square error of } \hat{\theta} \text{ under the new plan}}$$

The following is the efficiency table based on the exact distribution when n=30 and θ=1

r	t	Efficiency
5	0.5	1.8971
5	1.0	2.2353
5	1.5	2.1965
10	0.5	1.2707
10	1.0	1.3694
10	1.5	1.34510
15	0.5	0.9848
15	1.0	1.1084
15	1.5	1.0554
20	0.5	0.9950
20	1.0	0.9546
20	1.5	0.9158
25	0.5	0.9831
25	1.0	0.9119
25	1.5	0.8106

Surprisingly, from the above table it is clear that the Interval-censored type II plan is better for certain values of r and t. Intuitively one would expect the traditional Type II plan to be more efficient than the new plan as the data under the traditional plan is exact while that under the new plan is inexact.

We found two possible reasons for such a surprising result and we attempt to explain the same as follows.

1)   Let R be the number of failures that actually occurred at the conclusion of the experiment. The following is the table for expected number of failures when n=30, θ=1 and t is optimal.

r	Efficiency	E[R]
5	0.27037	20.9642
10	1.40457	20.9643
15	1.12668	18.9097
20	0.97826	22.0876
25	0.94982	26.0566

When n=30, θ=1 and optimal t=1.2, we terminate the data collection process as soon as we observe 5 failures, though in reality we have substantially more failures expected (E(R) ≅ 21). The loss of information caused by inexact data is compensated actually by obtaining more failures. This is one of the reasons why

the new plan is more efficient than the traditional plan, especially for lower values of r.

2)   Time taken to collect the data: As we discussed earlier, the data under the traditional plan consists of first r order statistics. The average time taken to execute the Type II plan is given by,

$$E_\theta T_{(r)} = \frac{1}{\theta} \sum_{i=1}^{r} \frac{1}{n-i+1}.$$

The average time taken to execute the Interval-censored type II plan is given by

$$t E_\theta(M)$$

It is found that the average time taken to execute the Interval-censored type II plan is greater than that of the Type II plan. The following table compares both the average times for n=30, θ=1 and t=optimal.

r	NP	TP
5	1.2	0.2
10	1.2	0.4
15	1.0	0.7
20	1.3	1.1
25	2.0	1.7

NP = Average time taken to execute the new plan.
TP = Average time taken to execute the traditional plan.

Since more time is spent in collecting the data under the new plan, one would expect more failures than required and this could be another contributing factor for the new plan to be more efficient than the traditional plan.

6. Conclusions: Though the Interval-censored type II plan results in generating inexact data unlike the traditional Type II plan, more failures occur (E(R)>>r) under the new plan. The Interval-censored type II plan is more efficient than the traditional plan especially when r<<n. It is obvious that one need not observe the units round the clock for lower values of r.

The Interval-censored type II plan is more cost- and time- efficient than the traditional plan. May be, the time has come to sound a death-knell to the traditional Type II plan!!

***

References

1.   Epstein, B., and Sobel, M (1953). "Life testing," J. Amer. Stat. Assoc., 48, 486-502.

2. Klein, J. P., and Moeschberger, M. L. (1997)."Techniques for Censored and Truncated Data," Springer-Verlag, New York.

3. Lawless, J. F. (1982). "Statistical Models and Methods for Lifetime data," Wiley, New York.

4. Lee, E. T. (1992). "Statistical Methods for Survival Data Analysis," Wiley, New York.

5. Mudivarthy, S., Rao, M.B., and Mitra, R. (1996). "Interval-censored Type II Plan," ASA Proceedings of the Physical and Engineering Sciences., 162-165.

APPLICATIONS AND CHALLENGES IN PROBABILISTIC & ROBUST DESIGN BASED ON COMPUTER SIMULATION

Nathan R. Soderborg, Ford Motor Co.
Concern Prevention and Advanced Reliability, Building 5, 20000 Rotunda, Dearborn, MI 48121-2053

Key Words: Robustness, Reliability, Simulation, Analytical Methods, CAE

ABSTRACT

Computer simulation has become an integral part of product design driven by the desire to reduce development time and cost and improve product quality. Effective simulation allows design optimization and refinement to occur early in the development process, before expensive prototypes are built and hardware testing is employed. However, the extent to which speed, cost, and quality improvements can be enjoyed is influenced by a number of factors including the capability of computer models and the context in which they are applied. This paper addresses the statistical context in which computer modeling is applied in product development. It calls for approaches that combine statistics and computer aided engineering (CAE) to shift model-based design decision making from a deterministic to a probabilistic and robust design perspective and provides examples of applications. It identifies challenges faced by industry in effectively carrying out this shift and outlines areas in which the statistical community can work to provide better solutions for industry.

THE PROMISE OF COMPUTER SIMULATION

For years the ability to analytically simulate the performance of engineered designs has steadily improved. This improvement is largely due to advances in computing power and speed which have enabled increasingly accurate modeling of increasingly complex systems and interactions. Thus the capability of engineers to use computer simulation to assess designs and identify opportunities for improvement has also improved dramatically. The automotive industry is still far from the day when vehicles can be designed, optimized, and verified completely "on a computer screen," yet continued improvement in computer simulation and modeling capabilities holds great promise for helping to reduce product development speed and cost and to increase product quality.

In the absence of effective simulation and analysis strategies, the product development process typically follows an iterative Design-Build-Test cycle. Designs may be developed using computer aided engineering and design (CAE, CAD) tools, and some aspects of their performance may be simulated using deterministic computer models. However, actual hardware must be built and evaluated before a decision can be made whether the design performs acceptably. If not, the product must be redesigned, rebuilt, and tested again.

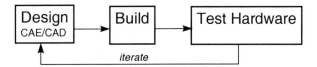

Figure 1. Traditional Product Development Process

Effective models coupled with intelligent probabilistic assessment and improvement strategies allow the cycle to be shortened. Iteration leading to an optimal design based on simulation can occur at the design stage without the necessity of building hardware until the final confirmation test.

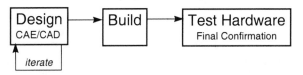

Figure 2. Improved Product Development Process

Under this scenario, product development and warranty costs will be reduced as effective modeling strategies allow designers to better avoid both over-design and under-design. Good models along with probabilistic and robust design strategies will aid in designing for a particular level of customer usage and in identifying significant low-cost parameters that influence product performance. Necessary hardware tests will be made more efficient by applying analytic assessment information to make the tests more statistically valid and require fewer test samples. Eventually the great bulk of design verification may be carried out "analytically," resulting in significant product development cost savings.

The ability to comprehensively analyze and assess designs before building hardware would also enable engineers to dramatically improve product quality. Design variables and settings that deliver reliable, robust performance would be identified up front, so design engineers would have direction on how to improve a design earlier in the product development process. They would be able to carry out an effective search for optimal designs analytically, identifying design combinations that may never have been found without these simulation capabilities and strategies.

CURRENT STATUS

What is the automotive industry's status toward realizing the promises of computer simulation? In reality, many design decisions are already being made based on analytical and CAE simulation. Such simulation has become an integrated and significant part of the product development process. Often, however, these studies are deterministic as opposed to probabilistic in nature. Moreover, the purpose of such simulation is often design assessment as opposed to design improvement. For example, using simulation to do robust design is not a standard practice. This is due partly to the fact that not all design and CAE engineers have training in designing experiments, and experimentation is perceived as difficult and time consuming—even when conducted using computer models!

FRAMEWORK FOR PROGRESS

This paper addresses two dimensions in which progress can be made in applying statistical methods to realize the full promise of computer simulation in the product development environment. One dimension-- that of "methods--" is straightforward. Progress in this dimension means moving from deterministic methods to probabilistic methods. A second dimension is the "purpose" for applying these methods. Often the purpose is to make assessments of designs. Another purpose is to make improvements to existing designs by identifying good settings for design parameters. This is the focus of robust design taught by Taguchi[1]. A third purpose is to improve a fundamental design concept to provide a new function or make a quantum improvement in performance or quality.

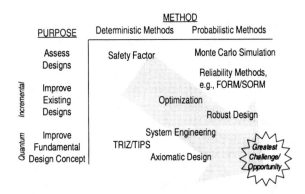

Figure 3. Pupose/Method Framework

Figure 3 shows these ideas arranged in a two dimensional table with Method on the horizontal axis and Purpose on the vertical axis. The items shown in the table are *examples* of the disciplines and tools that engineers use in assessing and improving designs. Some are easily classified by method. Those that fall between the two methods columns may be applied either deterministically or probabilistically.

Classifying these disciplines and tools by purpose is even more difficult than by method. Here they are classified by their most common purpose, but the classification is not intended to be exclusive. Some of the items listed may not be familiar to a statistical audience. For example, FORM and SORM are acronyms for First and Second Order Reliability Methods[2]. TRIZ is the Russian Acronym for Theory of Inventive Problem Solving (TIPS), an organized approach to using principles of invention for solving engineering and other types of problems[3]. Axiomatic Design is a structured method for developing good designs based on fundamental axioms[4].

The table is a conceptual framework intended to identify areas in which improvements in statistical methods and application integrated into computer simulation can make the biggest impact in product development. Certainly, the development and refinement of methods and tools that help engineers shift from deterministic to probabilistic methods will lead to progress. Development of statistical methods and tools that can be used in computer simulation to improve existing designs or improve fundamental design concepts will also lead to progress. Thus, the greatest opportunities and challenges in applying statistics to realize the full promise of computer simulation lie in the area represented by the lower right hand corner of the table in which the arrow represents the desired direction for making progress.

SHIFTING TO PROBABILISTIC METHODS

Engineers use many kinds of analytic models in design work. Examples include finite element analysis (FEA), regression equations based on empirical data, sets of equations that describe physical phenomena, etc. In a deterministic analysis, inputs are typically nominal or worst case values of design parameters—things such as part dimensions, material properties, cycle actuations, and loads. Outputs are typically *point estimates* of performance or life, sometimes expressed as safety factors or design margins.

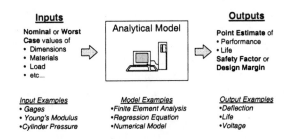

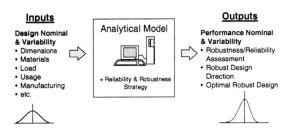

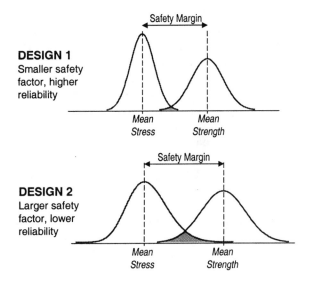

Figure 4. Typical Deterministic Analysis

Because deterministic analyses provide point estimates of performance values without accompanying estimates of variability, it is possible that a design calculated to have a higher safety margin than another design may actually have lower reliability. Using a simple stress-strength comparison, Figure 5 illustrates how this might occur.

Figure 5. Safety Margin and Reliability

In the figure, a larger interference region between stress and strength distributions indicates a higher probability of failure. Design 2 has a higher mean strength than Design 1, intended to provide a greater margin of safety. However, Design 2 has a greater probability of failure because the relative stress and strength distributions have greater variability. Deterministic analyses provide little understanding or capability to plan for the effects of real world variability. Results are often difficult to correlate to hardware test performance and provide no assessment of reliability or robustness or opportunity to conduct robust design. This may be all right if variability does not significantly affect a product's performance. But in the automotive industry our products are subject to

potentially significant manufacturing variation and used by a wide variety of customers under an extensive range of environmental conditions. Sensitivity to variability is key to product performance. What is desired are computer simulation strategies that allow truly probabilistic analysis.

Figure 6. Desired Probabilistic Analysis

PROBABILISTIC ASSESSMENT

Thanks to improvements in computer speed and computational capability, the ability to make probabilistic assessments of designs is steadily improving. For example, it is now possible to use Monte Carlo simulation to evaluate many designs. Also, probabilistic reliability methods such as FORM, SORM and others can be used in many cases to identify a design's "Most Probable Point" (MPP) of failure. Using the MPP, it is possible to estimate the design's reliability. A limiting factor for application of both these approaches is the number of computer runs required to complete the analysis. Evaluating complex systems may often require too many runs to make these approaches feasible to obtain a complete probabilistic assessment.

The perception that probabilistic methods are often too costly in CPU or set up time—especially for complex CAE models—is one reason that deterministic analysis is so much more prevalent in automotive engineering practice than probabilistic analysis. More fundamental, however, is a lack of deep understanding of probabilistic methods by many in the automotive engineering community. One barrier in particular is the misperception that the primary use for probabilistic methods ought to be to increase the accuracy of design *assessments* such as reliability or life predictions. But truly accurate assessments require precise statistical information about the distributions of sources of noise (variability) that affect the design's performance, and precise information of this type is very often either unavailable to engineers or extremely difficult to derive[5]. As a result, engineers become easily frustrated in trying to apply probabilistic methods.

What should be understood is that while probabilistic assessment is better than deterministic assessment, assessment alone is insufficient. Typical

Monte Carlo and reliability methods provide an evaluation vs. performance targets, but they do not give an engineer specific guidance on how to optimize a design for robustness. Paradoxically, this valuable guidance can often be obtained without knowing precise distributional information; often all that is required are rough approximations of key noise distributions or estimates of upper and lower limits. This means that trying to collect highly precise distributional information is often a waste of time that could be better spent working to make design improvements based on a fundamental understanding of product function and failure modes.

Getting engineers and statisticians to apply the most effective probabilistic methods may require a change of mindset. Rather than estimating or simulating performance and noise distributions and predicting reliability, engineers and statisticians should concentrate on finding the best experimentation and modeling strategies to identify robustness and functional improvements. This requires the belief that it is OK to do analysis that identifies a design improvement without being able to precisely quantify the improvement during the analysis.

PROBABILISTIC OPTIMIZATION

Various optimization techniques are used today for identifying how to improve designs. One natural step forward is to do a better job of incorporating probabilistic methods into these techniques. In a typical deterministic optimization process, an engineer sets up an experiment or search to determine the combination of design parameter settings that yields the best nominal response. Instead, in a probabilistic study, the experiment or optimization problem could be set up to identify how to minimize variability around a specified nominal response value, or maximize reliability subject to a set of constraints. From the point of view of robust design, the purpose of an experiment is to identify how to achieve the best *combination* of nominal response level and variability around that level. For example, if the response were a negative product attribute such as vibration, the objective would be to find a design that minimizes the "mean plus variation" ($\mu + \sigma$) of the response. Another measure that can used to quantify this combination is the signal-to-noise-ratio popularized by Taguchi.

ANALYTIC ROBUST DESIGN

The ideas of robust design are well known in the product development activities of the automotive industry. The goal is to design products that consistently deliver ideal function in the presence of noise. Parameter Design and Tolerance Design as developed by Taguchi, provide a structured approach to help engineers achieve this goal. The purpose of Parameter Design is to identify control factor settings that minimize performance variability in the presence of noise. The purpose of Tolerance Design is to identify economical tolerance specifications for control factors.

On the surface, it would seem straightforward to simply apply the standard approaches of Taguchi to computer simulation in order to do "analytic" robust design. However, there are several challenges that need to be considered and addressed before the full potential of robust design can be exploited through computer simulation. For example,

- Many computer simulations are expensive, both in preparation and computing time. Similar to the situation for hardware experimentation, it can be difficult for engineers to commit the necessary resources to experiments requiring many computer runs.

- Analytic models often involve a large number of design parameters and a large design space. As a result, highly nonlinear relationships between design parameters and response are common, but typical fractional factorial or orthogonal arrays can't capture this non-linearity.

- Many computer simulations model ."error states" (e.g., fatigue, vibration, wear), not ideal function. Multi-objective optimization may be required in order to ensure design changes that minimize one error state do not increase other error states.

- Early in product development when analytic robust design should be applied, design objectives and constraints have not been well defined. Design variable statistical data such as manufacturing variation, material properties, load variations, etc., may not be readily available to engineers.

- Finally, many computer models do not include important noise variables as inputs. This means that it will be important, if possible, to simulate the effects of this noise using variables that do exist in the model.

It is important that statisticians help develop methods and tools to address these challenges. Otherwise, engineers will continue to view robust design and reliability assessment as nice but impractical goals. At Ford Motor Company, we are working on developing systematic approaches for integrating reliability and robustness methods into CAE practices. Our focus is to improve the feasibility and accuracy of reliability assessment and robust design using computer simulation. Specifically, we are working on methods and tools to

- Conduct probabilistic assessments of designs much more efficiently than Monte Carlo simulation and provide information on how to improve designs.

- Create optimal sampling methods to set up experimental arrays well suited for computer-based experimentation (e.g., make better use of the ideas from DACE—Design and Analysis of Computer Experiments).
- Create response surface models that approximate more complex CAE models and can be run efficiently in large experimental arrays.
- Eliminate difficulties in selecting the "best" compound noise combinations in Parameter Design experiments.
- Identify critical parameter settings for MPP of failure to facilitate hardware test sample size reduction.

EXAMPLES

The results of applications of our methods have been promising. Included here are brief examples of results from three different types of application.

Example 1. Door "Sag" Reliability Assessment When a car door opens, it does not remain perfectly horizontal. A slight vertical sag occurs that is greatest at the door's outer edge and is measured when the door stands wide open. FEA models can be used to estimate this sag before the car is built. In this study, the goal was to check whether the sag requirement would be met by a particular vehicle design and explore opportunities to improve the design or reduce cost. The design requirement was a sag less than 1.5 mm. The FEA model included design variables such as number of missing welds, central gravity location, trim weight, and gages and material for door, hinge, reinforcement, and hinge pillar.

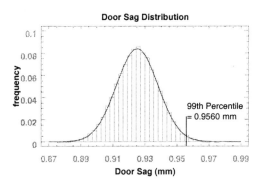

Figure 7. Door Sag Study Results

Based on nominal and variation inputs for these variables a distribution for Door Sag was created, showing the 99th percentile to be .96 mm. Also, the most important variables contributing to sag variability were identified. Of total sag variability, door reinforcement gage contributed 37%. Trim weight,

center of gravity, and hinge pillar reinforcement contributed 14%, 12%, and 11% respectively. From the study we concluded that

- Even when accounting for variation of the design variables, the design easily meets the requirement.
- Door hinge reinforcement is by far the most dominant factor for controlling door sag.
- It may be possible to reduce cost by reducing gage of door reinforcement; but before this is done it must be demonstrated that fatigue requirements are met.

Example 2. Engine Block Manufacturing Fixture Optimization Variation in deflection of an engine block machining fixture during the manufacturing process causes variation in the surface of the block being machined. In this study, the goal was to find optimal fixture design settings which minimize deflection and deflection variability in the presence of manufacturing noise. Design variables included four fixture locating positions, four clamp positions, and clamp force. Deflection was simulated using an FEA model.

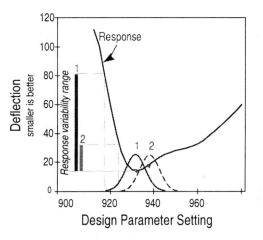

Figure 8. Optimization Results

Results are pictured in Figure 8. Based on the amount of response variability, it is clear that choosing Setting 1, which minimizes nominal deflection, is the wrong choice. Variability in deflection is much smaller for Setting 2 than Setting 1, even though Setting 2 yields a slightly higher nominal deflection value. This example illustrates how optimization can produce different results depending on whether variability *is* or *is not* considered.

Example 3. Vibration Robustness Study An irritating vibration phenomenon was discovered in a particular vehicle model during design verification testing. The phenomenon occurred on about 20% of vehicles that had been built when they were driven on the highway between 55 and 60 mph. Extensive computer simulation was conducted to identify a solution that could be implemented in the vehicles before they would be sold. These vehicle level simulations included settings for approximately 80 design variables as inputs: over 30 bushing stiffness parameters, over 20 engine mount damping and stiffness parameters, and over 20 other parameters. Using the methods developed at Ford for conducting analytic robust design studies, response surface models that approximated the original CAE model were created for several different speed levels. Several responses were measured at each of these speeds. The result of one analysis at 58 mph for a particular engine mount is shown in Figure 9.

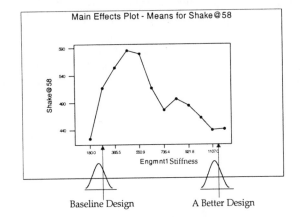

Figure 9. Engine Mount Stiffness vs. Shake

In the figure, the most robust design setting for the "shake" response is identified. The flat region of the response curve at the upper end of the stiffness range indicates an interval in which the response is insensitive to variability in stiffness. Changing the stiffness of the engine mount to fall in this interval provides a robust solution to the problem.

Example Summary Our work to make progress in conducting analytic robust design has been very successful. In spite of the fact that model quality varies and probabilistic data and assumptions are often tentative, we have found that applications of the methods and tools developed to do analytic reliability assessment and robust design make a big improvement over the status quo.

WHAT IS NEXT?

Being able to effectively apply probabilistic methods in computer simulation to improving existing designs is a big undertaking in itself. But further opportunity lies in the area of developing probabilistic

methods to improve fundamental design concepts. For example, we ought to be asking, "How do we integrate ideas from disciplines like Axiomatic Design and TRIZ into probabilistic computer simulation?" This is important because robust design and optimization are methods that search for the best possible performance from an existing set of design parameters. Inherent in these approaches is a necessity to make trade-offs based on existing design constraints. Approaches such as Axiomatic Design and TRIZ, on the other hand, focus on constructing the best design concept—a concept which delivers the desired functions without having to make fundamental trade-offs. It is in these areas that the role of statistics is least developed and which great promise lies for enhancing product development processes. There is great potential for experts from the reliability, optimization, robust design, TRIZ, and Axiomatic Design fields to work together to combine concepts from these fields to deliver more holistic, breakthrough methods and tools in this area.

REFERENCES

1. Taguchi, G., Introduction to Quality Engineering, American Supplier Institute, Inc., Dearborn. Michigan, 1986.
2. Madsen, H. O., Krenk, S., and Lind, N. C., *Methods of Structural Safety*, Prentice-Hall, Inc., 1986.
3. Altshuller, G., *And Suddenly the Inventor Appeared: TRIZ, The Theory of Inventive Problem Solving*, Technical Innovation Center, Inc., Worcester, Massachusetts, 1996.
4. Suh, N., *The Principles of Design*, Oxford University Press, New York, 1990.
5. Davis, T. P., "The Fallacy of Reliability Prediction in Automotive Engineering," *Quality Engineering*, Spring, 1998, pp. 19-21.

COLUMN-EMBEDDING METHODS FOR CONSTRUCTING CERTAIN FRACTIONAL-FACTORIAL DESIGNS

Joseph O. Voelkel, Rochester Institute of Technology

CQAS, 98 Lomb Memorial Dr., RIT, Rochester, NY 14623-5604 (jgvcqa@rit.edu)

Abstract:

We illustrate a new design-construction technique, column embedding. We show how it leads to good (but not necessarily optimal) solutions to two specific design problems: the construction of a 27-factor split-split-plot highly fractionated design, and the construction of a Plackett-Burman design for estimating all main effects and certain two-factor interactions.

Key Words: Split-Plot, Plackett-Burman, Optimal Design

1 A Column-Embedding Method for Constructing Split-Plot-Type, Highly Fractionated, Designs

In many experiments in industry, some factors are changed less frequently than others. In such cases, a split-plot design (e.g., Cochran and Cox (1957)) is the natural design to run. For example, in thin-film coatings, some factors, called main-plot factors (such as type of spatter, O_2 concentration, and temperature) must be held fixed for each coating batch, but other factors, called sub-plot factors (such as substrate type and substrate purity) can be varied within the batch. In this design, all treatment factors are crossed with each other, but experimental units for the sub-plot factors are nested under the experimental units for the main-plot factors.

We discuss a new method for creating good split-plot-type designs when (1) all factors are at two levels, (2) only a fraction of all the treatment combinations can be run, and (3) aside from the restrictions inherent in split-plot designs, complete randomization is performed, i.e. there is no blocking at the main-plot level and no additional blocking at the sub-plot level. For simplicity, we will refer to such designs as fractional-factorial split-plot (FFSP) designs. Restrictions (1) and (3) can be relaxed in a natural way. We provide an example of how the method can easily be extended to split-split-plot designs.

These designs have been discussed by Kempthorne (1952), and some FFSP designs were generated by Addelman (1964). An optimality criterion for fractional-factorial designs is minimum aberration (MA) (Fries and Hunter (1980), Chen (1992)). Although this criterion does not always lead to the most preferred designs—see Sun, Wu, and Chen (1997)—it is reasonable and widely used. Huang, Chen, and Voelkel (1998), using generating matrices (Franklin (1984)) and the MA criterion, created a catalog of MA FFSP designs for 16, 32, and 64-run experiments based on MA designs for experiments without the split-plot restriction—we refer to this as the *borrowing method* for creating such designs. They also illustrated how Chen's (1992) use of linear programming might be used, albeit awkwardly, to construct other MA FFSP designs (the *linear-programming method*). Bingham and Sitter (1999) calculated all non-isomorphic 8 and 16-run MA FFSP designs by use of a search algorithm—we will call this the *non-isomorphic-search method*.

We present a new method for generating good—although possibly not optimal—FFSP designs. For reasons that will soon become clear, we will call this a *column-embedding method*. To see the idea, consider constructing a $2^{(3+4)-(1+2)}$ design, where the notation means there are 3 main-plot factors with fractional element 1 and 4 sub-plot factors with fractional element 2. (This design was used as an example in Huang, et al., and was of course constructed by Bingham and Sitter.) Following Addelman's notation, let the main-plot factors be called A, B, C, and the sub-plot factors P, Q, R, S. Since an unrestricted MA FF design is resolution IV, but this design is resolution III, such a design cannot be found using the borrowing method. The MA FFSP design, found by either the linear-programming or non-isomorphic search methods, is based on the generators $C = AB$, $P = AR$, $Q = RS$. Here, the so-called *basic factors* are A, B, R, and S. The *added factors* are created using the following rules, the first two of which are required by the nature of split-plot designs:

1. *main-plot addition rule.* Each main-plot added factor can be added only by confounding it with main-plot basic factors.

2. *sub-plot addition rule.* Each sub-plot added factor can be added either by confounding it with sub-plot basic factors, or with main-plot basic factors and at least one sub-plot basic factor.

3. *main-plot MA rule.* We will require for our examples that the resulting design be MA at the main-plot level. This restriction is not mathematically required to create a MA FFSP design—see Huang, et al., for an example—but is virtually required in practice. Main-plot factors are typically exposed to more confounding than the sub-plot factors, and this rule attempts to minimize this confounding. When we follow this rule and find an MA FFSP design, we will say we have an MA FFMPSP design.

When both the main-plot addition and MA rules have been met, we will say that the *main-plot rules* have been met. Now, suppose the number of main-plot and sub-plot factors are given and, in notation consistent with the above, we want to construct a $2^{(n_1+n_2)-(k_1+k_2)}$ design. Let $n = n_1 + n_2$ and $k = k_1 + k_2$. We now show the proposed method with the $2^{(3+4)-(1+2)}$ example. First, find the MA 2^{7-3} design. Its generators are $e = abc$, $f = bcd$, $g = acd$, where we use lower-case letters for such *potential embedding designs*, whose name will soon become clear. In most of our examples we will use generators that are written in shortest-word-length format. (This can always be done—see Wu (1999).) We use this format to help see the how bad the confounding will look without having to resort to examining the defining contrasts. In this case, since our design requires a generator $C = AB$, whose length is 3, we cannot *embed* the main-plot factors in this preliminary design. However, at such a point, the following theorem may apply for other cases, so we state it now.

Theorem 1 *If a $2^{(n_1+n_2)-(k_1+k_2)}$ design can be embedded in a 2^{n-k} MA FF design where $n = n_1 + n_2$ and $k = k_1 + k_2$, then the design is an MA FFSP design. If the MA main-plot rules holds in this embedding, then the design is an MA FFMPSP design.*

Proof: This is simply the borrowing method of Huang, et al. ■

We next consider another potential embedding design, one that has one more factor but the same number of runs, a 2^{8-4} design. Here, $e = bcd$, $f = acd$, $g = abc$, $h = abd$, so once again we can still not embed $C = AB$ in this design. So we next investigate a 2^{9-5} design. We find that $e = abc$, $j = cd$, $f = aj$, $g = bj$, $h = abd$. (Note: MINITAB® provided correct generators, but they were not of shortest word

length; JMP® provided a design that was not MA.) Now a generator of the form $C = AB$ appears, so, by assigning $j \to C$, $c \to A$, $d \to B$, we have found a design that meets the main-plot rules. The generators are now $e = abA$, $f = aC$, $g = bC$, $h = abB$, $C = AB$. We call this the *mid-way design*.

We now need to see if the sub-plot addition rule can be met. We have nine factors in the mid-way design. Three have been selected as main-plot factors, so there are six remaining factors. However, we only require four sub-plot factors. So the next step is to see whether we can *embed* the four sub-plot factors while adhering to the sub-plot addition rule. If we let a and b be the sub-plot basic factors and f and g be the added factors, we have met the sub-plot addition rule. So we assign $a \to R$, $b \to S$, $f \to Q$, and $g \to P$, and drop out e and h, to get $Q = RC$, $P = SC$, $C = AB$. To compare this *embedding design* to the MA FFMPSP design, we must compare word-length pattern (WLP). Starting the WLP at words of length 3, we find that the embedding design has $WLP = (3, 3, 0, 0, 1)$, while the MA FFMPSP design has $WLP = (3, 2, 1, 1, 0)$. So the embedding design is reasonable, but is not as good as the MA FFMPSP design.

Of course, in practice, we do not know the MA FFMPSP design. So, in practice, we might stop at this slightly sub-optimal design. Or—and we recommend this—we could repeat the process with the next largest potential embedding design. Here this would be a 2^{10-6} design, whose shortest word-length generators are $e = ck$, $f = bcd$, $g = acd$, $h = kd$, $j = af$, $k = ab$, and a, b, c, and d are the basic factors. If we repeat the process, and use $j \to C$, $a \to A$, $f \to B$, then $c \to S$, $k \to R$, $e \to Q$, and $b \to P$, we arrive at $C = AB$, $P = AR$, $Q = RS$. This second embedding design is, in fact, the MA FFMPSP design. In practice, we could repeat this process to generate additional potential embedding designs based on the 2^{11-7}, the 2^{12-8}, and so on MA designs, and then use the best embedding design. If we use this method, but follow only the two addition rules, we will call this the *FFSP column-embedding method*. If we also follow the main-plot MA rule, we will call this the *FFMPSP column-embedding method*. The formal step for this latter method is

1. Find the $2^{n_1-k_1}$ MA design for the main-plot factors, which we will call the MA main-plot design. (We will assume this design can easily be found, since MA designs have been constructed for most reasonable cases.) Call the generators here the MA main-plot generators.

2. See whether this design is embedded in an MA 2^{n-k} design (again, we will assume such a design can easily be found). By *embedded*, we mean that a subset of n_1 factors exists in this design with generators equivalent to those of $2^{n_1-k_1}$ MA design. We call the larger design the *potential embedding design*. If such a subset exists, then the main-plot rules have been met—proceed to step 3.

3. See whether the sub-plot design can be embedded in this design, adhering to the sub-plot addition rule. If so, a MA FFMPSP design has been found, so stop. If not, let $n' = n$, $k' = k$, and proceed to step 4.

4. Modify $n' \rightarrow n'+1$, $k' \rightarrow k'+1$, and consider an MA $2^{n'-k'}$ design. See whether the MA main-plot design is embedded in this design. If such a subset exists, then the main-plot rules have been met, so proceed to step 5. Otherwise, return to step 4.

5. See whether the sub-plot design can be embedded in this design, adhering to the sub-plot addition rule. If so, an embedding design has been found. In this case, either (a) stop; (b) see whether other embedding designs can be generated from this potential embedding design; or (c) proceed to step 4 to search in the next largest potential embedding design. If an embedding design has not been found, proceed to step 4. (One can repeat step 4 until all embedding designs have been found. If there are many embedding designs, this may be difficult. We have not yet investigated the number of such designs. A modification of the approach used by Bingham and Sitter may be useful here. In the applications for which we have used this approach, we have found it relatively easy to generate reasonable embedding designs by hand.)

6. Using the MA criterion, select the best design from the embedding designs. If we wish to emphasize that all possible potential embedding designs are searched, we will preface the method with "*full*" .

Theorem 2 *If a FFSP design exists, the full FFSP column-embedding method will find the MA FFSP design. If a FFSP design exists for which the main-plot MA rule holds, the full FFMPSP column-embedding method will find the MA FFMPSP design.*

Proof: For a $2^{(n_1+n_2)-(k_1+k_2)}$ design, the search using the full column-embedding method will include all possible FF design arrangements from the

fully saturated design in $m = 2^{(n_1+n_2)-(k_1+k_2)}$ runs. But this design embeds all possible FF geometric designs of j factors in m runs, for $j = 1, 2, \ldots, m-1$, in particular for $j = (n_1 + n_2) - (k_1 + k_2)$. But FFSP designs are a (possibly null) subset of these designs. So, if a FFSP design exists, the MA FFSP design will be included in the embedding. Similarly if a FFSP design exists for which the main-plot MA rule holds, the MA FFMPSP design will be included in the embedding.∎

This proof is not particularly useful, however, because the effort using a direct search would normally be prohibitive. We have used this technique so far in practice to generate what appears to be a reasonably good design although mathematically we can only call it feasible. However, because the technique will produce an MA FFSP design if the first potential embedding design can be used, we believe it should produce reasonably good designs if we do not need to search potential embedding designs for which n' is much larger than n. This is the practical value of the method.

We now use the example that motivated this technique. In a large-scale testing area, a researcher wanted to study 27 factors. Of these, 7 were hard to change (H—two hours), 12 were moderate (M—20 minutes), and 8 were easy (E—5 minutes), so this naturally corresponds to a split-split-plot design. After discussions, it was decided that there would be 16 main plots, 4 sub plots per main plot and 2 sub-sub plots per sub plot, resulting in 128 runs. (Actually, to provide some protection, the main-plot design was run in two blocks. This is equivalent, for purposes here, to creating an eighth main-plot factor. This was successfully done, but is not reported here because it involved some techniques not discussed here.) Because this was a split-split plot design, the natural extension of the embedding technique was used.

The first potential embedding design was a 2^{27-20} design, but this did not satisfy the main-plot rules. A 2^{31-24} design was required to satisfy the main-plot rules and the sub-plot (and sub-sub-plot) addition rules. The embedding design was created in JMP, using its algorithm that attempts to construct a minimum-aberration design. The resulting design is shown in Table 1. To associate this directly with the JMP output, factor names X1, X2, ..., X31 were retained, and shortest word-length generators were not derived. For readability, the basic H factors have been bolded, the basic M factors have been italicized, and the basic E factor appears in a smaller font. This produces a feasible design. Because we did not have to embed this design too deeply, we believe this design is either MA or nearly MA.

2 A Column-Embedding Method for Constructing Designs from Designs with Complex Aliasing: a Plackett-Burman Example

There is no consensus on whether to use Plackett-Burman (PB) designs as screening designs. Many writers (e.g., Montgomery (1997)) take a very cautious stance and warn the user about the complex aliasing structure in these designs. However, this complex aliasing structure may be put to good use. For example, Hamada and Wu (1992), through a series of examples, showed that the analysis of data from these designs could uncover two-factor interactions in a way that is not possible with the geometric, 2^{k-p}, resolution III, designs. This occurs because, in a geometric design, every two-factor interaction is either totally confounded with, or orthogonal to, every main effect. In a PB design, every two-factor interaction is either partially confounded with, or orthogonal to, every main effect. For example, the AB interaction is orthogonal to the main effects of both A and B and partially confounded with all other main effects.

Hamada and Wu showed this usefulness in the *analysis* stage. We illustrate how this complex aliasing can be exploited in the *design* stage, by using a new method of creating fractional-factorial designs. This is another example of a column-embedding method, called a *PB column-embedding method*. We illustrate it with an example that arose in practice.

Thirteen factors, denoted by *A–M*, were under investigation. In addition to the main effects, the experimenter felt that *AB*, *BC*, *BG*, *CF*, and *DE* might be active. Using a geometric design would require 32 runs, but the experimenter felt this was an excessive number. Since there were 18 effects that were of potential interest *a priori*, a 20-run PB design could be considered. Other designs, such as a D-optimal design or a 24-run PB, could also be considered.

There are many ways to determine a "good" design here—we believe the best approach is to look at the possible designs from a number of viewpoints (and see Box and Draper (1975) and Box (1982).) But here, we are primarily interested in illustrating a new method for *searching* for a good design. We will use several standard measures, including the D-criterion, to measure the final goodness of the design.

When one searches for a D-optimal design, the usual strategy is to first fix the number of columns in

a design matrix that has more rows than actually required (a *row-embedding design*), and then to select an optimal subset of these rows using the D criterion. We will call such a strategy a *D-Row* strategy and the resulting design a D-row design. We adopt a different strategy. We use this strategy in the PB context, but note that it has broader applications. An outline of the strategy, which we will soon illustrate with our example, is:

1. Start with a *column-embedding design*, which we often refer to here simply as an *embedding design*. This is a main-effects-only design matrix that has more factors than are actually required in the experiment, but the correct number of rows. In our example, we choose a 19-factor, 20-run PB design. We denote this by $\mathbf{X}_{PB}$ and label the columns of this matrix with lower-case letters.

2. Select a preliminary subset of the columns equal to the number of factors required. Here, this would be 13 columns. However, do not yet equate these columns with any particular factors. Denote the associated matrix by $\mathbf{X}_{in}$. Denote the matrix of the remaining columns of $\mathbf{X}_{PB}$ by $\mathbf{X}_{out}$.

3. Bring an interaction into the design. To bring in an interaction, e.g., *AB*, will require that the A and B factors are *assigned* to columns—equivalently, the columns are assigned to the factors. Do these assignments using a myopically optimal strategy as shown below. The interactions brought into the design will be denoted by a matrix named $\mathbf{X}_{Inter}$. The full design matrix $[\mathbf{X}_{in} \ \mathbf{X}_{Inter}]$ will be denoted by $\mathbf{X}_{Des}$.

4. After each interaction is brought into the design, decide whether to *swap* any unassigned columns of $\mathbf{X}_{in}$ with columns from $\mathbf{X}_{out}$. This will also be done using a myopic strategy shown below. If another interaction needs to be added, return to step 3.

5. (Possibly) repeat this entire strategy—at some points, more than one decision is myopically optimal.

The keys to the strategy lie in steps 3 and 4. For this reason, we have written routines (in SPLUS®) that present information to help us make optimal decisions at these steps. The full algorithm could be automated, but we have not done so. We call such a strategy an R^2-*Column* strategy.

To begin the strategy, generate the 19-factor, 20-run, PB design in the standard way, and select the first 13 columns—a through m—as a preliminary subset. Begin the first iteration on step 3 by looking at the results shown in Table 2. (We refer to the design we want to find as "(13+5)//20" here to mean 13 main effects and 5 two-factor interactions in 20 runs.)

For now, only consider the first section of the output, labeled Rsq.Inter. This shows the R^2 obtained by regressing each of the $\binom{13}{2}$ two-factor interactions ab, ac, ..., lm on $\mathbf{X}_{Des}$ (which, at this point, is simply $\mathbf{X}_{in}$). So, for example, the ab interaction is fairly well predicted—$R^2 = 0.76$—by the $\mathbf{X}_{Des}$ columns.

The interactions to bring in to the design—AB, BC, BG, CF, and DE—are dominated by the B factor. Also, the C factor appears twice. No other factor appears more than once. Rather than bringing in these interactions one at a time, we decided to bring in the three interactions associated with B simultaneously. This means we need to assign the factors A, B, C, and G to columns. Because columns labeled c and d have the largest number of small correlations, we decide to assign $B \to c$ and to assign $C \to d$. We also assign $A \to a$ and $G \to b$, as suggested by the low R^2 values. At this point, step 4 is not useful—this will be clearer shortly—so we return to step 3. See Table 3. Because AB, BC, and BG are now part of $\mathbf{X}_{Des}$, the corresponding entries in Rsq.Inter are 1. The next natural factor to assign is F, since it interacts with the already-assigned C column. The myopic strategy requires that $F \to e$. But before we do this, consider the other two sections of Table 3.

The Rsq.XDesVIF section lists the R^2 for each column x in $\mathbf{X}_{Des}$ if it were to be predicted by all of $\mathbf{X}_{Des}$ except column x. The name includes "VIF" because this information is equivalent to the variance-inflation factor. The R^2's for the main effects are all quite low. The R^2's for the interactions are virtually identical to their values in iteration 1, indicating that the bringing in all three interactions simultaneously did not hurt us.

The Rsq.Xout section lists the R^2 values for each column in $\mathbf{X}_{out}$ when each column is regressed on $\mathbf{X}_{Des}$. If any R^2 value here is less than the R^2 value of an unassigned column in the Rsq.XDesVIF section, it suggests that a swap should be made. Here, a swap is not indicated. Based on these results, we assign $F \to e$. The next iteration appeals in Table 4.

We only need to add the DE interaction to $\mathbf{X}_{Des}$. However, in Table 4 the Rsq.Xout value of s is smaller than that of the Rsq.XDesVIF value of h.

This suggests a swap has the potential to lead to a better final design, so we perform $h \leftrightarrow s$. See Table 5.

Since neither D nor E have yet been assigned to the design columns, we look for the smallest value of R^2 in Rsq.Inter for unassigned columns. The value of 0.56 occurs for the gs interaction, so our swapping strategy appears to have worked. Assigning $D \to g$ and $E \to s$ leads to Table 6.

Several features appears in this table.

1. The Rsq.XDesVIF values of the main effects of B and C, whose associated interactions dominate the design, are the smallest values in this section. This is consistent with the aliasing structure of PB designs.

2. The Rsq.Inter and Rsq.Xout values associated with most of the unassigned columns are very high. This is not surprising, because we have used 18 of the possible 19 degrees of freedom in the experiment.

The final step in the design construction, which we do not perform here, is to assign the remaining factors to the assignable columns of $\mathbf{X}_{in}$. Since these remaining factors have no interactions associated with them, there is no need to examine additional tables. (A very precise designer might use the information in Rsq.XDesVIF to make these assignments. For example, if an unassigned factor was suspected of having a very large effect, if any, the assignment might be made to k, l, or m.)

We will shortly compare this design to several D-Row designs. But first, we point out three key features that are intrinsic to this R^2-Column design, but not to D-Row designs:

1. Because the design is simply a (well-chosen) PB design, all main-effect columns have an equal number of ± 1's and are orthogonal to each other. So, if no interactions are active, this design is optimal under any of the usual criterion.

2. For these same reasons, all interaction columns have an equal number of ± 1's.

3. If a subset of the interactions are active, the design will clearly still possess desirable properties. This is especially true if only the B-factor interactions are active, because its interactions were entered first into the design. (The swapping step may modify these properties; however, this effect should only be slight, since the swapped-in column was nearly orthogonal to the columns already in the design matrix.)

Now, D-Row refers to both the D criterion and to the algorithm for select such a design. What we will do to construct D-Row designs is typical:

1. Start with a row-embedding design. This design begins with a main-effects only "good" design matrix $\mathbf{X}$, one with more rows then the 20 actually required; has the correct number of main-effect columns, 13, whose associated columns are named A, B, ... M; and is augmented by additional columns corresponding to the 5 interactions of interest. Note that even after $\mathbf{X}$ has been chosen, there will typically be many distinct ways to assign factor names to the columns. We follow what we believe is the typical procedure of simply assigning them alphabetically.

2. Use an algorithm that selects a subset of 20 rows based on the D criterion. We have used JMP's algorithm for this selection.

We examined two $\mathbf{X}$ matrices for D-Row designs. The first matrix was a 2^{13-6}, 128-run, design, that was generated by JMP. The second matrix was a 48-run, PB, design, in which the first 13 columns were used, generated in MINITAB. We could have considered other designs, but these were the largest designs of these that could be automatically generated by these two packages. So we believe these results would indicate the best results a typical user could achieve.

The JMP D-optimal routine has parameters that can be set. We used the suggested "N random" value of 10 and the "K value" of 3.

For the 2^{13-6}, 128-run, row-embedded design, we ran the JMP algorithm for 30 "trips." We performed this entire process four times. This resulted in two distinct designs—the first three were identical. We then did the same thing for the 48-run PB design, and this resulted in four distinct designs. We then summarized our results in Table 7. The summary includes

1. The D-efficiency. To calculate the D efficiency, we first used the D criterion of $|\mathbf{X}'\mathbf{X}|^{1/p}/n$, where $p = 19$ includes the constant term in $\mathbf{X}$ and $n = 20$. This criterion is standardized by the number of factors and number of runs. We then found the efficiency as this D criterion divided by the D criterion for a theoretical orthogonal design (the D criterion for such a design is 1). This is the same definition used in JMP.

2. A summary of the balance for the main-effect columns, showing how many of the 13 columns

ranged from 10 +1's and 10 -1's down to an imbalance of 7 rows of one sign and 13 of the other.

3. A summary based on the maximum variance of sets of terms. These are efficiency values, comparing the maximum variance in each group to that theoretically obtained by an orthogonal design (0.05). The sets are factor B, the dominant factor; the 12 other main effects; the three interactions that included B; and the two other interactions.

The key features of this table are

1. On the D-criterion, our column-embedded design performed virtually the same as those derived from the D-Row 128-run design, and better than those based on the D-Row 48-run design.

2. The R^2-Column design was more balanced than any other design.

3. If we compare the R^2-Column design to the best D-Row design using the variance-efficiency measures, we see that the R^2-Column design concentrates it efforts in a natural descending order of which terms we suspect might be active. The best D-Row design, on the other hand, estimates some main effects with only a 40% efficiency, the efficiency obtained from an 8-run orthogonal design.

This refers to worst-case (all effects active) results. To illustrate what might occur in an actual scenario, we considered how these designs would perform if only the main effects of A, B, C, D, F, G, and the AB, BC, CF interactions were active. This selection is mostly arbitrary, but we did include in the selection two interactions that included B, and the two factors, D and G, that had the 8/12 amount of unbalance from the best D-Row design. Our results appear in Table 8. The R^2-Column design either outperforms the best D-Row design or performs essentially the same as it on all measure except for the last.

The inherent balance properties of the R^2-Column design would also allow it to perform better than the best D-Row design if any runs are missing at random. This balance will also no doubt be more appealing to the experimenter and should also result in better projective properties. So, for the variety of reasons we have cited, we believe that the R^2-Column strategy used in this example leads to a design that is preferable to the designs generated by the D-Row technique.

We finally note that the column-embedding and row-embedding features above could be combined. For example, a 2^{64-57}, 128-run, design could have been examined in which eventually 13 columns of main effects and 20 rows could have been selected. We have not examined such a possibility.

References

[1] Addelman, S. (1964), "Some two-level factorial plans with split-plot confounding," *Technometrics*, 6, 253-258.

[2] Bingham, D., and Sitter, R. R. (1999), "Minimum-aberration two-level fractional factorial split-plot designs" *Technometrics*, 41, 62–70.

[3] Box, G. E. P., and Draper, N. R. (1975), "Robust designs," *Biometrika*, 62, 347–352.

[4] Box, G. E. P. (1982), "Choice of response surface designs and alphabetic optimality," *Utilitas Mathematica*, 21, 11–55.

[5] Chen, J. (1992), "Some results on 2^{n-k} fractional factorial designs and search for minimum aberration designs," *Annals of Statistics*, 20, 2124-2141.

[6] Chen, J., and Wu, C. F. J. (1991), "Some results on s^{n-k} fractional factorial designs with minimum aberration or optimal moments," *Annals of Statistics*, 19, 1028-1041.

[7] Cheng, S., and Wu, C. F. J. (1999), "Finding defining generators with extreme lengths," Technical Report #332, Department of Statistics, University of Michigan.

[8] Cochran, W. G., and Cox, G. M. (1957). *Experimental Designs*. New York: John Wiley.

[9] Franklin, M. F. (1984), "Constructing tables of minimum aberration p^{n-m} designs," *Technometrics*, 26, 225-232.

[10] Fries, A., and Hunter, W. G. (1980), "Minimum aberration 2^{k-p} designs," *Technometrics*, 22, 601-608.

[11] Huang, P., Chen, D., and Voelkel, J. (1998), "Minimum-aberration two-level split-plot designs," *Technometrics*, 40, 314–346.

[12] Hamada, M., and Wu. C. F. J. (1992), "Analysis of designed experiments with complex aliasing," *Journal of Quality Technology*, 24, 130-137.

[13] Kempthorne, O. (1952), *The Design and Analysis of Experiments*. Huntington, New York: John Wiley.

[14] Montgomery, D. C. (1997), *Design and Analysis of Experiments*. Fourth edition. New York: John Wiley.

[15] Sun, D. X., Wu, C. F. J., and Chen, Y. (1997), "Optimal blocking schemes for 2^n and 2^{n-p} designs," *Technometrics*, 39, 298–307.

Table 1. Feasible 27-Factor Split-Split Plot Design.	
Factor Type	Generators
M	*X1*
E	X2
H	**X3**
M	*X4*
H	**X5**
H	**X6**
H	**X7**
	X8 = *X1**x2***X3***X4***X5*****X6*****X7**
M	*X9* = **X3***X4***X5*****X6*****X7**
	X10 = x2*X4***X5*****X6*****X7**
M	*X11* = *X1**X4***X5*****X6*****X7**
	X12 = x2***X3*****X5*****X6*****X7**
M	*X13* = *X1****X3*****X5*****X6*****X7**
E	X14 = *X1**x2***X5*****X6*****X7**
H	**X15** = **X5*****X6*****X7**
E	X16 = x2***X3***X4***X6*****X7**
M	*X17* = *X1****X3***X4***X6*****X7**
	X18 = *X1**x2*X4***X6*****X7**
M	*X19* = *X4****X6*****X7**
H	**X20** = **X3*****X6*****X7**
E	X21 = x2***X6*****X7**
M	*X22* = *X1****X6*****X7**
E	X23 = x2***X3***X4***X5*****X7**
M	*X24* = *X1****X3***X4***X5*****X7**
E	X25 = *X1**x2*X4***X5*****X7**
M	*X26* = *X4****X5*****X7**
H	**X27** = **X3*****X5*****X7**
E	X28 = x2***X5*****X7**
M	*X29* = *X1****X5*****X7**
M	*X30* = **X3***X4***X7**
E	X31 = *X1**x2***X3*****X7**

Key:
H **bold** font
M *italic* font
E small font

Table 2. (13+5)//20 PB Example. Iteration 1.

```
$Rsq.Inter:
      a    b    c    d    e    f    g    h    i    j    k    l    m
a 0.00 0.76 0.44 0.76 0.76 0.76 0.44 0.76 0.76 0.44 0.76 0.44 0.76
b 0.76 0.00 0.44 0.44 0.76 0.76 0.76 0.44 0.76 0.76 0.44 0.76 0.76
c 0.44 0.44 0.00 0.44 0.44 0.76 0.76 0.76 0.44 0.76 0.76 0.44 0.76
d 0.76 0.44 0.44 0.00 0.44 0.76 0.76 0.44 0.76 0.44 0.76 0.76 0.44
e 0.76 0.76 0.44 0.44 0.00 0.44 0.76 0.76 0.44 0.76 0.76 0.76 0.76
f 0.76 0.76 0.76 0.76 0.44 0.00 0.44 0.76 0.76 0.44 0.76 0.76 0.44
g 0.44 0.76 0.76 0.76 0.76 0.44 0.00 0.44 0.76 0.76 0.44 0.76 0.76
h 0.76 0.44 0.76 0.44 0.76 0.76 0.44 0.00 0.76 0.76 0.76 0.44 0.76
i 0.76 0.76 0.44 0.76 0.44 0.76 0.76 0.76 0.00 0.76 0.76 0.44 0.44
j 0.44 0.76 0.76 0.44 0.76 0.44 0.76 0.76 0.76 0.00 0.76 0.76 0.44
k 0.76 0.44 0.76 0.76 0.76 0.76 0.44 0.76 0.76 0.76 0.00 0.76 0.76
l 0.44 0.76 0.44 0.76 0.76 0.76 0.76 0.44 0.44 0.76 0.76 0.00 0.76
m 0.76 0.76 0.76 0.44 0.76 0.44 0.76 0.76 0.44 0.44 0.76 0.76 0.00


$Rsq.XDesVIF:
 a b c d e f g h i j k l m
 0 0 0 0 0 0 0 0 0 0 0 0 0


$Rsq.Xout:
 n o p q r s
 0 0 0 0 0 0
```

Table 3. (13+5)//20 PB Example. Iteration 2.

```
$Rsq.Inter:
      A    G    B    C    e    f    g    h    i    j    k    l    m
A 0.00 0.91 1.00 0.82 0.85 0.88 0.73 0.96 0.82 0.59 0.97 0.59 0.94
G 0.91 0.00 1.00 0.61 0.91 0.82 0.89 0.89 0.83 0.85 0.66 0.96 0.85
B 1.00 1.00 0.00 1.00 0.46 0.91 0.91 0.91 0.45 0.88 0.87 0.47 0.83
C 0.82 0.61 1.00 0.00 0.54 0.83 0.96 0.87 0.98 0.59 0.83 0.78 0.87
e 0.85 0.91 0.46 0.54 0.00 0.80 0.85 0.82 1.00 0.97 0.91 0.98 0.83
f 0.88 0.82 0.91 0.83 0.80 0.00 0.67 0.94 0.91 0.54 0.96 0.88 0.73
g 0.73 0.89 0.91 0.96 0.85 0.67 0.00 0.66 0.82 0.89 0.80 0.83 0.83
h 0.96 0.89 0.91 0.87 0.82 0.94 0.66 0.00 0.91 0.91 0.77 0.66 0.83
i 0.82 0.83 0.45 0.98 1.00 0.91 0.82 0.91 0.00 0.91 0.88 0.86 0.77
j 0.59 0.85 0.88 0.59 0.97 0.54 0.89 0.91 0.91 0.00 0.96 0.83 0.73
k 0.97 0.66 0.87 0.83 0.91 0.96 0.80 0.77 0.88 0.96 0.00 0.83 0.91
l 0.59 0.96 0.47 0.78 0.98 0.88 0.83 0.66 0.86 0.83 0.83 0.00 0.88
m 0.94 0.85 0.83 0.87 0.83 0.73 0.83 0.83 0.77 0.73 0.91 0.88 0.00


$Rsq.XDesVIF:
    A    G B    C    e    f    g    h    i    j    k    l    m   AB
 0.13 0.14 0 0.13 0.19 0.16 0.16 0.19 0.19 0.16 0.19 0.16 0.19 0.45
   BC   BG
 0.45 0.44


$Rsq.Xout:
    n    o   p    q   r   s
 0.81 0.74 0.2 0.86 0.2 0.2
```

Table 4. (13+5)//20 PB Example. Iteration 3.

```
$Rsq.Inter:
     A    G    B    C    F    f    g    h    i    j    k    l    m
A 0.00 0.93 1.00 0.98 0.86 0.97 0.80 0.99 0.85 0.59 0.97 0.66 0.98
G 0.93 0.00 1.00 0.62 0.92 0.85 0.94 0.94 0.96 0.88 0.75 0.99 0.86
B 1.00 1.00 0.00 1.00 0.52 0.91 0.91 0.98 0.74 0.88 0.92 0.69 0.96
C 0.98 0.62 1.00 0.00 1.00 0.84 0.99 0.88 0.98 0.59 0.95 0.78 0.88
F 0.86 0.92 0.52 1.00 0.00 0.80 0.88 0.98 1.00 0.97 0.93 0.98 0.84
f 0.97 0.85 0.91 0.84 0.80 0.00 0.78 0.98 0.92 1.00 0.99 0.88 0.74
g 0.80 0.94 0.91 0.99 0.88 0.78 0.00 0.75 0.85 0.89 0.80 0.96 0.95
h 0.99 0.94 0.98 0.88 0.98 0.98 0.75 0.00 0.91 0.93 0.86 0.75 0.84
i 0.85 0.96 0.74 0.98 1.00 0.92 0.85 0.91 0.00 0.91 0.88 0.86 0.78
j 0.59 0.88 0.88 0.59 0.97 1.00 0.89 0.93 0.91 0.00 0.99 0.95 0.99
k 0.97 0.75 0.92 0.95 0.93 0.99 0.80 0.86 0.88 0.99 0.00 0.96 0.92
l 0.66 0.99 0.69 0.78 0.98 0.88 0.96 0.75 0.86 0.95 0.96 0.00 0.88
m 0.98 0.86 0.96 0.88 0.84 0.74 0.95 0.84 0.78 0.99 0.92 0.88 0.00

$Rsq.XDesVIF:
    A    G    B    C    F    f    g    h    i    j    k    l    m   AB
 0.24 0.2 0.08 0.14 0.2 0.18 0.18 0.29 0.21 0.18 0.29 0.27 0.29 0.45
   BC   BG   CF
 0.45 0.54 0.54

$Rsq.Xout:
    n    o    p    q    r    s
 0.82 0.92 0.77 0.89 0.37 0.22
```

Table 5. (13+5)//20 PB Example. Iteration 3a.

```
$Rsq.Inter:
     A    G    B    C    F    f    g    s    i    j    k    l    m
A 0.00 0.94 1.00 0.92 0.83 0.86 0.88 0.87 0.74 0.69 0.88 1.00 0.96
G 0.94 0.00 1.00 0.51 0.93 0.89 0.99 0.72 0.90 0.99 0.99 0.98 0.83
B 1.00 1.00 0.00 1.00 0.90 0.56 0.91 0.96 0.75 0.88 0.91 0.56 0.90
C 0.92 0.51 1.00 0.00 1.00 0.87 0.96 1.00 0.94 0.87 0.84 0.83 1.00
F 0.83 0.93 0.90 1.00 0.00 0.80 0.93 0.88 0.89 0.85 0.98 0.98 0.96
f 0.86 0.89 0.56 0.87 0.80 0.00 0.94 0.94 0.93 1.00 0.98 0.84 0.80
g 0.88 0.99 0.91 0.96 0.93 0.94 0.00 0.56 0.78 0.91 0.80 0.90 0.92
s 0.87 0.72 0.96 1.00 0.88 0.94 0.56 0.00 0.89 0.98 0.64 0.99 0.93
i 0.74 0.90 0.75 0.94 0.89 0.93 0.78 0.89 0.00 0.88 0.84 0.99 0.98
j 0.69 0.99 0.88 0.87 0.85 1.00 0.91 0.98 0.88 0.00 0.96 0.92 0.72
k 0.88 0.99 0.91 0.84 0.98 0.98 0.80 0.64 0.84 0.96 0.00 0.90 0.72
l 1.00 0.98 0.56 0.83 0.98 0.84 0.90 0.99 0.99 0.92 0.90 0.00 0.88
m 0.96 0.83 0.90 1.00 0.96 0.80 0.92 0.93 0.98 0.72 0.72 0.88 0.00

$Rsq.XDesVIF:
    A   G    B    C    F    f    g    s    i    j    k    l    m   AB
 0.2 0.2 0.07 0.13 0.2 0.23 0.22 0.22 0.21 0.18 0.29 0.26 0.29 0.45
   BC   BG   CF
 0.45 0.5 0.5

$Rsq.Xout:
    n    o    p  q    r    h
 0.86 0.87 0.74 1 0.35 0.29
```

Table 6. (13+5)//20 PB Example. Iteration 4.

```
$Rsq.Inter:
    A    G    B    C    F    f    D    E    i    j    k    l    m
A 0.00 1.00 1.00 1.00 0.84 1.00 0.88 0.89 0.93 0.70 0.99 1.00 0.99
G 1.00 0.00 1.00 0.74 0.94 0.91 0.99 0.89 0.98 1.00 0.99 1.00 0.89
B 1.00 1.00 0.00 1.00 0.98 1.00 0.97 1.00 0.76 0.93 0.95 0.63 0.98
C 1.00 0.74 1.00 0.00 1.00 0.89 1.00 1.00 1.00 0.89 0.98 0.84 1.00
F 0.84 0.94 0.98 1.00 0.00 0.80 0.99 0.93 0.99 0.95 0.99 0.99 0.99
f 1.00 0.91 1.00 0.89 0.80 0.00 0.94 0.94 0.94 1.00 1.00 0.95 0.80
D 0.88 0.99 0.97 1.00 0.99 0.94 0.00 1.00 0.83 0.91 0.80 0.98 1.00
E 0.89 0.89 1.00 1.00 0.93 0.94 1.00 0.00 0.91 0.99 0.66 1.00 0.99
i 0.93 0.98 0.76 1.00 0.99 0.94 0.83 0.91 0.00 0.99 0.95 1.00 0.98
j 0.70 1.00 0.93 0.89 0.95 1.00 0.91 0.99 0.99 0.00 1.00 1.00 0.98
k 0.99 0.99 0.95 0.98 0.99 1.00 0.80 0.66 0.95 1.00 0.00 0.98 0.89
l 1.00 1.00 0.63 0.84 0.99 0.95 0.98 1.00 1.00 1.00 0.98 0.00 0.93
m 0.99 0.89 0.98 1.00 0.99 0.80 1.00 0.99 0.98 0.98 0.89 0.93 0.00

$Rsq.XDesVIF:
    A    G    B    C    F    f    D    E    i    j    k    l    m   AB
  0.28 0.28 0.1 0.2 0.29 0.33 0.24 0.24 0.22 0.19 0.33 0.32 0.34 0.46
    BC   BG   CF   DE
  0.45 0.52 0.56 0.56

$Rsq.Xout:
    n  o    p  q    r    h
  0.86 1 0.75 1 0.49 0.97
```

Table 7. (13+5)//20 Example. Comparison of Designs when all 18 Terms are in Model.

Design	% D-eff'y	Balance of Main-Effect Columns				% Efficiency of Max Var			
		10/10	9/11	8/12	7/13	B	Other main	B-inters	Other inters
Column-Embedded	84	13				90	66	48	44
Dopt1-3//128	85	9	0	4		86	40	59	68
Dopt4//128	83	9	0	4		90	32	53	46
Dopt1//48	79	5	5	2	1	64	43	55	48
Dopt2//48	78	5	5	2	1	57	40	39	31
Dopt3//48	79	3	8	1	1	82	40	58	45
Dopt4//48	78	4	6	3		60	45	49	44

Table 8. (13+5)//20 Example. Comparison of Designs when A, B, C, D, F, G, AB, BC, and CF are in Model.

Design	% D-eff'y	% Efficiency of Max Var			
		B	Other main	B-inters	Other inter
Column-Embedded	95	95	88	82	82
Dopt1-3//128	95	96	76	76	96
Dopt4//128	93	93	64	64	91
Dopt1//48	91	90	68	83	76
Dopt2//48	91	87	60	63	72
Dopt3//48	92	93	70	88	80
Dopt4//48	89	88	60	72	82

APPROXIMATE MAXIMIN DISTANCE DESIGNS

Michael W. Trosset[1], College of William & Mary
Department of Mathematics, P.O. Box 8795, Williamsburg, VA 23187-8795

Key Words: Computer Experiments, Space-filling Experimental Designs, Nonrectangular Experimental Regions

Abstract

Maximin distance designs are space-filling experimental designs that were introduced by Johnson, Moore, and Ylvisaker (1990). Unlike Latin hypercube and orthogonal array designs, the definition of maximin designs does not depend on the geometry of the design space; hence, maximin designs are potentially attractive for experiments that involve nonrectangular design spaces. Unfortunately, maximin designs are difficult to compute. To address this difficulty, we consider a framework in which approximations to maximin designs can be obtained as solutions to nonlinear programming problems in which the constraints define the design space. In this framework, our ability to calculate designs corresponds to our ability to manage the optimization constraints. Finally, we discuss some computational issues that arise when attempting to calculate designs in this manner.

1 Introduction

Many deterministic computer simulations are extremely expensive to evaluate. For example, Booker (1996) described a simulation of the aeroelastic and dynamic response of a helicopter rotor blade for which a single function evaluation requires approximately six hours of cpu time on a Cray Y-MP. In such situations, a common engineering practice (Barthelemy and Haftka, 1993) is to replace the expensive simulation, which we represent as a real-valued function f, with an inexpensive surrogate, $\hat{f}$. The surrogate is constructed from information obtained by evaluating f at a set of carefully selected *design sites*. This report considers the problem of how to choose a specified number of design sites in a sensible way.

The problem of choosing design sites is a problem of experimental design. This observation, together with the observation that various techniques from spatial statistics can be gainfully employed to construct the surrogates, led statisticians to begin to study the design and analysis of computer experiments. Surveys of the rapidly growing literature on this subject have been made by Sacks, Welch, Mitchell, and Wynn (1989), by Booker (1994), and by Koehler and Owen (1996).

Roughly speaking, there are two approaches to the design of computer experiments. The parametric approach is often supplied with a Bayesian interpretation. One assumes that f is the realization of a stochastic process and specifies a parametric family of possible processes. It is possible to extend familiar design criteria like D-optimality to this setting, then construct designs that are optimal with respect to the specified family. However, the practical difficulties of actually computing such optimal designs can be formidable.

The nonparametric approach to the design of computer experiments chooses design sites in a manner that is perceived to be "space-filling". When the experimental region $E \subset \Re^p$ is a bounded rectangle, Latin hypercube sampling (McKay, Beckman, and Conover, 1979) and orthogonal array sampling (Owen, 1992, 1994; Tang, 1993) are practical, inexpensive ways of generating space-filling designs. Design criteria that are explicitly space-filling include the minimax and maximin distance principles proposed by Johnson, Moore, and Ylvisaker (1990); however, these designs can be difficult to compute. The purpose of this report is to describe a method of approximating maximin designs that exploits conventional nonlinear programming algorithms.

The utility of Latin hypercube and orthogonal array sampling diminishes when, as is often the case, the experimental region is not rectangular. If E can be inscribed in a rectangle of the same dimension, then a plausible space-filling design can often be obtained by the simple *ad hoc* device of generating a space-filling design for the circumscribing rectangle and accepting the reulting design sites that fall in E.

---

[1] Supported in part by AFOSR Grant F49620-95-1-0210, by the Institute for Computer Applications in Science and Engineering (ICASE) under NASA Contract NAS1-19480, and by a gift from the Mobil Technology Company.

Our desire to improve such designs—and to generate plausible space-filling designs in regions that do not readily lend themselves to this device—motivates the work described in this report.

2　An Exact Formulation

Let N denote the specified number of design sites and suppose that $x_1, \ldots, x_N \in E$, where E is a compact subset of $\Re^p$. For convenience, we place x_i' in row i of the $N \times p$ design matrix $X = (x_{ik})$. We then abuse notation and write $X \in E$.

Let

$$d_{ij}(X) = \|x_i - x_j\|_2 = \left[\sum_{k=1}^{p} (x_{ik} - x_{jk})^2 \right]^{1/2}.$$

Then $X^* \in E$ is a maximin Euclidean distance design in E if and only if

$$\min_{i<j} d_{ij}(X^*) \geq \min_{i<j} d_{ij}(X)$$

for all $X \in E$, i.e. if and only if X^* is a global solution of the nonsmooth optimization problem

$$
\begin{aligned}
\text{maximize} \quad & \min_{i<j} d_{ij}(X) \\
\text{subject to} \quad & X \in E.
\end{aligned}
\tag{1}
$$

Maximin designs are intuitively appealing because they explicitly endeavor to spread the design sites as much as possible. Other metrics are certainly possible; in this report, we restrict attention to Euclidean distance.

In general, explicit formulae for X^* will not exist and maximin designs must be computed numerically. To do so, it may be helpful to reformulate Problem (1) as a smooth nonlinear programming problem. Imagine positioning each site in a prospective design X at the center of a closed ball of radius r, where r is small enough that the intersection of any pair of balls is either empty or a singleton point. To spread the sites in accordance with the maximin distance criterion, we seek a design for which r can be made as large as possible. Thus, X^* is a maximin Euclidean distance design in E if and only if it solves

$$
\begin{aligned}
\text{maximize} \quad & r \\
\text{subject to} \quad & X \in E, d_{ij}(X) \geq 2r;
\end{aligned}
$$

or, equivalently, if and only if X^* solves

$$
\begin{aligned}
\text{minimize} \quad & -r^2 \\
\text{subject to} \quad & X \in E, [d_{ij}(X)]^2 \geq 4r^2, r \geq 0.
\end{aligned}
\tag{2}
$$

There are interesting analogies between the problem of maximin distance design and certain problems in signal detection. Very recently, Gockenbach and Kearsley (1999) considered the problem of designing optimal sets of signals (finite time series) under amplitude constraints and in the presence of non-Gaussian noise. They considered a signal set optimal when the largest probability of mistaking any one signal for any other is minimal. This is a constrained minimax problem that the authors reformulated as a smooth nonlinear program that is strikingly similar to Problem (2).

Gockenbach and Kearsley (1999) noted that their nonlinear program has several unfortunate features, shared by our Problem (2), that typically confound standard nonlinear programming algorithms:

> "Difficulties arise when one tries to solve... with standard algorithms. The fact that there are far fewer variables than constraints results in three specific diffficulties:
>
> - There are 'almost' binding constraints at the solution.
> - The linearized constraints are often inconsistent.
> - The boundary of the feasible region is noticeably nonlinear."

Nevertheless, the authors were able to find solutions using a sequential quadratic programming algorithm that incorporates a constraint perturbation technique studied by Kearsley (1996). It seems likely that these methods can be adapted for finding maximin distance designs via Problem (2); however, this is a topic for future research. The purpose of the present report is to examine approximate formulations of the maximin distance design problem that can be solved by standard algorithms for nonlinear programming.

3　Approximate Formulations

Let ϕ denote any strictly decreasing function on $(0, \infty)$, e.g. $\phi(t) = 1/t$, and let $\phi_{ij}(X) = \phi(d_{ij}(X))$. Let $v(X)$ denote the vector of length $N(N-1)/2$ whose kth component is $\phi_{ij}(X)$, where

$$k = (j-1)(N - j/2) + i - j$$

for $j = 1, \ldots, N-1$ and $i = j+1, \ldots, N$. Then X^* is a maximin design if and only if it is a global solution of the constrained minimax problem

$$
\begin{aligned}
\text{minimize} \quad & \|v(X)\|_\infty \\
\text{subject to} \quad & X \in E.
\end{aligned}
\tag{3}
$$

We approximate $\|v(X)\|_\infty$ by the smooth objective function $\|v(X)\|_\sigma$, resulting in the more tractable optimization problem

$$\begin{aligned} \text{minimize} \quad & \|v(X)\|_\sigma \\ \text{subject to} \quad & X \in E. \end{aligned} \quad (4)$$

With $\phi(t) = 1/t$, this formulation was exploited by Morris and Mitchell (1995) in their study of distance designs that are maximin within the class of Latin hypercube designs.

Let X^σ denote a global solution of Problem (4). The following result justifies calling X^σ an approximate maximin design, although we submit that the plausibility of X^σ as a space-filling design does not depend on this justification.

Theorem 1 *Let $\sigma_k \to \infty$ as $k \to \infty$ and let X^∞ be any accumulation point of $\{X^\sigma\}$. Then X^∞ is a maximin distance design.*

Proof: For every $\sigma \in [1, \infty]$,

$$\|v(X^\sigma)\|_\sigma \to \|v(X^\infty)\|_\sigma$$

as $k \to \infty$. Hence, by diagonalization,

$$\|v(X^\sigma)\|_\sigma \to \|v(X^\infty)\|_\infty$$

as $k \to \infty$.

Let X^* denote any global solution of Problem (3). If X^∞ is *not* a global solution of Problem (3), then there exists $\epsilon > 0$ such that

$$\|v(X^\infty)\|_\infty \geq \|v(X^*)\|_\infty + 2\epsilon.$$

Choose K large enough that $k \geq K$ entails both

$$|\|v(X^\sigma)\|_\sigma - \|v(X^\infty)\|_\infty| < \epsilon$$

and

$$|\|v(X^*)\|_\sigma - \|v(X^*)\|_\infty| < \epsilon.$$

Then

$$\|v(X^*)\|_\sigma < \|v(X^\sigma)\|_\sigma,$$

which is a contradiction. $\square$

Once σ has been fixed, Problem (4) is equivalent to the following:

$$\begin{aligned} \text{minimize} \quad & \sum_{j>i} [\phi_{ij}(X)]^\sigma \\ \text{subject to} \quad & X \in E. \end{aligned} \quad (5)$$

How difficult it will be to solve Problem (5) will depend on how easily one can manage (a) the objective function, which in turn will depend on the choices of ϕ and σ; and (b) the constraints.

It is not always easy to choose ϕ and σ in such a way that Problem (5) is sensibly scaled. Because the scaling of Problem (5) will affect the performance of the algorithms that we use to solve, we endeavor to ameliorate this situation by assuming that E has been scaled so that $E \subset [0,1]^p$.

The transformation ϕ should be constructed in such a way that $\phi(t)$ is large when t is small. If ϕ has a singularity at $t = 0$, then the optimization algorithm will be strongly discouraged from considering designs in which sites are replicated. Thus, $\phi(t) = 1/t$ is a natural transformation. So is the computationally less expensive $\phi(t) = 1/t^2$. We have experimented with several transformations and currently prefer

$$\phi(t) = \log\left(1 + \frac{1}{t^2}\right). \quad (6)$$

For $\phi(t) = 1/t$, Morris and Mitchell (1995) tried $\sigma = 1, 2, 5, 10, 20, 50, 100$ until they were convinced that their simulated annealing algorithm had found a maximin design. Our goal is more modest, in that we are content to settle for a reasonable approximation to a maximin design. For $\phi(t) = 1/t^2$ and (6), we have found that $\sigma \in [5, 10]$ usually results in satisfactory designs.

Morris and Mitchell (1995) noted the following tradeoff: as σ increases, X^σ better approximates a maximin design, but Problem (5) is more difficult to solve. We have observed the same phenomenon, but generally we have been impressed with the performance of standard nonlinear programming algorithms on Problem (5). In contrast to the global search strategy employed by Morris and Mitchell (1995), we have only studied local search strategies. These strategies are sometimes trapped by nonglobal minimizers and do not always find X^σ. Nevertheless, for reasonable choices of σ, they consistently find designs whose minimal intersite distances are almost as great as those realized by X^σ.

There remains the matter of the constraints that define the experimental region, E. In order to treat Problem (5) as a nonlinear program, we assume that E is defined by finite numbers of equality and (more typically) inequality constraints and we hope that there are not too many highly nonlinear constraints. A plethora of nonlinear programming algorithms are available; the reader seeking to identify algorithms and software suited to specific applications is referred to Moré and Wright (1993). To take advantage of powerful computing platforms and state-of-the-art software with a minimal investment of time and effort, we recommend submitting jobs written in AMPL (A Mathematical Programming Language), documented by Fourer, Gay, and Kernighan (1993), to the NEOS (Network-Enabled Optimization System) server at Argonne National Laboratories:

http://www.mcs.anl.gov/neos/Server

Currently, our preferred solver is SNOPT.

4 An Example

We conclude by demonstrating an approximate maximin Euclidean distance design that comprises $N = 16$ sites in a simple nonrectangular region $E \subset [0,1]^2$. The decision variables are the $Np = 32$ elements of

$$X = [x_{ik}],$$

which of course must satisfy the bound constraints

$$0 \leq x_{ik} \leq 1.$$

To define E, we imposed summation constraints, requiring that

$$0.6 \leq x_{i1} + x_{i2} \leq 1.0.$$

We constructed an initial design $X^0 \in E$ by pseudorandom sampling from a uniform distribution on E. This was accomplished by using the S-Plus function runif to generate pseudorandom observations from a uniform distribution on $[0,1]^2$ and accepting those in E until $N = 16$ feasible sites had been generated. The initial design X^0 is displayed in Figure 1. Note that

$$\min_{i<j} d_{ij}\left(X^0\right) \doteq 0.03.$$

Problem (5), with ϕ specified by (6), $\sigma = 10$, and $X = X^0$, was coded in AMPL and submitted (by email to neos@mcs.anl.gov) to the SNOPT solver via NEOS. The text of this submission can be obtained from the author (trosset@math.wm.edu). The SNOPT solver took just 0.65 seconds to compute X^σ, the approximate maximin Euclidean distance design displayed in Figure 2. Note that

$$\min_{i<j} d_{ij}\left(X^\sigma\right) \doteq 0.18,$$

so that X^σ represents a six-fold improvement on X^0 with respect to the maximin distance design criterion.

Although much testing remains to be done, especially with more sites in higher-dimensional experimental regions defined by more challenging constraints, we believe that this example provides a proof of concept. Given access to the appropriate

algorithms and software developed by the numerical optimization community, approximate maximin distance designs offer considerable promise for constructing space-filling designs in nonrectangular regions.

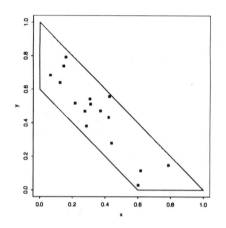

Figure 1: A Pseudorandom Sample from a Uniform Distribution on E

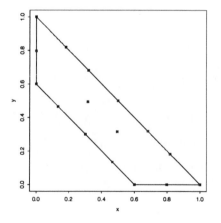

Figure 2: An Approximate Maximin Euclidean Distance Design in E

Acknowledgments

I am grateful for suggestions made by Andrew Booker and Max Morris, and for preliminary computational experiments performed by Laura Taylor.

References

Barthelemy, J.-F. M. and Haftka, R. T. (1993). Approximation concepts for optimum structural design—a review. *Structural Optimization*, 5:129–144.

Booker, A. J. (1994). DOE for computer output. Technical Report BCSTECH-94-052, Boeing Computer Services, Seattle, WA.

Booker, A. J. (1996). Case studies in design and analysis of computer experiments. In *Proceedings of the Section on Physical and Engineering Sciences*, pages 244–248. American Statistical Association.

Fourer, R., Gay, D. M., and Kernighan, B. W. (1993). *AMPL: A Modeling Language for Mathematical Programming*. Duxbury Press, Pacific Grove, CA.

Gockenbach, M. S. and Kearsley, A. J. (1999). Optimal signal sets for non-Gaussian detectors. *SIAM Journal on Optimization*, 9:316–326.

Johnson, M. E., Moore, L. M., and Ylvisaker, D. (1990). Minimax and maximin distance designs. *Journal of Statistical Inference and Planning*, 26:131–148.

Kearsley, A. J. (1996). *The Use of Optimization Techniques in the Solution of Partial Differential Equations from Science and Engineering*. PhD thesis, Department of Computational & Applied Mathematics, Rice University, 6100 Main Street, Houston, TX 77005-1892.

Koehler, J. R. and Owen, A. B. (1996). Computer experiments. In Ghosh, S. and Rao, C. R., editors, *Handbook of Statistics, Volume 13*, pages 261–308. Elsevier Science, New York.

McKay, M., Beckman, R., and Conover, W. (1979). A comparison of three methods for selecting values of input variables in the analysis of output from a computer code. *Technometrics*, 21:239–245.

Moré, J. J. and Wright, S. J. (1993). *Optimization Software Guide*. Society for Industrial and Applied Mathematics, Philadelphia.

Morris, M. D. and Mitchell, T. J. (1995). Exploratory designs for computational experiments. *Journal of Statistical Inference and Planning*, 43:381–402.

Owen, A. B. (1992). Orthogonal arrays for computer experiments, integration and visualization. *Statistica Sinica*, 2:439–452.

Owen, A. B. (1994). Lattice sampling revisited: Monte Carlo variance of means over randomized orthogonal arrays. *Annals of Statistics*, 22:930–945.

Sacks, J., Welch, W. J., Mitchell, T. J., and Wynn, H. P. (1989). Design and analysis of computer experiments. *Statistical Science*, 4:409–435. Includes discussion.

Tang, B. (1993). Orthogonal array-based Latin hypercubes. *Journal of the American Statistical Association*, 88:1392–1397.

TRENDS ACROSS SEVERAL SPATIAL PATTERNS

William D Heavlin, Advanced Micro Devices
P O Box 3453, MS 117, Sunnyvale, CA 94088-3453 <bill.heavlin@amd.com>

Key words: multidimensional scaling, nonparametric methods, rotations and reflections, semiconductor industry, spatial statistics, zero-one data.

Abstract. Semiconductor manufacturing operates on batches (lots) consisting of a few dozen circular wafers of silicon crystal. Scher et al (1990) consider the trends of univariate wafer statistics with respect to within-lot wafer sequence number. This work extends such considerations to the trending of wafer spatial patterns. Particular attention is given to identifying patterns that are similar save for mirror-image reflections and/or angular rotations.

The approach has five elements: (a) For each wafer, measured sites are transformed into a function of wafer site (x,y) coordinates. (b) For each pair of wafers, a relative rotation is calculated to minimize that pair's Euclidean distance. (c) To the resulting wafer-to-wafer distance and rotation angle matrices, multidimensional scaling methods represent the wafers as points in a low dimensional vector space. (d) These representations are trended against wafer sequence numbers, or (e) alternately, nonparametric methods can test the significance of any particular wafer sequence. A running example illustrates these methods.

1. Introduction

Integrated circuit manufacturing takes place on thinly sliced circular disks (wafers) of silicon crystal organized linearly into batches (lots) of 24 for multi-step processing. Both the diameter of the wafer and number of wafers per lot are sufficiently large for substantial spatial effects to emerge. For example, because of processing and equipment symmetry, radial effects commonly occur, and the probability that a circuit functions properly ("yields") is often higher for circuits manufactured near the wafer's center rather than near the wafer's edge.

Contemporary semiconductor manufacturing processes approach 400 distinct steps. Whether single-wafer or batch, certain effects can induce ultimately different results on wafers within the same lot: a batch furnace may induce a gradient of thickness across its length; a single wafer etcher may suppress the yield of initial wafers as its environment ramps to a desirable equilibrium.

To identify such wafer sequence patterns, Scher et al (1990) develop "Wafer Sleuth." The basic idea is at several processing steps to randomize wafer order, always taking care to record the wafer sequence both before and after. When end-of-line electronic measurements are taken — voltages, currents, yield — these are compared to all the recorded sequences.

An example is in figure 1. The y-axes consist of the wafer yields for a particular lot, the x-axes the wafer sequences recorded at steps A, B, and C. The scatters associated with the latter two steps appear random, but that with step A demonstrates suppressed yields for early wafers.

***Figure 1. Wafer Sleuth Example.** Y-axis: yield, x-axes: wafer sequences, plotted: one lot's 24 wafers.*

Wafer Sleuth is typically applied lot by lot. This makes it convenient to use even in development environments, where the volume of lots is low and the planned variations among lots are many. Wafer Sleuth implementation details vary among manufacturing operations, but typical practice includes between 10 and 50 randomization points, lot by lot analysis, and a combination of yield and transistor-level measurements. The result is that the volume of Sleuth data quickly mounts — the number of lots, randomization points, and responses all multiply — and for that reason the automated screening of interesting scatterplots is desirable. Screening methods include Pearson correlations, control charts with Western Electric rules, and runs tests.

In spite of such volume, in one important sense Wafer Sleuth remains underapplied. This is because it uses only wafer-level summary statistics, and does not exploit the patterns inherent in each wafer. (One attempt to deal with this divides wafers into 8-9 sectors, and applies Wafer Sleuth to each sector separately. This approach is both ad hoc, and further multiplies the number of Sleuth scatters to consider.) Incorporating the details of the wafer patterns is one of two defining themes of this paper.

This paper's second theme is based on an obvious but neglected weakness of Wafer Sleuth: that it cannot detect the source of a spatial pattern, if that pattern is the same on every wafer. To address this problem, a modification to Wafer Sleuth's standard sequence randomization is proposed.

Section 2 motivates and describes this modified randomization. In section 3 is a brief survey the spatial statistics literature, a description of the within-wafer model subsequently used. Section 4 applies the within-wafer model to form the data structures suitable for multidimensional scaling and subsequent graphical data analysis. Section 5 complements section 4 by developing nonparametric tests of statistically significant wafer sequences, and section 6 makes some closing comments and conclusions.

2. Motivating Problem (and Digression)

Applied to any wafer-level summary statistic, Wafer Sleuth is insensitive to detecting the sources of spatial patterns that affect all wafers equally.

Suppose, however, that the Sleuth randomizations that reorder the wafers' sequences induce also angular rotations. Let a denote the order of a given wafer after randomization at a particular step. Then we propose to rotate wafer a by the angle $\theta(a)=360(a-1)/n$, where n is the number of wafers in the lot. For a typical value of $n=24$, each wafer is rotated $15°$ more than the previous one, thereby sweeping out all $360°$.

By inducing angular rotations while reordering wafer sequence, Wafer Sleuth becomes sensitive to detecting particular process steps that induce static patterns. There are two caveats: (a) angular rotations do not improve sensitivity to radially symmetric static patterns, and (b) certain manufacturing steps — masking, pattern-based inspection, electronic structure testing — must reset all wafer angles to zero to accomplish their tasks.

Inducing Sleuth angles represents an additional complication, but on the whole this is of smaller scale than the initial introduction of Wafer Sleuth random resequencing. Section 4 returns to this concept of rotating one wafer's spatial pattern until it resembles another's.

3. Within-wafer Spatial Model

In classifying kinds of spatial data, Cressie (1993, section 1.2) distinguishes among (a) geostatistical, (b) lattice, and (c) spatial point data. The latter two both are common in semiconductor manufacturing. When the response is continuous and recorded over a domain that is either continuous or discrete, kriging (Ripley, chapter 4.4) and related methods are applicable; when the response is binary, the model due to Besag (1974) can be appropriate. Spatial point data has received the special attention of Diggle (1983).

The present example studies the yield pattern on wafers from a particular lot: it is binary response lattice data, and a small variation of Besag's (1974) model is applied. Let the lattice consist of the coordinates $\{s_i: i\}$ and denote the associated zero-one response by $\{z(s_i): i\}$,

or more briefly, $\{z_i: i\}$. Besag asserts this conditional lattice model:

$$\Pr[z(s_i)|\{z(s_j), j<>i\}] \ \alpha \ \exp\{\textstyle\sum_j F(z(s_j), \| s_j - s_i\|^2)\},$$

where the sum is over sites s_j near s_i, and $F(.,.)$ affords some generality. The version of this model used here is

$$\text{odds}[z(s_i)|\{z(s_j), \text{all } j\}] = \text{odds}_0 \cdot \textstyle\prod_j[\text{ odds}(\ j)^{(2z(i)-1)}],$$

where $\text{odds}(\ j) = 1/[\ \text{off}_0 + (\ s_j-s_i)^T \mathbf{D} (\ s_j-s_i)\ ]$. In this notation, the free parameters are odds_0 and elements comprising the diagonal matrix $\mathbf{D}$; they are estimated by maximum pseudo-likelihood. The parameter off_0 is analogous to kriging's nugget parameter, and controls the contrast between the zeros and ones.

In our applications, we fit this modified Besag model to each wafer separately. The result is that each wafer is represented by a function, and this function can be evaluated not only at lattice points $\{s_i: i\}$ but also over a continuous domain.

For wafer a denote this function by $W_a(\mathbf{s})$, where $\mathbf{s}$ is typically the 2-vector (x,y). Further, denote the reflection of wafer a by $R_a((x,y))=W_a((x,-y))$; this notation assumes the center of gravity of the lattice $\{s_i:i\}$ is $(0,0)$.

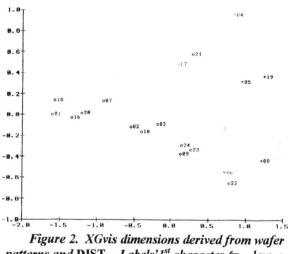

Figure 2. XGvis dimensions derived from wafer patterns and DIST$_2$. Labels' 1st character (=,– low, o neutral, + high) implies 3rd dimension.

4. Multidimensional Scaling and Wafer Sleuth

The distance between two sets of points is the minimum distance between all possible point pairs. This basic principle underlies the following definitions: Define

$$DIST_{\theta,1}(W_a, W_b)=$$

$$[\textstyle\sum_{i \in G} \|W_a((x_i, y_i)) - W_b(\text{rot}(x_i, y_i), \theta)\|^2]^{1/2}$$

where rot(**s**,θ) denotes the rotation of point **s** by angle θ, and G represents a set of coordinates (a grid) over which the wafer functions' distances are evaluated. Define

$$DIST_1(W_a, W_b) = \min\{DIST_{\theta,1}(W_a, W_b): \text{all } \theta\}.$$

Let $\theta_1(W_a, W_b)$ denote the corresponding rotation angle at which the minimum $DIST_{\theta,1}(W_a, W_b)$ is realized. Because the choice of rotating W_a or W_b is arbitrary, $DIST_1(W_a, W_b) = DIST_1(R_a, R_b)$ and $DIST_1(R_a, W_b) = DIST_1(W_a, R_b)$. This motivates an additional distance metric:

$$DIST_2(W_a, W_b) = \min\{DIST_1(W_a, W_b), DIST_1(W_a, R_b)\}.$$

$DIST_1$ represents the distance between wafers when one is *rotated* to best resemble the other; $DIST_2$ when one is rotated and perhaps *reflected* as well.

As before, let n denote the number of wafers. By including reflected wafers, $DIST_1$ (alternately θ_1) comprises a $2n \times 2n$ matrix of interwafer distances (angles), while $DIST_2$ comprises an $n \times n$ matrix.

Multidimensional scaling (MDS) algorithms derive points in low-dimensional vector spaces that approximately reproduce distance matrices such as $DIST_1$ and $DIST_2$. MDS has an extensive literature, surveyed, for example, in Young (1987) and Kruskal and Wish (1978). Buja et al (1998) implement an MDS algorithm, called XGvis, that works in conjunction with the dynamic graphical display environment XGobi.

For $DIST_2$, XGvis' MDS options are set to 12-dimensional space and Euclidean distance, and results are displayed in figure 2.. The two plotted dimensions result from the rotating these 12 dimensions to maximize the variances of the lower indexed dimensions (dimension 1, 2, 3 and so on) in preference to the higher indexed dimensions (e.g. dimensions 10, 11, 12); in the context of eigen-analysis, such variance-maximizing rotations are familiar. Plotted symbols and colors are used to indicate the magnitude of the third dimension.

The configuration in figure 2 has a funnel shape, with lower-yielding wafers forming an almost one-dimensional stem, and higher-yielding ones blooming at one end. This description is reinforced when dynamically rotated in XGobi. The first dimension tracks closely with yield, the second with the degree of contrast between low- and high-yielding wafer portions.

Because of the special nature of the matrix θ_1, a special MDS algorithm is appropriate. In radians, angles are the lengths of arc on the great circles connecting points on a unit sphere. Therefore, any scaling algorithm should constrain all points to such a sphere.

Observe that if x_a and x_b are points on a unit sphere, their dot product equals $\cos(\theta_{ab})$. This invites the following algorithm: (1) Define the matrix **C** whose elements are the cosines of θ_1. (2) Compute the first k eigenvectors **P**$_k$ eigenvalues λ_k of **C**, and the factor scores

$\mathbf{X}_k = \mathbf{P}_k \cdot \text{diag}(\lambda_k)^{1/2}$. (3) Normalize $\mathbf{X}_k$ so that its rows are all unit vectors. (4) Similar to before, apply the variance-maximizing rotations to $\mathbf{X}_k$.

For $k=3$, the results are in figure 3. The circle represents the plane in which the third dimension zero. With the exception of the (red) wafers W09 and W24, all points (blue) have a positive third dimension, and can be visualized as lying on the half-sphere projecting toward the viewer. Because the reflected wafers consistently lie at coordinates exactly on the other side of the (dimension 1, dimension 2)-origin (with the same value for the third dimension), they are not plotted here.

Whether derived from $DIST_2$ or θ_1, in application one can plot these newly computed dimensions against various wafer processing sequences. Their evaluation of significance can proceed as for univariate wafer statistics. One can respond to subjectively interesting patterns, or apply formal methods such as correlation coefficients, control chart rules, or runs tests.

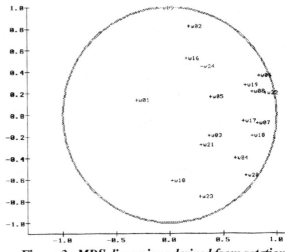

Figure 3. MDS dimensions derived from rotation matrix θ_1; points all lie on unit 3-sphere. Labels' 1st character (– minus, + positive) implies 3rd dimension.

5. Statistically Significant Wafer Sequences

Section 1 notes that Wafer Sleuth creates many scatterplots, and that it is desirable to screen these to discern the more important ones. This section is devoted to tests of significance in a manner consistent with matrices such as $DIST_2$. The approach resembles that of Friedman and Rafsky (1979).

Let ω denote a given sequence of wafers, such that if wafer a is first in the sequence, then $\omega(a)=1$, if second, then $\omega(a)=2$, and so on. Define the $n \times n$ matrix SEQ_ω such that $SEQ_\omega(a,b)=|\omega(a)-\omega(b)|$; effectively, SEQ_ω is a distance matrix based on the one-dimensional ordering implicit in ω.

Define RD_2 as an nxn matrix in which the elements of $DIST_2$ are replaced with their ranks. Finally, let YD denote an nxn matrix such $|\eta(a)-\eta(b)|$, where $\eta(a)$ is the rank of wafer a in yield.

An index of similarity between RD_2 and SEQ_ω is $\sum_{a,b}RD_2(a,b)\cdot SEQ_\omega(a,b)$, where the sum is over all wafers a and b, and larger values indicate greater similarity. For brevity, let us denote this double sum by $RD_2\cdot SEQ_\omega$. Similarly, an index of similarity between wafer yield ranks and a given sequence ω is $YD\cdot SEQ_\omega$.

From three randomized 19-symbol orthogonal arrays from Owen (1998), each of strength 3, are computed 20577 pseudo-random permutations of 19 wafers. Each such permutation implies a different value of ω, for which are computed both $RD_2\cdot SEQ_\omega$ and $YD\cdot SEQ_\omega$. The scatter of these 20577 values of $RD_2\cdot SEQ_\omega$ against $YD\cdot SEQ_\omega$ is in figure 4. Between these two statistics is substantial correlation, about 0.63 for this example.

More appropriately, $RD_2\cdot SEQ_\omega$ can be applied conditionally. This is because any analysis based on spatial patterns would very likely be used to supplement and complement, rather than reproduce, the simpler analysis based on wafer yields alone. In figure 4, the median value of $YD\cdot SEQ_\omega$ is about 841, and the red vertical stripe highlights the roughly 1000 sequences with values of $YD\cdot SEQ_\omega$ within ±2 of 841.

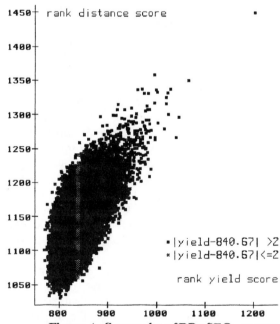

Figure 4. Scatterplot of **$RD_2\cdot SEQ_\omega$** *versus* **$YD\cdot SEQ_\omega$,** *for random sequences ω. Vertical (red) stripe demarks cases near median value of* **$YD\cdot SEQ_\omega$.**

Of these, for that ω with the highest $RD_2\cdot SEQ_\omega$, figure 5 shows the wafer maps. This sequence shows some clustering of similar patterns, but on the whole is not obviously visually arresting. Experience shows that in such displays as figure 5, viewers respond to gradients in pixel density (yield), but less so to patterning per se.

6. Comments and Conclusions

This approach represents wafer patterns as points, rather than planar objects, in a low-dimensional space. It consists of five steps: (a) For each wafer, measured sites are transformed into a function of wafer site (x,y) coordinates. (b) For each pair of wafers, relative rotations (and perhaps reflections) are calculated to minimize that pair's Euclidean distance. (c) To the resulting wafer-to-wafer distance and rotation angle matrices, multidimensional scaling methods represent the wafers as points in a lower dimensional vector space. (d) These representations are trended against wafer sequence numbers, or (e) alternately, are used to calculate nonparametric tests of significance for particular wafer sequences.

Relatively novel is the combination of (b) with (c). Buja et al (1998) observe that when used for dimension reduction, MDS is essentially identical with principal components. But in the present application, because the distances are defined as minima over a family of transformations, this equivalence does not quite hold, and MDS emerges as the more flexible alternative. In other contexts, the greater flexibility afforded by MDS is well recognized.

The example presented here does not have the deliberately randomized Sleuth angle rotations proposed in section 2. Nonetheless, many steps induce slight-to-moderate rotations, enough that the definitions of distances $DIST_1$, $DIST_2$, and θ_1 remain appropriate.

These definitions offer some opportunities to accelerate computations: One can exploit symmetry relations — $DIST_1(W_a,W_b) = DIST_1(R_a,R_b)$ and $DIST_1(R_a,W_b) = DIST_1(W_a,R_b)$ — to reduce the number of wafer pairs to consider by 25 percent. And, as done here, one can calculate the distances $DIST_1(W_a,W_b)$ over a set of lattice points smaller than the lattice on which the data resides. Finally, the computationally intensive tasks (pairwise distance calculations, variance-maximizing rotations) involve repeated applications of one-dimensional minimizations; these have natural algorithms exploiting parallel processing.

R68_50

Figure 5. Wafers' yield maps ordered by sequence ω with the most significant value of $RD_2 \cdot SEQ_\omega$, among those near median value of $YD \cdot SEQ_\omega$.

Presented here is a particular set of choices for identifying trends among spatial patterns with little attention to alternatives, richly described, for example, in Hansen, Nair, and Friedman (1999). Let us close with an attempt to gather open issues together. First, further refinements of the within-wafer model are both possible, and probably justified. Complementing the application here with continuous response lattice data, and with spatial point data should provide further insights. Likewise, alternative parameter fitting algorithms can be explored. New data collection schemes often present their own intrinsic complications; quite likely randomly induced wafer angle rotations will prove no exception. For example, the proposal in section 2 links wafer rotation to wafer sequence, but this can be de-coupled so that wafer angle and slot position are completely uncorrelated.

MDS is applied here to $DIST_2$, but $DIST_1$ remains a viable option, and the advantages and disadvantages of each needs to be better understood. Similarly, MDS encompasses many algorithms and options, and the best choice among these merits attention. Mainstream MDS algorithms can be applied also to θ_1, and the practical consequences of doing so could be explored more fully. In at least one important sense, the algorithmic separation of $DIST_1$ and θ_1 is unsatifying. For a radially symmetric pattern, θ_1 becomes indeterminate, but no accommodation has been made for this. Note also that generalizations beyond planar objects are possible. For example, for rotations in 3-space, the angles θ_{12} and θ_{13} give rise to an inner product of $\cos(\theta_{12})\cos(\theta_{13})$, and the rest applies.

A basic tension exists between the rather visual MDS approach of section 4 and the hypothesis testing approach of section 5; there could be more compatibility.

The statistic $RD_2 \cdot SEQ_\omega$ is not the only one conceivable, and alternatives may prove both viable and ultimately more worthy. In particular, chamber tools offer adjacent wafers physically different processing chambers. For this reason, contemporary Wafer Sleuth practioners pay considerable attention to even-odd and every-third wafer effects, and such issues offer a physical basis for considering the lower tails of distributions of test statistics such as $RD_2 \cdot SEQ_\omega$. Finally, displays such as figure 5 may be optimum neither for pattern discovery nor for pattern verification. Other forms may enhance visual pattern recognition in ways that meld better with formal significance testing.

7. References

Scher, G, Eaton, DH, Fernelius, BR, Sorenson, J, and Akers, J (1990), "In-line statistical process control and feedback for VLSI integrated circuit manufacturing," *IEEE Transactions on Components, Hybrids, and Manufacturing Technology*, **13**:3, pp 484-489.

Cressie, NAC (1993), *Statistics for Spatial Data*, Wiley, New York.

Ripley, B (1981). "Stochastic process prediction," chapter 4.4, pp 44-75 in *Spatial Statistics*, Wiley: New York.

Besag, JE (1974), "Spatial interaction and the statistical analysis of lattice systems," *Journal of the Royal Statistical Society, B*, **36**, pp 192-225.

Diggle, PJ (1983) *Statistical Analysis of Spatial Point Patterns*, Academic Press, NY, 157pp.

Young, FW (1987), *Multidimensional Scaling: History, Theory, and Applications*, RM Hamer, ed, Lawrence Erlbaum: Hillsdale.

Kruskal, JB, and Wish, M (1978), *Multidimensional Scaling*, EM Uslaner, ed, Sage: Newbury Park.

Buja, A, Swayne, DF, Littman, ML, and Dean, N (1998), "XGvis: interactive data visualization with multidimensional scaling," *Journal of Computational and Graphical Statistics*, to appear. Preprint available from http://www.research.att.com/areas/stat/xgobi/index.html#xgvis-paper.

Friedman, JH and Rafsky, LC (1979), "Multivariate generalizations of the Walf-Wolfowitz and Smirnov two-sample tests," *Annals of Statistics*, 7:4, pp 697-717.

Owen, AB (1998). "Orthogonal arrays in c," accessed from http://lib.stat.cmu.edu/designs/oa.c, June 3.

Hansen, MH, Nair, VN, and Friedman, D (1999) "Process improvement through the analysis of spatially clustered defects on wafer maps," submission to *Journal of the American Statistical Association*. Preprint available at http://cm.bell-labs.com/cm/ms/who/cocteau/papers/pdf/overview.pdf.

THE VALUE OF PRE EXPERIMENTAL SCIENTIFIC INFORMATION IN THE DESIGN OF ACCELERATED TESTS

M.J. LuValle, Lucent Technologies
25 Schoolhouse Rd., Somerset, NJ, 08873

Keywords: Experiment design, accelerated tests, kinetics, demarcation maps.

INTRODUCTION

An important question for both educators and employers of applied statisticians is "How important is the ability to understand and integrate subject matter theory with statistical concepts?". This question can probably only be answered through experience. The purpose of this paper is to offer a quantitative comparison of the effect of such knowledge on experiment design for a simple accelerated testing example drawn from my experience (LuValle et. al., 1998). After some introductory material, we compare the asymptotic relative efficiency of 3 designs.

• A typical design used in the problem, similar to standard designs for model estimation.

• The physics based design, using a non-intuitive form of step stress (this is not statistically optimized).

• The (intuitively) closest design to the physics based one in the space of constant stress designs.

THE PROBLEM

A relatively new device in optical networking is the fiber grating. To make this device a piece of optical fiber is exposed to ultraviolet light in such a way so as to generate corrugations in the index of refraction of the core (the part of the fiber that carries the light) of the fiber. This pattern of index change can act in several ways, some patterns are wavelength specific mirrors. Other patterns couple light from the core to the cladding. Each of these effects has a use in optical communications networks. Unfortunately this optically induced change in index is meta stable. So over time it vanishes. Fortunately (Erdogan et. al., 1995) the effect can be modeled as a set of independent sites each with an individual activation energy. Thus baking the device for a while eliminates low activation energy sites, while retaining the high activation energy (highly stable) sites.

More explicitly, each microscopic site where index change occurs decays at a rate:

$$rate = v \exp\left(\frac{-E_a}{kT}\right) \qquad (1)$$

where k is Boltzmann's constant, and T is absolute temperature. Each site has it's own value of E_a, but all sites are assumed to have a common value of v. This model has been fit to many gratings over wide temperature ranges, and it was assumed that if there were a second value of v, it would be obvious through the method of fit. This assumption has been shown incorrect, and a special experiment design was developed using a physics based tool, a demarcation map, which greatly increased the ability to see this effect (LuValle et. al. 1998).

SOME PHYSICAL THEORY

Both the modeling method usually used, and the design are based on a simple approximation. Typically, the rate given above in this particular problem is associated with an exponential probability of decay. The approximation is to replace this exponential decay with a step function. Place the step at the time t such that $rate \times t = 1$. Using this approximation in (1) allows us to derive:

$$E_a \equiv E_d = kT \ln(vt) \qquad (2)$$

The accuracy of the approximation is shown in the Figure 1 below. The straight lines are the approximation, while the points show actual probability of occupancy. For this plot, v is 10^{15} Hertz, temperature is 25 C, and time is 100 hours.

For changing temperature, at any value of v we can calculate equivalent time by setting E_d under one temperature equal to E_d under another, and solving:

$$kT_1 \ln(\nu t_1) = kT_2 \ln(\nu t_2)$$

or

$$t_1 = \frac{(\nu t_2)^{T_2/T_1}}{\nu}$$

So the demarcation energy after time t_1 at temperature T_1, and then time t_2 at temperature T_2 takes the form:

$$E_d = kT_2 \ln\left(\nu\left(t_2 + \frac{(\nu t_1)^{T_1/T_2}}{\nu}\right)\right) \quad (3)$$

For model fitting, E_d is a universal time for collapsing data from different temperatures onto a single master curve. For experiment design, the plot of E_d vs $\log_{10}(\nu)$ for different temperature trajectories allows one to track the parameters corresponding to "completed" reactions during that test.

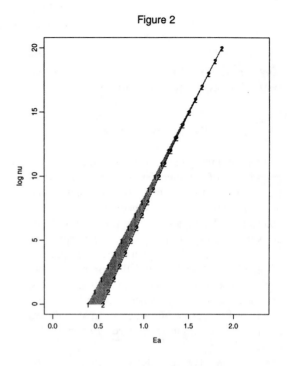

Figure 2

Figure 1

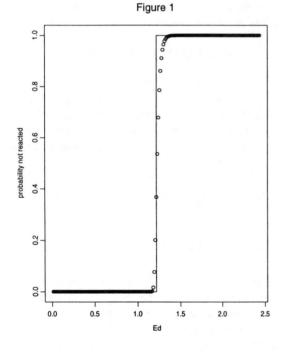

For example suppose that as part of manufacturing, a device is exposed to 100 C for 40 hours, and the life stress is 38 C for 25 years. Then the demarcation map takes the form:

The region to the left and above the figure indexes the set of reactions exhausted by the burn-in. The shaded region is the set exhausted by life.

THE STATISTICAL MODEL

The original analysis procedure (LuValle et. al, 1998) was a semiparametric procedure based on (3). The data takes the form of a time series on each device, one device per condition. The error in the time series is often complicated, but we assume for the purpose of this paper that it is iid between measurements. The parameters of interest are those of the structural model. For the purpose of this paper the full model has the form:

$$y_t = c_1 \int_0^{kT\ln(\nu_1 t)} d\mu_1(E_a) + c_1 \int_0^{kT\ln(\nu_2 t)} d\mu_2(E_a) + \varepsilon_t$$

Each distribution is uniform with endpoints to be chosen from the data. The model is a nonlinear model with eight parameters of interest.

The asymptotic covariance matrix for this structural model is $(X'X)^{-1}$ where X is the derivative matrix of the model (calculated numerically). The asymptotic relative efficiency for each parameter is the ratio of the appropriate diagonal element from the matrix evaluated for each experiment.

In the calculations I use the following values for the true values of the parameters:

$$v_1 = 10^3, v_2 = 10^{16}$$
$$E_{a11} = 0.2, E_{a21} = 0.8$$
$$E_{a12} = 0.7, E_{a12} = 2.4$$
$$C_1 = 1, C_2 = 5$$

THREE EXPERIMENT DESIGNS

The first design is the one first used in the real experiment that motivated this, although it is plainly not optimal, it was considered good enough for estimating the single V model (LuValle et. al. Ibid).

<u>Design 1</u>
1 device at 220 C for 80 hours
1 device at 180 C for 120 hours
1 device at 160 C for 30 hours

Each device is assumed to go through the 100 C 40 hour burn-in. Figure 3 above uses a demarcation map to compare cell 3 of the design above to life. The life region is the shaded region. The experimental region is the one between lines marked with "1" and "2". We see that under our model, a major portion of the degradation observed arises from the second integral, so the signal from the first integral is swamped.

However, the major contribution during life is the contribution from the first integral. Figure 4 below shows a demarcation map for an experimental cell which gets around this problem by exposing the device to a very short (25 second) very high temperature (300 C) temperature excursion. (These devices could stand this).

This exposure essentially removes the effect of the large value of V with nearly 0 aging going on at the smaller value. Subsequent aging at 120 C for 200 hours provides an opportunity to see much lower values of V.

This is illustrated in figure 4 by the part of the shaded region (aging during life) that the region between the line labeled 2 and the line labeled 3 in figure 4 covers.

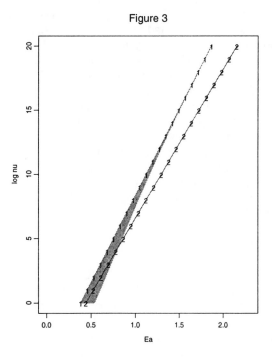

Figure 3

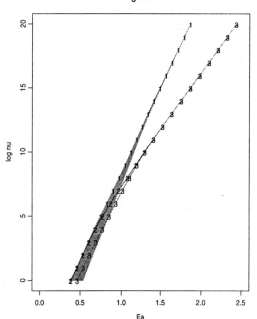

Figure 4

The experiment design based on this strategy suggested by physical insight takes the form:

Design 2
1 device at 220 C for 80 hours
1 device at 180 C for 120 hours
1 device at 160 C for 30 hours
1 device at 300 C for 25 seconds followed by 200 hours at 120 C
1 device at 300 C for 25 seconds followed by 200 hours at 100 C.

In order to provide a fairer comparison than just design 1 to design 2, a third design that looks like design 2 without the exposure at 300 C is evaluated.

Design 3
1 device at 220 C for 80 hours
1 device at 180 C for 120 hours
1 device at 160 C for 30 hours
1 device at 120 C for 200 hours
1 device at 100 C for 200 hours

RESULTS

The asymptotic relative efficiency for design 2 and 3 with respect to design 1 is provided for each parameter in the model in the table below. Conceptually, the improvement can be motivated by the fact that the step stress experiment moves the X matrix closer to an orthogonal matrix.

Asymptotic relative efficiencies of the designs

Parameter	ARE 2:1	ARE 3:1
v_1	4.64	1.64
E_{a11}	5.12	4.53
E_{a12}	3.16	1.60
v_2	19.98	1.79
E_{a21}	2.71	1.99
E_{a22}	19.53	1.86
c_1	9.66	4.41
c_2	3.45	2.55

The step stress experiment clearly dominates both the original experiment and the 2nd experiment designed to be close to the step stress.

CONCLUSION

This paper provides a quantitative comparison of the effectiveness of incorporating pre-experimental subject matter knowledge into design of experiments. A sample size savings of nearly a factor of ~5 is found comparing the naïve and the physically motivated design. factor of ~2 is found comparing the physically motivated design, and the closest constant stress design. The application of optimal design methods may change this, however I would conjecture that physically appropriate step stress designs will always beat constant stress designs in this type of hidden failure mode problem. For more complex kinetic models using simple design practices, I have seen the ARE of the physically motivated design approach 1000.

In the real problem, the sample size savings translated into the ability to detect a potentially important second degradation mode originally masked by the obvious one observed in the accelerated tests.

Returning to our question, How important is the ability to understand and integrate subject matter theory with statistical concepts?" This is only one anecdote, but in my experience, combined physical and statistical insight generally provides an ability to see around the problem and identify new ways to approach it that purely statistical thinking, and purely physical thinking won't see. This can result in dramatic improvements in performance of experiments, in particular in detecting important departures from assumed models.

Employers must consider how much they are willing to live with missing this knowledge. Educators must consider with how much more marketable their students would be with this extra training, and the cost of giving it to them.

REFERENCES

1. LuValle M.J., Copland L.R., Kannan, S., Judkins, J.B., and Lemaire, P.J., "A strategy for extrapolation in accelerated testing", Bell Labs Technical Journal, 1998, V3, #3, pp 139-147.

2. Erdogan, T., Mizrahi, V., Lemaire, P.J., and Monroe, D., "Decay of ultraviolet induced fiber Bragg gratings", Journal of Applied Physics, 1994, V76, (1), 73-80

ENUMERATING NONEQUIVALENT ORTHOGONAL ARRAYS VIA MINIMUM G-ABERRATION

Yingfu Li, Boxin Tang, Lih-Yuan Deng, University of Memphis

Boxin Tang, Dept. of Mathematical Sciences, University of Memphis, Memphis, TN 38152

Key Words: Orthogonal array, Hadamard matrix, Plackett-Burman design, nonequivalence of orthogonal arrays, nonregular factorial design, minimum G-aberration, confounding frequency vector, J-characteristics.

Abstract:

Deng and Tang (1999) introduced the criterion of minimum G-aberration for assessing nonregular factorials. In this paper, we explore the power of this criterion in discriminating nonequivalent orthogonal arrays. For many combinations of run size n and the number m of factors, collections of nonequivalent orthogonal arrays are obtained. Nontrivial orthogonal arrays are found that cannot be embedded into Hadamard matrices. Our findings among other things disprove a conjecture of Vijayan (1976) on the embedding problem of Hadamard matrices.

1 Introduction

A two level orthogonal array of size n, strength d and having m constraints is an $n \times m$ matrix with entries $+1$ and -1 such that the 2^d possible level combinations in any submatrix with d columns occur equally often. We will concentrate on orthogonal arrays of strength two in this paper. When used as a factorial design, an orthogonal array allows orthogonal estimation of the main effects. We will speak of orthogonal arrays and factorial designs interchangeably in this paper. The most studied orthogonal arrays are the regular series of 2^{n-k} designs.

Deng and Tang (1999) recently introduced the criterion of minimum G-aberration. By searching through orthogonal arrays given by taking columns

from the five Hadamard matrices of order 16, Deng and Tang (1998) demonstrated that minimum G-aberration not only provides a fruitful criterion for ranking different orthogonal arrays but also is very powerful in discriminating nonequivalent orthogonal arrays. In this paper, we will further explore the discriminating power of minimum G-aberration for run size greater than 16. A Hadamard matrix H is a square matrix of order n with entries -1 and $+1$ such that $H^T H = E$, where E is the identity matrix. A Hadamard matrix can be *normalized* so that its first column consists of all $+1$'s. Hadamard matrices considered in this paper are assumed to be normalized. Two orthogonal arrays are said to be equivalent if one can be obtained from the other by permuting rows, permuting columns, switching the signs of a column, or a combination of the above.

Two objections naturally arise in association with the approach of taking orthogonal arrays from Hadamard matrices. First, there exist orthogonal arrays that cannot be embedded into Hadamard matrices. Vijayan (1976) considered the embedding problem of Hadamard matrices and showed that an orthogonal array can be embedded into a Hadamard matrix if there are at most five missing columns. He conjectured that if $m \geq n/2 - 1$, then an orthogonal array can be embedded into a Hadamard matrix. Second, the complete set of nonequivalent Hadamard matrices is only available for order 28 or lower. It is well known that for order 12 or lower, there is precisely one Hadamard matrix up to equivalence. For orders 16, 20, 24, and 28, the numbers of nonequivalent Hadamard matrices are 5, 3, 60, and 487, respectively (Spence 1995, Table 1). It is clear that even if the complete set were available for order 32 or higher, it would not be practical to examine all of them.

The research of Boxin Tang is supported by the National Science Foundation DMS-9971212.

In this paper, we use minimum G-aberration to search for nonequivalent orthogonal arrays without restricting to arrays from Hadamard matrices. For many combinations of run size n and the number m of factors, collections of nonequivalent orthogonal arrays are obtained. Minimum G-aberration designs and many other nontrivial orthogonal arrays are found that cannot be embedded into Hadamard matrices. Moreover, our findings discredit the conjecture of Vijayan (1976).

Section 2 provides a brief review of minimum G-aberration and describes our search method. The results and accompanying discussions are given in Section 3.

2 Method and Preliminary results

We provide in subsection 2.1 some background material related to the present paper. The central concepts here are minimum G-aberration and its associated confounding frequency vector. In subsection 2.2, orthogonal arrays of 3 and 4 factors are discussed. Then in subsection 2.3, our method for searching for orthogonal arrays of more than 4 factors is described.

2.1 Minimum G-aberration and confounding frequency vector

Consider an orthogonal array $D = (c_{ij})_{n \times m}$ with $c_{ij} = \pm 1$, where each row of D represents a run and each column corresponds to a factor. For a nonempty subset s of the index set $Z_m = \{1, \ldots, m\}$, we define

$$J_s(D) = \sum_{i=1}^{n} \prod_{j \in s} c_{ij}.$$

For $s = \phi$, the empty set, define $J_\phi(D) = n$. The collection of J_s values are called J-characteristics. We note that the definition of and the notation for J-characteristics in this paper are slightly different from those given in Deng and Tang (1999) and Tang and Deng (1999). Design D is regular if $|J_s| = 0$ or n for all $s \subseteq Z_m$. It is nonregular if there exists an $s \subseteq Z_m$ such that $0 < |J_s| < n$. For orthogonal designs, it is obvious that $J_s = 0$ if $|s| = 1, 2$.

For simplicity, we use both $|J_s|$ and $|s|$ in this paper, where the former denotes the absolute value of J_s and the latter the number of elements in s. The meaning of $|\cdot|$ should be clear from context. Note further that J_s must be a multiple of 4 for orthogonal designs (Deng and Tang 1999, Proposition 3). We know that the run size n of an orthogonal array must be a multiple of 4. Let $n = 4t$ and f_{kj} be the frequency of k column combinations that give $|J_s| = 4(t + 1 - j)$ for $j = 1, \ldots, t$. The confounding frequency vector of design D is then defined to be

$$F(D) = [(f_{31}, \ldots f_{3t}); (f_{41}, \ldots f_{4t}); \ldots; (f_{m1}, \ldots f_{mt})].$$

Let $f_l(D_1)$ and $f_l(D_2)$ be the l-th entries of $F(D_1)$ and $F(D_2)$, respectively. Let l be the smallest integer such that $f_l(D_1) \neq f_l(D_2)$. If $f_l(D_1) < f_l(D_2)$, then D_1 has less G-aberration than D_2. If no design has less G-aberration than D_1, then D_1 has minimum G-aberration.

If two orthogonal arrays are equivalent, obviously they have the same confounding frequency vector. In other words, two orthogonal arrays are nonequivalent if their confounding frequency vectors are different. Each confounding frequency vector corresponds to at least one nonequivalent orthogonal array. Deng and Tang (1998) found many orthogonal arrays from Hadamard matrices of order 16, and their results showed that confounding frequency vectors provide a simple yet powerful tool for identifying nonequivalent orthogonal arrays.

2.2 Orthogonal arrays of 3 and 4 factors

For convenience, we refer to the first t elements of a column vector of length $n = 4t$ as the first quarter, and the second t elements as the second quarter, and so on. Consider two orthogonal columns c_1 and c_2. Without loss of generality, we assume that for c_1, the first and second quarters consist of $+1$'s and the third and fourth quarters -1's, and for c_2, the first and third quarters consist of $+1$'s and the second and fourth quarters -1's. The following lemma provides a quich check if a third column c_3 is orthogonal to c_1 and c_2.

Lemma 1. *Let c_1 and c_2 be as given above. Let x_i*

be the number of $+1$'s in i^{th} quarter of a column c_3, for $i = 1, 2, 3, 4$. Then column c_3 is orthogonal to c_1 and c_2 if and only if $x_2 = t - x_1$, $x_3 = t - x_1$, and $x_4 = x_1$.

Lemma 1 says that x_2, x_3, and x_4 are completely determined by x_1 if c_3 is to be orthogonal to c_1 and c_2. Since $0 \leq x_1 \leq t$, we conclude that there are $t + 1$ different choices of x_1. Note that the array with $x_1 = x_1^*$ can be obtained from the array with $x_1 = t - x_1^*$ by switching the signs of c_3. So we need only to consider $0 \leq x_1 \leq [t/2]$ where $[t/2]$ denotes the largest integer not exceeding $t/2$. Note further that $J_s = 4|x_1 - x_2| = 4(t - 2x_1)$ for $s = \{1, 2, 3\}$. Thus different values of x_1 give different confounding frequency vectors. This shows that the number of nonequivalent arrays is precisely $[t/2] + 1$. This result is a special case of a more general result of Seidan (1986) who provided a complete enumeration of orthogonal arrays of strength d and having $d + 1$ factors.

Lemma 2. *Let c_1 and c_2 be as given above. Suppose that c_3 and c_4 are two columns orthogonal to c_1 and c_2. Let y_i be the number of $(+, +)$ pairs in the ith quarter of c_3 and c_4. Then c_3 is orthogonal to c_4 if and only if $y_1 + y_2 + y_3 + y_4 = t$.*

Let x_i and t_i be the numbers of $+1$'s in the ith quarters of c_3 and c_4, respectively. Let y_i be as given in Lemma 2. We may assume $0 \leq x_1 \leq t_1 \leq [t/2]$. In addition, we have $0 \leq y_i \leq \min(x_i, t_i)$. Note that x_i and t_i for $i = 2, 3, 4$ are completely determined by x_1 and t_1. Now for each pair (x_1, t_1) with $0 \leq x_1 \leq t_1 \leq [t/2]$, we examine $y_i = 0, 1, \ldots, \min(x_i, t_i)$ to see if they satisfy $y_1 + y_2 + y_3 + y_4 = t$. Then we compute the confounding frequency vector via

$$|J_{123}| = 4|x_1 - x_2|$$

$$|J_{124}| = 4|t_1 - t_2|$$

$$|J_{134}| = 4|y_1 + y_2 - y_3 - y_4|$$

$$|J_{234}| = 4|y_1 + y_3 - y_2 - y_4|$$

$$|J_{1234}| = 4|y_1 + y_4 - y_2 - y_3| - 4|x_1 - x_2| - 4|t_1 - t_2|,$$

for the cases with $y_1 + y_2 + y_3 + y_4 = t$ met. Collections of orthogonal arrays are obtained by applying

the method to run size $n \leq 100$. The number M of nonequivalent orthogonal arrays are given in Table 1.

n	8	12	16	20	24	28	32	36
M	2	1	5	3	10	7	19	15
n	40	44	48	52	56	60	64	68
M	31	28	52	48	79	79	119	123
n	72	76	80	84	88	92	96	100
M	171	184	243	268	332	379	452	523

Table 1. The number of nonequivalent orthogonal arrays of 4 factors for run size $8 \leq n \leq 100$.

We suspect that the results in Table 1 are exact in that they provide complete enumeration of orthogonal arrays of 4 factors, but are unable to provide a rigorous proof at the moment. This problem will be looked into in the future.

2.3 Searching for orthogonal arrays of more than 4 factors

The basic idea of our searching for nonequivalent orthogonal arrays is simple. For each orthogonal array in the given set of nonequivalent orthogonal arrays with k columns, we examine an additional column to see if it is orthogonal to the existing k columns. Lemma 1 is useful in this regard. If this additional column is orthogonal to the existing k columns, we then proceed to compute the confounding frequency vector for the whole array. Results in subsection 2.2 allow us to get jump-started by working with $k = 4$ first.

In the following, we look at some computational issues. Computation of the entire confounding frequency vector can be very time-consuming for large m since it involves evaluating $2^m - 1 - m - m(m-1)/2$ J-characteristics. One way to alleviate the computational burden is to use the following abridged version

$$F^*(D) = [(f_{31}, \ldots f_{3t}); (f_{41}, \ldots f_{4t}); (f_{51}, \ldots f_{5t})].$$

In fact, this is the approach adopted in Deng and Tang (1998) and Deng, Li, and Tang (1999). Deng and Tang (1998) call the criterion using $F^*(D)$ MA-5 classfier, we refer to that paper for a detailed discussion on MA-4 and MA-5 classfiers.

Since two nonequivalent orthogonal arrays may have the same confounding frequency vector, we record up to 10 orthogonal arrays with k columns for each confounding frequency vector. Based on this enlarged set of orthogonal arrays of k columns, we search for the set of orthogonal arrays with $(k + 1)$ columns. Our studies show that this modification does make a difference, especially for large k.

Finally, calculation of confounding frequency vector for a candidate orthogonal array with $k + 1$ columns can be done by simply calculating the J-characteristics involving the $(k + 1)$th column. This is achievable because the confounding frequency vector of the parent orthogonal array with k columns has already been computed.

3   Results and Discussions

Deng and Tang (1998) considered the problem of ranking and classifying orthogonal arrays from Hadamard matrices of orders 12 and 16 using the criterion of minimum G-aberration. For early work using other approaches, the reader is referred to Lin and Draper (1992), Sun (1993), and Wang and Wu (1995). We have applied our method described in Section 2 to orthogonal arrays of 12 and 16 runs. No new orthogonal array is found except that the trivial 12 run array given by $x_1 = 0$ in Lemma 1 cannot be embedded into a Hadamard matrix. In subsection 3.1, our results on orthogonal arrays of 20 runs are presented and discussed. We discuss in subsection 3.2 our findings on orthogonal arrays of 24, 28, 32, and 36 runs.

3.1   Orthogonal arrays of 20 runs

Our results on orthogonal arrays of 20 runs are given in Table 2, where m denotes the number of factors, M the total number of orthogonal arrays found by our method, M_0 the number of orthogonal arrays that cannot be embedded into Hadamard matrices of order 20. To check if an array is from a Hadamard matrix, we compare its confounding frequency vector with those of the designs in Deng, Li, and Tang

(1999) where designs from Hadamard matrices of orders 20 and 24 are catalogued.

m	5	6	7	8	9	10
M	10	42	71	94	130	127
M_0	0	8	20	14	5	2

Table 2. The numbers of nonequivalent orthogonal arrays of 20 runs.

Vijayan (1976) studied the problem of embedding orthogonal arrays into Hadamard matrices. He showed that a Hadamard submatrix of n rows and m columns can be embedded into a Hadamard matrix if $m \geq n - 4$. He further conjectured that embedding is achievable if $m \geq n/2$. A Hadamard submatrix of n rows and m columns means that its columns are orthogonal. Adjoining a column of all $+$'s to an orthogonal array of n runs and m factors gives rise to a Hadamard submatrix of n rows and $m + 1$ columns. Thus Vijayan's (1976) conjecture can be stated as an orthogonal array of n runs and m factors can be embedded into a Hadamard matrix if $m \geq n/2 - 1$. Table 2 gives five nonequivalent orthogonal arrays of 9 factors and two nonequivalent orthogonal arrays of 10 factors unattainable from Hadamard matrices of order 20. This disproves the conjecture of Vijayan (1976). In the appendix, the two nonequivalent orthogonal arrays of 10 factors are presented.

3.2   Orthogonal arrays of 24, 28, 32, and 36 runs

Results on orthogonal arrays of 24, 28, 32, and 36 runs are given in Table 3.

	$n = 24$	$n = 28$	$n = 32$	$n = 36$
$m = 5$	49	59	215	254
$m = 6$	423	1055	5368	7294

Table 3. The numbers of nonequivalent orthogonal arrays of 24, 28, 32, and 36 runs.

Compared our designs with those in Deng, Li, Tang (1999), we found that for $n = 24$ and $m = 6$, there are 15 designs that cannot be embedded into Hadamard matrices of order 24.

Finally, we want to mention that more detailed information on the orthogonal arrays obtained in this paper is available from the authors.

References

Deng, L.Y., Li, Y., and Tang, B. (1999). Catalogue of nonregular designs with small runs from Hadamard matrices based on generalized minimum aberration criterion. Submitted for publication.

Deng, L.Y., and Tang, B. (1999). Generalized resolution and minimum aberration criteria for Plackett-Burman and other nonregular factorial designs. *Statistica Sinica*, in press.

Deng, L.Y., and Tang, B. (1998). Design Selection and Classification for Hadamard Matrices Using Generalized Minimum Aberration Criterion. Submitted for publication.

Lin, D.K.J. and Draper, N.R. (1992). Projection properties of Plackett and Burman designs. *Technometrics* **34** 423-428.

Spence, E. (1995). Classification of Hadamard matrices of order 24 and 28. *Discrete Mathematics* **140** 185-243.

Sun, D. X. (1993). Ph.D. Thesis, University of Waterloo, Canada.

Seidan, E. (1986). On the existence, construction, and enumeration of some 2-symbol orthogonal arrays. In *Statistical Design: Theory and Practice*, Proceedings of a conference in honour of Walter T. Federer.

Tang, B. and Deng, L.Y. (1999). Minimum G_2-aberration for nonregular fractional factorial designs. *Annals of Statistics*, to appear.

Vijayan, K. (1976). Hadamard Matrices and Submatrices. *J. Austral. Math. Soc.* **22A** 469-475.

Wang, J.C. and Wu, C.F.J. (1995). A hidden projection property of Plackett-Burman and related designs. *Statistica Sinica* **5** 235-250.

Appendix

In this appendix, we present the two nonequivalent orthogonal arrays of size 20 and 10 factors (Table 2) that cannot be embedded into Hadamard matrices of order 20. In our presentation, each run is represented by an integer. The coding scheme is described as follows. For a run which consists of +'s and −'s, changing − to 0 and + to 1 gives rise to a binary sequence. This binary sequence can equivalently be represented by an integer in our number system of base ten. For example a run $(- - - + -)$ can be represented by integer 2 and a run $(- - + + ++)$ by 15. It is obvious that the integer codes of runs are between 0 and $2^{10} - 1 = 1023$, since we have 10 factors. The integer codes of runs are given in ascending order for convenience. The following are the two arrays.

Array 1 : consists of runs 3, 20, 41, 228, 250, 322, 383, 397, 439, 472, 607, 625, 662, 682, 717, 792, 814, 868, 945, and 963.

Array 2 : consists of runs 1, 18, 47, 232, 244, 328, 373, 415, 442, 455, 606, 635, 657, 685, 710, 780, 820, 867, 930, and 985.

INFRASTRUCTURE DEGRADATION: AN APPLICATION OF CENSORED REGRESSION MODELS

Hanga C. Galfalvy, Douglas G. Simpson, University of Illinois
Hanga C. Galfalvy, 725 S. Wright Street, Champaign, IL 61820 (galfalvy@stat.uiuc.edu)

Key Words: Tobit regression,variance function estimation, t distribution.

Abstract:

In civil engineering applications inspectors often use bounded condition scores to evaluate infrastructure units. When analyzing the relationship between these scores and the age and other attributes of the units, least squares methods are not appropriate, although they are widely used. This paper considers a generalization of the Tobit censored regression model to solve the problems of censoring and heteroscedasticity that characterize these condition scores. We apply the method to a cross-sectional study of factors influencing roof condition as a function of age.

1 Introduction

Several times in our consulting experience, we met the problem of modeling infrastructure degradation with age, expressed as a decrease in condition scores given by inspectors. These datasets are collected in cross-sectional or a mixture of cross-sectional and longitudinal observational studies, and present a common challenge: the condition scores are on a bounded scale, with a sizeable proportion of the observations exactly on the boundaries.

Least squares methods are not appropriate in this case, and have been shown (see Amemiya, 1984) to result in biased estimates. We propose instead to view the semicontinuous, bounded condition scores as the realization of an underlying, continuous latent variable. We will model this latent variable and use it to draw conclusions about the condition score, which can be viewed as the censored value of the latent variable.

The models we will consider have two parts: a regression model connecting the latent, unobservable, variable to the predictor(s), and a censoring mechanism that connects the latent variable to the observed response. While in survival models the censoring mechanism is often random and complicated, in our models the censoring is at fixed, pre-specified values.

We will apply the censored regression methods to a cross-sectional dataset on roof conditions, called ROOFER, which is the result of a large cross-sectional study conducted by the U.S. Army Construction Research Laboratories (USACERL). It contains inspection data on the condition of roofs measured on three separate scales, and several explanatory variables. Bailey et al.(1997) analyzed the data using multiple linear regression analysis and quantile regression. The least squares regression ignored the essentially bounded nature of the response: on all three scales, scores can take values only between 0 and 100, not on the whole real line. In the original dataset, most of the censoring occurs on the upper end (meaning that there are a lot of 100 scores), but there is some evidence of lower censoring as well.

2 Censored Regression Approach

To describe our model, we adopt the following notations:
z = uncensored (latent) random variable
L = lower censoring point
U = upper censoring point
y = censored response, $y = min(max(z, L), U)$.
Let us assume that the regression model for the uncensored variable is:

$$z_i = \mathbf{x_i}^T \boldsymbol{\beta} + \sigma u_i, i = 1, \ldots, n, \qquad (1)$$

where the u_i are independent and identically distributed with distribution function F, continuous density function f, and σ is the scale parameter. Then the likelihood function for the observed (censored) variable y can be written as

$$L(\boldsymbol{\beta}, \sigma) = \prod_{y_i = L} F\{\sigma^{-1}(L - \mathbf{x_i}^T \boldsymbol{\beta})\}$$

$$* \prod_{y_i = U} [1 - F\{\sigma^{-1}(U - \mathbf{x_i}^T \boldsymbol{\beta})\}]$$

$$* \prod_{L < y_i < U} \sigma^{-1} f\{\sigma^{-1}(y_i - \mathbf{x_i}^T \boldsymbol{\beta})\} \qquad (2)$$

For a continuous f, the MLEs can be found by solving the maximization problem for this likelihood, using an iteration algorithm like Newton-Raphson, or an EM algorithm.

Allowing L to take up the value $-\infty$ and U to be $+\infty$, the model can be applied to data censored at only one point (with a lower or an upper limit) or with no censoring.

In econometrics, one such censored regression model has been widely used. The Tobit model, developed by Tobin (1958), assumes that the error terms u_i are normally distributed. The standard model is for data that is censored at a lower censoring point (usually 0). There are several extensions to the standard Tobit model. A good review of the generalized Tobit models can be found in Maddala (1983) or in Amemiya (1984).

Using the normal distribution for the error terms is a common choice, although computations in the censored case are more complex than the least squares method. However, the Tobit MLE is not robust to violations of the assumptions for the latent variable model: normality and homoscedasticity. Goldberger (1980) showed that, in contrast to the uncensored case, the MLEs are asymptotically biased when the original data was non-normal or heteroscedastic; for example, if it had Student's t distribution. Amemiya (1984) lists several robust estimators for the Tobit model.

Degradation data from infrastructure studies are likely to present outliers, consequently a heavy-tailed distribution is more suitable than the normal distribution. Student's t distribution works well as a robust alternative to the normal for non-censored data (see Lange et al., 1989), so one alternative is substituting the t distribution for the normal. He et al. (1999) established breakdown points for t regression. Since the t distribution can be viewed as a normal with random scale parameter, this gives flexibility to the model. The resulting conditional percentile estimates will be more robust.

This Tobit-like model, which we call t-tobit, is more general. The normal distribution is a limit-case of the family of t distributions indexed by the degrees of freedom parameter. T-tobit models, then, include the t-regression, least squares regression and the Tobit model as special cases. This enables us to compare these models to the general t-tobit.

Amemiya (1984) proved the consistency and asympotic normality of the Tobit MLEs. The proof can be extended to the case of t distribution with fixed number of degrees of freedom, but it is quite lengthy, so it is omitted. This means that the t-tobit regression gives consistent estimates if the error terms in the latent variable model have, in fact, a t distribution, while in this situation the Tobit estimates are asymptotically biased.

3   Estimating the scale function

What if the data shows evidence of heteroscedasticity? Even in the case of least squares regression, we cannot ignore it. One approach is to try to transform the response or the predictors. Another is to try to include a model for the variance function in the regression model. Variance function models have been discussed in Davidian and Carroll (1987). In this section, we will extend their methods to the heteroscedastic t-tobit model.

As previously mentioned, the tobit regression model is not robust to heteroscedasticity: the resulting estimates will be asymptotically biased. Thus it is necessary to use a method that results in consistent estimates.

Consider the heteroscedastic t regression model for the latent (uncensored) variable z:

$$z_i = \mathbf{x}_i'\boldsymbol{\beta} + \sigma\psi(\boldsymbol{\theta}, \mathbf{x}_i)e_i \qquad (3)$$

where e_i are iid and have t distribution with ν degrees of freedom; and ψ is a function of the predictors with parameter $\boldsymbol{\theta}$ to be estimated from the data. The response z cannot be observed, instead the data includes y, which is censored at a lower value L and upper value U.

The variance determining or scale function ψ (estimating the standard deviation instead of the variance) can be modeled as a parametric function of the predictors, whose form we have to decide on before the analysis. Some possibilities include:

$$\psi(\boldsymbol{\theta}, \mathbf{x}_i) = 1 + \theta_1(\mathbf{x}_i'\boldsymbol{\beta})^{\theta_2} \qquad (4)$$

which is linear in a power of the mean, or, if we assume that the variance only depends on one of the predictors (the kth), the quadratic scale function will look like

$$\psi(\boldsymbol{\theta}, x_i^k) = 1 + \theta_1 x_i^k + \theta_2(x_i^k)^2. \qquad (5)$$

Similarly, we can use cubic, exponential etc. functions, all including parameters that have to be estimated from the data. Davidian et al.'s method consists of three steps.

First, we run a homoscedastic regression, and compute the residuals $r_i, i = 1, \ldots, n$. In the second step, the absolute value of the residual is regressed on the design matrix corresponding to the scale function, with weights that are inversely proportional to the current variance function estimates. The goal is to estimate the parameter θ and compute the estimated scale function. Finally, divide the variables by the regression equation with this estimated function, run the (now approximately homoscedastic) regression again, and compute updated estimates of β and σ.

When the response is censored, the least squares regression in the first step becomes an ordinary t-tobit regression. We compute the censored residuals from the latent variable regression:

$$r_i = y_i - \mathbf{x}_i'\hat{\beta}^{(1)} \qquad (6)$$

where the y values are censored from below at L and above at U, and the second term is the predicted value for the uncensored response z. If we view these residuals as estimators for the values of the errors e_i from the latent variable models, they are censored from above/below, whenever the corresponding y_i is censored from above/below. This means that the residuals are not censored at the same values, they have individual L_i and U_i censoring points.

The second step would take the absolute residuals and regress them on the basis for the scale function. In the first iteration the current scale function is a constant, so we do not have to use weights at all. If we start with an uncensored response and normal errors for the latent variable model, the least squares regression here makes sense, even though the distribution of the absolute residuals is not normal. In the censored case, the absolute residuals will be censored at individual censoring points; some from above, some from below, so using a LS regression model is not appropriate. The correct alternative is to use the generalized t-tobit model that allows different censoring points for different observations. In our experience, final results are similar whether LS or t-tobit was used in the second step.

After computing $\hat{\theta}$ from the second step, the final t-tobit regression will be based on the homoscedastic model:

$$\frac{z_i}{\psi(\hat{\boldsymbol{\theta}}, \mathbf{x}_i)} = [\psi(\hat{\boldsymbol{\theta}}, \mathbf{x}_i)]^{-1}\mathbf{x}_i\beta + \sigma e_i \qquad (7)$$

The asymptotic properties of $\hat{\beta}$ and $\hat{\sigma}$ will depend on the properties of $\hat{\beta}^{(1)}$ and that of $\hat{\theta}$, but iterating to convergence by using the updated variance function as weights in the second step will remove this dependence. See Davidian et al. for the details.

4 Applications

To illustrate our ideas, we selected a subset of the large ROOFER dataset, focusing on the condition, as measured by the membrane condition index (MCI) score, of roofs with age less than 20 years, that have no embedded edge metal and have interior drains. As explanatory variables, we selected the membrane type of the roof and age. The membrane type is a binary variable, taking up the value 1 if the roof membrane type is EPDM, and 0 otherwise. A summary of the dataset can be found in Table 1. About 30% of the data is 'censored from above'.

Variable	Min.	Mean	Max.
MCI	30	89.52	100
Age	0.5	9.37	19.01
Membrane	0	0.23	1
N	270		

Table 1: Summaries for ROOFER

The regression model for the "Membrane Condition Index" will have the age of the roof and the membrane type as predictor. The condition score of a new roof does not depend on the membrane type, and since differences between the membrane types show up only later, it is logical to use the interaction (product) of age and membrane type rather than the membrane type alone. So the model for the latent variable is

$$MCI_i = \beta_0 + \beta_1 * AGE_i +$$
$$+\beta_2 * AGE_i * MEMB_i + u_i, \text{ for i=1, \ldots, 270} \qquad (8)$$

Our goal is to study the effects of the failure to account for censoring on the conclusions of the analysis. We assume a t distribution for the errors of the linear regression model for the true condition score, with unspecified degrees of freedom.

The results of the homoscedastic t-tobit and LS regressions are presented in Table 2. The "best" degree of freedom estimate is 2, which argues that the tobit model is not a good fit, we need the heavy-tailed t distribution to fit this data.

For the normal regression model, which is an analysis of covariance, the effect of the membrane type is clearly significant on the $\alpha = .05$ level. However, for the censored t regression, the approximate p-value for this coefficient is about .16, and that is not significant. The Tobit model assumes an infinitely large value for degrees of freedom instead of the "best" estimate 2, overestimates the magnitudes of the the intercept and the slope. The differences in the scale parameter estimates are hard to interpret, because in the case of the t distribution it is related to the degrees of freedom.

Effects	LS method Estimate	P-value	Tobit Estimate	P-value	T-tobit Estimate	P-value
Intercept	96.09	0.000	105.07	0.000	103.60	0.0000
AGE	-0.81	0.000	-1.33	0.000	-1.06	0.0000
AGE*MEMB.	0.60	0.013	0.65	0.057	0.36	0.158
Scale	13.08	-	17.35	-	11.9	-

Table 2. Homoscedastic MLEs

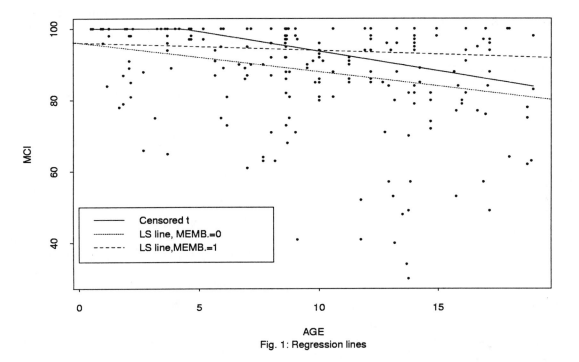

Fig. 1: Regression lines

Provided that our assumptions are met, we detected a difference in the conclusions one draws from the least squares and the t-tobit analyses. Based on the normal model, we would conclude that membrane type is significant, when in fact it is not.

Figure 1 plots the two different regression lines for the LS regression - one for each material type - together with the censored t-tobit regression line with only age as a predictor. Looking at the plot, we would conclude that the LS and t-tobit models are not very different, and they both do a good job summarizing the data. Note that the LS lines suggest a very gradual decline in the response, whereas the t-tobit line accurately portrays the the median quality (perfect score) of almost-new roof sections.

Looking at the figure, however, we also notice that the response variable is clearly heteroscedastic, the

Effects	Estimate	P-value	Var.fct.
Intercept	102.97	0.0000	3.21
AGE	-1.02	0.0000	0.59
AGE*MEMB.	0.33	0.128	-

Table 3. Heteroscedastic t-tobit MLEs

variance of MCI changes with age. We decide to look at the residuals from the first regression and use the techniques from the last section to estimate the variance function.

Table 3 displays the results of the heteroscedastic t-tobit regression. The variance function estimate was assumed to be a polynomial in the age variable, and a linear fit was found to be adequate; the coefficients are in the last column of the table. Comparing the two sets of t-tobit coefficients shows that they differ very little; in this case, the bias caused by heteroscedasticiy in the t-tobit model is negligible. The membrane type is still not significant.

In conclusion, although the difference between the normal and t-tobit models is statistically significant, the normal model provides a qualitatively similar conclusion to the t-tobit regression. And while the data is heteroscedastic, in this case the homoscedastic coefficient estimates were very close to the ones resulting from a heteroscedastic model with variance function estimation.

5   Remarks on computation

The likelihoods for the tobit and t-tobit analysis are highly nonlinear and require an iterative algorithm to solve them. In the appendix, we provide the likelihood and the likelihood equations for the t-tobit model, and one way of computing the MLEs is to use a function minimizer or an equation solver algorithm.

The major statistical packages include a function for parametric survival analysis, which is censored regression with random censoring. They give a choice of possible distributions for the error term. Also, they allow the user to choose the type of censoring - lower, upper or interval censoring. This means that these functions are fitting the same censored regression models discussed in this paper.

The SAS manual has an example on doing a Tobit regression using proc lifereg. Unfortunately, the choice of the distributions is rather limited for our purposes, because they include only the distributions commonly used in survival regression - the exponential, the extreme value distribution, the Rayleigh and the normal distribution. Thus, the Tobit model can be run, but the t-tobit model can not.

In S-plus, the function survreg, run with the normal distributions, will perform the Tobit regression. However, the t distribution is not among the possible choices. With some changes in the source code and the addition of a C function for computing tail probabilities of the Student's t distribution, t-tobit regression can be run with fixed degrees of freedom. The degree of freedom corresponding to the largest value of the maximum likelihood will be the "MLE" of the degrees of freedom. Estimating the degrees of freedom together with the coefficients is not a good method: the likelihood is no longer concave, and degrees of freedom smaller than 1 present robustness problems. In fact, since the variance for the t distribution with 2 degrees of freedom is not defined, we often want to restrict the degrees of freedom to

values higher than 2, and, for simplicity of presentation, use only integer values.

6   Conclusions

In this paper we have shown that in cross-sectional studies for evaluating infrastructure with a bounded condition score, censored regression methods make more sense than the classical methods based on the normal distribution. We propose a new model, the t-tobit, which uses the t distribution for the latent (unobserved) response variable. This is a generalization of the Tobit model used in econometrics. The predicted value for the censored response from a linear model will be a broken line, quantile-based prediction intervals can be easily computed. We also proposed a method for estimating the variance function when the model is heteroscedastic.

An alternative to the parametric censored regression models is censored quantile regression. Based on the work of Powell, it estimates quantiles of the response rather than expected value. An important advantage is that they are distribution-free.

An area for future research would be to study models for correlated data with censored response. In the case of longitudinal studies, objects are re-measured at different time points, and the challenge is to identify the common aging trend of infrastructure objects. Another important direction is spatial correlations, which often exist among objects included in an infrastructure evaluation study.

Acknowledgements: The authors thank David Bailey, USA-CERL, for making the data available and for providing background information about the data, and Xuming He for his suggestions as the peer-reviewer of this paper. Research supported in part by NSF Contracts DMS-95-05290 and by the University of Illinois Research Board.

7   References

Amemiya, T.(1984). Tobit models: a survey. *Journal of Econometrics* **24**, 3-61.

Bailey, D. M., Simpson, D. G., He, X., Geling, O., Lau,S. and Trachtenberg, F. (1997). Statistical Analysis of ROOFER database from 21 army installations. USACERL technical report 97/83.

Davidian, M. and Carroll, R. J.(1987). Variance function estimation. *Journal of the American Statistical Association*, **82**, 1079-1091.

Goldberger, A.S., (1980). Abnormal selection bias,

SSRI workshop series no. 8006 (University of Wisconsin, Madison, MA).

He, X., Simpson, D. G. and Wang, G. (1999). Breakdown points of t-type regression estimators. In revision for *Biometrika*

Lange, K. L., Little, R. J. A. and Taylor, M. G. (1989). Robust Statistical Modeling Using the t Distribution. *Journal of the American Statistical Association*, **84**, No. 408, 881-896.

Maddala, G. S. (1983). Limited-dependent and qualitative variables in econometrics. Cambridge University Press.

Powell, J. L. (1986). Censored regression quantiles. *Journal of Econometrics*, **32**, 143-155.

Tobin, J. (1958) Estimation of relationships for limited dependent variables. *Econometrica*, **26**, 24-36.

Empirical Dynamic Modeling of Process Systems with Output Multiplicities

Jeffrey DeCicco and Ali Cinar*
Chemical and Environmental Engineering Department
Illinois Institute of Technology
Chicago, IL 60616

Key Words: Nonlinear, Multiplicities, Additive

Abstract

Nonlinear multivariable time series modeling of process systems with exogenous manipulated input variables is presented. The model structure is similar to that of a Generalized Additive Model (GAM) and is estimated with a nonlinear Canonical Variate Analysis (CVA) algorithm called CANALS. The system is modeled by partitioning the data into two groups of variables. The first is a collection of future outputs, the second is a collection of past input and outputs, and future inputs. This approach is similar to linear subspace state space modeling. An illustrative example of modeling is presented based on a simulated continuous chemical reactor that exhibits multiple steady states in the outputs for a fixed level of the input.

1 Introduction

In most process systems such as continuous chemical reactors, linear models such as ARX, ARMAX or state space models are at best an approximation that only perform well over a small region of operation. Most chemical reaction systems exhibit steady state behavior in the outputs (zero frequency response) that is not a linear function of the inputs.

A typical example of this situation is the relatively simple physical model of a continuous exothermic reaction carried out in a continuous stirred tank reactor (CSTR) where reactant A is converted to product B ($A \rightarrow B$) [Uppal et al., 1974] . For specific operating conditions the outputs, (dimensionless reactor temperature and conversion) exhibit output multiplicities for a fixed level

of the input, (dimensionless cooling water temperature) (Figure X and Y). The dynamic behavior of the process exhibits what is referred to as extinction as outputs shift from the upper steady state to lower as the input level is decreased, or ignition when the outputs jump from the lower to upper steady state when the input level is increased. Figure 3 shows the dynamic nature of the data.

The empirical dynamic model is composed of a dynamic representation of a set of latent variables, and a function linking these latent variables to the (original) observed physical outputs. In the case of linear state space modeling, a collection of future outputs is regressed, by means of reduced rank regression on a collection of past inputs and outputs, and future inputs. This approach gives an estimate of the approximate state variables [Van Overschee and De Moor, 1994]. In the case of nonlinear model development nonlinear transformations of the regressors are found by means of CANALS algorithm [van der Burg and de Leeuw, 1983]. This leads to a model structure for a set of latent variables that are then linked to the outputs with a linear model.

The next section describes the basic model structure, and the steady analysis of that model structure. Section 3 describes the model estimation. Section 4 gives a detailed study of modeling simulated data from the chemical reactor.

1.1 Previous Work

GAM modeling and nonlinear CVA estimation of dynamic processes has been proposed before. The ACE algorithm [Breiman and Friedman, 1985] has been used fit nonlinear dynamic models [Chen and Tsay, 1993]. They only considered single output models and inferred a parametric structure from the nonparametric graph of the function. This work considers multiple outputs and directly uses the nonparametric functional estimate in the final

---

*Author to whom correspondence should be addressed (*cinar@charlie.iit.edu*)

model. The general theory for using nonlinear CVA to fit dynamic model is outlined [Larimore, 1990a]. The examples and application presented were limited to cases without exogenous inputs.

2 Model Structure and Analysis

The model structure consists of two sub-models. The first is a dynamic model describing a set of latent variables. The second relates the latent variables to the outputs.

Let $\mathbf{y}_t \in \mathcal{R}^l$, $\mathbf{u}_t \in \mathcal{R}^m$ and $\mathbf{x}_t \in \mathcal{R}^n$ represent the outputs, inputs and latent variables, respectively. The outputs, inputs and latent variables $\mathbf{y}_t$, $\mathbf{u}_t$ and $\mathbf{x}_t$ are collections of individual variables y_t^i, u_t^i and x_t^i,

$$\mathbf{y}_t = \begin{bmatrix} y_t^1 \\ \vdots \\ y_t^l \end{bmatrix}, \ \mathbf{u}_t = \begin{bmatrix} u_t^1 \\ \vdots \\ u_t^m \end{bmatrix}, \ \mathbf{x}_t = \begin{bmatrix} x_t^1 \\ \vdots \\ x_t^n \end{bmatrix}. \quad (1)$$

The model structure for a single latent variable is

$$x_{t+\beta}^i = \sum_{j=1}^{\beta} \sum_{k=1}^{m} h_{i,j+k-1}^* \theta_{k,j}^p \left(u_{t+\beta-j}^k \right)$$

$$+ \sum_{j=1}^{\beta} \sum_{k=1}^{l} h_{i,\beta m+j+k-1}^* \phi_{k,j}^p \left(y_{t+\beta-j}^k \right), \quad (2)$$

where h^* are scalar coefficients, θ^p and ϕ^p are nonlinear functions, and β is the past window length that determines the model structure. The model structure linking the latent variables to the outputs is

$$\mathbf{y}_t = \mathbf{y}_{ss} + \mathbf{C}\mathbf{x}_t, \quad (3)$$

where $\mathbf{C}$ is an $l \times n$ coefficient matrix, and y_{ss}^i is some centering point for the observed data. The centering point might be the mean of the observed data series or some known steady state operating condition.

For sake of simplicity consider a single-input single-output (SISO) model. A linear model could have the structure

$$y_t = a_1 y_{t-1} + a_2 y_{t-2} + a_3 y_{t-3} + b u_{t-1}. \quad (4)$$

Let y_{ss} represent the steady state solution of (4) corresponding to u_{ss} or

$$y_{ss} = \frac{b u_{ss}}{1 - a_1 y_{ss} - a_2 y_{ss} - a_3 y_{ss}}.$$

The steady state output is clearly a linear function of the input. Now consider the nonlinear additive model

$$y_t = a_1 \theta_1 (y_{t-1}) + a_2 \theta_2 (y_{t-2}) + a_3 \theta_3 (y_{t-3}) + b\phi(u_{t-1}).$$

The steady solution y_{ss} then satisfies

$$y_{ss} - a_1 \theta_1 (y_{ss}) - a_2 \theta_2 (y_{ss})$$
$$- a_3 \theta_3 (y_{ss}) - b\phi(u_{ss}) = 0. \quad (5)$$

The solution of (5) can be found through numerical continuation techniques by using the software package **AUTO** [Doedel et al., 1998]. For this simple example it possible to model a SISO system with output multiplicities with the above model structure. The chemical reactor example presented is a two output single input system. The steady state solution follows in an analogous manner as the SISO case.

3 Model Estimation

In the case of linear state space models it is possible to recover an estimate of the approximate Kalman filter states from a reduced rank regression of a collection of future outputs on past collection of inputs and outputs, and future inputs [Van Overschee and De Moor, 1994]. The terms "past" and "future" refer to a discrimination of previously observed historical data that is used in the estimation of a causal dynamic model.

Consider the collection of past and future observed outputs:

$$\mathbf{y}_\beta(t) = [\mathbf{y}_t^T \ \mathbf{y}_{t+1}^T \ \cdots \ \mathbf{y}_{t+\beta-1}^T]^T, \quad (6)$$
$$\mathbf{y}_\gamma(t) = [\mathbf{y}_{t+\beta} \ \mathbf{y}_{t+\beta+1}^T \ \cdots \ \mathbf{y}_{t+\alpha-1}^T]^T,$$

where $\alpha = \beta + \gamma$, and γ is the future window length. The parameters γ and β determine model structure. The collections of past and future inputs are defined in the same manner as the outputs as $\mathbf{u}_\beta$, and $\mathbf{u}_\gamma$ respectively. Let $t+\beta-1$ represent the present time. The observations of the above collections are

$$\mathbf{Y}_\beta = [\mathbf{y}_\beta(1) \ \ldots \ \mathbf{y}_\beta(N)]$$
$$\mathbf{Y}_\gamma = [\mathbf{y}_\gamma(1) \ \ldots \ \mathbf{y}_\gamma(N)]$$

The collections of observed past and future inputs are defined in the same manner as the outputs as $\mathbf{U}_\beta$ and $\mathbf{U}_\gamma$ respectively. The reduced rank regression that leads to the approximate states variables is

$$\min_{\mathbf{L_1}, \mathbf{L_2}, \mathbf{L_3}} \| \mathbf{Y}_\gamma - [\mathbf{L_1}\mathbf{U}_\beta + \mathbf{L_2}\mathbf{Y}_\beta + \mathbf{L_3}\mathbf{U}_\gamma] \|^2 \quad (7)$$

such that $[\mathbf{L}_1\ \mathbf{L}_2] = \mathbf{\Gamma H}$ and $\mathbf{\Gamma}$ and $\mathbf{H}$ both have rank n, where n are the number of states (latent variables). The approximate state variables sequence follows as

$$\mathbf{X}(1) = \mathbf{H} \left[\begin{array}{c} \mathbf{U}_\beta \\ \mathbf{Y}_\beta \end{array} \right], \qquad (8)$$

with

$$\mathbf{X}(t) = [\hat{\mathbf{x}}(t + \beta) \ldots \hat{\mathbf{x}}(t + \beta + N - 1)].$$

With the estimated state variables sequence (8) and the observed inputs and outputs it is possible to estimate the model coefficients of the linear state space model including the Kalman filter gain.

The nonlinear model will extend the reduced rank regression (7) to allow for nonlinear transformations of the regressors namely the lagged inputs and outputs. This is done through a modified version of the CANALS algorithm. The algorithm implemented allows for nonparametric regression in the form of the supersmoother [Friedman, 1984]. The loss function becomes

$$\min_{\tilde{\mathbf{L}}, \mathbf{L}_4, \mathbf{G}, \mathbf{Z}_\gamma} \| \mathbf{L}_4 \mathbf{Z}_\gamma - \tilde{\mathbf{L}} \mathbf{G} \|,^2 \qquad (9)$$

where $\mathbf{Z}_\gamma$ and $\mathbf{G}$ are nonlinear transformations of $\mathbf{Y}_\gamma$ and $\left[\mathbf{U}_\beta^T\ \mathbf{Y}_\beta^T\ \mathbf{U}_\gamma^T \right]^t$. CANALS finds the nonlinear transformation that make up the latent variable model and CVA reduced rank regression [Larimore, 1990b] is used to determine the number of latent variables and model coefficients of (2). Least squares is used to estimate the link function (3).

4 Modeling of CSTR

To illustrate the modeling procedure, data were simulated from a physical model of an exothermic CSTR [Uppal et al., 1974]. The system consists of two outputs, (conversion, y_1, and dimensionless reactor temperature, y_2), and a single input (dimensionless cooling water temperature The input was manipulated around an operating region of output multiplicities.

First the steady state performance of different model structures was compared. Different model structures consisted of changing β and setting $\beta = \gamma$. For each case the maximum possible number of latent variables was chosen (In this example the maximum number of latent variables is 2γ).

The input-output data used to develop the model structures are shown in Figure 2. The actual and predicted steady state diagrams for the different model structures are shown in Figure 3 for the conversion and Figure 4 for the temperature. The dashed box represents the domain and range of the observed input and outputs. Outside the dashed box the model extrapolates the systems behavior by assuming a linear model structure. The sum of squared error (SSE) of prediction for different structures for one-step-ahead prediction and pure simulation is calculated. Pure simulation is the case where only known inputs are used in predicting future values of the outputs. To do this predicted outputs are recursively used in the model to predict future outputs. One-step-ahead prediction utilizes observed outputs in the model to predict the next step into the future.

A visual inspection shows that the steady state performance of the mode improves inside the dashed box as $\beta = \gamma$ increase. The one-step-ahead and pure simulation prediction error decreases as β and γ increase. This suggests a model order selection procedure where $\beta = \gamma$ is increased to minimize some criteria such as AIC.

Figure 2 shows the pure simulation prediction from the black-box model with $\gamma = \beta = 3$. The one-step-ahead predictions are indistinguishable from the data. A few erroneous transitions are unavoidable in the case of black-box modeling because it will never be possible to exactly reproduce the steady state behavior of the process. Without exact steady state performance conditions will always exists where the model will have erroneously transitions.

To illustrate the variability of the estimated steady state diagrams five models were developed from five separate data sets. The structure of the model was kept constant with $\gamma = \beta = 3$. Figure 1 shows the actual steady state curve and the five estimated steady state and actual diagrams. As expected, there exists variability but each model captures the general trajectory of the actual process.

5 Summary and Conclusions

A nonlinear dynamic modeling approach was presented that can accurately reproduce nonlinear steady state behavior. An example was presented that illustrates the approach on a CSTR model that exhibits output multiplicities. The primary focus of this paper was the illustration of steady state performance however, it is important to note the utility of the dynamic model. The dynamic model performance is especially important in the application of automatic feedback controllers. This model works well in multi-step ahead prediction as a recursive filter.

Acknowledgments

Financial support provided to Jeffrey DeCicco from the National Center for Food Safety and Technology and USDA is gratefully acknowledged.

References

L. Breiman and J. H. Friedman. Estimating Optimal Transformations for Multiple Regression and Correlation. *J. Amer. Statist. Assoc.*, 80:580–598, 1985.

R. Chen and R. S. Tsay. Nonlinear Additive ARX Models. *J. Amer. Statist. Assoc.*, 88(423):955–967, 1993.

E. J. Doedel, A. R. Champneys, T. F. Fairgrieve, Y. A. Kuznetsov, B. Sandstede, and X. Wang. AUTO 97: Continuation and Bifurcation Software for Ordinary Differential Equations (with HomCont), 1998.

J. H. Friedman. A Variable Span Smoother. Technical Report No. 5, Dept. of Statistics, Stanford University, 1984.

W. Larimore. Identification and filtering of nonlinear systems using canonical variate analysis. In *Nonlinear Modeling and Forecasting,: Proceedings of the Workshop on Nonlinear Modeling and Forecasting Help September, 1990 in Santa Fe, New Mexico Vol 12*. Addison-Wesley, 1990a.

W. Larimore. Order-Recursive Factorization of the Pseudoinverse of a Covariance Matrix. *IEEE Trans. Automat. Contr.*, 35:1299–1303, 1990b.

A. Uppal, W. H. Ray, and A. B. Poore. On the Dynamic Behavior of Continuous Stirred Tank Reactors. *Chem. Eng. Sci.*, 29:1–2, 1974.

E. van der Burg and J. de Leeuw. Nonlinear Canonical Correlation. *British J. Math. Statist. Psychol.*, 36:54–80, 1983.

P. Van Overschee and B. De Moor. N4SID: Subspace Algorithms for the Identification of Combined Deterministic-Stochastic Systems. *Automatica*, 30(1):75–93, 1994.

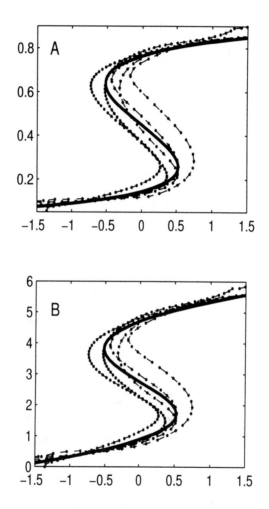

Figure 1: Actual (solid line -) and predicted steady state diagrams (dash-dotted -.) for conversion (A) and dimensionless reactor temperature (B). Five models were developed from five separate data sets to illustrate the variability in the estimated steady state diagrams.

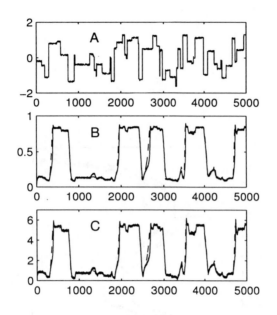

Figure 2: Dimensionless cooling water temperature (A), conversion (B), and dimensionless reactor temperature (C). Actual outputs (solid line -) and predicted outputs (dashed - -).

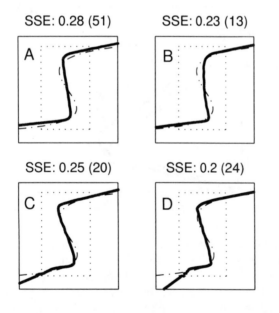

Figure 3: Actual and predicted steady state conversion curves for different empirical model structures. The sum of squared error (SSE) for one-step-ahead and (pure simulation) is calculated. The dotted box (:) represent the domain and range of the inputs and outputs. Outside the box observed data is not available. Four model structures with $\gamma = \beta = 1$ (A), $\gamma = \beta = 2$ (B), $\gamma = \beta = 3$ (C), and $\gamma = \beta = 4$ (D).

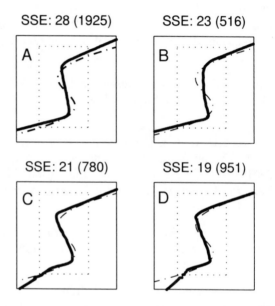

Figure 4: Actual and predicted steady state dimensionless reactor temperature curves for different empirical model structures. The sum of squared error (SSE) for one-step-ahead and (pure simulation) is calculated. The dotted box (:) represent the domain and range of the inputs and outputs. Outside the box observed data is not available. Four model structures with $\gamma = \beta = 1$ (A), $\gamma = \beta = 2$ (B), $\gamma = \beta = 3$ (C), and $\gamma = \beta = 4$ (D).

BIAS REDUCTION IN TOMOGRAPHIC GAMMA SCANNING: EXPERIMENTAL AND SIMULATION RESULTS

Tom L. Burr, Thomas H. Prettyman, Los Alamos National Laboratory
Tom Burr, MS E541, Los Alamos National Laboratory, Los Alamos NM, 87545

Key Words: tomographic gamma scanning, bias reduction

1. Abstract

Tomographic Gamma Scanning (TGS) is one of the first nondestructive assay methods for nuclear material that is designed to correct for sample-specific attenuation effects. Performance results have confirmed that the transmission-corrected emission spectra can largely correct for attenuation effects, resulting in good overall assay performance.

However, recent efforts to minimize the bias in simulated data led to the discovery that the effective number of unknown parameters (from the unknown background counts at each emission image) must sometimes be reduced. Otherwise, the maximum likelihood solution to the transmission corrected emission data leads to detectable assay bias in some cases.

In this paper we present performance results for both experimental and simulated data for a modified likelihood-based approach to reducing the effective number of unknown parameters. We also discuss the connection between related penalized likelihood approaches and Bayesian or empirical Bayesian approaches.

2. Introduction

Tomographic Gamma Scanning (TGS) is a γ-ray nondestructive assay (NDA) method to assay special nuclear material (SNM) in heterogeneous samples, particularly residues and waste. The principle of the method is that the rate of γ-ray emission is roughly proportional to the total SNM mass T. However, sample-specific attention of the γ-rays complicates the relation between the γ-ray emission rate and T. Furthermore, because the samples could be heterogeneous, both the γ-ray attenuation and source rate vary within the volume of the sample. Therefore, TGS uses tomography to form 3-dimensional images of the γ-ray attenuation. In effect, the attenuation coefficient is estimated in each of many small volume elements (voxels) of the sample. An isotopic transmission source that emits more than one γ-ray (usually Se^{75}) is used to obtain attenuation images as a function of energy. The emission images are then corrected for the attenuation of γ-rays by using the linear attenuation coefficient images. The amount of radioactivity, or the mass, in any region of interest in the sample can then be estimated by integrating the transmission-corrected emission image over the volume of the region. In this paper, the region of interest is the entire sample. The goal is to study the performance of candidate analysis methods in estimating total mass T. See [1] and [2] for more details and caveats about where TGS is applicable.

3. TGS Image Reconstruction

The volume of a 55-gal drum is typically divided into $N = 1600$ 3-dimensional voxels. In a standard scan protocol, data is collected at 150 individual points in polar coordinate (displacement-angle) space for each of 16 vertical layers, giving a total of $M = 2400$ measurement positions (bins) [2]. During an initial scan, transmission measurements are made using an external Se^{75} source to characterize the γ-ray attenuation of the drum. This allows reconstruction (estimation) of the so-called system matrix $A^{M \times N}$. Because of interactions that affect γ-ray energies, the γ count rate at a given energy channel includes the effects of γ-rays that originated with higher energy but appeared at the given energy channel. The simplest way to account for this underlying background is to measure the (background) γ-rays in energy channels near the channel(s) of interest. The net γ count rate $n = g - c_1 * b$ where g is the observed gross counts in the energy region of interest (ROI) energy channels, b is the observed background counts near the ROI, and c_1 is the ratio of the number of peak ROI channels to the number of background ROI channels. Also, the detection rate of γ-rays must be corrected to a full-energy interaction (FEI) rate that accounts for losses due to deadtime and pulse pileup (detector response issues). The FEI can be estimated by using a Cd^{109} source that emits an 88 keV γ-ray [2], and defined by $FEI = CF(RL) \times n$, where $CF(RL)$

is the estimated correction factor for rate loss. Following [1] we will include $CF(RL)$ in the definition of the $A^{M \times N}$ matrix, which means that when we consider estimation errors in $A^{M \times N}$ we must include estimation errors in $CF(RL)$. With i indexing bins and j indexing voxels, the image reconstruction problem can then be cast most simply as

$$g_i \sim \text{Poisson}(\sum_j A_{ij} x_j + c_1 \mu_{b,i}), \qquad (1)$$

where x is a vector of nonnegative values that describes the distribution of γ emitting material within the drum, and $\mu_{b,i}$ is the true background count rate from bin i. The observed background counts b_i are well modeled by a $\text{Poisson}(\mu_{b,i})$ distribution. For our purposes here we can assume that $T = \sum_{i=1}^{N} x_i$ is the mass of sample. Therefore, our problem is to estimate total mass $T = \sum_{i=1}^{N} x_i$.

4. Statistical Issues

We assume that $M > N$ in Eq. (1) so ordinary least squares (OLS), or weighted least squares (WLS) are options for estimating each x_i and therefore also T. In fact, Eq. (1) is essentially the same as what commonly appears in a typical 2-stage calibration experiment. Stage 1 is the equivalent of our "estimate A stage" and stage 2, which concerns us here, uses the estimated A to estimate $T = \sum_{i=1}^{N} x_i$. The unique features of our application of Eq. (1) are:

1) The dimension of A is very large (M-by-N is approximately 2400-by-1600) and A is often ill-conditioned.

2) There can be significant spatial correlation among neighboring x_i. This is the main reason for the introduction of Bayesian methods in image analysis. Further, it is the main reason for the recent popularity of Markov Chain Monte Carlo (MCMC) methods [3]. MCMC methods are useful for interpreting the nonstandard, large dimensional (N dimensional in this case) posterior probability distributions that arise from using nonstandard prior probability distributions for x^N.

3) The error structure of the *net* response is nonstandard, being $n \sim \text{Poisson}(\mu_g) - c_1 \times \text{Poisson}(\mu_b)$. When g is small, it might be important to use the Poisson distribution rather than an approximating Gaussian.

4) There can sometimes be non-negligible errors in the A matrix. One TGS system [2] deliberately collapses some bins to reduce variance at the expense of slightly increased bias. The usual bias-variance tradeoff suggests that this is a good idea. To date

we have ignored the possibility of bias in the A matrix. That is, errors in A are all modeled as random errors [4] (with standard deviation denoted σ_A here) so future work must include both (a) treatment of possible bias in some entries of the A matrix, and (b) a plan for dealing with dynamically changing dimension of the A matrix due to bin collapsing.

5) The number of unknowns is M background means and N voxel emission rates. The number of observations is $2M$ (M (g_i, b_i) pairs), so there are possibly too few observations per unknown. Therefore, asymptotic results for maximum likelihood (ML) estimates do not necessarily apply. In particular, there is no guarantee that ML is asymptotically (as $M \to \infty$) unbiased. However, we have empirically observed that ML appears to be essentially bias-free provided μ_{net} is reasonably large. Bias appears to be more prevalent for small μ_{net} values.

This paper focuses on issue 5 for small μ_{net} values ($\mu_{net} \approx 1$). Other publications [4-6] have considered issues 1-4. Our main performance measure is $PM_1 = E\{\hat{T} - T\}^2$, where E is the expected value with respect to the distribution of $\hat{T}$. A more typical performance measure in multivariate calibration is $E\{\sum_{i=1}^{N} (\hat{x}_i - x_i)^2\}$. Note that $PM_1 = E\{[\sum_{i=1}^{N} (\hat{x}_i - x_i)]^2\}$, so that covariances among the $\hat{x}_i$ can potentially degrade or improve performance, depending on their sign. Very few image analyses are concerned with this particular global performance measure (image analyses tend to try to "sharpen features" or detect edges). It is well known that there is guaranteed to be a biased solution vector $\hat{x}$ that has lower (better) PM_2 than does the OLS solution vector $\hat{x}$ [7]. We expect there is a similar result for PM_1 but we are unaware of it. Our second performance measure is the absolute bias, $PM_2 = |E\{\hat{T} - T\}|$.

We modify Eq. (1) in [1] to work with g so that the joint probability of g given x, A, μ_b, and c_1 is

$$p(g|x) = \prod_i \frac{\exp(-\mu_{g,i}) \mu_{g,i}^{g_i}}{g_i!}, \qquad (2)$$

where $\mu_{g,i} = \sum_j A_{ij} x_j + c_1 \mu_{b,i}$ is the mean of the gross counts at bin i due to all voxels. We note here that reference [8] presented a way to view the transition from maximizing the likelihood in Eq. (2) to maximizing a suitable posterior probability for x (in a Bayesian approach) that involved a temporary assumption that we could see the contribution at bin i from each individual voxel j. Note however that the mean for g_i at bin i is generally affected by more than one voxel j.

5. Modified Likelihood Method

This paper focuses on one potential bias source (the "too many parameters" effect). We note that this effect does **not** apply to WLS. However, we believe [4-6] that ML-based methods generally outperform WLS and other "matrix-inversion" based methods (including "biased-regression" methods), largely in cases where A is ill-conditioned. We also note that another potential bias source is errors in A. Again, ML-based methods appear to be relatively insensitive to errors in A and there is no guarantee that "errors-in-variables" methods [9] can reduce the bias caused by errors in A.

We have empirically discovered that the estimate of μ_b is important. The ML method that uses the Estimation-Maximization method (MLEM [1]) jointly estimates μ_b and μ_g by maximizing their joint likelihood. A modified version of MLEM uses $\hat{\mu}_{b,i} = b_i$ (which [1] calls the MLEM-FB (fixed background) method). An "empirical Bayes" argument could justify using $\hat{\mu}_{b,i} = \alpha b_i + (1 - \alpha)\bar{b}$ where the weight α could be selected according to the relative variances of the prior and likelihood. But, to our knowledge Bayesian methods for this type of problem have to date [4,6] only used $\hat{\mu}_{b,i} = b_i$ (as does MLEM-FB).

To address the "too many parameters" issue, here we consider one way to reduce the effective number of parameters. Ultimately, any reduction in the number of parameters has a Bayesian justification (we use prior knowledge to penalize "rough" solution vectors for example). Any Bayesian analysis will need to specify a prior probability for x and for μ_b. And, "penalized likelihood" methods, which typically constrain the solution vector $\hat{x}$ to be smooth, or force $\hat{\mu}_b$ to be parametrically related to $\sum_j A_{ij} x_j$, are formally justified via the assumed prior probability for the true x or μ_b vector.

Because we have noticed a tendency for $\mu_{b,i}$ to increase when $\mu_{n,i}$ increases, in this paper, we assume that $\mu_{b,i} = \mu_b + c_2 \mu_{n,i}$. Also, the scanning protocol suggests that the source term $\mu_{n,i}$ will contribute to the background in other (but generally not all) bins. Therefore, we anticipate that $\mu_{b,i} = \mu_b + c_2 \mu_{n,i}$ is an adequate (but not exactly correct) assumption.

Figure 1 is a plot of (a) the estimated mean net count rate versus bin number and (b) the estimate mean background count rate in a typical sample item. Note that (b) is approximately proportional to (a), so the assumption $\mu_{b,i} = \mu_b + c_2 \mu_{n,i}$ is reasonable.

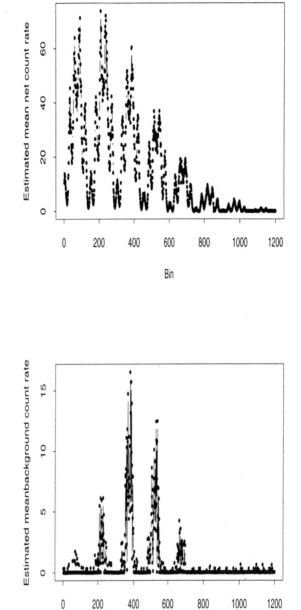

Figure 1: Estimated (a) mean net count rate ($\mu_{n,i}$) and (b) mean background rate ($\mu_{b,i}$) for a sample item. Bin number corresponds to a scan-position (angle, displacement, elevation) determined by the scanning protocol.

We will present results for two methods:
(1) MLEM and (2) MMLEM (modified MLEM). Let
$\hat{g}_i = \sum_j A_{ij}\hat{x}_j + c_1\hat{\mu}_{b,i}$ and $\hat{b}_i = \hat{\mu}_{b,i}$.

The MLEM solution [1] ($\hat{x}_j$ and $\hat{\mu}_{b,i}$) satisfies

$$\sum_i A_{ij} = \sum_i A_{ij}g_i/\hat{g}_i \text{ for all } j, \text{ and} \qquad (3)$$

$$c_1 + 1 = c_1 g_i/\hat{g}_i + b_i/\hat{b}_i \text{ for all } i. \qquad (4)$$

Let $\hat{g}_i = (1 + c_1 c_2)\sum_j A_{ij}\hat{x}_j + c_1\hat{\mu}_b$, and $\hat{b}_i = \hat{\mu}_b + c_2\sum_j A_{ij}\hat{x}_j$.

The MMLEM solution satisfies

$$\sum_i A_{ij} = \sum_i A_{ij}g_i/\hat{g}_i \text{ for all } j, \text{ and} \qquad (5)$$

$$c_1 + 1 = c_1/M \sum_i g_i/\hat{g}_i + 1/M \sum_i b_i/\hat{b}_i. \qquad (6)$$

We note that Eq. (5) is the same as Eq. (3), which both hold for $j = 1, 2, \ldots N$. However, Eq. (4) holds for $i = 1, 2, \ldots M$, while Eq. (6) is a single equation.

6.   Simulation Study

Here we give assay results for simulated data for a $2 \times 3 \times 3$ full factorial using $M = 8$ and $M = 64$, noise-to-signal ratio NSR = average of $c_1\mu_{b,i}/\sum_i A_{ij}x_j$= L (.1) , M (1) , or H (10), and $T = .1$, 1, or 2 gms. Elsewhere [4,6] we report simulated assay results of several classical and Bayesian methods for a 2^6 full factorial experimental design varying the following six factors with $N = 6$ (using values that agree reasonably well with those of real containers except that we use very small M and N):

(1) condition of A (High or Low),

(2) $M = 8$ or 16,

(3) $\sigma_A = 0$ or $\approx 0.2 \times A$,

(4) $T = \sum_{i=1}^N x_i = 1$ or 10,

(5) $\sigma_x = 0$ or > 0 (σ_x^2 is the variance of the x_i), and

(6) noise to signal ratio, $\frac{c\mu_b}{Ax} = .2$ or 1.

In [4,6] we reported a severe increase in performance measure PM_1 (large values are bad) as the condition (ratio of largest to smallest singular value) of A increased for all "matrix-inversion" based methods (including biased regression methods such as ridge regression) compared to MLEM (and Bayesian methods that approximated the posterior distribution for x). We also observed a bias in MLEM for small T values, but did not attempt to reduce the

bias. Because of the strong performance of MLEM across a range of conditions, we are now considering the utility of bias-reduction in MLEM for certain (especially low T) situations.

In Table 1 we present $\overline{\hat{T}}$, $\sigma(\overline{\hat{T}})$, and $PM_1 =$ MSE for each run for MLEM and MMLEM for the 18 cases, varying first NSR, then T, and then M. That is, run 1 is LLL for the 3 factors, run 2 is MLL, run 3 is HLL, run4 is LML, ..., and run 18 is HHH. For example, run 2 is NSR=1 (M), $T = .1$ (L), and $M = 8$ (L).

Table 1 entries are based on 100 simulations per run, so generally, reported PM_1 values are within approximately 20% of their true values. The estimated standard deviations of the $\overline{\hat{T}}$ are given in Table 1, and we see that the $\overline{\hat{T}}$ values are also within approximately 20% of their true values. The values in bold are the ones for which MMLEM has lower observed PM_1 than MLEM. Note that in the case where MLEM is clearly biased (runs 10,11,12), MLEM actually has lower PM_1 than MMLEM. Averaged over the 18 runs, the MLEM PM_1 value is .33 and the MMLEM PM_1 is .17.

Figures 2 and 3 display the $\overline{\hat{T}}$ and $\sigma(\overline{\hat{T}})$ values in Table 1. The main features to notice in Figures 2 and 3 are: (1) when we increase the dimension of A from 8-by-2 to 64-by-2, the bias in MLEM for small T (.1g) is obvious; (2) when we increase the dimension of A from 8-by-2 to 64-by-2, there is no apparent bias in MMLEM for small T (.1g); and (3) the NSR appears to affect the bias for both MLEM and MMLEM.

At present, we have experimental results for one real sample. The MLEM estimate for T was 470 units (we are missing the information needed to convert the units to gms) and the MMLEM estimate for T was 534 units. This is an encouraging result because the MLEM bias [4-6] tends to be low for the low T values (in both real and simulated data). This negative bias causes a difficulty in waste disposal because we often need to demonstrate that the true mass does not exceed some limit, say 10 gms. If we have a negative bias, then it is more difficult to defend claims that T does not exceed 10 gms.

7.   Summary

We have presented a comparison of MLEM (state of the art) to MMLEM (new) for a small designed experiment using simulated data and one real sample. MLEM demonstrated both positive and negative bias relative to the MMLEM result. In real data, MLEM has demonstrated both positive and negative

Table 1: $\overline{T}$, $\sigma(\overline{T})$, PM_1 (100 simulations per run).

Run (NSR,T,M)	MLEM			MMLEM		
	$\overline{T}$	$\sigma(\overline{T})$	PM_1	$\overline{T}$	$\sigma(\overline{T})$	PM_1
1 (LLL)	.66	.15	.47	.37	.11	**.18**
2 (MLL)	.08	.04	.04	.24	.07	.09
3 (HLL)	.32	.05	.10	.22	.03	**.05**
4 (LML)	1.07	.14	.15	1.05	.13	**.14**
5 (MML)	1.33	.20	.30	1.21	.17	**.21**
6 (HML)	1.87	.25	1.0	1.17	.17	**.20**
7 (LHL)	2.08	.38	.38	1.97	.34	**.34**
8 (MHL)	2.56	.35	.66	2.3	.28	**.37**
9 (LHL)	3.0	.30	1.29	2.3	.25	**.31**
10 (LLH)	0.0	0.0	.01	.08	.03	.02
11 (MLH)	0.0	0.0	.01	.14	.06	.02
12 (HLH)	0.0	0.0	.01	.11	.04	.02
13 (LMH)	.93	.29	.23	1.07	.30	**.10**
14 (MMH)	1.56	.32	.43	1.38	.28	**.13**
15 (HMH)	1.26	.34	.44	.95	.23	.46
16 (LHH)	2.04	.28	.13	2.01	.29	.13
17 (MHH)	2.2	.26	.18	2.19	.25	.24
18 (HHH)	1.77	.33	.18	1.54	.23	**.11**

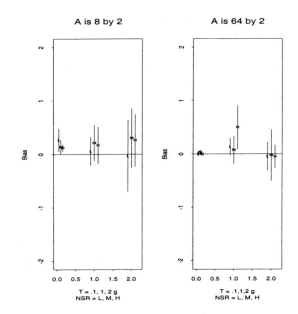

Figure 3: Bias and error bars ($\pm 2\hat{\sigma}$) for MMLEM for (a) A an 8 by 2 matrix, and for (b) A a 64 by 2 matrix

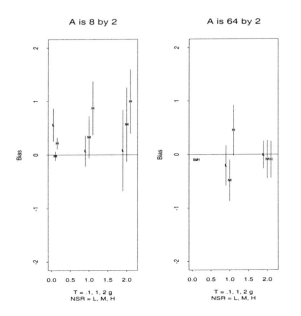

Figure 2: Bias and error bars ($\pm 2\hat{\sigma}$) for MLEM for (a) A an 8 by 2 matrix, and for (b) A a 64 by 2 matrix

bias, but most often demonstrates negative bias for small T values. This negative bias makes it difficult to defend claims that T does not exceed a prescribed limit. Also, Eq. (1) is not always adequate for real data, so we have considered the "cleanest" situation that TGS might encounter. Departures from Eq. (1) could be modeled with "equation error" following the "error-in-variables" literature. One such departure can occur in TGS if the nuclear material is "lumpy." If the nuclear material is concentrated in lumps, then significant self-attenuation of γ flux within each lump can occur. Methods to diagnose this phenomenon involve interrogating with several γ energies, because the self-shielding effect decreases as γ energy increases. We have not addressed the "sample-lumpiness" issue in this paper. In cases with low T, sample lumpiness is less likely to be an issue.

Generally, any source of bias in the MLEM method is of interest, as are ways to reduce any TGS bias. Reference [10] presents some methods for reducing bias in maximum likelihood methods that might be effective (without "reducing the number of parameters"). This paper considered one way to reduce the number of parameters that was motivated both by physical reasoning and by empirical observations such as Figure 1. However, we assumed that $\mu_{b,i} = \mu_b + c_2\mu_{n,i}$ held exactly, and in practice, it will at best only hold approximately. Therefore, a next step would be to study the sensitivity of MMLEM to modest and strong departures from the $\mu_{b,i} = \mu_b + c_2\mu_{n,i}$ assumption. Also, we noted in Table 1 that PM_1 (MSE) does not necessarily decrease when we decrease the bias, although the average PM_1 for MMLEM is lower than that for MLEM.

Another paper [9] considered the "errors in A" issue. Although results in [4,6] and [9] indicate that ML-based methods are the least sensitive to errors in A, it is of interest to modify MLEM to correct for errors in A. However, the success of any modification for errors in A will probably depend on the dimension of A (or the effective number of free parameters). It is tedious but important to characterize the bias of MMLEM-like methods because of the need to include the effect of departure from the assumptions that are used to reduce the effective number of unknowns. Therefore, future work will: (1) consider the utility of including a probability model for the true A matrix as a way to handle the "errors in variables" aspect, and (2) characterize the bias of MMLEM-like methods as a function of departure from the assumptions that are used to reduce the effective number of unknowns. Finally, as mentioned, an additional source of error in A and/or in Eq. (1) arises from lumpiness of the nuclear material, so current work is aimed at characterizing all error sources in A.

8.   References

[1] Prettyman, T., Cole, R., Estep, R. and Sheppard, G. (1995), "A Maximum-Likelihood Reconstruction Algorithm for Tomographic Gamma-ray Nondestructive Assay," Nuclear Instruments and Methods in Physics Research A 356, 470-475.

[2] Prettyman, T. and Mercer, D. (1997), "Performance of Analytical Methods for Tomographic Gamma Scanning," LA-UR-97-1168 (Los Alamos National Laboratory technical report).

[3] Besag, J., Green, P., Higdon, D., and Mengersen, K. (1995), "Bayesian Computation and Stochastic Systems," Statistical Science 10, No. 1, 3-66.

[4] Burr, T., Mercer, D., and Prettyman, T. (1998), "Comparison of Bayesian and Classical Reconstructions of Tomographic Gamma Scanning for Assay of Nuclear Materials," American Statistical Association Proceedings of the Section on Physical and Engineering Sciences, 195-200.

[5] Mercer, D. Prettyman, T., Abhold, M, and Betts, S. (1997), "Experimental Validation of Tomographic Gamma Scanning for Small Quantities of Special Nuclear Material," Proceedings of the 38th annual meeting of the Institute of Nuclear Materials Management, Phoenix, Arizona, July 20-24.

[6] Burr, T., Mercer, D. and Prettyman, T. (1998), "A Study of Total Measurement Error in Tomographic Gamma Scanning to Assay Nuclear Material with Emphasis on a Bias Issue for Low-Activity Samples," Proceedings of the 39th annual meeting of the Institute of Nuclear Materials Management, Naples, Florida, July 26-30.

[7] Frank, I., and Friedman, J. (1993), "A Statistical View of Some Chemometrics Regression Tools," Technometrics, 35, No. 2, 109-148.

[8] Green, P. (1990), "On Use of the EM Algorithm for Penalized Likelihood Estimation," Journal of the Royal Statistical Society B, 52, No. 3, 443-452.

[9] Burr, T., and Knepper, P. (1999), "A Study of the Effect of Measurement Error in Predictor Variables in Nondestructive Assay," 4th Topical Meeting on Industrial Radiation and Radioisotope Measurements and Applications, October 4-7, 1999, Raleigh, North Carolina.

[10] Kendall, M. and Stuart, A. (1979), "The Advanced Theory of Statistics," Volume 2, Griffin and Company.

AN APPLICATION OF GRAECO-LATIN SQUARE DESIGNS
IN THE SEMICONDUCTOR INDUSTRY

Diane K. Michelson, SEMATECH, Steve Kimmet, Harris Semiconductor
Diane K. Michelson, SEMATECH, 1206 Montopolis Dr., Austin, TX 78741-6499

Key Words: wafer fabrication, design of experiments, particulates

Introduction

In semiconductor fabrication, wafers are processed in a cleanroom environment. Transistors and other electronics are formed by photolithography to form the structures. Cleanroom measures, such as HEPA filters and laminar airflow, are responsible for keeping contamination off the wafers. However, particulates do enter the cleanroom and may contaminate wafers. Certain types of particulates may cause Electrical Over Stress (EOS) failures when the die are probed. The purpose of this experiment was to evaluate the relationship between the size and density of defects introduced in the wafer fab at specific masking levels versus the resultant failures of the individual die at circuit probe for EOS related tests.

Root cause failure analysis of EOS failures is complicated by the nature of the EOS mechanism. An EOS failure mode in the customers' application is characterized by a large concentration of energy at the fail site and subsequent massive silicon damage which vaporizes the defect or anomaly.

Due to the fact that there is a relatively low incidence of failures (~50 ppm) and there is difficulty in determining root cause of those failed devices, it was determined that a controlled experiment should be performed which would intentionally place a large quantity of known defects on die during specific levels of manufacturing. These known defects could then be detected at circuit probe EOS testing and definitively linked to the specific process levels. Thereupon, improvements could be implemented in the wafer fab at the critical levels to reduce those occurrences of defects.

A common EOS fail site is depicted in the two photographs in Figure 1. Note the massive damage to the circuit area which usually prevents analysis of the root cause of the failure.

The goals of the experiment are to evaluate the relationship between defect size and density and EOS failures at probe, determine which masking levels are critical for defect-related EOS failures, and build an EOS-susceptible population of die to help in perfecting a test screen for the failures.

0.145mm EOS damage site

0.300mm EOS damage site

Figure 1. Photograph of two die with EOS fail sites in the power output.

Experimental Design

The experiment was designed to estimate the effect of defect size and density on parameters which may cause EOS failures. To minimize extraneous variability, a 4×4 Graeco-Latin square design was employed. Note that the Graeco-Latin square was used as a setup tool rather than an analysis tool. The layout of the design is shown in Figure 2. Defects were scattered randomly throughout each cell as can be seen in Figure 3. Within each of the sixteen cells resides a total of sixteen whole die to be included in the experiment. If any portion of a die fell upon a border of an adjacent cell it was excluded from the experiment.

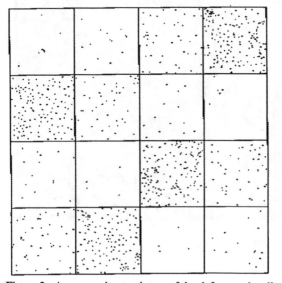

1um 1 d/cm	2um 3 d/cm	4um 9 d/cm	8um 27 d/cm
2um 27 d/cm	1um 9 d/cm	8um 3 d/cm	4um 1 d/cm
4um 3 d/cm	8um 1 d/cm	1um 27 d/cm	2um 9 d/cm
8um 9 d/cm	4um 27 d/cm	2um 1 d/cm	1um 3 d/cm

Figure 2. Defect mask sizes and densities in Graeco-Latin square arrangement.

Figure 3. An approximate picture of the defect mask cells.

The defect was created on the wafer using an Anti-Reflective Coating (ARC) to place a defect representing an "island" of material into the circuit. This island of material portrayed a real defect as commonly experienced in the fab. Engineering selected a group of masking levels to study.

As an example of how the defect is produced, we will illustrate the Poly Gate process. The first step at Poly Gate requires the deposition of polysilicon material onto the wafer. Next a layer of ARC was applied to the wafer which was followed by a photo and etch step using the defect mask. This step removed all the ARC except where a defect was to be formed. Subsequently, the wafer would be processed through the normal Poly photolithography and etch steps which would form the Poly Gate regions. The leftover areas of ARC would protect the removal of the "defect" areas of polysilicon during that etch process. Thus, upon completion of these steps, the remaining features on the die would include the normal geometrical shapes of polysilicon material along with small circles of "defect" polysilicon randomly places throughout the circuit area.

The details of the experiment were conducted as follows. Two lots of eighteen wafers each were created. Three wafers from each lot received ARC treatment with the defect mask at each of the levels under consideration and three wafers did not receive the mask at all so as to provide a control cell of the experiment.

Shown in Figure 4 is a wafer map of a wafer which received the defect mask at the Poly Gate level. The map is a representation of the results obtained during 100% wafer probing. Each colored square represents one integrated circuit (die) which after being tested is classified as either a "good die" (Bin 1) or a "bad die" (Bins 4 through 16). A green square indicates a good die, while any other color indicates a failure. To aid in seeing them in black and white, bad die have been marked with a dot. If a die fails one of the four EOS tests, it is classified as a bad die in either Bin 11, 12, 13, or 14. Notice that the bottom left and top right of the wafer has large quantities of failures. These two regions contain the largest size of defects (8 micron) as well as the highest density of defects (9 and 27 defects per square centimeter). The analysis of the experimental data is summarized in the next section.

The responses analyzed were the yield of each cell, that is, the count of good die divided by 16, and the count of EOS failures in each cell. Models were found for each response for each masking level of the experiment.

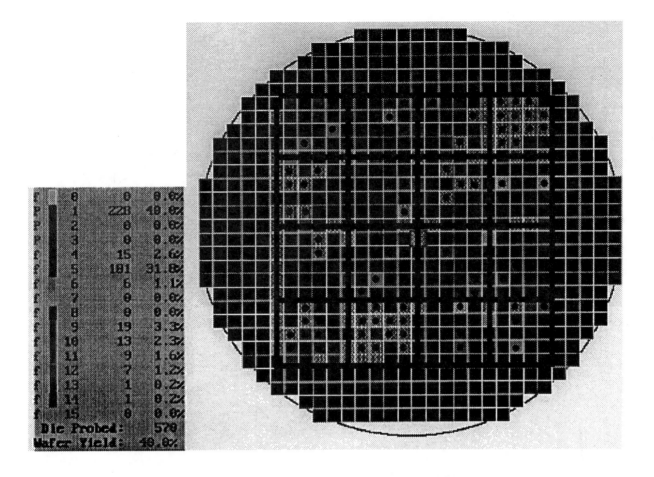

Figure 4. Wafer map test results of an experimental wafer with overlaid grid of experiment cells.

Analysis

The yield of each test cell (of sixteen die) was subsequently analyzed as a function of defect size and density. Since the distribution of yield is non-Gaussian, and the yield is constrained between 0 and 1, a logit transformation was applied to the response. The transformation is $log(yield / (1 - yield))$.

The independent variables *row* and *column* do not have the same meaning as the same variables in a traditional Latin square design. Traditionally, row and columns in a Latin square are used by blocking variables. Here, we used them simply as a grid. In any case, they were not ever significant and were removed from the models.

The level with the best-fitting model had an R^2 of .64. The model is

$$logit(yield) = 1.26 - .0993*size - .0119*density - .0135*size*density.$$

The root mean squared error is 0.6766.

Figure 5 is a contourplot of yield by size and density. The circled point had one cell with a yield of 0%. There were no cells with 100% yield. The 0% yield cell was not included in the analysis.

Good models were found for three of the five levels studied. A good model with the variables size and density could not be found for the control cell, as is to be expected, since the control cell did not receive the treatment mask. The reason good models were not found for other masking levels is the amount of circuit area that is printed on the die is limited for those levels and the structures on those levels are typically separated from adjacent structure by fairly great distances so that defects are not usually a contributor to yield loss.

Similar contourplots were found for each of the other well-fitting models. All showed that as size of defects and density of defects increase, yield decreases, as expected. For designed experiments, R^2 values of above 90% are usually expected. The results of this experiments show that significant factors were not accounted for by looking only at defect size and density. However, it is gratifying to see a model for the relationship between yield and defect size and density

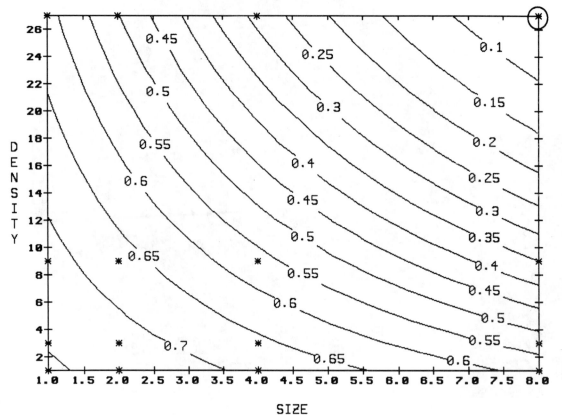

Figure 5. Contourplot of yield vs defect size and density.

coming from a designed experiment rather than a regression on historical data.

In addition to yield, the other response examined was the quantity of die in each cell which failed for the EOS tests. For those levels for which the count of EOS failures was very skewed, we used the response *log*(EOS+1). The constant is added so that we are not calculation logarithms of zero for the cells without EOS failures.

The models for this analysis did not have high predictive capability. The highest R^2 values was 23%. Once again, significant factors are not accounted for in the experiment. However, some results can be obtained by examining the contourplots of the predicted surfaces.

The model for the masking level with the best-fitting model is:

$$EOS = .2529 + .1161*size + .0652*density - .0147*size*density.$$

The error standard deviation is 0.8157.

Figure 6 is a contourplot of EOS counts by size and density. Notice that high defect density and large defect size contribute to more EOS failures. This masking level did not have EOS failures for the combination of the highest defect density and largest defects, hence the predicted surface decreases as it heads to the upper right corner of the contourplot. However, it is clearly understood that to minimize EOS failures at this masking level, it behooves us to decrease both the size and amount of particulates on the wafers.

262

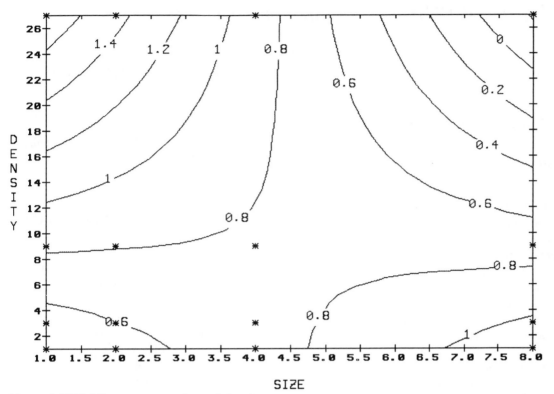

Figure 6. EOS failure counts vs size and density.

Conclusion

We have shown an innovative way to study the effect of defect size and density on wafer yield and yield loss due specific failures. Being a controlled, designed experiment, this method is preferable to regression methods which measure particle counts at certain levels and correlates particle counts to yield or failures at the end of the line. This method requires full mask photolithography and Anti-Reflective Coating.

The Graeco-Latin Square design was used to arrange the cells on the wafer so that the impact of across-wafer variability was minimized.

The information has been given to the wafer fab for inclusion in their continuous improvement processes. It is expected that the two most significant levels for EOS will be added to the real-time inline-monitoring program for added particles. In addition, new studies will be undertaken to determine the potential contributors of defects ranging in the 2 to 8 micron size.

Improved PCB Inspection: Spatial Design Issues

Dee Denteneer, Philips Research

Prof. Holstlaan 4, 5656 AA Eindhoven, the Netherlands (Dee.Denteneer@philips.com)

Key Words: PCB inspection, Loess, spatial design

1 Introduction

Process control on Printed Circuit Boards (PCB's) involves measurement of the volume of a large number of small heaps of solder paste deposit, that are used to fix components, such as IC's, on the board. A crucial prerequisite for accurate measurement is a good estimate of the warpage of the PCB as it is clamped into the machine and for this purpose surface estimation is needed. In this paper, we will review the use of local fitting to estimate board warpage and particularly show that the principles of Loess as described in Cleveland and Grosse (1991) can be successfully applied.

There are also some relevant differences between surface fitting as described in Cleveland and Grosse and the requirements of the current application. These derive mainly from the fact that, usually, surface estimation is invoked as an exploratory tool to investigate a single data set. In the current application, however, surface estimation is used as an automatic tool to be applied to a sequence of PCB's under the time constraints inherent to a production process. Apart from a greater focus on computational speed, the current application differs in that some preprocessing can be performed, which would not be useful in the standard context. These differences are more fully explored in Denteneer and Kuepers (1999), with particular focus on the speed up of the computations.

In this paper we will focus on the spatial design issues that arise in surface estimation with local polynomials. These issues concern both the selection of the fit locations, where the local polynomials are estimated, and the selection of the reference locations, where the board height is measured. Two distinct versions of this latter issue are discussed. First, we will consider the selection of reference locations from well defined areas on the surface of the PCB. Next, and in much greater detail, we will discuss the selection of a subset of a given collection of reference locations.

This paper is organized as follows. In section 2, we describe the background to our research: process control on solder paste deposits on PCB's and the

relevance of surface estimation in this context. In section 3, we will focus on the design issues. Finally, in Section 4, we summarise and draw some conclusions for future research.

2 Process control on solder paste deposits

Manufacture of printed circuit boards (PCB's) involves the deposition of a large number of small heaps of solder paste that are used to fix the components on the PCB. Roughly, the process is as follows. Firstly, heaps of moist solder paste are deposited on the board. Secondly, the components (e.g. IC's) are placed on the board and each pin of each component must be positioned in such a heap of solder paste. Finally, the paste is dried to fix the components.

Typically, the number of heaps of solder paste deposit is in the order 1000-5000, the areas of the heaps range from $1mm^2$ to $.1mm^2$, and their heights range from $.1mm$ to $.2mm$.

The volume of solder paste in each heap is crucial to board quality: too much paste may cause components to short-circuit and too little paste may result in loose components. Hence, accurate measurement of these volumes is essential for process control in PCB manufacture. Typically, the measurement is performed after the deposition of the paste and before the placement of the components, so that the results can also be employed in a reject (and re-wash) policy that prevents expensive components from being placed on defective boards.

One way to carry out the measurements is optically. The spot on the PCB where the solder paste is located is illuminated with a laser beam and the height of the top of the heap is computed from the amount of reflected light. Thus one measures height within a coordinate system imposed by the machine. However, to estimate the volume in each heap of solder paste deposit requires measurement of the height relative to the PCB surface. Now the PCB is warped as it is clamped into the machine and board warpage may exceed the target height of the heaps of solder paste by a factor 10 to 20. It follows that the estimation of PCB surface height is as essential an ingredient for accurate volume measurement as the

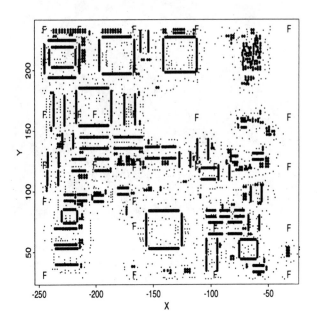

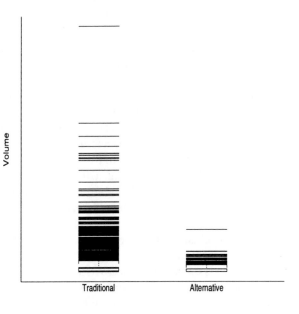

Figure 1: Example layout of PCB: shown are 1998 target locations (big dots), 1697 reference locations (small dots), and 34 fit locations (denoted by F)

Figure 2: Distribution of estimated volumes on one PCB obtained by means of the traditional approach and by means of the proposed approach

measurement of the top height of each heap. Thus volume measurements essentially involve estimation of the surface height of the PCB at a number of target locations, i.e. at the locations were the heaps of solder paste are deposited.

To compute this surface height, the following procedure was traditionally used at each target location. The height of the surface of the PCB was measured at a number of reference locations, near the target location. Then a, local, quadratic polynomial was fitted to these reference measurements and the height of the surface was estimated by evaluating this polynomial at the target location. This procedure, however, was subject to occasional problems. Both noise peaks in the measurement of the height at the reference locations and unfortunate spatial arrangement of these reference locations can lead to substantial errors. To remedy these problems, the following changes were proposed to this original procedure. Essentially they extend a rather direct implementation of local polynomial fitting with the sophisticated principles pioneered in Cleveland and Grosse (1991).

- For each target location, employ a large number, rather than a small number, of reference measurements. This can be implemented without taking extra reference measurements by borrowing reference locations from neighboring

target locations.

- Reduce bias by employing distance weights in the fitting procedure.

- Increase robustness, by using an iterative fit which down-weights reference measurements with large residuals.

- Fit at a limited number of *fit* locations only, as fitting at each target location would be computationally prohibitive.

- Compute the heights at the target locations by 'blending' these fits. Blending is the spatial generalization of interpolation by means of a cubic polynomial, see Cleveland and Grosse for details.

Figure 1 illustrates the set-up. It displays the target locations (big dots), the reference locations (small dots), and the fit-locations (denoted by F) on the surface of the PCB. As usual, there are a large number of target locations and reference locations: 1998 and 1697, respectively. The number of fit-locations is small by computational necessity and equals 34 in figure 1.

An illustration of the results that can be achieved in this way is given in figure 2. It displays the results obtained in a test procedure with an empty board:

as the board had no paste on it, so that all volumes are zero, there is an easy ground truth available for objective comparison in this case. Shown are volume measurements obtained via the traditional procedure for surface estimation and volume measurements obtained via a procedure for surface estimation that incorporates the changes suggested above. The measurement machine can switch between the traditional and the proposed procedure so that the only difference between the measurements consists of the routines for surface estimation. Clearly, the measurement obtained with the traditional procedure are more scattered and deviate more from zero (the truth) than the measurements obtained via the proposed procedure.

The figure illustrates two relevant findings, that generalize to other boards that were measured. Firstly, there is a statistical improvement in that the inter quartile range of the empirical error distribution, i.e. the distribution of the difference between true and estimated volume, is reduced. In this case, the reduction was by one third. This was found to be the case in all other boards investigated and appears to hold quite generally. Secondly, there is a much improved resistance to outliers. In the results as shown, the largest outlier is reduced by a factor of 6. Clearly, the occurrence of outliers and their size is highly dependent on the type of PCB tested. However, this figure is typical for what can be achieved with difficult PCB's.

Substantial experiments were carried out that corroborate these findings. We will not dwell much further on the comparison, however. Theoretically at least, it is fairly obvious that the new procedure will outperform the traditional approach and this was borne out in practice.

3 Spatial design issues

There are several design issues that are relevant to surface estimation concerning both the selection of reference locations and the selection of fit locations. These are addressed in this section.

3.1 Selection of fit locations

Strictly speaking, the current application does not involve surface estimation as we are interested in the height of the surface in a fixed number of target locations only. As this number of target locations is typically very large, tools for surface estimation are appropriate. However, this difference poses a design issue: in standard local fitting the fit locations, where the fitting is performed, are derived from the

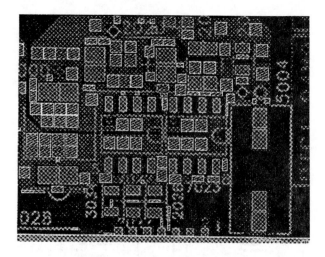

Figure 3: Detail of a PCB, showing a number of target locations (check board pattern) and a number of reference locations (diagonal stripes)

measurement locations. In the current application, they can be derived from either the reference locations or from the target locations. Intuitively, the former approach has the advantage of greater accuracy: fitting is done closer to the measurements at the reference locations. However, the latter approach has the advantage of greater relevance: fitting is done closer to the target locations where the estimates are actually required. So with both approaches one may argue for improved accuracy at the target locations. We have opted for the latter approach, but comparisons carried out by means of simulations revealed minor differences only.

In our implementation, the fit locations were based on the kd-tree based on the set of target locations. Again, the differences with using a simple grid were minor.

3.2 Selection of reference locations

Aim of the selection of the reference locations is to have accurate estimates at the target locations with the lowest number of reference measurements: less measurements results in a faster machine. Note that there are two benefits to a reduction of the number of reference measurements. Firstly, it reduces measurement time for the machine. Secondly, it reduces the computational complexity of the algorithms.

Strictly speaking, the design issue involves the selection of areas on the PCB, rather than locations: any area on the bare copper on the surface area, that is larger than a given minimal area, suits as a reference location. The traditional selection algorithm works roughly as follows. For each target location divide up the space around this location into four quadrants. Then these quadrants are searched in turn for the closest area of bare copper that exceeds the minimum size. These areas are then used as reference locations. Usually, 8 to 10 reference locations are associated with each target location. The set of reference locations then is the union of the reference locations found for all target locations.

This yields a large set of reference locations, and reduction can be necessary for reasons of speed. Moreover, as fitting is done only at a modest number of fit-locations, it seems more appropriate to derive reference locations from fit-locations rather than from target locations. Our implementation then reduces the given set of reference measurements by using this intuition.

To this end, we use the Delaunay triangulation of the reference locations; see e.g. Ripley (1981) for an excellent introduction to Delaunay triangulation and Voronoi (or Dirichlet) tesselation. Very briefly, the Voronoi tesselation associates a tile with each point in a collection of spatial points. The tile is the area closer to this point than to any other point in the collection. The Delaunay triangulation then connects the points that are neighbors in the Voronoi tesselation. This defines a neighboring relation on a collection of planar points and we can speak of first neighbors: points that are directly connected in the triangulation, second neighbors: points that are indirectly connected in the triangulation via one intermediary point, etc.

Our algorithm first computes the Delaunay triangulation of the set of reference locations. Then for each fit-location, it selects the reference location closest to it, as well as the reference locations that are upto k-th neighbors of this initial point. All points so selected are retained in the collection of reference locations, all other points are deleted. Thus it is possible to reduce the collection of reference locations by an appropriate choice of k

In our experiments we made use of Baddeley's contribution to StatLib, which was also used to create Figure 4. Figure 4 illustrates our approach: it displays a fit-location (denoted by F), the reference location closest to it (denoted by 0), as well as the 1- and 2-neighbors (denoted by 1 and 2, respectively). Figure 5 presents results obtained in a small simulation study. In this study, the coordinates of an actual board are used: there are 1600 reference locations in the original set and there are 34 fit-locations. The heights at the reference locations were assumed known, and derived from an actual fit, and these were contaminated with Gaussian noise. Then

estimation based on the original set was compared to estimation on reduced sets obtained by incorporating upto third and fourth neighbors. The results were then compared with respect to maximum error and error sum of squares at the 1900 target locations.

The results show that taking the reference locations that are upto third and fourth neighbors substantially reduces the number of reference locations: taking up to third neighbors reduces the set of reference locations to approximately 40% of the original size, taking upto fourth neighbors reduces this set to 65% of the original size. The results indicate that such reductions can be achieved without sacrificing much accuracy. However, we attribute the improvement in accuracy obtained by reducing the set to 65% of the original size to the use of Gaussian noise in the simulation, and do not expect this to carry over to the application.

Also, it can be seen that the approach suffers from a lack of granularity. Taking third neighbors reduces the set of reference locations to 40% and taking fourth neighbors reduces to 65%. However, there is nothing in between so that it is difficult to achieve a prescribed number of reference locations. We have currently resolved this issue using randomization: use all up to k-th nearest neighbors, and select randomly from the $k + 1$-th nearest neighbors until the target number of reference locations is attained. Clearly, alternative approaches that select from the set of $k + 1$-th neighbors based on variance measures will be more appropriate and are currently being investigated.

3.3 Selection of fit locations, revisited

After selection of the reference locations, it is possible to define the collection of reference locations associated with each fit location: the set of reference measurements that will be used to fit the local polynomial at that particular fit location. If we ignore the re-fitting with the robustness weights, the variance of the estimated coefficients of the local polynomial, $\hat{\beta}$, can be expressed simply in terms of the design matrix, X, and a weight matrix D_w:

$$\text{var}(\hat{\beta}) = \sigma^2 (X^t D_w X)^{-1} X^t D_w^2 X (X^t D_w X)^{-1}. \quad (1)$$

Here the rows of X are associated with the reference locations. They consist of a 1 for the intercept term and the reference locations centered with respect to the fit location. In addition, the rows can contain terms that are powers of these centered reference locations if higher order polynomials are used, but here attention is confined to local linear polynomials.

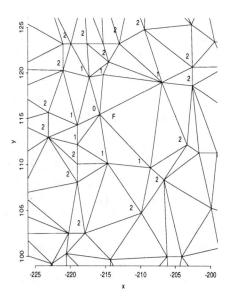

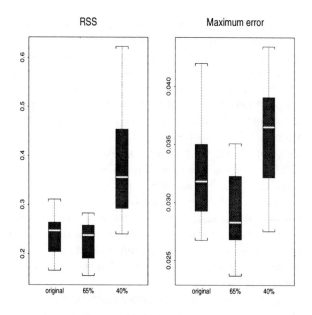

Figure 4: Illustration of Delaunay triangulation as a tool to reduce the collection of reference locations: displayed are a fit location (denoted by F), the reference location closest to it (denoted by 0), as well as the 1- and 2-neighbors (denoted by 1 and 2, respectively)

Figure 5: Accuracy of surface estimation at target locations obtained with the original collection of reference measurements and with reduced collections of reference measurements

The diagonal weight matrix contains the distance weight associated with each reference location. In the current application, we have made use of the tricube weight function, but the exact form of the weights is not important here.

Then, for local linear polynomials, the elements of $\hat{\beta}$ contain exactly the ingredients that are used in predicting the height at the target locations: the value of the estimated local polynomial and the values of both first derivatives, all of these evaluated at the fit location. So, generally, best overall prediction will be obtained if the variance, (1), is minimized.

Now, (1) is a function of the fit location, both via the centering and via the distance weights. Hence, is appears to be possible to improve the overall prediction at the target locations by minimizing (1) with respect to the fit location. Experiments to this end are currently being performed.

However, the reduction of (1) can also be approached in a very simple and pragmatic matter. To this end, refer to figure 1. From the image it is clear that the fit locations at the edges of the PCB are located outside the reference locations that will be associated with them. Intuitively, to obtain the coefficients of the local polynomial, we must extrapolate from the data. It is well known that extrapolation inflates variances. Hence it makes sense to shift the

fit locations slightly towards the center of the data to reduce the variance of the estimated coefficients.

A simulation experiment to this end was carried out. In it, two procedures were used to estimate height at the target locations. The first procedure used the standard kd-tree. The second procedure used an alternative set of fit locations, based upon the standard kd-tree, but with the locations at the edges moved slightly toward the associated reference locations. In fact, we shifted the fit locations on the edges by a prefixed, non-optimized, amount of $10mm$ toward the center of the PCB. Then for each procedure, mean squared error of the prediction at the target locations was computed. This experiment was replicated 20 times for four distinct boards.

In figure 6, the results so obtained are displayed. The four panels correspond to the four distinct PCB's that were included in the experiment. Within each panel, the left boxplot corresponds to the results with the standard kd-tree, and the right boxplot corresponds to the results with the shifted fit locations.

For each board, the average mean squared error was reduced to about 90% by using the shifted fit locations. Overall, these results substantiate the intuition that considerable gains in precision can be achieved by shifting the fit locations as computed from the standard kd-tree to the center of the reference locations.

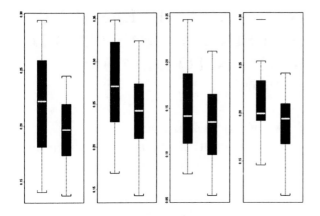

Figure 6: Accuracy of surface estimation at target locations using fit locations based on kd-trees (left boxplots) and fit locations shifted toward the reference locations. The four panels corresspond to four distinct boards.

4 Conclusion

In this paper we have shown that surface estimation is important for process control on PCB's and that the sophisticated principles for local fitting as pioneered in Cleveland and Grosse (1991) constitute a natural and successful extension of the engineering approach. Although this approach was found to be very successful, questions concerning the relative performance of this approach as compared to other techniques for surface estimation, such as kriging, have been left unanswered. This is a relevant subject for further research.

The focus of the paper has been on spatial design issues for surface estimation, i.e. accurate estimation of the surface with the least number of measurements. This involved both the selection of fit locations and the selection of reference locations. The information flow in our approach is displayed in figure 7. It can be seen that the selection of the final, shifted, fit locations depends on the selected reference locations. The selection of the reference locations, in turn, depends on the selected fit locations. In the current approach, we have only marginally exploited this mutual dependency by shifting the fit locations after selecting the final set of reference locations. However, there is no compelling reason to stop here. It is also possible to iterate the algorithm in order to further improve upon the selected fit and reference locations. Again, this could be a fruitful subject for further research.

From the applied point of view, however, the thing to go after is undoubtedly the measurement noise. A better understanding of the noise associated with

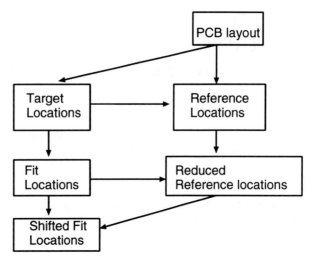

Figure 7: Flow of information for design issues

the reference measurements, and its dependence on size and material, pays off immediately in both selection of reference measurements and fitting procedures: just avoid the places that you know will cause trouble. Moreover, a better understanding of the noise will give further structure and direction to the research questions to be investigated.

Acknowledgements

We thank Mike Keane for suggestions concerning the Delaunay triangulation; we used Adrian Baddeley's contribution to StatLib for our experiments with the Delaunay triangulation. The software by Loader, see http://cm.bell-labs.com/cm/ms/departments/sia/project/locfit/index.html, was used in some of the experiments.

References

[1] Cleveland, W.S. and Grosse, E. (1991), Computational Methods for Local Regression, *Statistics and Computing* 1, pp. 47-62.

[2] Denteneer, D. and Kuepers, C. (1999), Extensions of Local Fitting for Improved PCB Inspection, In Proc. Interface'99.

[3] Friedman, J. H. and Rafsky, L. C. (1981). Graphics for the Multivariate Two-Sample Problem. *Journal of the American Statistical Association* 76, pp. 277-287.

[4] Ripley, B.D. (1981). *Spatial Statistics*, J. Wiley, New York.

A STUDY ON THE ESTIMATION OF PARAMETERS OF THE NORMAL DISTRIBUTION FROM SINGLY TRUNCATED SAMPLES, WITH APPLICATION TO DIATOM COMMUNITIES.

John A. Hendrickson, Jr., Metropolitan Health Plan,
822 S. 3rd St., Suite 140, Minneapolis, MN 55415 (john.hendrickson@co.hennepin.mn.us)

Key Words: Environment, Maximum Likelihood, Minimum Modified Chi-Square.

Abstract:

The Maximum Likelihood Estimator (MLE) of the mean and variance of singly truncated normal distributions was considered by A. C. Cohen, Jr. [Technometrics 1 (1959):217–237; 3 (1961):535–541]. Another application occurs in environmental studies by Ruth Patrick, et al. [Notulae Naturae 259 (1954):1–12]. Hansen and Zeger [ASTM STP 652 (1978):29–37] compared the MLE with a regression-based estimator; the MLE had smaller Mean Squared Error (MSE) in the plausible parameter space for the environmental application. Subsequently, many samples examined by Patrick and her colleagues yielded implausible MLEs or MLEs were inestimable using tabled values of Cohen's auxiliary function. This paper reports a comparison of MLE and a three-moment estimator (3M) with Minimum Chi-Square, Minimum Modified Chi-Square (MMCS) and Minimum Distance estimators over the same parameter space. In simulations, the last three had smaller MSEs than MLE or 3M does. MMCS seems best overall for this application.

1 Introduction

1.1 Statistical setting

It is useful to begin by noting that in truncation, all the observed values exceed (for left truncation) or are less than (for right truncation) a known truncation point. The number of missing observations is unknown. In contrast, in censoring, all the fully observed values exceed (for left censoring) or are less than (for right censoring) a known censoring point. The number of censored observations is known. For singly truncated samples, all the truncated values are on one side of the distribution.

Some early applications of singly-truncated normal distributions were in industrial quality control settings.

1.2 Biological setting

Preston (1948) postulated that in some communities of organisms, the logarithms of the abundances of the species would be approximately normally distributed, were the full community present in the sample. But, some of the rarer species were likely, on average, to be absent in a given sampling effort. Hence, the observations would be truncated between 0 representatives of a species and 1 representative of that species. Since the logarithm of 1 is 0, this provides for an estimation situation in which the truncation point is 0 for the distribution of the logarithms. For various reasons, $\log_2$ is used both in theory and practice in ecology.

From a sample, with n observed species, one may, through statistical estimation of the mean and variance of the parent distribution, also estimate N, the number of species in the full community.

Patrick, Hohn and Wallace (1954), using Hald's (1949) estimation methods, applied Preston's theory to communities of diatoms, which are microscopic algae with silica frustules; the frustules, once cleaned and mounted on a microscope slide, are readily identified to species by an experienced diatomist. Much can be inferred about the health of an aquatic ecosystem by study of the communities of aquatic organisms, such as the diatom community. Studies of diatom communities have been conducted on many river systems, not only in North America, but throughout the world. In that process, a great deal of evidence on the empirical plausibility of these lognormal distributions has been accrued by Ruth Patrick and her colleagues over the past 45 years. In particular, the use of Maximum Likelihood estimation has led to some implausible estimates of N.

The present study was begun in an effort to find

Work done for Dr. Ruth Patrick, Academy of Natural Sciences, Philadelphia, PA, in conjunction with studies for the Savannah River Plant.

alternatives to Maximum Likelihood Estimation for the sampling situations encountered with diatom collections. An earlier study, by Hansen and Zeger (1975, 1978), had led to the recommendation that MLE was preferable on statistical grounds over a regression-based algorithm proposed by Gauch and Chase (1974). This study is closely related to the methods employed by Hansen and Zeger.

For general discussions of the statistical and ecological aspects of the problem of estimating the number of species in an ecological community, the reader might turn to McIntosh (1976), or Hendrickson (1979). A primarily statistical review appears in Bunge and Fitzpatrick (1993).

2 Objectives of this study

We wish to get estimates of the following quantities for the parent, non-truncated normal distribution from a sample with a known truncation point: $\hat{\mu}$, $\hat{\sigma}^2$, and $\hat{N}$.

We also wish to apply appropriate statistical criteria to selecting not only an estimation method, but also to evaluating the candidates presented by the chosen method. This may, sometimes, involve a 'stopping rule.'

3 What we have available

The observations (or their logarithms) are usually tabulated in a frequency distribution prior to analysis. This makes it convenient to use estimation methods and goodness of fit criteria which involve class frequencies.

T: The truncation point.

x_i: The class mark for the ith class. The class with the largest class mark is the Lth class.

f_i: The frequency in the ith class.

n: This is given by $\sum_{i=1}^{L} f_i$, the number of non-truncated observations. [In tabulating the frequency distribution, a diatom species with an abundance of an exact power of two is scored as 0.5 in each of the two adjacent classes; those with an abundance of one individual in the sample are scored only as 0.5 in the interval between 1 and 2, which is between 0 and 1 on the $\log_2$ scale.]

m_1: The sample mean of the non-truncated observations, $(1/n)\sum_1^L (f_i x_i)$

s^2: The sample variance of the non-truncated observations.

ν_j': The jth sample moment about the truncation point, for $j = 1$ to 3.

4 Available estimators

For singly truncated normal distributions, there are two published point estimators, namely the Maximum Likelihood Estimator (Cohen, 1949, 1950, 1959, 1991; Hald, 1949; Akhter, 1996), which is identical to the Method of Moments Estimator, and the Three-Moment Estimator (Cohen, 1959, 1991).

From the section on estimation in most statistics texts, (e.g., Mood, Graybill and Boes, 1974), there are several other candidates, especially the Minimum Chi-Square Estimator (MCS), the Minimum Modified Chi-Square Estimator (MMCS), and the Minimum Distance Estimator (MD). Each of these is based on the empirical frequency distribution, rather than just the sample moments. Thus, each is likely to be more sensitive to the shape of the exposed part of the parent normal distribution than are estimates based only on moments.

4.1 Maximum Likelihood Estimator

For the MLE, we need to deal with a tabled, auxiliary function, $\hat{\theta}(s^2/m_1^2)$ (Cohen, 1961, 1991)

The point estimates of the parameters are as follows:

$$\hat{\mu} = m_1 - \hat{\theta}(m_1 - T) \qquad (1)$$

$$\hat{\sigma}^2 = s^2 + \hat{\theta}(m_1 - T)^2 \qquad (2)$$

For each method, we then estimate $\hat{N}$ as

$$\hat{N} = \frac{n}{0.5 - erf\left((T - \hat{\mu})/(\hat{\sigma}\sqrt{2})\right)/2}$$

Cohen (1959,1991) also provides confidence regions for these parameter estimates.

4.2 Three-Moment Estimator

Cohen (1959,1991) provided a set of point estimates of the parameters based on the first three sample moments about the point of truncation. He speculated that this method might do better than the MLE as the percentage of truncated data grew large.

$$\tilde{\mu} = T + \frac{\nu_3' - 2\nu_1'\nu_2'}{\nu_2' - 2(\nu_1')^2}$$

$$\tilde{\sigma}^2 = \frac{(\nu_2')^2 - \nu_1' \nu_3'}{\nu_2' - 2(\nu_1')^2}$$

This method does not tie the parameter estimates directly to the sample mean and variance of the truncated observations.

4.3 Minimum Chi-Square Estimator

For this and the remaining two estimators, in order to maintain a close relation between the observed moments and the estimated parameters of the normal distribution, estimates for μ and σ in this study were computed by selecting, successively, values of θ and using and then computing the corresponding parameter estimates as above.

We seek to minimize, over a set of $(\tilde{\mu}, \tilde{\sigma})$ pairs,

$$\chi^2 = \sum_{i=1}^{L} \left((\tilde{f}_i - f_i)^2 / \tilde{f}_i \right)$$

when $\tilde{f}_i$ is computed for a normal distribution with $\tilde{\mu}$ and $\tilde{\sigma}^2$, where, analogously to equations (1) and (2),

$$\tilde{\mu} = m_1 - \tilde{\theta}(m_1 - T)$$

$$\tilde{\sigma}^2 = s^2 + \tilde{\theta}(m_1 - T)^2$$

and we allowed $\tilde{\theta}$ to take a sequence of values from 0 to 1.5. (A larger limit on this range was tried in some simulations, but 1.5 seemed to give results which were more focused about the true distribution.) The minimum χ^2 was selected from that sequence.

4.4 Minimum Modified Chi-Square Estimator

We seek, here, to minimize, over the same set of $(\tilde{\mu}, \tilde{\sigma})$ pairs,

$$\chi^2_{Modified} = \sum_{i=1}^{L} \left((\tilde{f}_i - f_i)^2 / \min(f_i, 1) \right)$$

as proposed by Cramér (1951).

4.5 Minimum Distance Estimator

Here, we seek to minimize, over the same set of $(\tilde{\mu}, \tilde{\sigma})$ pairs, the weighted maximum difference between observed and predicted cumulative distributions,

$$D = \max_{1 \le k \le L} \left| \sum_{i}^{k} f_i - \sum_{i}^{k} \tilde{f}_i \right| / (\Pr(z > z_0))$$

where the cumulative distribution in the truncated area is assumed to be identical to its theoretical counterpart and z_0 is the standardized truncation point.

The divisor is needed to avoid a tendency to shift the mean further and further into the non-truncated tail, with correspondingly smaller probabilities attached to the observed portion of the distribution.

4.6 Criteria

We seek an estimator (or a sequence of estimators) with minimal values of Mean Squared Error (MSE) over the plausible parameter space.

$$MSE = (\hat{\mu} - \mu)^2 + var(\hat{\mu})$$

Rao (1965) noted that MLE is not likely always to be the best MSE estimator over the full range of possible parameters.

As a potential evaluation of goodness of fit, the empirical distribution of the Kolmogorov-Smirnov one-sample statistic was compared with percentage points of its theoretical distribution for intrinsic hypotheses. These results are also summarized.

5 Distributions investigated

Simulations were done in Fortran-IV using G-floating point values on a MicroVAX-3100, Model 20e. The pseudo-random normal deviates were computed using the RNOR algorithm of Kahaner and Marsaglia, available from the Statlib website at Carnegie Mellon.

We held constant $\sigma = 5$ and used $N = 150$ in all the simulations. We considered the following group of means, μ:

$$8.75, 7.5, 6.25, 5, 3.75, 2.5, 1.25, 0, -1.25, -2.5, -3.75$$

The table captions include the values for μ, σ, N, and the average observed sample size, $\bar{n}$. The average percentage of missing data can be computed as $100(N - \bar{n})/N$.

Hansen and Zeger had considered as means

$$2.5, 0, -2.5$$

in a study comparing MLE to a regression-based estimator. MLE was much better in all three cases. In

Table 1: μ=7.5, σ=5, N=150, $\bar{n} \approx 140.0$

	MLE	3M	MCS	MMCS	MD
μ					
mean	7.46	7.50	7.17	7.56	7.44
MSE	0.34	0.41	0.63	0.56	0.36
σ					
mean	4.99	4.96	5.22	4.90	5.01
MSE	0.19	0.23	0.35	0.33	0.20
N					
mean	150.1	149.8	153.4	149.3	150.4
MSE	37.8	47.7	84.4	66.2	41.9

Table 2: μ=5, σ=5, N=150, $\bar{n} \approx 126.2$

	MLE	3M	MCS	MMCS	MD
μ					
mean	4.91	4.94	4.48	4.76	4.92
MSE	0.81	0.97	1.53	1.37	0.80
σ					
mean	5.01	4.98	5.26	5.08	5.00
MSE	0.32	0.39	0.57	0.53	0.32
N					
mean	152.1	151.9	159.1	154.9	152.0
MSE	238.2	275.6	472.8	392.5	221.6

Table 3: μ=2.5, σ=5, N=150, $\bar{n} \approx 103.6$

	MLE	3M	MCS	MMCS	MD
μ					
mean	2.17	2.22	2.01	2.07	2.47
MSE	3.67	4.19	2.29	2.23	1.58
σ					
mean	5.05	5.03	5.17	5.14	4.96
MSE	0.67	0.75	0.60	0.59	0.48
N					
mean	159.7	160.1	162.3	160.6	152.7
MSE	2521	3499	1082	1016	716

Table 4: μ=0, σ=5, N=150, $\bar{n} \approx 75.1$

	MLE	3M	MCS	MMCS	MD
μ					
mean	-0.51	-0.50	0.40	0.28	0.98
MSE	15.3	24.3	2.19	1.92	2.60
σ					
mean	5.01	4.97	4.76	4.82	4.53
MSE	1.67	1.96	0.57	0.50	0.71
N					
mean	240.3	2724	144.7	147.6	131.5
MSE	355339	3×10^9	1133	1084	1225

Table 5: $\mu = -2.5$, σ=5, N=150, $\bar{n} \approx 46.3$

	MLE	3M	MCS	MMCS	MD
μ					
mean	-2.57	-4.18	-0.51	-0.82	0.70
MSE	23.26	195.54	6.16	4.40	11.56
σ					
mean	4.96	5.30	4.25	4.37	3.79
MSE	1.38	18.2	1.11	0.84	1.91
N					
mean	285		107	113	83
MSE	340491		2830	2267	4982

their simultations, $n = 150$, as they simulated until 150 non-truncated observations had accrued in the sample.

The present study is examining four alternatives to MLE and comparing them to MLE by MSE.

6 Results

Simulation results are all for sample sizes of 1000 pseudo-samples. The observed mean sample sizes are recorded for each of the simulated populations.

In Table 1, the MLE is clearly best by the MSE criterion.

In Table 2, the MLE is still a very good estimator, although approximately tied with MD.

In Table 3, 3M has become the worst of the estimators, a trait it will retain throughout the shift to fewer observations being non-truncated. Cohen (1991) had speculated that 3M might become the better estimator as MLE began to deteriorate. MD is the best of the estimators here.

In Table 4, MMCS has become the best of the estimators.

In Table 5, MMCS is still the best estimator, while 3M has become very prone to numerical instability.

When $\mu = -3.75$, the sample sizes are small

Table 6: Average sample sizes for distributions not reported in Tables 1–5.

μ	$\hat{n}$
8.75	144.0
6.25	134.1
3.75	115.9
1.25	89.8
−1.25	60.0
−3.75	34.0

enough that it is difficult to get good numerical behavior for MLE or for 3M.

The best approximation to the percentage points of the one-sample Kolmogorov-Smirnov statistic for intrinsic hypotheses was from the MMCS distributions. The approximation was reasonable for the upper 20% for means as small as 1.25 and close to nominal for the upper 5% for means as small as 3.75.

7 Summary

For data losses up to about 16%, MLE is a good estimator.

For data losses from about 16% to about 40%, Minimum Distance is the best esimator considered.

For data losses exceeding about 40%, Minimum Modified Chi-Square is the best estimator considered.

As an overall strategy for diatometer studies, we have recommended the use of Minimum Modified Chi-Square estimator.

8 References Cited

Akhter, Ahmad Saeed (1996). Cohen's method of parameter estimation in truncated normal distribution using moments corrected by quadratic approximation. *Statistica*, 56:167–173.

Bunge, J., and Fitzpatrick, M. (1993). Estimating the Number of Species: A Review. *Journal of the American Statistical Association*, 88:364–373.

Cohen, A. C., Jr. (1949). On estimating the mean and standard deviation of truncated normal distributions. *Journal of the American Statistical Association*, 44:518–525.

Cohen, A. C., Jr. (1950). Estimating the mean and variance of normal populations from singly and doubly truncated samples. *Annals of Mathematical Statistics*, 21:557–569.

Cohen, A. C., Jr. (1959). Simplified Estimators for the Normal Distribution When Samples Are Singly Censored or Truncated. *Technometrics*, 3:217–237.

Cohen, A. C., Jr. (1961). Tables for Maximum Likelihood Estimates: Singly Truncated and Singly Censored Samples. *Technometrics*, 3:535–541.

Cohen, A. C., Jr. (1991). *Truncated and Censored Samples: Theory and Applications*. New York: Marcel Dekker.

Cramér, H. (1951). *Mathematical Methods of Statistics*. Princeton. 575pp.

Gauch, H. G., Jr., and Chase, G. B. (1974). Fitting the Gaussian Curve to Ecological Data. *Ecology*, 55:1377–1381.

Hald, A. (1949). Maximum likelihood estimation of the parameters of a normal distribution which is truncated at a known point. *Skandinavisk Akturietidskrift*, 8:65–88.

Hansen, J., and Zeger, S. (1975). A Comparison of Two Methods for Estimating the Parameters of a Truncated Normal Probability Distribution, Draft Technical Report, Biometry Section, Division of Limnology and Ecology, The Academy of Natural Sciences, Philadelphia. 29pp.

Hansen, J., and Zeger, S. (1978). A Comparison of Two Methods for Estimating the Parameters of a Truncated Normal Distribution. In *Biological Data in Water Pollution Assessment: Quantitative and Statistical Analyses. ASTM STP 652*. K. L. Dickson, J. Cairns, Jr., and R. J. Livingston (eds), 29–37. Philadelphia: American Society for Testing and Materials.

Hendrickson, J. A., Jr. (1979). The Biological Motivation for Abundance Models. In *Statistical Distributions in Ecological Work, Statistical Ecology Volume 4*, J. K. Ord, G. P. Patil, and C. Taillie (eds), 263–274. Fairland, MD: International Co-operative Publishing House.

McIntosh, R. (1976). Ecology since 1900. *Issues and Ideas in America*. B. W. Taylor, and J. White (eds), 353–372. Norman: University of Oklahoma Press.

Mood, A. M., Graybill, F. A., and Boes, D. C. (1974). *Introduction to the Theory of Statistics*. Third Edition. New York: McGraw-Hill.

Patrick, R., Hohn, M. H., and Wallace, J. H. (1954). A New Method for Determining the Pattern of the Diatom Flora. *Notulae Naturae*, 259. 12pp.

Preston, F. W. (1948). The Commonness and Rarity of Species. *Ecology*, 29:254–283.

Rao, C. R. (1965). *Linear Statistical Inference and its Applications*. New York: Wiley.

STATISTICAL DESIGN AND ANALYSIS OF EXPERIMENTS
AS APPLIED FOR WEAR OF ENGINEERING MATERIALS STUDIES

Surapol Raadnui, King Mongkut's Institute of Technology North Bangkok
Surapol Raddnui, KMITNB, 1518 Pibulsongkram Road,Dusit, Bangkok, 10800, Thailand

Key Words: Wear, Design of Experiment (DOE)

ABSTRACT

Starting from the technical and economic need to evaluate the operating variables that significantly affects wear of engineering materials in detail, statistical methods have been employed. The most important advantages of the application of mathematical statistics in the study of complicated wearing systems are, among others, the complete and reliable information obtained for both the effects of the main factors and the existing interaction between factors and also the reduction in experimental expenditure as a whole.

In this particular work, two typical wear tests, namely, adhesive and abrasive wear tests were conducted and statistically assessed. A fractional factorial experimental design was implemented and analyzed using ANOVA table. The experiments have been realized according to special factorial designs that have varied the levels of five important cause variables in each typical wear test series. Within the confines of the test rigs, effective factors and interaction effects were quantified and discussed.

INTRODUCTION

Wear of machine elements or tools is one of the most common problems encountered in industry today. Of the different mechanisms of wear, namely, adhesive, abrasive, fatigue and tribochemical reaction. All wear processes, it is necessary to know the affect of the different wear parameters. Important parameters affecting the abrasive wear behavior are thought to be abrasive media characteristics, applied load, speed, hardness and characteristics of engineering materials. The relationship between these parameters and their effects on the wear process are not clearly known due to the complexity and randomness of the process (Dasgubta et. al, 1997). Similarly, Important parameters affecting the sliding wear behavior are thought to be materials characteristics, contact configuration, applied load, speed, hardness and characteristics of environments. In addition, several studies have been focussed on the influences of wear variables on friction and wear characteristics of bearing materials (Begelinger and De Gee, 1972, 1974; Habig, et. al, 1981 and Bayer, et. al, 1975). However,

the relationship between wear variables and their effects on the wear process are not clearly reported.

The existence of variables, time spent and random characteristics of wear processes requires an efficient experimental design study to analyze the process. The design of the experimental method is a typical approach to identify any complex process characteristics efficiently and logically by experiment (Spuzic et. al, 1997). A fractional factorial experimental design which much information is extracted using a small number of experiments is utilized. The experimental results are analyzed using ANOVA (analysis of variance).

EXPERIMENT

The objective of this study was to determine which factors were affecting the characteristics of abrasive and adhesive wear processes. A careful study of the wear process identified five variables that were thought to be important and were candidates for possible control procedures. It was determined that five factors, for each case of fundamental wear processes, should be varied over the ranges in Tables 1 and 2.

Table 1 Experimental levels for abrasive wear tests: five-factor experiment.

Factor	High (+)	Low (-)
A:Load	20 N	10 N
B:Speed	1.0 m/s	0.5 m/s
C:Test duration	2 min.	1 min.
D:Mean grain size	425 μm (#60)	250 μm (#40)
(Al_2O_3 abrasive media)		
E:Electrode type	$\leq 3\%$Cr.	$\geq 30\%$Cr.

Table 2 Experimental levels for adhesive wear tests: five-factor experiment.

Factor	High (+)	Low (-)
A:Load	30 N	10 N
B:Speed	0.50 m/s	0.20 m/s
C:Test duration	3 min.	1 min.
D:Surface roughness (R_a)	6 μm	1 μm
E:lubricant	Dry	Lubricated

The characteristics of specimens, testing procedures and wear testers used in this work were reported elsewhere (Raddnui, 1999). Two–level fractional–factorial designs are usually constructed assuming that it is of interest to estimate linear effects free and clear of two-factor interactions and that three and four-factor interactions rarely exist in nature (Mason et. al, 1989). Hence, little effort should be spent to estimate them, particularly at the beginning of an experimental program. Two-factor interactions, on the other hand, occur frequently and a good design is the one that estimates them directly or separates them from the linear effects. As it was pointed out above, three (or more) factor interactions rarely take place. Therefore, in this paper, only five main effects (A, B, C, D and E) and ten two factor interaction effects (*i.e.* AB, AC, AD, AE, BC, BD and so on), in each case, were considered in the experimental design. All the other effects were considered as trivial effects.

EXPERIMENTAL RESULTS AND DISCUSSION:
Abrasive Wear Tests

Analysis of variance (ANOVA), was used to assess the significance of the factor main effects and factor interactions, the conclusions for abrasive wear experiments were that:

1. Factor E affected the average response at the 1 percent level of statistical significance.
2. Factors A, B, and C affected the average response at the 5 percent level of statistical significance.
3. Factor D did not have a significant effect on the average response value.
4. Interactions AB, AD, AE, BC, BD, BE, CD and CE affected the average response at the 5 percent level of statistical significance.
5. Interactions AC and DE did not have a significant effect on the average response value.

Plots of the factor main effects are in Fig. 1. Selected plots of the two factor interactions are in Figs.2 and 3.

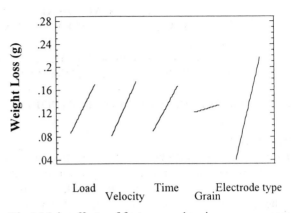

Fig.1 Main effects of factors on abrasive wear process.

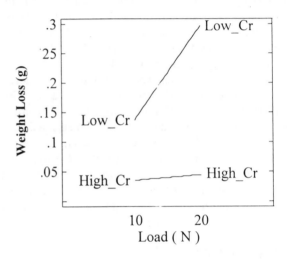

Fig.2 Two-factor Interaction plot (AE).

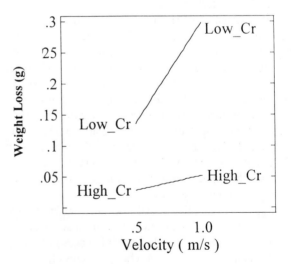

Fig.3 Two-factor Interaction plot (BE).

From the above figures, it can be concluded that:

1. LOAD X ELECTRODE TYPE – Since the aim was to identify factors and settings that resulted in the most wear, setting factors A (load) and E (electrode type) at their high levels would produce the largest average response.

2. SPEED X ELECTRODE TYPE – Similar wear behavior to the above was observed. However, the effects of changing speed levels on wear were greater when load level was changed in general.

EXPERIMENTAL RESULTS AND DISCUSSION:
Adhesive wear tests

Similar to the previous discussion, utilizing ANOVA table, the conclusions were that:

1. Factor E and B affected the average response at the 5 percent level of statistical significance.

2. Factors A and D affected the average response at the 10 percent level of statistical significance.

3. Interactions AE, BE, CD and DE affected the average response at the 10 percent level of statistical significance.

4. Interactions AB, AC and BD had a significant effect on the average response value at the 25 percent level of statistical significance.

Plots of the factor main effects are in Fig.4. Typical plots of the two factor interactions are in Figs.5 and 6.

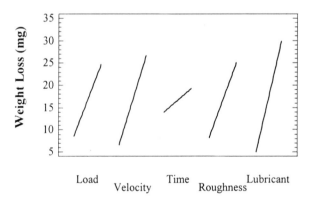

Fig.4 Main effects of factors on adhesive wear process.

Interpretation of the two larger interaction terms is as follows.

1. SURFACE ROUGHNESS X LUBRICANT– Since the aim was to identify factors and settings that resulted in the most wear, setting factors D (surface roughness) and E (lubrication condition) at their high levels would produce the largest average response.

2. SPEED X LUBRICANT – Similar wear behavior to the above was observed. However, the effects of changing speed levels (from low to high) on wear were greater when speed level was changed (from low to high) in general.

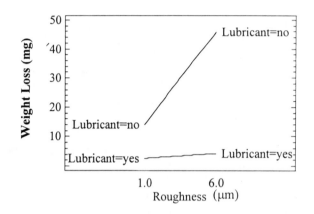

Fig.5 Two-factor Interaction plot (BD).

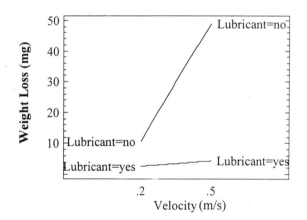

Fig.6 Two-factor interaction plot (AD).

CONCLUSIONS

The conclusions that could be drawn from these experiments included:

1. DOE allow us to measure the influence of several wear factors on a response.

2. Statistically designed wear experiments are economical *i.e.* the total number of required test conditions can be successfully reduced.

3. DOE allow the estimation of the magnitude of experimental error to be made.

REFERENCES

Bayer, R.G. (1975), The influence of lubrication rate on wear behaviour, Journal of Wear, 35, pp. 35-40

Begelinger, A. et. al (1972), Boundary lubrication of sliding concentrated steel contacts, Journal of Wear, 22, pp. 337-357

Begelinger, A. et. al (1974), Thin film lubrication of sliding point contacts of AISI 52100 steel, Journal of Wear, 28, pp. 103-114

Dasgubta, R. et. al (1997), Slurry erosive wear characteristics of hard faced steel: effect of experimental parameter, Wear 213, pp.41-46

Habig, K.H. et. al (1981), Friction and wear tests on metallic bearing materials for oil-lubricated bearing, Journal of Wear, 1981, pp. 43-54

Mason, R.L. et. al (1989), Statistical Design and Analysis of Experiments, John Wiley & Sons, Inc.

Raadnui, S. (1998), Evaluation of operating variables that significant affect wear of hard facing surfaces utilizing fractional factorial experimental design, Proceedings of the 11th International Conference on Condition Monitoring and Diagnostic Engineering Management (COMADEM'98), pp.713-720, Tasmania, Australia

Raadnui, S. (1999), Statistical approach for experimental analysis of sliding wear behaviour of bearing materials, Proceedings of the 12th International Conference on Condition Monitoring and Diagnostic Engineering Management (COMADEM'99), pp.153-157, Sunderland, England

Spuzic, S. et. al (1997), Fractional design of experiments applied to a wear simulation, Journal of Wear, 212, pp.131-139

COMMERCIAL QUALITY:
THE NEXT WAVE IN STATISTICAL THINKING?

Rekha Agrawal, Roger Hoerl, GE Corporate Audit Staff
Rekha Agrawal, GE Audit Staff, 3135 Easton Turnpike, Fairfield, CT 06431

Key Words: Six Sigma; Non-Manufacturing

A statistician working in the private sector today is facing a dramatically changing environment. These changes are universal rather than local and are likely to affect the entire profession. Its ability to respond rapidly to these changes will determine the degree to which the profession can impact the organizations in which statisticians work. In this paper, we will discuss our perceptions of this new environment, and our views on the appropriate response.

The organization of this paper is as follows: (1) motivation for why statisticians should respond proactively to the new environment, (2) the specific environment in which we, the authors, work (GE Corporate Audit Staff), (3) background on GE's Six Sigma initiative, (4) how Six Sigma has impacted our roles, (5) some specific examples of the work we've been doing in these roles, (6) our view of the future of the statistics profession, and finally some conclusions.

Motivation

Important challenges are surfacing in the current business environment for statisticians. Some may view these changes as threats to the job security of professional statisticians. For example, as easy-to-use statistical software proliferates, statisticians no longer find themselves the exclusive "owners" of statistical analysis. At GE, for example, most employees have Minitab pre-loaded whenever they obtain a new computer. Coupling this with their Six Sigma training (which we will discuss in further detail later), these employees are able to do statistical analysis at their desks whenever they want. This obviously reduces the need to call a statistician when, for example, a DOE is required or regression analysis is called for.

Also, traditional markets for industrial statisticians are becoming saturated. Many manufacturing environments, for example, have been using statistical tools for years. The engineers are very comfortable with our roles in such environments. It can be argued that there is little room for the creation of more statistical consulting roles in such environments.

Also, there are several professions that are encroaching on our core competencies. For example, at GE and elsewhere "data-miners" have taken on a unique identity, apart from the statistics profession. Often, these people have expertise in areas such as operations research, simulation and optimization, computer science, or even marketing and financing. It is not clear how the work that these people do is unique from statistical analysis of large data sets, but there is certainly a perception of a unique discipline here. Modeling using neural nets is another example of other disciplines, e.g., chemistry and chemical engineering, taking a lead role in what could be considered statistical methodology.

In today's environment, it is critical that statisticians contribute to short term, bottom line results. This need has always existed to some degree, but in today's environment the pressure is much greater. What we do as statisticians needs to be seen as clearly adding immediate value. If it isn't, our organizations will quickly find other professions and people that can.

Other factors are also having a profound impact on the traditional role of an industrial statistician. World wide, there is a general shift to a service economy. This is true not only in the United States (Dow Jones, November 3, 1999), but also in places such as Europe (Wall Street Journal, November 8, 1999) and Australia (Dow Jones, November 9, 1999). Much more of the energy spent is directed towards services, as opposed to manufacturing. Specifically, industries such as E-commerce and other high technology areas are expanding very rapidly. Since these are areas that industrial statisticians have not traditionally played in, such a shift in the world economy may appear to threaten our job security. The positive aspect of this change, however, is that there is a lot of growth opportunity in these areas, since they are relatively uncombed. Further, if the profession can demonstrate added value to these new and expanding areas, then it can grow as this sector does. In fact, as we will elaborate on shortly, we view these changes as tremendous opportunities.

GE Corporate Audit Staff

The environment in which we work reflects a lot of the challenges that are listed above. The Audit Staff is a part of the Corporate Finance organization of GE. It supports all of the different GE businesses around the world. GE has 12 very diverse businesses, including such things as aircraft engines, medical imaging, consumer appliances, power generation, broadcasting (NBC), and a huge financial services business (GE Capital). The GE audit staff works with all of these businesses, at all of their different locations around the world.

Its primary objective is one of controllership, ensuring the integrity of GE's financial reporting, and compliance to all federal and local regulations. However, another critical role that it plays is that of leadership development. Many of the future leaders of the company are likely to be on the audit staff early in their careers. For example, the current CEO of GE Capital, and also the Chief Financial Officer of GE are Audit Staff graduates. Finally, the audit staff plays a key role of supporting operational excellence. In other words, auditors often support efforts to investigate and improve key business initiatives, such as Six Sigma, supply chain management, globalization, and E-commerce.

The Audit Staff tends to be composed of young, talented, ambitious people. By design, they are from very diverse backgrounds, not only in terms of their formal education, but also gender, ethnicity, and citizenship. Significant (and growing) portions of auditors are citizens of European and Asian countries, and are based out of these regions. While they are on staff they travel all over the globe, switching project assignments and locations every four months. They work extremely long hours, and typically only return home every other weekend, or once a month if they are working outside their home region. In exchange for this commitment, successful auditors are rapidly promoted, and receive very important jobs upon leaving the Audit Staff.

Clearly, the environment in which we are working is very different from that of a traditional manufacturing environment. In what follows, we will describe what our role is in this environment, and some of the challenges and opportunities we face. In order to understand this role, however, we will first give the context of the Six Sigma initiative at GE, and the impact this initiative has had on how statistics is applied.

Six Sigma at GE

Six Sigma is a quality and business improvement initiative that GE began in 1995. It focuses on eliminating defective products and services by improving the processes that generate them. Six Sigma was originally developed at Motorola. See Harry (1994) for more details on the origins of Six Sigma.

In our opinion, one of the key differences between Six Sigma and previous quality initiatives is the level of upper management support that it has received. Jack Welch has been extremely passionate in driving it through the organization, taking measures such as tying 40% of the business CEO's incentive compensation to Six Sigma results. One of the striking messages that Welch has sent is insisting that people with Six Sigma experience will be rewarded by being given "big jobs" and leadership positions. This has generated tremendous enthusiasm, and attracted people viewed as top performers to Six Sigma.

We believe that another key reason for the impact of Six Sigma is the fact that it has been tied to financial benefits. Figure 1 shows a chart of the costs and benefits of Six Sigma since 1996 (1998 GE Annual Report). This chart clearly shows that the investment the company has made in this quality initiative has paid off financially – we are projected to achieve more than *two billion dollars* in benefits this year (1999).

Six Sigma Progress

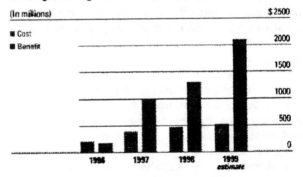

Figure 1. Six Sigma Costs vs. Benefits.

As part of this initiative, virtually all exempt employees at GE, in all of the different businesses, have been trained in Six Sigma. This has involved a massive amount of statistical training, perhaps the most massive statistical training effort in history. In most businesses, for example, SPC and design of experiments are a part of the standard training. A broad base understanding of some of the basic statistical concepts, terminology, and tools has been the result. See Hoerl (1998) and Hahn et al. (1999) for further discussion.

People in the company have different titles and roles depending on what part they play in this initiative. For example, Green Belts are people who are trained in the Six Sigma tools, and who apply Six Sigma to their jobs. Black belts, on the other hand, are fully dedicated to Six Sigma. This is their only job. They lead projects, and they implement process improvements using the appropriate tools. Master Black Belts mentor and train Black Belts, and guide the overall effort. Finally, champions are the process owners who promote the Six Sigma mindset, and eliminate barriers to implementation.

The evolution of Six Sigma at GE has been an interesting one. The initiative started off very focused on manufacturing improvements, on reducing defects in internal processes, and saving money in re-work and defect avoidance. It's probably safe to say that the projects done at this time were not all aligned strategically. Gradually, Six Sigma evolved to the design phase of a product or process life cycle – helping to build a better product the first time. Next, the initiative moved to general business processes, what GE calls Commercial Quality, applying the statistical tools and concepts to non-manufacturing processes. At about this time, the nature of Six Sigma projects also changed to become more strategically aligned, where individual projects were mapping to larger efforts. The final stage of this evolution is incorporating Six Sigma into designing better commercial quality products and processes, such as call centers, consumer credit processes, E-commerce, and so on. Most of the GE businesses have begun on this last stage fairly recently.

To understand the strategic importance of Commercial Quality to GE, it is important to keep in mind that GE Capital provides about 45% of the profits of the company. Continuing to focus on manufacturing would miss an extremely large portion of the total opportunity. In addition, as Deming often noted, even in a manufacturing business, a small percentage of the people actually build the product. The rest of the workforce is involved in non-manufacturing processes (Commercial Quality), such as sales, marketing, HR, IT, logistics, purchasing, and so on.

Our Role

On the Audit Staff, we form a separate Quality organization, which reports directly to the Vice President of the Audit Staff. The VP of Audit Staff is a corporate officer of the GE Company, one of only about 125 in the entire company, which has about 300,000 total employees. The fact that the Quality Leader reports to her is indicative, we think, of the importance that quality, and Six Sigma, is given.

Our role involves defining how Six Sigma can help the Audit Staff improve its performance,

developing the auditor's skills to do this, and then mentoring and supporting the auditors in implementation. Specifically, we have given ourselves the following objectives:

To drive a Six Sigma mindset in the organization, as defined by the following behaviors:
- Creative understanding/application of the appropriate Six Sigma tools, for all audits.
- Widespread approach of quality to all audits.
- The institutionalization of metrics that measure auditors' approach to quality, and tie to their appraisals.
- Tying all efforts to the defined customer Y, for each review.
- More rigorous, data-driven approach to all audits.

We have succeeded in making progress on all of these objectives. In particular, the auditors' appraisals now include a section that incorporates Six Sigma know-how. We would like to feel that these things are indicative of our vitality and integral role within the organization.

Examples

In this section we'll briefly review some examples of the projects that we've been involved in. We hope to show that while the application areas are very different from what "industrial statisticians" have traditionally been involved in, they are very amenable to statistical thinking.

1. Data Quality on a Website

The first example we'll discuss is a situation in which our company was providing continuous information over a website. The accuracy of the information given on this site is critical, and could have significant implications to many people using the site. Therefore, this project had as a goal to improve the accuracy of information on the site.

Many interesting questions came up as we were investigating this project. For example, there is no "true" or "standard" source for the data that we were providing. Therefore, how do we assess accuracy? How do our customers assess accuracy?

Another interesting feature that came up in this investigation had to do with the fact that the data is essentially supplied to us from another source. This raised a variety of questions – How do we drive improvement from the supplied source? What are the implications of the fact that some of our competitors are using the same data source, while others are using a different data source?

We were able to make a significant improvement in the accuracy of this site as a result of applying statistical thinking and some basic statistical tools.

2. Sales Margin

Another project in which we were involved was with one of GE's manufacturing businesses. This is a manufacturing business in which a wide range of products is sold, and so the amount of margin from a sale depends heavily on which type of product is sold.

The business observed that there had been a decline in sales, when comparing sales for a given quarter to sales in the same quarter for the previous year. A team was put together and asked to investigate reasons for why this might have occurred. Further, the team was asked to look at whether or not the business was losing high end product mix. In other words, was there reason to think that the business' market for high margin products was decreasing?

When we got involved in this project, the first question we asked was "Does the goal make sense?" We were concerned about whether the year over year comparison accurately reflected a decreasing trend, or whether, for example, the previous year didn't just happen to be particularly high in sales.

There were other questions that the project team was having difficulties answering. For example, this business had a variety of different customers, whose size ranged from very small to very large. The project team was trying to understand how customer size impacted the conclusions about sales.

This project was an example of a situation in which some fairly simple statistical tools, such as appropriate graphs and summaries, enabled the team to communicate very effectively with upper level management about what the data were saying. Essentially, the data allowed us to ask the right questions about the sales process, and helped us to focus resources in addressing the solutions. Interestingly, the sales data had not been "mined" prior to the application of Six Sigma, to better understand the sales process and profitability, although it certainly existed previously.

3. Contingency Planning

One of GE's businesses abroad was under some serious cost constraint pressures, and it was decided that to deal with these pressures, some changes to the organization were required. In particular, it was felt that there was a need to reduce headcount. This was an unfortunate situation, but one deemed necessary in light of the business environment. We were not asked to agree or disagree with this decision, only to help apply Six Sigma to implement the decision as effective as possible.

In the country in which this GE business was located, there were particular laws on how such a headcount reduction was to be executed. Specifically, an announcement of the intended reduction was required months before it was actually executed. This was very different from the way such a change might be done in the United States.

The goal of the project that we were involved in was to assure maintenance of the operational quality levels of the business during the transition period. There was a fear that while the employees of the company were under notice that there would be reductions, they may not be motivated to perform as they had been. Given the length of the transition period, this could have serious negative consequences for the business.

There were many issues that became apparent during the course of this project. There was clearly a need to react quickly, if the performance of the business began deteriorating. Lack of a timely response would put the remaining employees' jobs at risk. At the same time, given the sensitive nature of the changes, overreaction would also affect the organization adversely. This implied a need for timely information, and this information needed to be communicated frequently to all the right constituents. Finally, there was a need for clear interpretation of the data.

At the end of this project, it was felt that statistical thinking had helped in resolving this critical business issue, to make the business decision as effective as possible.

4. Other examples

There are many other fascinating issues that we have been involved with. We name a few below:
1. Optimal pricing strategies
2. Understanding variable cost productivity
3. Reserve analysis at an insurance company

4. Understanding what drives the amount of time required to qualify a new supplier
5. Improving an accounts payable process.

The above list should indicate that statistical thinking can be applied in any area or business processes. The breadth of applicability is astonishing. In addition, we have found that significant improvements are often possible by applying simple statistical methods on existing data. In other words, the rate of improvement in Commercial Quality is often greater than in manufacturing and engineering, and the payoff is often greater.

What about Statistical Consulting?

We would like to emphasize a point here about statistical consulting. We feel very strongly that what we do is not statistical consulting. Consultants in general are a dying breed at GE, and the word has negative connotations associated with it. The reasons for this are clear: consultants are perceived to be passive givers of advice on other people's projects, which is not what our business culture values. We feel that we need to take ownership of results, and personally execute, as opposed to being consultants. We need to be leaders on the critical path of the important initiatives, and we cannot do that as consultants. Hence, we make a major distinction between statistical consulting and what we do. This is not to say that statistical consulting is "bad", or that we never consult. The point is that our role is significantly changed in the new environment, and we have to use our previous experience and credibility as consultants as a base in order to move on to more strategic roles.

The Look of Statistics in the Future

Our experience suggests that our profession is changing in dramatic and exciting ways. Further, since the root causes of these changes lie in the business environment, rather than within the profession, we do not believe that statisticians have the option of "just saying no". Below are some of the changes that we perceive:

1. *Blurring between industrial and business statisticians.* We see the distinction between these titles becoming obsolete. Statisticians in the private sector will need to be flexible in applying statistics to a whole variety of fields – everything from manufacturing and designing new products to working on marketing, sales, and service processes and products.

2. *Fundamental change in role.* The statistician's role is changing from one of a technical consultant to more of a strategic role. Rather than performing statistical analyses, the role involves such things as

ensuring that others have the right tools and training to do their own analyses, defining the company's strategy for the use of statistics, auditing where the company is doing well in the use of appropriate methods, determining when and how external consultants should be used, and so on. While this has exciting implications and ramifications, it inherently has more risk. These tasks require a completely different set of skills, and most of us statisticians find them harder than data analysis!

3. *Importance of role is elevated.* Since the role is becoming more strategic, we think that this change will be reflected in the organizational structure. There will be more pressure to perform consistently with good results, and more people will be watching.

4. *Technical skills are still key, but a variety of other skill are also required.* While technical skills will always be our core competency, statisticians in the future will also need to exceed in organizational effectiveness, be able to think strategically about an entire business model, quickly learn new application areas, and be able to drive tangible business results. Thus, the bar is being raised on the necessary qualifications to be successful in this environment.

5. *Breadth of learning opportunity is huge.* For statisticians who have the curiosity and the energy to learn about the application areas in which they are working, there is a huge opportunity. This creates new career opportunities.

6. *New dilemma: Success most likely implies promotion out of the role.* As a result of the combination of factors given above, it is quite conceivable that the statisticians of the future will not stay statisticians for very long. By being broader and more business focused versus technique oriented, they will most likely be so valuable to their organizations that they will be offered other roles. This can only be good news for our profession in general, but has some consequences in terms of the pipeline that we need to keep replacing those people. This is what we are seeing at GE. In the company's largest group of statisticians, the people within the group are frequently promoted to other roles, and the group is constantly having to replace them.

While these types of changes are taking place, our profession needs to take a really good look in the mirror and ask ourselves some hard questions:
1. Do we want to change – and are we energized?
2. Are we prepared to step up to the new role? *If we don't, others will!*
3. Are we attracting the right people for the job?

4. Are we educating people in an appropriate manner for the challenges they will face?

Until we are comfortable that we as a profession can handle the magnitude of the impending change, all statisticians should be feeling some apprehension. Our concern is that too many in our profession are too complacent to be giving serious thought to the future.

Conclusions

We hope that we've clearly demonstrated that statisticians are in a new and unfamiliar environment, dealing with new problems and challenges. There are a plethora of opportunities to apply statistical thinking to almost every business process – financial, service, marketing, just to name a few. However, a dramatic mindset change will be required if we as a profession are going to be successful at dealing with the new environment. There is a huge opportunity to make an impact on our organizations – if we don't do it, others will.

Are you ready?

References

1. Dow Jones, "US NAPM Non-Mfg Oct Business Index 60; Sept 61", Brian Blackstone, November 3, 1999.

2. Dow Jones, "Australia Watch: 2000 Econ Set For Growth And Distortion", Iain McDonald, November 9, 1999.

3. Harry, M.J. (1994), The Vision of Six Sigma: Roadmap for Breakthrough, Sigma Publishing Company, Phoenix, AZ.

4. Hahn, G.J., Hill, W.J., Hoerl, R.W., and Zinkgraf, S.A. (1999), "The Impact of Six Sigma Improvement", The American Statistician, 53, 3, 1-8.

5. Hoerl, R.W., (1998), "Six Sigma & the Future of the Quality Profession", Quality Progress, June, 35-42.

6. Wall Street Journal, "Deregulation Could Trim Structural Unemployment", Sarah Ellison, November 8, 1999.

BURN THE BROWNIES

Effective Experimental Design through Statistical Thinking

Lynne B. Hare
Nabisco, Inc.

Key Words: Statistical Thinking, Experimental Design

Sarah's Story

Sarah was in sixth grade when she came home from school and said to her mother, "Mom, the teacher says we have to do a science project and write a paper about it. What should I do?"

Her mother said, "Ask your father. He's a scientist."

Sarah came to me, somewhat reluctantly I think, because she knew that I'd suggest something weird.

"Well, you like brownies, don't you?"

Softly, and not revealing an abundance of parental trust, "Yes."

"What is it you like most about brownies?"

Somewhat less reserved, "The chocolate flavor."

"Let's bake brownies. You could make some with extra chocolate, and you could see how closely you have to follow the baking destructions." I know it's directions, but our family always calls them destructions. It might have something to do with my facility around the house.

Recognizing that her mother had abandoned her to her father's care and that while she might suffer her father's idiosyncrasies for a while, at least she'd get to eat a few brownies, Sarah agreed.

Off to the grocery store we went. Sarah and I (mostly I) bought 8 boxes of brownie mix and a package of bakers' chocolate. Then we went home to attack the kitchen. For our control sample, we used the manufacturer's formulation, baking time and baking temperature. Then we modified these directions with increased levels of chocolate, baking time and baking temperature according to the 2-level factorial design shown below.

Batch	Chocolate	Baking Time	Baking Temperature
1	Regular	Regular	Regular
2	High	Regular	Regular
3	Regular	High	Regular
4	High	High	Regular
5	Regular	Regular	High
6	High	Regular	High
7	Regular	High	High
8	High	High	High

We discussed (Daddy went off on) the importance of randomizing the order of baking, and finally, with just enough time to clear the kitchen for supper, we had baked all 8 packages of brownies.

Sarah announced happily to her older siblings that we were having brownies for dessert. They revealed the usual sibling non-objectivity by saying something to the effect of "Big whoop," and retreating to the family room until the prospect of being asked to help clean the kitchen passed. (Their father had spoiled them. They've since recovered nicely.)

After dinner, because Mom said we had to wait until then, we had a tasting party. Our friends, Ed and Frances, came over to help. That made Sarah happy because Ed and Frances tended to be more objective about brownies and other things than her brother and sister. It made me happy because it increased the size of the panel of judges, and I had better jolly well find something worth saying as a result of this experiment or we'd never do science Daddy's way again.

We asked everyone to taste the brownies, being careful not to eat too much (until later) and to check boxes on the form we made up to reflect how well each judge liked the brownies and how strong they thought the chocolate flavor was. Our rating scale was a typical food industry intensity scale.

After Sarah's bedtime, I crunched the numbers and came up with a peculiar interaction involving the amount of chocolate flavor and the oven temperature. This is not what I had expected, but then science is all about the discovery process. The next day, Sarah and I sat down to go over the results. I explained the interaction to her. "With the recommended oven temperature, an increase in chocolate increases the perception of chocolate flavor slightly, but with a higher temperature, increasing the amount of chocolate has a much more pronounced effect on chocolate flavor perception."

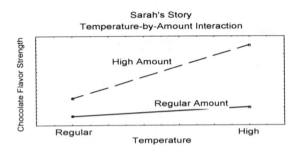

Sarah had no trouble understanding this notion, especially after seeing the hand-drawn version of Figure 1, and she started writing up her science project. I had some nagging doubts about the results, however, and I discussed these findings with food science colleagues at work. They drew long chain polymers on my board for a half-hour, giving me a dose of what I had been giving them, and then said, "Yes, it stands to reason." It is always a good idea for conclusions to have some basis in subject matter theory. I relayed this confirmation to my daughter.

Sarah has written beautifully almost as long as I have known her, and I had no doubt that she would explain her project accurately and concisely. A few days later, I asked her if the teacher had returned her paper, because I wanted to know how I did. "Dad, it's too soon for that. The teacher never gives us our grades back until a couple of weeks."

So I waited a couple of weeks and asked again. This time Sarah's response was unusually sheepish. "The teacher said that what I did *isn't science.*"

I came unglued. I started barking. Sarah's siblings took great pleasure in watching me flail my arms, and rant and rave. But to Sarah's absolute horror, I said I was going to have a discussion with the teacher. "No," her mother interceded, much to Sarah's relief, "You

should not have a discussion with the teacher. Not until you calm down."

My wife, you see, is a teacher, too. And she knows what they're up against. It's not an easy job. At the end of my mandatory cooling off period, I went to the school to express my views. "I'm an industrial statistician," I explained. "I work with scientists and engineers to maximize the information obtained from experimentation and to improve products and processes by using statistical methods. What Sarah did was, in miniature, what I do at my company on a daily basis. Further, some of the greatest advances in medicine and in industrial productivity and quality have come about as a result of the application of statistical methods. I'd like you to reconsider your position on Sarah's paper."

The teacher agreed to think about it, but said that it was certainly different from any science in her experience. Of course it is, I thought. It has only been around for 75 years. What should I expect?

The school year was coming to a close, and it had been months since Sarah turned in her paper. I had already caused her enough strain, so with some reluctance I asked her about her paper.

"Oh, that," She replied. "The teacher said that the science project didn't work, and she threw out all the papers."

Ouch!

My editor friends have commented that it is a shame that Sarah's paper is lost. They would have published it to show that children as young as Sarah can understand the principles of experimental design.

Statistical Thinking as a Key to the Use of Experimental Design

Why is it, I have often wondered, that there is such reluctance to understand and use planned experimentation? At one company, I spent literally years building a clientele of scientists and engineers who would actually design experiments using the techniques they learned from me and from the statisticians in my group. I did this via educational sessions, by one-on-one consulting sessions and by providing successful collaborators an opportunity to give testimony of their successes gained through the use of the discipline. Almost all the established culture crumbled when a key research leader retired and was replaced by someone who lacked the understanding or appreciation of designed experimentation. I had to rebuild, and sadder but wiser, I started by seeking root causes of motivation.

My strategy is different now and, I like to think, more successful. Now, I take the process view, and I work to learn what motivates people. If I can understand that, perhaps I can appeal to their values. It seems that most people are irritated by variation, at least to the extent that they realize that if they could reduce variation, things would run more smoothly. Manufacturing processes would run with fewer interruptions, and research would take place with greater clarity.

So, I push Statistical Thinking. Statistical Thinking is a philosophy of learning and action based on the following fundamental principles:

- All work occurs in a system of interconnected processes,
- Variation exists in all processes, and
- Keys to success are understanding and reducing variation.[1]

These are not my thoughts. Several of us from the Statistics Division of ASQ have been promoting Statistical Thinking for 5 years.[2][3][4] Also, others have expressed similar ideas under different banners. See Brian Joiner's book, for example.[5]

By *philosophy*, we mean literally, the "love of wisdom" as the Greek implies. This is our basic mind-set, our attitude as we approach the application of our discipline. And as we apply this statistical discipline, learning grows.

There seems to be little argument with the notion that *"all work occurs in a system of interconnected processes."* Its application to manufacturing is almost a no-brainer. The transition to other kinds of work processes such as conducting meetings and matching checks to invoices in Finance is a little less obvious, but people generally understand it.

That *variation exists in all processes* is easy to grasp in most situations, especially as it involves processes that belong to others. It is important that we understand the difference here between variation and variety. Certainly, in the creative process, *variety* of thinking is to be desired, while in the course of conducting many business processes, the reduction of *variation* will lead to improvement.

Lastly, there can be little real process improvement without an understanding of variation. For understanding variation is essential for its reduction, and without reducing variation, we cannot make

processes run more smoothly, with higher levels of throughput and acceptance.

Figure 2 is a diagram that I have found helpful to illustrate the potential of the process view and the way statistical thinking leads into more frequent and more effective uses of statistical methods.

Statistical Thinking and Methods

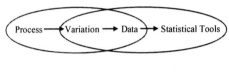

Notice that it is the process view that drives the effort. By process view, I mean the examination of a particular process in its context as it meshes with other, interrelated processes. Such examination opens the opportunity for understanding of the ways variation can creep into the process. This leads to a need to acquire the "right amount of the right kind of data" to provide assessment of the process variation. This sequence, process – variation – data (PVD), sets the stage for sensible use of statistical tools. With the PVD sequence, the use of statistical tools can be proactive and value-rich. Without it, the application of statistical tools is passive, not valued and certainly not understood.

When seeking reasons for the lack of appreciation of the potential of statistical methods, I look at formal education: K-12, college and graduate school. Students claim that statistics was their worst course ever. A colleague, David Moore of Purdue University, tells of a sign posted, we assume, by a student. It said:

If it moves, it's biology;
If it stinks, it's chemistry;
If it doesn't work, it's physics;
If it puts you to sleep, it's statistics!

My good friend, David Banks of the National Institute of Standards and Technology, provided a telling anecdote from a university in the northeast. The statistics professor droned on to such boring lengths that a student yawned sufficiently widely to dislocate his jaw. He screamed in agony, fell to the floor, passed out, and had to be taken from the lecture hall by the university medical staff. I conclude that bad statistics teaching kills.

Seriously, when statistics is taught, the focus is on statistical tools, the right hand end of Figure 2. Relevance is lost because of the missing process view. Teaching focuses on the wrong end of the diagram! No wonder students say that statistics was their worst course ever. The strategy for teaching statistics would benefit from a healthy dose of Statistical Thinking. For more on this subject, see Hoerl and Snee.[6]

OFAAT

When students are not taught well or when, forewarned by their colleagues, they avoid the discipline altogether, they go out into the world to seek their fortunes, missing the opportunity to use Statistical Thinking and methods to advance their careers and make better decisions for their organizations. Their inclination is to make decisions by differences that seem right and to advocate "one factor at a time" (OFAAT) experimentation. I've seen this happen among many that I consider the brightest and best. The temptation to do something "quick and dirty" gives the OFAAT lure great appeal. Besides, they learned from their chemistry teacher that, in order to measure the effect of one variable, they must hold everything else constant and only change that one variable. The chemistry teachers learned it from their chemistry teachers who learned it from theirs, and so on, all the way back to Merlin.

The problems with OFAAT are, of course, that:

- It fails to take advantage of the power of numbers by balancing factor levels among treatment combinations,
- It ignores the existence of interactions or coupled effects among factors, and
- It does not build knowledge.

The Experimentation Process

Instead, experimentation, like all other work, must be viewed as a series of interconnected processes. This is Statistical Thinking applied to experimentation. The late Horace Andrews[7] used to regale his design students with Figure 3 very early in the semester. It speaks to the iterative nature of experimentation, and it is entirely consistent with the message of the opening chapter of Box, Hunter and Hunter,[8] which we all know and love, or should.

Sound experimentation begins with a review of subject matter knowledge coupled with a creative tension that generates new ideas. This leads to the development of

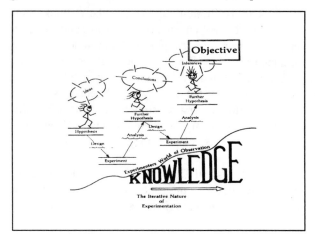

a fundamental plan for experimentation to be carried out carefully and concisely, generating data whose analysis and interpretation will lead to conclusions *and* point the experimenter in the direction of further experimentation.

Experimenters often go to great lengths in the data generation process only to turn their backs on the data derived. As much effort should go into the analysis and interpretation of data as goes into their generation. Understandably, experimenters may be reluctant to apply statistical tools because they feel unprepared to do so. The tools, however, do not have to be sophisticated. The first step in the analysis of any data set should be a simple graph. Data should be examined graphically and in many ways. A good starting point is a plot in the order that the data were generated.

Always, always, always, without exception, plot the data!

I encourage my colleagues to "*Always, always, always, without exception, plot the data.*" Some have actually listened and done it. On one occasion, a young researcher attached a graph of his data to his plant trial report, but he said nothing in his conclusions about an obvious downward time trend in his responses. I've had to modify the "Always, always, always, without exception, plot the data." rule to include "… and look at the plot!"

One experiment is rarely adequate: it takes a sequence of experiments to be fully successful. That is to say, the *process* of experimentation is necessarily iterative. It cycles through inductive and deductive reasoning, while it builds knowledge until an objective is reached. Merely conducting a screening design and playing the winner will not lead to progress. I have had difficulty discouraging that strategy. Some believe that I should be happy that experimenters are, at least, doing something "statistical." I object, contending that it is the *process* - the iterative process of experimentation - that leads to success because it aids in building knowledge. Recently, Box and Liu provided great support.[9] They said:

An industrial innovation of major importance, such as the development of a new drug or the development of a new engineering system, comes about as a result of an investigation requiring a sequence of experiments. Such research and development is a process of learning: dynamic, not stationary; adaptive, not one-shot. The route by which the objective can be reached is discovered only as the investigation progresses, each subset of experimental runs supplying a basis for deciding the next. Also, the objective itself can change as new knowledge is brought to light. To catalyze such innovation, the statistician must be part of the investigational team. Unfortunately, many statistics students (and their professors) have little or no training or experience to qualify them for this important role.

Consider a statistician who is analyzing some data coming from, say, a factorial arrangement that has been designed in collaboration with an experimenter. The statistician knows that, even though the data are subject to observational error, certain probability statements can be made and conclusions drawn. However, it may not be realized how such conclusions are conditional on the experimental environment. For example, if the statistician had been working on the same problem with a different experimenter, then

a) *the design would almost certainly have contained somewhat different factors,*
b) *different ranges for the factors would have been chosen,*
c) *different transformations for the factors might be used, ..., and/or*
d) *different tentative models might have been considered.*

Such differences would affect the conclusions drawn from this single experiment far more than would observational error.

However, in an investigational sequence of experiments, although different experimenters will take different routes and begin from different starting points, they nevertheless can arrive at similar solutions. Like

mathematical iteration, scientific iteration tends to be self-correcting. The concept of iterative investigation requires a mind set unfamiliar to many students of statistics. It is difficult to teach and to illustrate, and it needs to be experienced to be appreciated.

Amen and amen, but how do we acquire this mind set and how do we get others to do these things?

Statistical Education in Industry

Education, like any other industrial activity is a series of interrelated processes. We tend not to think of it that way, but we do so at our peril. The process view, that is, starting from the left end of Figure 2, will increase the probability of success. It is by taking the process view that we understand the importance of setting the stage for education. We understand who the customers are, what they are up against and what their needs are.

There is much more that can be said about statistical education in industry, and I would encourage readers to examine the Hoerl and Snee article (Reference 6). I will add only two ideas, one on education (to meet all learning styles) and one on useful techniques.

From my school teacher wife, I learned about four learning styles.[10] They are easy to remember if you remember "Forms, Facts, Feelings and Futures." The notion is that some people learn best when they understand how something works (Forms). People in this category weigh theory against common sense, and they are skills oriented. Others learn best if they are simply told what the thing is (Facts). They integrate their observations into the known, and they want the details. Others learn best if they are told why something is important to be learned (Feelings). They take time to reflect, and they seek harmony. And still others learn best if they can answer the question "What if ..." They learn by trial and error. For example, what if I took this technology and applied it to financial analysis? These are Futures learners. No one is 100% in any one of these categories. Instead, there is a little of each category in each of us. During our education process at Nabisco, I try to include some facet of each learning style in each lesson. I usually start out with feelings because it is the most fun, especially these days when the workplace is tightening down and so much is expected of everyone. "These techniques will increase your chances of project success and make you look better" usually gets everyone's attention. The order of facts and forms doesn't seem to matter very much, but I usually do the futures last because some knowledge through facts and forms is required.

The useful technique is an educational device we use: the paper helicopter exercise described in the Box and Liu paper (Reference 9). We limit the variables to 5, and we use a half-replicate of the 2^5 factorial design. People learn by doing, especially when the needed tool is introduced at the time of the need. During our introductory sessions, we examine mean flight times only, but we do talk about analyzing variation as a response, and we use that discussion to introduce the idea of product and process ruggedness. All concepts are explained using examples from our own or a related industry. We are careful not to include examples that are so current that we upset the political apple cart. And we avoid examples that will draw focus to the underlying science and away from the statistical thinking and methods that went into the design, analysis and interpretation of the results.

When people see that "this stuff works," that it can lead to success and that it can even be fun, they usually want more. Only then can we engage in major value-added activities using appropriate statistical methods, and only then can we use these powerful techniques strategically.

Applying Designs Strategically in Industry

Go where the big bucks are. At the end of the day, there is usually money left on the table in manufacturing. Look at scrap, rework, wasted time and lost opportunity. There is money, usually money that is not easily quantified, in research. Look at major projects where investigators are engaging in OFAAT experimentation, and look at experimentation that has taken place without benefit of a careful examination of the resulting data. You might even want to plot these data, strange as it seems.

In all of these endeavors take the process view. Look at the big picture with regard to sources of process variation. Focus on the sources of variation and work collaboratively with experimenters to help get the right amount of the right kind of data to quantify the variation so you know where you itch the most.

Experiment boldly. It will be expensive because you will end up making bad stuff, and bad stuff doesn't sell. Sometimes you have to make bad stuff to learn how to make good stuff. Experimentation, done right, builds knowledge, and knowledge is an investment yielding competitive advantage. The value of the competitive advantage far outweighs the cost of the bad stuff.

Done right, the use of Statistical Thinking in experimentation will contribute greatly to the bottom line … no matter what Sarah's teacher said. To do it right, sometimes you have to "Burn the Brownies."

---

[1] Glossary of Statistical Terms, ASQ Quality Press, 1996.

[2] Britz, G., D. Emerling, L. Hare, R. Hoerl and J. Shade, "How to Teach Others to Apply Statistical Thinking," *Quality Progress*, June, 1997.

[3] Hare, L., R. Hoerl, J. Hromi and R. Snee, "The Role of Statistical Thinking in Management," *Quality Progress*, February, 1995.

[4] Special Publication, Statistical Thinking, ASQ Statistics Division, available from the Quality Information Center, American Society for Quality, PO Box 3005, Milwaukee, WI 53201-3005, 1996.

[5] Joiner, B., *Fourth Generation Management*, McGraw-Hill, New York, 1994

[6] Hoerl, R., and R. Snee, "Redesigning the Introductory Statistics Course," *Technical Report No. 130*, Center for Quality and Productivity Improvement, University of Wisconsin, Madison, WI., 1995

[7] Snee, R., L. Hare. and J.Trout, eds., *Experiments in Industry – Design, Analysis and Interpretation of Results*, Quality Press, Milwaukee, WI., 1985

[8] Box, G., W. Hunter, and J. Hunter, *Statistics for Experimenters*, John Wiley and Sons, New York, 1978.

[9] Box, G. and P. Liu, "Statistics as a Catalyst to Learning by Scientific Method, Part I – An Example," *Journal of Quality Technology*, Vol. 31, No. 1, January, 1999, pp. 1-15.

[10] McCarthy, B., (1988) "4Mat Cycle of Learning," EXCEL, Inc., Barrington, IL 60010

OUTLIERS AND THE USE OF THE RANK TRANSFORMATION TO DETECT ACTIVE EFFECTS IN UNREPLICATED 2^r EXPERIMENTS

Víctor Aguirre-Torres, Ma. Esther Pérez-Trejo, Instituto Tecnológico Autónomo de México (ITAM)
Víctor Aguirre-Torres, Departamento de Estadística, ITAM, Río Hondo # 1, México, D.F. 01000, MEXICO

Key Words: Faulty Observations, Industrial Experimentation, Nonparametric Analysis, Rank Transformation.

Introduction

The use of statistically designed experiments for quality and productivity improvement has become a major force throughout the world. Unreplicated two level fractions are commonly used experimental plans in early stages of industrial applications; they are relatively easy to set up and economical. Of course these plans are also found in many other fields. However, to detect the significant factors, traditional ANOVA methods can not be easily applied since there are no degrees of freedom available for estimation of the variance of the experimental error. For this reason the normal plot of the effects, or Daniel plot (Daniel (1959)), has been the most widely used tool for detecting significant effects. One important assumption of this procedure, called factor sparsity, is that only a small proportion of the effects is significantly different from zero. The basic idea is that the null effects tend to line up on a straight line, while active effects are expected to depart from this line. A thorough discussion on the use of the Daniel plot can be found in Box, Hunter and Hunter (1978) or Montgomery (1997).

Given an experiment, the possibility that one or more outliers appear is always present. They may be due to oversights in the experimental procedure, gross measurement errors, mistakes in recording the observations, or a true state of the phenomenon under study. Daniel (1976) suggests a way to detect one outlier in his plot, by looking for a gap between groups of points falling along distinct parallel lines. It is interesting to note that, as far as the author knows, it is not discussed in the literature the possible effects of two or more outliers on the Daniel plot.

More recently, Box and Meyer (1987) addressed this same problem. Using a Bayesian approach, they analyzed unreplicated factorial experiments allowing for faulty observations. Their method consists of an iterative procedure that computes first p_i, the probability that the i-th effect is active, assuming that there are no outliers. Temporarily choose the active effects as those with $p_i > P$. Compute q_j, the probability that the j-th observation is faulty, with the active effects held fixed by the above choice. Recompute p_i assuming all observations with $q_j > Q$ to be outliers, and so on. The authors indicate that usually convergence is achieved in one or two iterations with $P = Q = 0.5$. The computations required for each iteration are fairly complex and require very specialized software. The authors do not discuss the statistical risk involved with this iterative procedure. Further discussion about this point of assessing the error rate of the Bayesian method may be found in the Monte Carlo study given in Aguirre-Torres and Pérez-Trejo (1999).

As an alternative, this paper presents an effective and simple method for analysis of unreplicated two level factorials allowing for faulty observations. It is based on the idea of applying the rank transformation to the original observations, this is done with the purpose of deflating the impact of the outliers on the analysis, and then assess the significance of the contrasts computed on the ranks. This evaluation of significance will be done employing the Daniel plot and a more formal and less subjective method to declare significant effects. In particular, from the Monte Carlo study mentioned before, it is recommended to use the rank transformation in conjunction with a modified version of Benski's (1989) procedure. The simulation study also shows that combining the rank transformation with this test, control of the experimentwise error rate is established even in the presence of outliers.

A word of caution should be mentioned, this paper does not propose to apply blindly the rank transformation to the observations of an unreplicated design, without much thinking about the problem and a careful analysis of the original data. However, it is well known that for data suffering the presence of outliers, the analysis of ranks is a good approach. See for example Conover (1980) page 337, or Montgomery (1997) page 146. Faulty observations is just one model, but other possibilities should be considered, like a power transformation of the original response, Box, Hunter and Hunter (1978) page 334; or dispersion effects, Box and Meyer (1986a).

The Proposed Method

The method consists of three steps.

First step. Analyze the original response, using Daniel Plot to assess what contrasts may be significant and to check for possible patterns in this plot, also use a more formal method to detect significant effects. In particular, consideration will be given here to a modification of Benski's (1989) method. This choice is justified from the Monte Carlo study mentioned before.

Benski's approach to assess significance of the effects is based on the idea of applying Olsson's (1979) normality test (W') on the estimated contrasts. If the test is rejected, identify the outlying effects as active. We will take as a rule for rejection when the associated P-value of the test is small, say less than 0.05. To identify the active effects, consider d_F the interquartile range of the effects. A contrast outside $[-2d_F, + 2d_F]$ is considered to be associated with a significant effect, this is the so-called fourth-spread test. The formulas to compute W' and its corresponding P-value are given in the appendix of Benski (1989). They are very easy to program on any spreadsheet.

Second step. Apply the rank transformation to the data and compute the contrasts as usual. Find out what contrasts, if any, are significant using the Daniel plot and also the more formal procedure mentioned before.

Third step. Compare the results of both analyses, the usual practice is as follows: if both procedures give similar results then this approach gives no evidence of an abnormal or faulty observation, check for the need of a possible power transformation. When the two procedures differ substantially, then the results from the second analysis should be preferred since it is less likely to be affected by unusual observations.

The experimenter may make a residual analysis fitting the effects that were significant from the rank analysis, and try to find the outlying observations. If the experimenter is successful in this task, then he or she should try to determine the reason why this happened. Ideally these faulty experimental runs should be repeated and incorporated into the analysis accordingly. If this is not possible due to time and budget constraints, then the results obtained using the ranks procedure may be taken as a more reliable basis for action.

At first sight, one may wonder why it is appropriate to check normality, on the vector of contrasts computed from the rank transformation of the data. The justification of this fact is given in Macdonald (1971), in that paper it is shown that under the null hypothesis that no effect is active, effect estimates based on the ranked observations are asymptotically normal with zero mean and equal variance. The author also mentions that the normal approximation is very close for at least eight contrasts. Since Benski's procedure is based on checking normality of the contrasts, this result justifies its application as proposed.

The idea of comparing the analyses of the original observations and the ranks, for unreplicated factorial experiments, is given in Aguirre-Torres (1993). However in that paper the comparison was made only on the basis of the Daniel plots, in this work we combine the results of those plots with a more formal test method to declare significant effects. The P value associated with this testing procedure helps establish a more objective comparison of both analyses. It is shown in Aguirre-Torres and Pérez-Trejo (1999), that with the procedure based on the ranks it is possible to control the error rate even in the presence of outliers.

Illustration of the Method

To illustrate the method, consider the following examples.

Example 1

This is example IV from Box and Meyer (1986b). Table 1 contains, the data as well as the effects for the original response and the rank transformation. Figure 1 shows the Daniel plot of the effects from the original data. It is hard to tell whether there are any active contrasts. In fact Box and Meyer claim that "it is impossible on the evidence of these data alone to draw reliable inferences about active and inert contrasts" and conclude that it is possible that as many as 5 contrasts are active.

Label	Data	Effects (data)	Ranks	Effects (ranks)
{1}	.08		2	
A	.04	-.19	1	-3.500
B	.53	-.02	12	0.250
AB	.43	.00	10	-0.250
C	.31	-.08	7	-0.750
AC	.09	.03	4	0.750
BC	.12	-.07	5	-1.500
ABC	.36	.15	8	2.000
D	.79	.27	16	4.750
AD	.68	-.16	13	-2.750
BD	.73	-.25	14	-5.000
ABD	.08	-.10	3	-1.500
CD	.77	-.03	15	-0.500
ACD	.38	-.01	9	0.000
BCD	.49	.12	11	2.750
ABCD	.23	.02	6	0.250

Table 1. Data, Example 1.

Looking at this same Daniel plot, one may wonder whether the contrasts form patterns of parallel lines due maybe to outliers. Figure 2 shows the normal plot of the contrasts after applying the rank transformation to the data, This picture looks similar to the previous one.

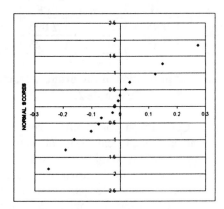

Figure 1. Daniel Plot. Original Data.

Table 2 shows the application of the formal method to the original response, the P-value for the normality test is quite high and hence there is no evidence of significant effects. This table also displays the results of test of significance on the contrasts, computed after the rank transformation is applied to the data. The P-

value for the normality test is large and similar to the previous one, therefore there is no evidence of active effects.

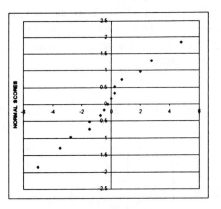

Figure 2. Daniel Plot. Ranks.

Comparison of both analyses. Both exercises resulted in basically the same conclusion, graphically and statistically, hence in this case there is no evidence of outlying observations.

Data	W'	P value	dF	± 2 dF
Original	0.9687	0.7435	0.115	± 0.23
Ranks	0.9715	0.8106	2.0	± 4.00

Table 2. Application of the proposed method, example 1.

Before assuming that the analysis on the original response is right, one may wonder if a power transformation should be used. Particularly, observe that for this experiment

$$Y_{max} / Y_{min} = 19.75 > 3$$

Box, Hunter and Hunter (1978), page 241, suggest the use of a variance stabilizing transformation of the form Y^λ in this situation. For this experiment, using the procedure given in Box, Hunter and Hunter (1978) page 232, to estimate the power of the transformation, leads to a value of $\hat\lambda = 0.5$, that is the square root of the data. Analyzing the transformed data in a similar fashion as above, leads to exactly the same kind of results.

In summary, for this example, the results are as follows: there is no evidence of significant factors, there is no evidence of outliers or the need for a variance stabilizing transformation.

Example 2

This example appears in Box and Draper (1986, pp. 131-134), it is an unreplicated 2^4 experiment. Table 3 contains, the data as well as the effects for the original response and the rank transformation.

Label	Data	Effects (data)	Ranks	Effects (ranks)
{1}	47.46		8	
A	49.62	-0.80	13	0
B	43.13	-4.22	1	-6.25
AB	46.31	0.91	5	-.75
C	51.47	3.71	15	4.75
AC	48.49	-2.49	10	-3.25
BC	49.34	-0.80	12	.5
ABC	46.10	1.20	4	1
D	46.76	1.01	6	0
AD	48.56	-.58	11	1
BD	44.83	-1.18	3	-.25
ABD	44.45	0.72	2	.25
CD	59.15	1.49	16	1.25
ACD	51.33	0.40	14	2.25
BCD	47.02	-1.58	7	-1
ABCD	47.90	1.52	9	1.5

Table 3. Data, Example 2.

Figure 3 shows the Daniel plot of the effects, original response. It is hard to say if there are significant effects. Also in this example, the central points split into two parallel lines, suggesting the possibility of an outlier.

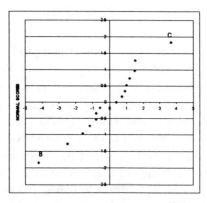

Figure 3. Daniel Plot. Original Data.

Figure 4 shows the Daniel plot of the contrasts after applying the rank transformation to the data. In this case this plot gives a clear indication that the effects B and C are significant, it also shows some evidence that the AC interaction is active. Notice that in this instance the two Daniel plots look different.

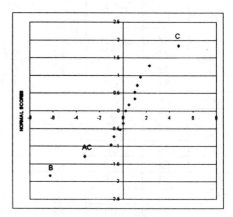

Figure 4. Daniel Plot. Ranks.

Table 4 shows the application of the lack of normality test to both the original response and the rank transformation of the data. In the first case the P-value is, by far, above 0.05, and hence there is no evidence of significant effects. For the rank transformation, the P-value is below 0.05, this gives evidence of active effects. Using the fourth spread test to detect outliers one finds that effects B and C are significant, interaction AC is in doubt since it is right on the borderline.

Data	W'	P value	dF	± 2 dF
Original	0.9536	0.4703	2.095	± 4.19
Ranks	0.8757	0.0429	1.625	± 3.25

Table 4. Application of the proposed method, example 2.

Comparison of both analyses. The graphical and statistical tests of significance differ substantially, this is evidence that something does not fit the usual assumptions for the original response. In particular, given the pattern in the original Daniel plot one may suspect the existence of an outlier. Figure 5 displays the normal plot for residuals after fitting a model with effects B and C as active, there is a clear indication that observation 13 is an outlier.

The ratio

$$Y_{max}/Y_{min} = 1.37 < 3$$

from Box, Hunter and Hunter's recommendation, there does not seem to be the need for a variance stabilizing transformation.

In summary, the results after examining the original response an the rank transformation were significantly different, in particular, the analysis on the rank transformation gave evidence that the main effects B and C are active, both from a graphical and an statistical test point of view. An outlying observation was detected using the above information.

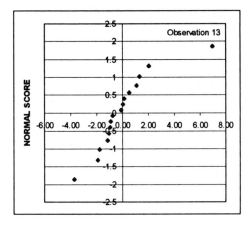

Figure 5. Normal Plot of Residuals. Example 2.

Notice that, for both examples, the P-value obtained from the lack of normality test, helps to compare the results from both analyses on a more objective ground.

Summary

Unreplicated two level factorials have played a major role in the current movement of quality and productivity improvement. The Daniel Plot, which makes use of the factor sparcity principle, is an important tool for analyzing this type of test plans. However, in practice, it is not too rare that outlying observations may occur in an experiment of this kind without the perception of the experimenter. The practitioner is then faced with the problem of how to analyze such designs under that situation.

It is therefore important to have methods of analysis that deal with outliers. A procedure that is relatively easy to use, that makes use of graphical and more formal statistical tools has been proposed. It has the advantage of keeping control of the error rate even in the presence of outliers, and also shows a consistently good power, relative to a broad set of methods, for finding significant effects when there are faulty observations present in the data set.

Acknowledgements

Part of this research was performed while the first author was on sabbatical leave at the Departamento de Probabibildad y Estadística, Centro de Investigación en Matemáticas, Guanajuato. This research was also partly sponsored by Asociación Mexicana de la Cultura, A. C.

References.

Aguirre-Torres, V. (1993). "A Simple Analysis of Unreplicated Factorials with Possible Abnormalities". *Journal of Quality Technology*, 25, 183-187.

Aguirre-Torres, V.; and Pérez-Trejo, M. E. (1998). "Outliers and the use of the Rank Transformation to Detect Active Effects in Unreplicated 2^f Experiments". Technical Report No. I-99-13 (PE/CIMAT), Centro de Investigación en Matemáticas.

Benski, H. C. (1989). "Use of a Normality Test to Identify Significant Effects in Factorial Designs". *Journal of Quality Technology*, 21, 174-178.

Box, G. E. P., and Draper, N. R. (1986). *Response Surfaces and Empirical Model Building*. John Wiley & Sons, New York, NY.

Box, G. E. P.; Hunter, W. G.; and Hunter, J. S. (1978). *Statistics for Experimenters*. John Wiley & Sons, New York, NY.

Box, G. E. P., and Meyer, R. D. (1986a). "Dispersion Effects From Fractional Designs". *Technometrics*, 28, 19-27.

Box, G. E. P., and Meyer, R. D. (1986b). "An Analysis for Unreplicated Fractional Factorials". *Technometrics*, 28, 11-18.

Box, G. E. P., and Meyer, R. D. (1987). "Analysis of Unreplicated Factorials Allowing for Possible Faulty Observations". *Design, Data and Analysis*, C. Mallows (ed.), John Wiley & Sons, New York, NY.

Conover, W. J. (1980). *Practical Nonparametric Statistics*. Second edition. John Wiley & Sons, New York, NY.

Daniel, C. (1959). "Use of Half Normal Plots in Interpreting Factorial Two-Level

Experiments". *Technometrics*, 1, 311-341.

Daniel, C. (1976). *Applications of Statistics to Industrial Experimentation*. John Wiley & Sons, New York, NY.

Macdonald, P. (1971). "The Analysis of a 2^n Experiment by Means of ranks". *Applied Statistics*, 20, 259-275.

Montgomery, D. C. (1997). Design and Analysis of Experiments. Fourth edition. John Wiley & Sons, New York, NY.

Olsson, D. M. (1979). "A Small Sample Test for Non-Normality". *Journal of Quality Technology*, 11, 95-99.

COMPUTER DECISION SUPPORT SYSTEM
FOR QUALITY PLANNING

Soumaya Yacout, École Polytechnique de Montréal
Jacqueline Boudreau, Université de Moncton
Soumaya Yacout, Director of Studies, École Polytechnique de Montréal,
C.P. 6079, Succ. Centre-Ville, Montréal, Quebec, Canada H3C 3A7

Key Words : Computer simulation, Quality Management, Partially Observed Markov Decision Process

Introduction

For many years simulation technique has been used as a powerful tool for decision making. It allows a decision maker to model a system, and to verify the impact of different actions on the system's performance, before actual implementation of these actions. It also allows for the building of complex models where many random and deterministic factors are interacting simultaneously. A simulation model allows the decision makes to observe the system's performance over a long period of time without having to actually wait for that time.

Simulation techniques have been used since the fifties. Robinson and Higton (1995) explained the close relation between the development of simulation techniques and the evolution of the computer. In the eighties, the relatively easy access to personal computers helped increasing the use of simulation techniques. Since that time, simulation has been used to model many systems, including quality systems. In 1991, Abdel Hamid and Leidy developed a simulation model that analyzed the attribution of resources to quality control activities related to software development. Their model performed a cost/benefit analysis for different quality assurance policies. Different scenarios for the allocation of resources were analyzed. In 1997, Gautreau and Yacout built a simulation model of a Partially Observed Markov Decision Process. Quality assurance actions were simulated and their effects on product cost and quality were analyzed.

In this article we extend the work of Gautreau and Yacout. A generic simulation model for a decision support system for quality planning is constructed. Four scenarios and four different policies are simulated. The system is constructed based on the well developed theories of a POMDP. Sampling inspection, statistical process control and Taguchi's loss function are simulated as discrete events.

Constructing The Model

The simulation model is constructed as in figure 1.

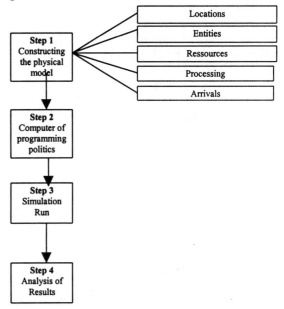

Figure 1. Steps of constructing the simulation model.

In step 1, the physical model describes a department where working units perform different tasks. In this article, these tasks are called processes. The tasks may require human interaction with machines, raw material and products. This interaction is described in the function called "locations". Input and output to the location are described in the function called "entities". The number of machines, of operators and all other resources needed to achieve the tasks are described in the function "resources". The way the tasks are performed is described in the function "processing". This function may need to be programmed. A simple programming language similar to BASIC is used. The arrival of raw material or of a product to the simulated location is described in "arrivals". The company's politics concerning quality control and quality assurance are programmed in step 2. Four possible scenarios and four politics are programmed. Step 3 consists of running the simulation model for a specified period of time. Since the results of a single simulation run represent only one of several possible outcomes, multiple replications are run to test the reproducibility of the results and to obtain a specified confidence interval for the results. Since simulation utilizes a pseudo-random number generator for generating random numbers, running the simulation multiple times simply reproduces the same sample. In order to get an independent sample, the starting seed value for each run was different for each replication, thus ensuring that the random numbers generated from replication to replication are independent. In step 4, the results of different policies are compared. The comparison is based on two measures of performance, the mean cost (mcost) and the average outgoing quality (AOQ) calculated from equations 1, 2 and 3 as follows:

$$mcost = \Sigma \ (cost)/Rep, \qquad (1)$$

$$AOQ = mnconf/N, \qquad (2)$$

$$mnconf = \Sigma \ nconf/Rep, \qquad (3)$$

where nconf is the number of nonconforming units per run, (cost) is the cost calculated at the end of each run, and Rep is the number of replications, N is the total number of units produced per run, and mnconf is the mean of nonconforming units.

The Simulation Model

The constructed model simulates three types of events:

1.The external causes of variation

These are discrete events causing abnormal variations in the process performance. The arrival of these causes is modeled as a Homogeneous Poisson Process with means, λ_i, where i is the number of types of external causes.

2.The Process Performance

The process may be in one of many states. These states are unobserved, and depend on the arrival of the causes of variation, i. The output of the process is observed and is a random variable having a normal distribution $N(\mu, \sigma_i)$ that depends on the states of the process, i , where μ is the mean of the ouput, and σ_i , is its standard deviation.

3.The Suggested Quality Policies

Four quality policies are simulated: a "do-nothing" policy, an appraisal policy, a prevention policy, and an appraisal-prevention policy. Two alternative appraisal policies are simulated: a 100% inspection policy, and a Dodge Romig inspection plan. The prevention policy uses Statistical Process Control (SPC), where a search for the causes of variation is done. If these causes are found, they are eliminated. The number of arrivals of subsequent causes decreases and the process performance is improved. Since absolute validation of the degree to which the simulation model corresponds to the real process is difficult, different scenarios about how the process performs are made. The process is simulated with and without periodic maintenance that is done after every production cycle. The outcomes of each scenario and each policy are calculated. These are the costs and the average outgoing quality. Costs are calculated by using Taguchi's loss function, or by assuming a constant cost to every nonconforming unit produced. External and internal costs, inspection costs, sampling costs are calculated. More details concerning the calculation of costs can be found in Yacout and Chang (1997), Yacout and Gautreau (1999), and Yacout and Boudreau (1998).

Numerical Example

A process of packing fish in vacuum-sealed plastic bag is simulated. The input to the chosen location is processed fish and empty plastic bags. These arrive according to a steady rate of 1 bag/min. An operator performs the task of packing fish in a vacuum-sealed plastic bag and weighing the bag. The nominal weight of a bag is 142g. The specification limits are equal to USL=153.6g and LSL=130.4g . The times between the arrival rates of the causes of variation are supposed to be known and are equal to $\lambda_1 = 0.10437$, $\lambda_2 = 0.07395$, $\lambda_3 = 0.02458$, $\lambda_4 = 0.02463$, where i = 4. The weight, w_i , of the bag varies according to the type of causes present, and has a normal distribution $N (\mu = 142, \sigma_i)$, where $w_i = 142 \pm \sigma_i$, $\sigma_i = s_i\sigma = 4.341$, 5.85, 7.39, 10.35, respectively for i=1,2,3,4. All inspections and sampling costs are known. Figures 2,3,4, and 5 show the results obtained for the four policies when periodic maintenance is considered, and when it is not considered. A steady state behavior is found after almost 200 production cycles when periodic maintenance is assured. It is quickly found after 10 cycles when periodic maintenance is not considered. This state corresponds to an out of control state under policies 1 and 2, and to an in-control state under policies 3 and 4.

Conclusion

The simulation technique is used to help the decision maker analyze the effects of four quality policies on the costs and the outgoing quality of a process. The model simulates the arrivals of the external causes of variations, a POMDP, and the alternative policies which include sampling inspection and SPC. The costs are calculated by using Taguchi's loss function. The user must specify the rate of arrival of external causes of variations, the process performance under normal and abnormal conditions, and the costs associated with sampling inspection and to the use of SPC. The results are analyzed and compared. A decision may be made based on scientific analysis.

References

Abdel Hamid, T.K. and Leidy F.H. (1991). An Expert Simulator for Allocating the Quality Assurance Effort in Software Development Simulation, vol. 56, no. 4, p. 233-240.

Gautreau,N., Yacout S., and Hull, R. (1997). Simulation of Partially Observed Markov Decision Process and Dynamic Quality Improvement. Computers and Industrial Engineering, 32, p. 691-700.

Robinson, S. and Higton N. (1995), Computer for Quality and Reliability Engineering. Quality and Reliability Engineering International. Vol. 11, p. 37-377.

Yacout, S., Chang, Y. (1997). Modeling Process Quality Costs For Alternative Quality Plans. Quality Engineering. Vol. 9, no. 3, p. 419-431.

Yacout, S., Boudreau, J., (1998). Computer Simulation: A Decision Support Tool for Quality Planning. Proceedings of the American Statistical Association.

Yacout, S., Gautreau, N. (1999). A Partially Observable Simulation Model for Quality Assurance Policies. Accepted for Publication in the International Journal of Production Research.

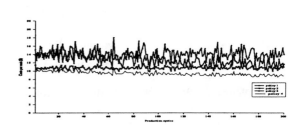

Figure 2: costs of quality per unit for scenario 2

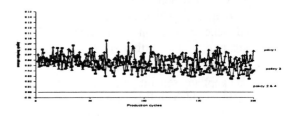

Figure 3: average outgoing quality for scenario 2

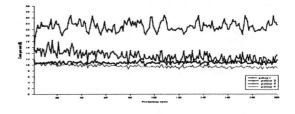

Figure 4 : costs of quality for scenario 4

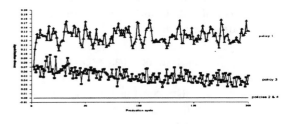

Figure 5 : average outgoing quality for scenario 4

A LIKELIHOOD APPROACH TO CONTROL CHARTS

Moshe Pollak, The Hebrew University of Jerusalem
Department of Statistics, The Hebrew University of Jerusalem, 91905 Jerusalem, Israel

Key Words: SPC, Cusum, Quality Control

ABSTRACT

Some of the basic problems underlying the application of classical SPC methods are:

a) Often one does not know in advance what the direction and magnitude of a change will be, ignorance of which may seriously affect the efficiency of a scheme. This is especially true in a multivariate setting.

b) The process being observed is multivariate, and applying separate surveillance schemes to each variable is impractical.

c) Baseline pre-change parameter values are not really known, and a large enough learning sample cannot be obtained readily. Consequently, true ARL's may deviate considerably from nominal ones.

d) The distribution of the observations may be non-normal, perhaps even unknown, suggesting a need for non-normal parametric procedures as well as nonparametric methods.

e) Change of level is not always the object of surveillance; at times other parameters are of interest (such as a change of the slope of a regression).

f) A change may be gradual, whereas classical schemes are designed for detecting an abrupt change.

g) The observations may be dependent, whereas classical procedures are designed for independent variables.

The last decade has seen advances on these fronts. A central idea underlying these advances is the construction of appropriate likelihood ratios and the exploitation of their statistical properties. Extension of this view to settings more complicated than the classical simple changepoint problem yields reasonably efficient procedures for the more complicated problems. Here, the basic idea is illustrated, and its implementation to some of the problems mentioned above is given.

1. Introduction

The goal of control charts is: 1) to ascertain that a process is in control (if that's the case) 2) to detect a change quickly (if one has taken place). Usually, the evidence is a set of observations $X_1, X_2, X_3, \ldots$ taken sequentially. One wants to avoid false alarms as well as large delays between a change and its detection. A control scheme is deemed efficient if it detects a change as quickly (on the average) as possible, subject to a constraint on the rate at which false alarms are tolerated. The usual constraint is

$$A_{\text{verage}} R_{\text{un}} L_{\text{ength to}} F_{\text{alse}} A_{\text{larm}} \geq B$$

where B is a pre-specified constant.

The two popular control schemes are: 1) Shewhart, which regards observations one by one and raises an alarm if the present observation is extreme, and 2) Cusum, which regards the sequence of partial sums of observations (past to present). The advantage of Shewhart is that it is simple; its disadvantage is that it's inefficient; it doesn't let information accumulate. The advantage of Cusum is that it is efficient if the observations are independent and one is on the alert for an abrupt change from one completely known distribution to another completely known distribution. The disadvantage is that rarely are these conditions met. Both methods suffer from the problems outlined in the abstract above.

The changepoint problem has at least a superficial similarity to the hypothesis testing problem. In hypothesis testing, one asks whether H_0 is the case, or whether H_1 is (immediately) in effect. In the process control context, one allows H_0 to be in effect initially, and one wonders whether H_1 ever takes over. So, it may pay to take a cue from hypothesis testing: there, the main tool is the likelihood ratio (or generalized likelihood ratio). So, it may be worthwhile to try the same for changepoint detection. It is this that is proposed here.

2. Notation

ν = changepoint

$X_1, X_2, ..., X_{\nu-1}$ are pre-change
$X_\nu, X_{\nu+1}, ...$ are post-change

P_ν = probability function when X_ν is
the first post-change observation
E_ν = expectation when X_ν is
the first post-change observation

$\nu = \infty$ means that there's never a change;
the process is in control forever

LIKELIHOOD RATIO:

$$\Lambda_k^n = \frac{f_{\nu=k}(X_1, X_2, ..., X_n)}{f_{\nu=\infty}(X_1, X_2, ..., X_n)} .$$

As an example, consider the classical problem of detecting a change of mean of independently distributed normal variables:

$X_1, X_2, ..., X_{\nu-1}$ are iid $N(0,1)$
$X_\nu, X_{\nu+1}, ...$ are iid $N(\theta,1)$.

Here

$$\Lambda_k^n = \Lambda_k^n(\theta) =$$

$$= \frac{\prod_{i=1}^{k-1}((1/\sqrt{(2\pi)}) e^{-X_i^2/2}) \prod_{i=k}^{n}((1/\sqrt{(2\pi)}) e^{-(X_i-\theta)^2/2})}{\prod_{i=1}^{k-1}((1/\sqrt{(2\pi)}) e^{-X_i^2/2}) \prod_{i=k}^{n}((1/\sqrt{(2\pi)}) e^{-X_i^2/2})}$$

$$= \exp(\theta \sum_{i=k}^{n} X_i - (1/2)\theta^2(n-k+1))$$

$$= \exp(\theta [\sum_{i=k}^{n}(X_i - (1/2)\theta)]) .$$

3. Using the Likelihood Ratios

What should one do with the likelihood ratios? The popular cusum is basically a maximum likelihood method. Define

$$W_n = \max_{1 \le k \le n} \Lambda_k^n$$

and stop the first time n that $W_n \ge$ constant. In the normal example:

$$M_A = \min\{n \mid \max_{1 \le k \le n} \exp(\theta[\sum_{i=k}^{n}(X_i-(1/2)\theta)]) \ge A\}.$$

Note that this is the same as stopping the first time that

$$\max_{1 \le k \le n} ([\sum_{i=k}^{n}(X_i-(1/2)\theta)]) \ge (\ln A)/\theta ,$$

which is the usual form of the cusum with $d=(1/2)\theta$ and $h=(\ln A)/\theta$.

Another scheme, known as the Shiryayev-Roberts procedure, uses the sum of the likelihood ratios rather than their maximum. The efficiency of this procedure is similar to that of the cusum, but it has a few technical properties which make it an attractive alternative. Formally, the Shiryayev-Roberts procedure defines

$$R_n = \sum_{k=1}^{n} \Lambda_k^n$$

and calls for stopping the first time n that $R_n \ge$ constant. In the normal example:

$$N_A = \min\{n \mid \sum_{k=1}^{n}\exp(\theta[\sum_{i=k}^{n}(X_i-(1/2)\theta)]) \ge A\}.$$

In order to satisfy the constraint {ARL to False Alarm $\ge$ B}, obviously one needs a way to connect between B and the stopping threshold A. This can be done in a number of **ways**:

1. Where available, use tables to set A so that ARL to false alarm = B.

2. Where available, use mathematical formulae to set A so that ARL to false alarm = B (or $\approx$ B).

3. A useful fact which is true under completely general circumstances - even if the observations are dependent - is:
FACT: $ARL_\infty(CUSUM(A))$
$\ge ARL_\infty(SHIRYAYEV\text{-}ROBERTS(A))$
$\ge A$.

Therefore, one can simply set A=B to at least guarantee that ARL to false alarm $\ge$ B.

4. Do a Monte Carlo study with a number of different A's to obtain an A so that ARL to false alarm = B.

So,

WE'RE ALL SET:
In order to define a control scheme, it suffices to:
a) WRITE DOWN A LIKELIHOOD RATIO.
b) CHOOSE A LEVEL A USING ONE OF THE WAYS 1-4 LISTED ABOVE.
c) RUN A CUSUM OR A SHIRYAYEV-ROBERTS CONTROL SCHEME.

4. Examples

The problem of detecting a change in the mean level of independent normal observations with known baseline is described in Section 2. In this section, more complicated cases are described. For ease of exposition, the normal context is kept, though the same ideas can be applied to other parametric families. For more explicit details of the procedures described below, see the References section at the end of the article.

a) Unknown Post-Change Parameter Values

When the post-change parameter value is unknown, the classical solution is to choose a representative value - call it θ_1 - and insert it instead of θ in $\Lambda_k^n(\theta)$. The disadvantage of this is that if the true value of θ is different from θ_1 then the resulting control scheme will be slower in detecting a change than it might have been had θ been guessed correctly. Furthermore, if the change can be either above or below the initial value, a single representative won't do.

One solution is to choose a prior over the parameter space. If g is the density of the prior, then $\int \Lambda_k^n(\theta) g(\theta) d\theta$ is a bona fide likelihood ratio. Although there is no guarantee that g is the real prior or even that a prior really exists, it still is valid as a technical gimmick to choose a prior arbitrarily and apply cusum or Shiryayev-Roberts, and, at least when A is large, the resulting control scheme will be close to optimal. (The reason for this is that the prior "washes out" when A is large. For small values of A, choice of prior admittedly makes more of a difference.) For example, in the normal case of Section 2, suppose one chooses a $N(0,1)$ prior over θ. Then

$$\int \Lambda_k^n(\theta) g(\theta) d\theta$$
$$= \int_{-\infty}^{\infty} \exp\{\theta \sum_{i=k}^{n} X_i - (1/2)\theta^2 (n-k+1)\} (1/\sqrt{(2\pi)}) e^{-\theta^2/2} d\theta$$
$$= (1/\sqrt{(n-k+2)}) \exp\{(\sum_{i=k}^{n} X_i)^2/[2(n-k+2)]\} \quad .$$

For example, the Shiryayev-Roberts statistic will be

$$R_n = \sum_{k=1}^{n} (1/\sqrt{(n-k+2)}) \exp\{(\sum_{i=k}^{n} X_i)^2/[2(n-k+2)]\}$$

and after choice of an appropriate value of A, one raises an alarm the first time that R_n exceeds A.

Another solution is to estimate the parameter instead of putting a prior on it. This procedure seems to be slightly less efficient than the former procedure, but in situations where integration with respect to a prior is difficult, estimation is an excellent alternative. However, the estimation requires care: for technical reasons, the estimator should depend on the first $n-1$ observations only, and not on the last observation. For example, again using the detection-of-a-normal-mean problem of Section 2 as an illustration, define

$$\theta_k^n = \sum_{i=k}^{n-1} X_i / (n-k)$$

and substitute this for θ in $\Lambda_k^n(\theta)$. The Shiryayev-Roberts statistic becomes

$$R_n = \sum_{k=1}^{n} \exp\{\theta_k^n [\sum_{i=k}^{n} (X_i - (1/2)\theta_k^n)]\} \quad .$$

b) Multivariate Process Control

There is no essential difference between univariate spc and multivariate spc. Again, all that's required is a likelihood ratio. For example, suppose that the observations are independent random vectors (X_i, Y_i, Z_i) such that the three components are independent, that prior to a change each has a $N(0,1)$ distribution and after a change the means become $E(X_i, Y_i, Z_i) = (\theta_1, \theta_2, \theta_3)$ where $\theta_1, \theta_2, \theta_3$ are unknown and at least one of them differs from zero. (It is assumed that the distribution remains normal with unit variance.) As in part a), one can choose a prior over $\theta_1, \theta_2, \theta_3$ or estimate the parameters. For example, if one chooses a standard normal prior independently for each of $\theta_1, \theta_2, \theta_3$ then the joint likelihood ratio Λ_k^n becomes the product of the separate likelihood ratios:

$$(1/\sqrt{(n-k+2)})^3 \exp\{[(\sum_{i=k}^{n} X_i)^2 + (\sum_{i=k}^{n} Y_i)^2 + (\sum_{i=k}^{n} Z_i)^2]/[2(n-k+2)]\}$$

and the Shiryayev-Roberts statistic R_n is the sum $(k=1,...,n)$ of Λ_k^n. Note that this enables simultaneous tracking of the three parameters by a single control scheme; there is no reason to track them separately. (As a matter of fact, simultaneous tracking leads to faster detection, especially if the aggregate parameter change is nonnegligible but each parameter changes only by a small quantity.) As for checking which parameter changed after having raised an alarm at time N, one can use a maximum likelihood approach; find the index k which maximizes Λ_k^N and base an estimate of the post-change parameters on $(X_k, Y_k, Z_k), ..., (X_N, Y_N, Z_N)$. See Section 5.

Alternatively, one can estimate the unknown parameters, and proceed in a manner analogous to that described in part a).

c) Unknown Baseline Parameters

The operating characteristics of control charts are notoriously sensitive to misspecification of baseline parameters. For instance, consider the cusum designed to detect an increase of one standard deviation of a normal mean at an ARL to false alarm equivalent to a Shewhart one-sided control scheme. The true ARL to false alarm of this cusum scheme is less than 33% than its nominal value if the standard deviation is underspecified by 5%. Even a 1% underspecification will cause the true ARL to false alarm to be 10% less than its nominal value. In other words, if one relies on a learning sample to estimate the unknown standard deviation, the size of the sample must be huge - much larger than often feasible. Hence, a procedure which is insensitive to ignorance of baseline parameters can be useful.

If the problem possesses an invariance structure, a likelihood ratio approach can be helpful. To illustrate this, consider the problem of detecting a change in a normal mean described earlier, only that now the standard deviation (assumed to be constant throughout) and the pre-change mean as well as the post-change mean are assumed to be unknown. This can be represented as

$$X_1,X_2,...,X_{v-1} \text{ are } N(\mu\sigma,\sigma^2)$$
$$X_v,X_{v+1},... \quad \text{ are } N((\mu+\theta)\sigma,\sigma^2)$$

where σ,μ,θ are unknown.

For the sake of simplicity, suppose for starters that one is interested in detecting a change of (at least) δ standard deviations should one take place. (In other words, δ is a representative of θ.) If one defines

$$Y_i = (X_i-X_1)/(X_2-X_1) \text{ for } i=3,4,...,n$$

then the joint distribution of $(Y_3,Y_4,...,Y_n)$ depends neither on σ nor on μ. Therefore, the likelihood ratio of $(Y_3,Y_4,...,Y_n)$ for $v=k$ vs. $v=\infty$ (for $k\geq3$) can be explicitly worked out, without its entailing any unknown quantities. In other words, assuming that a change does not take place in the first two observations, a cusum or a Shiryayev-Roberts procedure based on the sequence of invariant statistics $(Y_3,Y_4,...,Y_n)$ can be worked out, and its actual ARL to false alarm will equal its nominal one. (Admittedly, the likelihood ratio is not a simple formula, but it can be worked out into an expression which is readily programmable.) If the change does not occur early, then the efficiency of this procedure (in terms of speed of detection) will be very close to that of the analogous procedure in the context of known baseline parameters.

Of course, one may object to taking a representative for the post-change parameter of interest. In that case, it is possible to go the route described in part a); i.e., either to put a prior on the unknown post-change parameter or to estimate it. Obviously, this adds a further complication, but it is feasible to write out a programmable expression for the likelihood ratios in this case, too.

d) Nonparametric Control Charts

There are contexts where the distribution of the observations is completely unknown, and nevertheless one is interested in detecting a change of level or a change of dispersion. Such a situation calls for a nonparametric approach. The natural candidates on which to base surveillance are the sequential ranks,

$$r_n = \sum_{i=1}^{n} I(X_i\leq X_n)$$

where I is the indicator function. The challenge is to find a way to write a likelihood ratio for the sequence $r_1,r_2,...,r_n$, in such a way that the resulting procedure will be efficient (relative to a case where the distribution is known). This is usually feasible, though the details are too lengthy to be described here.

e) Detecting a Change in Regression

Invariance considerations can be applied also to detecting a change in regression. For example, suppose one is interested in detecting a change in the slope of a regression over time, where the standard deviation of the error is constant over time but its value is unknown, the pre-change slope and post-change slope are both unknown and the observations are independent and normally distributed. Thus

$$X_i \sim N((\alpha+\beta i+\theta(i-v+1)^+)\sigma,\sigma^2).$$

For ease of exposition, consider the case that one is satisfied with taking a representative for the change θ in slope; say the representative change in slope is δ standard deviations. Then, assuming the change does not occur prior to the fourth observation, the joint distribution of

$$Z_i = [(X_i-X_1)-(i-1)(X_2-X_1)]/[(X_3-X_1)-2(X_2-X_1)]$$

does not depend on α,β,σ. Therefore, one can base the control chart on the sequence $Z_4,Z_5,Z_6,...$. Again, one can take a prior on θ if taking a representative δ is not desired. Due to the length of exposition, the details are omitted.

f) Detecting a Gradual Change

Again consider the change-of-normal-mean problem of Section 2, only now assume that the change is gradual. A possible formulation of this is that

$$X_1, X_2, ..., X_{v-1} \sim N(0,1)$$
$$X_i, i=v,v+1,..., v+m-1 \sim N([\theta(i-v+1)/m], 1)$$
$$X_{v+m}, X_{v+m+2}, ... \sim N(\theta,1)$$

It is easy to write out a likelihood ratio Λ_k^n. If one does not want to fix representatives for θ,m then it is possible to choose a prior over θ,m or to estimate them, as in part a).

g) Dependent Observations

The same approach can be applied to dependent observations - time series, Markov chains, etc. The fact that the observations are dependent wreaks havoc with the classical control charts, but the approach proposed here is indifferent to whether the observations are or are not independent; as long as a likelihood ratio can be written down, the approach works. (The control charts based on $Y_3, Y_4, Y_5,...$ in part c) and on $Z_4, Z_5, Z_6,...$ in part e) are examples of this; note that both of these are sequences of dependent random variables.)

5. A Simulated Example

In this section, a simulated example of the multivariate problem described in part b) of Section 4 is presented. The (arbitrary) choice of parameters is $v=21$, $\theta_1=.5$, $\theta_2=.9$, $\theta_3=0$. Figure 1 depicts the three sequences. Figure 2 describes the Shiryayev-Roberts control chart. With a choice of A=1000, an alarm is raised at $N_{A=1000} = 35$. Finally, Figure 3 is a plot of the likelihood ratios Λ_k^{35}. The maximum likelihood is at k=19. The maximum likelihood estimates of $(\theta_1, \theta_2, \theta_3)$ are (.39, 1.04, .02).

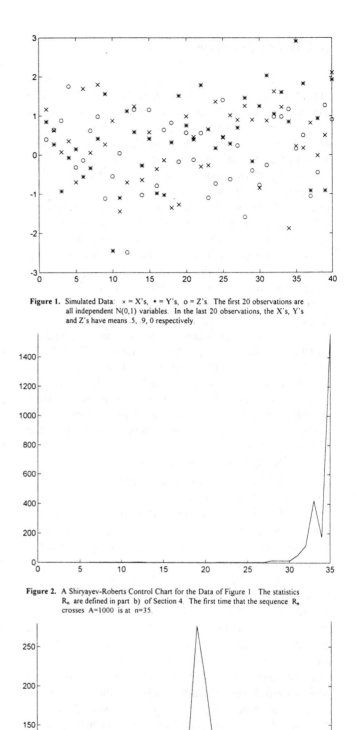

Figure 1. Simulated Data: × = X's, * = Y's, o = Z's. The first 20 observations are all independent N(0,1) variables. In the last 20 observations, the X's, Y's and Z's have means .5, .9, 0 respectively.

Figure 2. A Shiryayev-Roberts Control Chart for the Data of Figure 1. The statistics R_n are defined in part b) of Section 4. The first time that the sequence R_n crosses A=1000 is at n=35.

Figure 3. The Likelihood Ratios Λ_k^{35}, k=1,...,35. These are the likelihood ratios of the data for v=k vs. v=∞ at the time the alarm is raised. The maximum of the likelihood ratios is attained at k=19.

SELECTED REFERENCES
(See also references in the following sources.)

[i] (A review which is relatively easy to follow. Contains material on normal parametric detection of change of mean and parametric detection of change of standard deviation with unknown nuisance parameters, and nonparametric detection of a change of mean.)

[1] Pollak, M., Croarkin, C. and Hagwood, C. (1993).
Surveillance schemes with applications to mass calibration.
NISTIR 5158, Technical Report of the Statistical Engineering Division, The National Institute of Standards and Technology, Gaithersburg, MD 20899.

[ii] (Optimality and OC's of the Cusum and the Shiryayev-Roberts procedures.)

[1] Lorden, G. (1971).
Procedures for reacting to a change in distribution.
The Annals of Mathematical Statistics 42 1897-1908.

[2] Moustakides, G.V. (1986).
Optimal stopping times for detecting changes in distributions.
The Annals of Statistics 14 1379-1387.

[3] Pollak, M. (1985).
Optimal detection of a change in distribution.
The Annals of Statistics 13 206-227.

[4] Pollak, M. (1987).
Average run lengths of an optimal method of detecting a change in distribution.
The Annals of Statistics 15 749-779.

[5] Pollak, M. and Siegmund, D. (1985).
A diffusion process and its application to detecting a change in the drift of Brownian motion.
Biometrika 72 267-280.

[6] Yakir, B. (1995).
A note on the run length to false alarm of a changepoint detection policy.
The Annals of Statistics 23 272-281.

[7] Yakir, B. (1997).
A note on optimal detection of a change in distribution.
The Annals of Statistics 25 2117-2126.

[iii] (Theory of procedures based on reduction by invariance.)

[1] Gordon, L. and Pollak, M. (1997).
Average run length to false alarm for surveillance schemes designed with partially specified pre-change distribution.
The Annals of Statistics 25 1284-1310.

[2] Yakir, B. (1998).
On the average run length to false alarm in surveillance problems which possess an invariance structure.
The Annals of Statistics 26 1198-1214.

[iv] (Unknown baselines, nonparametric procedures.)

[1] Pollak, M. and Siegmund, D. (1991).
Sequential detection of a change in a normal mean when the initial value is unknown.
The Annals of Statistics 19 394-416.

[2] Siegmund, D. and Venkatraman, E.S. (1995).
Using the generalized likelihood ratio statistic for sequential detection of a change-point.
The Annals of Statistics 23 255-271.

[3] Gordon, L. and Pollak, M. (1995).
A robust surveillance scheme for stochastically ordered alternatives.
The Annals of Statistics 23 1350-1375.

[v] (Procedures based on estimation of post-change parameters.)

[1] Lorden, G. and Pollak, M. (1999).
Changepoint detection via Robbins-Siegmund estimation.
Technical report.

[vi] (Detecting a change in regression.)

[1] Yakir, B., Krieger, A.M. and Pollak, M. (2000).
Detecting a change in regression: first order optimality.
To appear in The Annals of Statistics.

[2] Krieger, A.M., Pollak, M. and Yakir, B. (1999).
Surveillance of a simple linear regression.
Technical report.

[vii] (Other references of general interest.)

[1] Basseville, M. and Nikiforov, I.V. (1993).
Detection of Abrupt Changes: Theory and Application.
Prentice Hall.

[2] Lai, T.L. (1995).
Sequential changepoint detection in quality control and dynamical systems.
The Journal of the Royal Statistical Society B 57 613-658.

Index of Authors